PERGAMON INTERNATIONAL LIBRARY
of Science, Technology, Engineering and Social Studies

The 1000-volume original paperback library in aid of education, industrial training and the enjoyment of leisure

Publisher: Robert Maxwell, M.C.

ENERGY AND ECONOMIC MYTHS
Institutional and Analytical Economic Essays

Publisher's Notice to Educators

THE PERGAMON TEXTBOOK INSPECTION COPY SERVICE

An inspection copy of any book published in the Pergamon International Library will gladly be sent without obligation for consideration for course adoption or recommendation. Copies may be retained for a period of 60 days from receipt and returned if not suitable. When a particular title is adopted or recommended for adoption for class use and the recommendation results in a sale of 12 or more copies, the inspection copy may be retained with our compliments. If after examination the lecturer decides that the book is not suitable for adoption but would like to retain it for his personal library, then our Educators' Discount of 10% is allowed on the invoiced price. The Publishers will be pleased to receive suggestions for revised editions and new titles to be published in this important International Library.

PERGAMON TITLES OF RELATED INTEREST

Balogh, Lord. *Fact and Fancy in International Economic Relations: An Essay on International Monetary Reform*

Cumes, J.W. *The Indigent Rich: A Theory of General Equilibrium in a Keynesian System*

Stillwell, F.J.B. *Normative Economics. An Introduction to Microeconomic Theory and Radical Critiques*

ENERGY AND ECONOMIC MYTHS
Institutional and Analytical Economic Essays

Nicholas Georgescu-Roegen

PERGAMON PRESS INC
New York / Toronto / Oxford / Sydney / Frankfurt / Paris

Pergamon Press Offices:

U.S.A.	Pergamon Press Inc., Maxwell House, Fairview Park, Elmsford, New York 10523, U.S.A.
U.K.	Pergamon Press Ltd., Headington Hill Hall, Oxford OX3, OBW, England
CANADA	Pergamon of Canada, Ltd., 207 Queen's Quay West, Toronto 1, Canada
AUSTRALIA	Pergamon Press (Aust) Pty. Ltd., 19a Boundary Street, Rushcutters Bay, sN.S.W. 2011, Australia
FRANCE	Pergamon Press SARL, 24 rue des Ecoles, 75240 Paris, Cedex 05, France
WEST GERMANY	Pergamon Press GmbH, 6242 Kronberg/Taunus, Frankfurt-am-Main, West Germany

Copyright © 1976 Pergamon Press Inc.

Library of Congress Cataloging in Publication Data

Georgescu-Roegen, Nicholas.
    Energy and economic myths.

    Includes index.
    1. Economics -- Addresses, essays, lectures.
2. Production (Economic theory) -- Addresses, essays, lectures.   3. Institutional economics -- Addresses, essays, lectures.   I. Title.
HB34.G38   1976   330   76-10265
ISBN 0-08-021027-9

All Rights Reserved. No part of this publication may be reproduced, stored in a retrieval system or transmitted in any form or by any means: electronic, electrostatic, magnetic tape, mechanical, photocopying, recording or otherwise, without permission in writing from the publishers.

Printed in the United States of America

To
Paul Anthony Samuelson

Contents

    Foreword   ix

    Acknowledgments   xxvii

Part I: Natural Resources and the Economics of Production

1. Energy and Economic Myths (1972)   3
2. Process Analysis and the Neoclassical Theory of Production (1971)   37
3. The Entropy Law and the Economic Problem (1970)   53
4. The Economics of Production (1969)   61
5. Process in Farming Versus Process in Manufacturing: A Problem of Balanced Development (1965)   71
6. Economic Theory and Agrarian Economics (1960)   103

    Postscript (1966)   142

    Postscript (1975)   146

Part II: Institutional Economics

7. Structural Inflation-Lock and Balanced Growth (1968)   149
8. The Institutional Aspects of Peasant Communities: An Analytical View (1965)   199

Part III: Epistemology and Methodology

9. Dynamic Models and Economic Growth (1974)   235
10. Further Thoughts on Corrado Gini's *Dellusioni dell'econometria* (1966)   255
11. Measure, Quality, and Optimum Scale (1964)   271
12. Toward Partial Redirection of Econometrics (1952)   297

Part IV: Pure Theory

13. Vilfredo Pareto and His Theory of Ophelimity (1973)   307
14. A Diagrammatic Analysis of Complementarity (1952)   351

    Index   371

Foreword

With three exceptions the essays reprinted in this volume were published initially after 1964, the year when the volume *Analytical Economics: Issues and Problems* (which includes a selection of my essays published between 1935 and 1960) was put together. The exceptions are the essays reprinted here as Chapters 6, 12, and 14. The reasons for their inclusion will be explained at the proper places later.

The eight essays of Part I and II are the product of a fundamental change in my orientation as an economist. However, this reorientation did not cause me to abandon my interest of long standing in theoretical economics and in some general epistemological issues. The essays of Part III and IV as well as some of Part I bear witness to this. The reader will thus know the reason for the exceptionally broad spectrum of topics covered by this volume. The retrospective comments offered below emphasize the highlights of each of these essays and add a few afterthoughts, but they are also intended to serve as a perusal guide.

Economists, as the pundit Percy W. Bridgman once argued, are the most opportunistic scholars. However, the opportunism Bridgman had in mind differs from the bandwagonning of the last two or three years when works on the energy problem came from writers of almost any profession, most of them speaking of the Entropy Law with only a superficial understanding of it. With this clarification, economists must plead guilty to Bridgman's charge, but we must reject the implicit castigation. For the economic profession would have failed in its fundamental mission if it had not swung its attention from one issue to another as issues changed with the continuously changing economic process. What would physicists themselves have done had the properties of matter been subject to a change as rapid as that of the economic process? The economist's predicament is that this process (especially in recent times) seems to change faster than a student can learn about it. Modern economics, in fact, has shifted its focus of attention even faster than its predecessors. At the 1973 annual meeting of the American Economic Association I happened to chair a session on the lexicographic ordering of utility. The session was scheduled in a small conference room of some forty seats. In my opening remarks I pointed out that had a session on utility theory been on the program some time in the 1950s, the session would have had to be held in a Grand Ball Room, and even so the latecomers would have found only standing room! But by 1973, economists had shifted their attention to other subjects—to input-output systems, linear programming, economics of education, social welfare, health, poverty, urban economics, and so on. The shifting has usually been so swift that by now many of those topics are no longer in vogue.

But the economic process changes not only with time; it changes also with place—a truth borne out by history from the dawn of civilization to our own day. At the time when the Pharaohs' economic system was building the pyramids, which after more than four thousand years continue to be the object of general admiration, the tribes of Central Europe were living in an economy of the Cro-Magnon type. Equally great, if not greater, differences exist even today; one need only compare the economy of the United States with that of the tribes of Kalahari. Some authors—I in perhaps the strongest vein—have argued that those economists who have been interested in studying the actual economic process, not some wholly imaginary abstraction, have come out with a theoretical edifice molded on the particular economic reality in which they have happened to live. Witness the fact that some economists have now set up courses on the economics of crime. The long chain of evidence goes back in time to the earliest preoccupations with economic problems. And the case of my own reorientation is only one small link in it.

Even though modern economists have shifted rapidly from one topic to another, especially in the last 40 years, in one particular respect they generally have kept to one and the same furrow. They have unflinchingly (and in general unconsciously as well) clung to the mechanistic epistemology inherited from the stalwart founders of the Neoclassical school. Directly as well as indirectly, this fact has been responsible for most of the drawbacks of standard economics that have culminated in the current display of impatience, at times, and disorientation, at others, by some of the most highly regarded representatives of the profession.

W. Stanley Jevons, one of the two authentic Neoclassical pioneers, *did* write a book in 1865, *The Coal Question*, in which he examined the problem of Great Britain's exhaustible coal deposits in relation to that country's economic prosperity. Yet, soon thereafter, the mechanistic epistemology, which was still prevalent among philosophers and scientists, got the best of him, and he set out to found a new economics as *"the mechanics of utility and self-interest"* (1871). It may be that Jevons, like all later disciples of the new doctrine, abandoned any interest in the issue of natural resources because of the mineralogical bonanza of the last one hundred years or so that has benefited especially the few countries which have been the front runners in industrial progress. It is, however, highly symptomatic that later on, in his *Principles of Science* (1874), Jevons took care to reveal his utter fascination with the mechanistic dogma of Laplace. And because in mechanics everything consists of locomotion, which is completely reversible, the new school came to equate the economic process with an isolated, self-sustaining roundabout movement between "production" and "consumption," as we find it portrayed graphically in almost every introductory manual.

With the mechanistic epistemology, the mathematical tool also was introduced into the economist's workshop, a mutation that ought to have constituted a blessing. However, the mechanistic epistemology by its very nature encouraged an uncontrolled use of that tool. The result has been an outgrowth of paper-and-pencil (PAP) exercises, most of them having no relation with actuality. Unfortunately, the outgrowth is far from benign. It has given standard economics its present tonal mode, so much so that PAP endeavors, even when empty, now represent the highest aspiration of professional performance. The hardest task of any special science—that of coming to grips with facts—has been relegated to a secondary level of importance and of professional respect. And so have the preoccupations with institutions and qualitative change which are the essence of evolution.

These aspects of standard economics are discussed repeatedly and pinpointedly in the essays of this volume. However, my past experience tells me that at this juncture one point needs unparsimonious emphasis in order to prevent (as much as may be possible) misinterpretations of my position. I would be among the last servants of science to deny the indispensable role of theory, which must necessarily aspire to be quantitative and hence mathematical, provided "theory" is not separated completely from fact. But, as my master Joseph A. Schumpeter did so poignantly, I would also be among the first to defend the absolute necessity of historical and institutional studies in social sciences, hence in economics. To stress the need for a change of venue of the economic profession in this direction, I have placed the pieces concerned with natural resources and institutional aspects at the beginning of this volume and those of a theoretical or methodological nature in the last parts.

PART I: *Natural Resources and the Economics of Production*

What caused me to look at the economic process from an unorthodox viewpoint is the particular nature of the economy of my native country, Romania, at the time when I returned from my training in the Western schools with a formidable armamentarium of mathematical standard economics. To begin with, I despaired at discovering that that armamentarium could hardly help me penetrate the economic problems of that country. Romania's institutions were not adapted to the Walrasian principle of profit maximization, a fact which at first appeared to me and, most certainly, to any Western observer as crass organizational ineptitude. Gradually (and not without some intellectual distress) I came to realize that those institutions—imperfect though they were as all institutions inevitably are—were nonetheless marvelously adapted to an agrarian economy in which a large part of the population (about fifty percent) was wholly superfluous relative to the available land and capital.

I had to search for a new theoretical scaffold in order to explain the internal logic of these institutions. I was thus led, first of all, to the conclusion that Walrasian economics completely ignores the issue of its applicability. As I pointed out in a 1955 essay (Chapter 10 in *Analytical Economics*), the elegant Arrow-Debreu proof of the existence of a Walrasian equilibrium is idle in this respect. The reason is that it rests on the assumption that every member of the community is endowed *ab initio* with a real income sufficient for life—which means that the economic problem of the community has already been solved by some means or other. But a definite expression of my new message appeared only in a 1960 essay, which for this very reason is reprinted below (Chapter 6).[1]

In a country in which an economically "superfluous" population presses on the scarce land, as in Romania's case before World War II, a wage system based on the marginal productivity of labor would be utterly inapt; it would decrease the real product and increase *unwanted* leisure. I was confronted with "a reality without theory," i.e., without a logically ordered description of its mode of functioning. In my endeavor to fill this gap, I brought to light the reason why the standard framework for consumer behavior cannot fit the conduct within peasant communities. The "utility" function of a peasant contains not only the amounts of commodities available to him, but also his actions as judged by the cultural matrix of the village. This result prompted me to analyze the institutional aspects of the peasant communities from a broader viewpoint. Most of the results in this direction, completed over the years even after I settled in the United States, formed the object of another essay included here as Chapter 8 (to which I shall return later on).

The main result of the 1960 essay is the analytical framework developed for production and distribution in overpopulated agrarian economies. This framework lends support to the platform of the Agrarian movement. To recall, Agrarians claimed that only a structure of family farms supplies the answer to the misery of such economies, of which there still are numerous examples in the world. The validity of my conclusions has been vindicated by the agrarian policies of the Communist governments of Yugoslavia and Poland. The logic of my argument has also provided the basis for some works by prominent agricultural economists from Poland and even from the USSR ("Postscript 1966" to Chapter 6). I should however add here that an agrarian reform cannot be a solution to overpopulation beyond any limit. The only human solution to the population explosion is a steady decrease in fertility.

Signs of the dreadful consequences of the pressure of population on the limited carrying capacity of the earth—hunger and starvation—were visible (for whoever was willing to see them) long before the recent racket about energy began. Already in the fifties the production of food needed to be increased by at least thirty percent in order to satisfy the basic nutritional needs of everyone. The situation is now much worse, nay, desperate. The explosive pressure of population within this

---

[1] Not that I had not tried long before 1960 to bring some of my colleagues to see my position. I had tried as much as I could, but without success. Sixteen years since, I still harbor a sense of gratitude to the editorial board of the *Oxford Economic Papers*, to George Richardson in particular, for their enthusiastic reception of my unorthodox piece.

finite planet—on which the Classical economists based their pessimistic prognosis—can no longer be subject to doubt. There is no longer another America or Australia to provide a safety valve for population pressure. World population has recently exceeded the four billion mark. Last year, it grew by eighty million people, as many as the populations of East and West Germany combined. To feed only these eighty million additional people (which usually consist of people of all ages) the production of grains must be increased by at least twenty million tons. And the sad story is that the present rate of production does not suffice to maintain even the present population at the floating nutritional line.

Sadder still, the situation is much worse than such global figures show. In the early fifties, again, two-thirds of the world population lived on the verge of starvation or actually were starving. Suffering in the greatest part of the world is badly aggravated by the unequal distribution of population relative to the production capacity of one sort or another. As already pointed out in my 1960 essay, even if the difference in soil productivity is taken into account, "the average Yugoslav would still have only one quarter as much food as the average Dane," simply because the density of people with respect to agricultural land was, at the time, four times as great in Yugoslavia as in Denmark. To apprehend how grim the consequences of the uneven access to resources as well as of the unchecked population growth can be, one need only imagine (if anyone could) the United States populated as densely as Bangladesh today. It is hardly thinkable that even the United States could feed that population, which would then be five billion people!

From what the received doctrine had first taught me, I thought, together with all standard and Marxist economists, that there is no need to distinguish between the laws governing agriculture and those governing industry. The same contact with the agrarian problems of Romania soon convinced me that that view is utterly wrong. To find out the irreducible reasons for the difference between the two activities, I proceeded to analyze the nature of the farming activity in great detail, greater than even agronomists had done. From the very beginning I ran into a major obstacle: the absence of any explicit analytical definition of "process," a notion so frequently used in all disciplines. Even the production function by which standard theory represents a productive process—whether in farming, mining, or manufacturing—proved under strict querying to be mainly a vapid verbalism translated into a colorless formula.

To remove this difficulty, in an essay prepared for the 1965 meeting of the International Economic Association and reprinted here as Chapter 5, I developed a general analytical representation of any material process. Although the foundation of this new conception was laid out in that paper, some aspects of it were clarified and some additional consequences were reached in two subsequent papers (Chapters 2 and 4 below).

A first point made obvious from the outset by the new approach was the necessity of dividing the factors involved in a material process into two analytical categories: the *funds* (i.e., the agents) and the *flows* (i.e., the materials undergoing transformation). The application of this idea uncovered an issue theretofore ignored—the issue of capital utilization (which has since induced several follow-through contributions). In a practical way, capital utilization bears on development planning; in a theoretical way, on the determination of the minimum total cost of production. An extensive technical analysis of this last problem was later offered in the essay reprinted here as Chapter 2.

The new analytical conception immediately brought to light the immutable reason for the essential difference between the process of a farm (almost everywhere on the globe) and the process of a factory. The immense economic advantage of the factory over a farm lies in the possibility of eliminating the idleness of all agents. A glaring illustration is the economy achieved by the transformation of the chicken farms in the United States into "chicken factories," which now produce chickens *in line* (as in a factory), instead of *in series*.

The idea that the economic process is not a mechanical analogue, but an entropic, unidirectional transformation began to turn over in my mind long ago, as I witnessed the oil wells of the Ploesti field of both World Wars' fame becoming dry one by one and as I grew aware of the Romanian peasants' struggle against the deterioration of their farming soil by continuous use and by rains as well. However, it was the new representation of a process that enabled me to crystallize my thoughts in describing for the first time the economic process as the entropic transformation of valuable natural resources (low entropy) into valueless waste (high entropy). I may hasten to add, as I did in Chapter 5 and *Analytical Economics*, that this is only the material side of the process. The true product of the economic process is an immaterial flux, the enjoyment of life, whose relation with the entropic transformation of matter-energy is still wrapped in mystery.

Following a suggestion of the biologist Alfred A. Lotka, I developed the idea that the economic process is an extension of the biological evolution, a fact by which the human species is differentiated in an essential way from all other species. It took forty-five million years for the Eocene *Eohippus*—an animal not bigger than a beagle—to be changed by biological mutations into the powerful horse of today. The human species found a far speedier way of becoming more powerful in numberless directions. It began to produce detachable limbs—exosomatic organs — instead of waiting to acquire them by biological mutations. This idea, expounded in the Introduction of my *Analytical Economics* and in greater detail in *The Entropy Law and the Economic Process*, revealed the close connection between mankind's exosomatic mode of existence and the Entropy Law. With the exosomatic evolution, the human species became addicted to the comfort provided by detachable limbs, which, in turn, compelled man to become a geological agent who continuously speeds up the entropic degradation of the finite stock of mineral resources.

The highlights of this new viewpoint together with a few consequences were presented in a Distinguished Lecture at the Economics

Department of Alabama University (Chapter 3). In that lecture, among other things, I exposed the concept of optimum population and insisted on the point that in discussing the population problem one must also take into account how long a given population may survive on the finite stock of the earth's resources. A deeper point was that the "amount of life" of the human species, measured in man-years, depends not only upon the size of that stock but also upon the speed of its use. It led to a result apt to cause general frowning: technological progress is not always a blessing in regard to the economy of energy. The weightiest example is the mechanization of agriculture, which substitutes the scarcest kind of energy (of terrestrial origin) for the free solar energy. Why, this last energy reaches us in the prodigious flow of 10^{20} BTU per week and will also last for at least another four billion years!

Solar energy had been discussed by several previous authors, who focused however on its *direct* use. Also, over the last one hundred years preoccupation with the conservation of the environment has been spurred on and off by deforestation, soil erosion, fish depletion, and the pollution of lakes and rivers. But it was only about ten years or so ago that some of us—Georg Borgstrom, Kenneth E. Boulding, Harrison Brown, Ezra J. Mishan, Hugh Nicol, and Joseph J. Spengler to mention only those writers best known to me now—began independently and each one in his own way to challenge the idea of continuous economic growth, which formed at that time (and still forms) the proudest article of standard economic faith. Other writers competent to speak on the problem of mankind's continuous existence within a finite environment joined our ranks. However, by now almost anyone—from preachers to tax assessors—expatiates on the energy crisis. In Garrett Hardin's sense, this is a new "tragedy of the commons." Not surprisingly, such headlong interventions involve many myths, myths that can only harm the understanding of the gravity of the problem and the search for a solution compatible with the laws of matter-energy.

One commentator from a Communist country thus claimed that in my recent works I prove how capitalism struggles for its life in the claws of the Entropy Law—as if Communism were immune to that law. Symmetrically, our Washingtondom must believe that capitalism, too, is so immune, for otherwise it would not have come out with Project Independence, in which we are still invited to have faith. Economists set aside the issue with the opiate "come what may, we shall find a way." Apparently, we will never stop overselling technology even when it moves against the economy of mankind's dowry of energy. To wit, in a paper prepared for the *Symposium on Technology and Public Policy* (Vanderbilt University, November 6-7, 1975), John Sawhill, a former head of the Federal Energy Administration, still cried victory over the discovery of how to produce food protein from crude oil. A landslide of specific recipes for saving energy also confronts us nowadays; but many remind us of the story of Nasreddin Hodja in which the holes in his coat sleeves were patched with a piece of cloth cut from the bottom of his pants.[2]

[2] Nasreddin Hodja is a legendary Islamic character in numberless short, witty stories, everyone with a valuable object lesson. See *Encyclopedia Britannica*, Volume 9, page 966.

For reasons such as these I have insisted on the necessity of projecting every scheme and every analysis of the environmental problem against a comprehensive background consisting of the laws of thermodynamics and a proper inventory of mankind's dowry of accessible matter-energy. We must continuously bear in mind that, in a plastic image, our accessible environment is like an hourglass which cannot, however, be turned upside down and in which the useful matter-energy from the upper half turns irrevocably into waste as it continuously pours down into the lower half.[3]

In the lecture delivered in 1972 at the School of Forestry and Environmental Studies of Yale University (reprinted here as Chapter 1) I surveyed the most popular economic myths and also touched upon a few myths of larger currency. In addition, I stressed some points of my previous works and offered, in conclusion, a minimal bioeconomic program.

Throughout my discussions of the Entropy Law I took the position that matter, not only energy, is subject to entropic degradation. The illustrative systems considered in thermodynamics do not deal in a specific way with this obvious truth although the most general formula for entropy includes the appropriate terms. We may burn a piece of coal more slowly or more quickly. But, once burned, its available energy is irrevocably degraded into the unavailable form. Similarly, we may wear out a penny more slowly or more quickly, but the molecules dissipated through use are irrevocably unavailable for another use. As I insisted repeatedly, both available energy and available matter can be used only once. Curiously, however, technology, instead of economizing available matter and energy, has often encouraged people to squander them. "When the blades are finally dull, you just toss the whole thing [rasor] away," boasts a recent advertisement of Gillette Company. Why not toss the whole automobile away when the ashtrays are finally full?

If we speak nowadays only of the energy crisis, it very likely is because when the oil embargo came we had plenty of matter—in the form of plants, airplanes, and automobiles—but we lacked energy. A sophisticated argument for omitting matter might invoke the Einstein equivalence between mass and energy, $E = mc^2$. However, under the conditions in which terrestrial life is possible the formula works mostly in one direction, from mass to energy. Only in the laboratory and for a few atoms of the smallest weight can we produce matter from energy alone. The copper used in electrical items, for example, existed as copper ever since the formation of the earth (at least) and will continue to exist as such regardless of its entropic dissipation.

Matter must therefore be kept as a separate item from energy in any analysis of needs and resources. The bare truth is that we need both matter and energy to obtain either matter or energy. And since matter

---

[3] Nicholas Georgescu-Roegen, "Economics or Bioeconomics?", paper read at the 1975 AEA Meeting in Dallas, and "A Different Economic Perspective," paper read at the AAAS Boston Meeting, February 21, 1976. (Both papers are incorporated in a forthcoming book.)

and energy cannot be brought to a common denominator, there is no way to reduce our economic balance sheet to a single coordinate—even if we were to ignore the other important factor, the disutility of labor.[4] There is no such a thing as net energy, any more than there is net matter. We can speak only of *accessible matter* and *accessible energy* (as I did in Chapter 1).

The above considerations help to clear up the confusion surrounding the scarcity of matter. Since our spaceship is a closed system (i.e., it exchanges with outer space only energy), in the end matter may prove even more crucial than energy for the maintenance of industrial activity at the level now prevailing in the advanced economies. What about recycling?, one may ask, invoking the cycles of carbon dioxide, oxygen, nitrogen, etc., so dear to all ecologists. Here again, those who extol recycling ignore the sting of the idea. Dissipated matter cannot be recycled; we can recycle only what we may call "garbojunk"—scrap paper, used aluminum cans, used cables, etc. Actually, we can recycle only part of this "garbojunk," as evidenced by "nuclear garbage." Just because the entropic loss that accompanies the cycles extolled by ecologists is not conspicuous over a short time interval does not mean that the loss is not there. To see this, one need only compare the present conditions of our planet with those prevailing in earlier geological eras. A closed system, such as the earth, cannot therefore be an everlasting steady state unless all entropic transformation of matter has ceased[5] —as is roughly the case with the moon.

Today we turn our hopes toward energies other than that of fossil fuels. Systems for harnessing solar energy, for example, are being offered continuously. In judging the economic importance of these systems, however, we must bear in mind that the materials used in their construction were produced mainly with the aid of fossil fuels. It is therefore in perfect order to ask whether such systems could reproduce themselves (an essential condition for any *viable* structure!) without the support of fossil fuels. In this respect we have only some indirect indications. The cost of the most recent installation of solar energy (at the George A. Towns Elementary School in Atlanta) was close to one million dollars, although the edifice is small and the installation provides only sixty percent of the needed energy.

This suggests that the tremendous amount of matter necessary to capture the ultrafine mist of solar energy may constitute a fantastic burden in real terms for an economy based on that energy alone. The same observation applies to other forms of energy. The Fermilab accelerator, for instance, has a diameter of one mile and one quarter and includes several thousand magnets.

Matter dissipates as a consequence of natural laws, through chipping, dissolving, wind blowing, etc. But material entropy is further increased by the actions of almost all living creatures, especially man's.

---

[4] The point can be proved easily by an input-output representation of all the operations. *Ibid.*

[5] "A Different Economic Perspective." The point bears on Sec. viii of Chapter 1.

When we cut down a tree and burn it far away from the forest, we cause an additional material dissipation. The same is true of the glass of milk drunk away from the farm as well as of the graphite of the pencil with which I am now writing. Some vital chemicals may thus become dangerously scarce in the future. With the increasing shortage of fossil fuels and the rejection of the Faustian deal with plutonium (if we are wise), the logical panorama for the future of mankind is a radical deurbanization with most people practicing organic agriculture on family farms and relying on wood for fuel and many materials, as in the traditional villages.[6] The positions of the Agrarians of Eastern Europe or of the famous group of Vanderbilt University no longer deserve to be scoffed at as flights of fancy.

Naturally, world population must be lowered to the level at which it could be fed by organic agriculture. To set a higher level by counting on mechanized agriculture and heavy-feeding high-yield varieties means to turn one's back to the present handwriting on the wall.

Naturally also, the nations with the highest population densities have to reduce their birth rates more drastically. This idea, unfortunately, has given rise to heated controversies that have paralyzed the work of several international conferences. To wit, the chorus of the representatives from heavily populated countries argued, in opposition, that the solution to the present hunger and misery is a redistribution of the means of production now concentrated in the advanced economies. Obdurate partisanship has blinded both sides from recognizing that they are both right.

Unequal distribution tends to maintain itself and, possibly, to become more unequal. The reason, apparently overlooked, is that powerful economies command an immense purchasing power, which, among other things, puts them in a dominant position on all markets of essential resources. Underdeveloped nations can thus have only a small share of those markets, far too small to sustain development. The conclusion is that the world must subscribe to a policy epitomized by "Factories, not food, for the hungry." The factories should enable the hungry nations to produce in the first place their own agricultural implements and fertilizers—not luxury consumer goods for their "elites," as has been the result of much of the international aid.

We must not ignore, however, the purely technical difficulties of this program. Just as endosomatic organs divide creatures into classes, species, etc., exosomatic organs have divided mankind since of old into exosomatic species. *Homo Indicus* is a distinct exosomatic species from *Homo Americanus.* The upshot is that our exosomatic organs are ill-fitted for the masses of the Asian subcontinent. Those people use

---

[6] In Chapter 1 I suggested that thermonuclear energy may remain forever beyond our control, as is the case with other "bombs"—gunpowder and dynamite. Highly symptomatic in this respect is the fact that Edward Teller, who in 1960 was fiercely against nuclear reactors and entertained sanguine hopes for controlled fusion, has recently reversed himself fully. Be this as it may, we should not erect edifices without staircases and elevators by counting on the future possibility of screening gravitation.

primitive cooking contraptions which burn dried dung; we are using electric, self-starting, self-stopping, self-cleaning stoves. Our R & D is mainly interested in innovations above our exosomatic level, whereas the economic development of the Indian people must necessarily proceed from their own level. Our R & D (as well as theirs!) must be reoriented accordingly. Ignorance of this exosomatic stumbleblock is the main cause for the conspicuous failure of the immense financial aid to develop the underdeveloped countries. A counter-proof is the striking success of the USA aid in putting back on their feet nations that already belonged to the same exosomatic species as *Homo Americanus.*

All this calls for a radical change of values *everywhere.* Only economists still put the cart before the horse by claiming that the growing turmoil of mankind can be eliminated if prices are right. The truth is that only if our values are right will prices also be so. We had to introduce progressive taxation, social security, and strict rules for forest exploitation, and now we struggle with anti-pollution laws, precisely because the market mechanism by itself can never heal a wrong. Some economists even speak of the cost of pollution or of resource depletion, thus proving that they are completely unaware of the physics of economic value, i.e., of thermodynamics. Prices are only parochial elements of the economic struggle of the human species. In any case, they cannot reflect the needs of future generations for the simple reason that those generations cannot bid on the present day market. In defense of standard economics, we are frequently told that the algorithm by which income is distributed according to the preference between the present and the future will take care of the intergenerational allocation of resources. The rebuttal of this position was offered in the essay reprinted here as Chapter 1. Here I wish to add only that it is utterly inept to transpose to the entire human species, even to a nation, the laws of conduct of a single individual. It is understandable that an individual should be impatient (or myopic), i.e., to prefer an apple now over an apple tomorrow. The individual is mortal. But the human species or a nation has no reason to be myopic. They must act as if they were immortal, because within the immediate horizon they are so. The present turning point in mankind's evolution calls for the individual to understand that he is part of a quasi immortal body and hence must get rid of his myopia.

PART II: *Institutional Economics*

Although Lord Keynes' *General Theory* is a seminal masterpiece, his analysis stopped short of a truly physiological examination of the relationship between the monetary system and the real coordinates of the economic process. His followers retreated further from this particular program, so that Keynesianism was gradually reduced to a mechanico-descriptive piece of knowledge limited to simple arguments about the ultra familiar graph with the $45°$ straight line. More recently, the attention was shifted entirely to an even simpler mechanico-descriptive formula, the Phillips curve.

Keynes' advice concerning "government spending" was misunderstood from the very beginning.[7] For a convenient analogy, from the idea that a shot of insulin is beneficial to any person who happens to be diabetic, Keynes' followers concluded that such a shot will make even a healthy person healthier. The introduction of the Phillips curve provided an even stronger incentive for latter-day Keynesians to preach government by inflation instead of government by taxation. The few opposing voices stood no chance of gaining their case in the public court, first, because all politicians are naturally happy to have the support of scholars for the easiest, yet the most perverse way of governing, and second, because they fought with equally blunt weapons. That is, they also stopped short of any structural analysis of the effects of inflation.

In all periods of history, inflation has been a legal, albeit abusive money forging. The cost of one ton of steel is always one ton of steel, of an egg an egg, but the cost of a forged dollar is much less than one dollar. The forged money goes immediately into the pockets of some groups, which vary with the political interests of each government. There are other groups who pay for this, for in real terms inflation operates a covert transfer of income from people who live on contractual incomes and also are politically powerless—they cannot strike, for example. In every country, they are the masses of retired persons, in some countries, the earners of controlled wages and salaries.

It is, therefore, clear that inflation changes the income distribution and this, in turn, changes the general profile of industrial activity. The rub is that if inflation is stopped or only slowed down, the profile of demand will change again. But this time, some industries will discover that inflation has caused them to build an excess capacity. Unemployment will result and spread to other industries by the usual multiplier effect.

In Chapter 7, the reader will find this kind of physiological analysis of inflation applied to Latin American countries, with the accent on Brazil, a part of the world where money forging by the government seems to be an endemic disease. However, even in a country in which all employees would be entitled to *instantaneous* escalation of their remuneration, they would still be losers; their pensions and life insurance policies will nevertheless depreciate, and (if taxation is progressive) their real taxes will automatically increase.

The essay included as Chapter 8 stems out of Chapter 6. It is an attempt to uncover the internal, economic reasons behind the institutions that have been observed in traditional villages in diverse parts of the world. The traditional village has amazed every observer by its power to meet adversities and to maintain a social stability not encountered in other communities. Since no other social organization has had as tremendously long a life in the entire known social evolution

---

[7] As proved by a very recent lament of one of the most distinguished partisans, Lawrence Klein, who admitted that "we Keynesians probably underestimated the severity of the problem." See his "Intractability of Inflation," *Methodology and Science*, VII, No. 3 (1974), p. 158.

of mankind, all its object lessons acquire a unique importance for any social scientist.

Among the salient points of my analysis is a fact with a topical flavor. The exceptional viability of the village community derives from an interesting, albeit overlooked feature of the village itself. Every village has been (or still is) a truly organic slice of the environment; it includes sufficient plow land, some woodland, some orchards, a body of water, and a convenient "hearth" of the village, i.e., a place where homes surrounded by gardens can be built. In a traditional village community an almost self-sufficient economy moves along smoothly under the power of the sun and with as much local material recycling as possible (a point to which I alluded in Part I). It was only under the pressure of population and with the subjugation of the countryside by the towns armed by science that the traditional village disintegrated in many parts of the world, although it still survives, not surprisingly, in some.

The village community teaches us another important fact about social economics. During its immensely long life, that community was always concerned only with the distribution of income, not with the distribution of property. Property in the modern sense did not even exist in it. The regimen of private property is a recent development; at bottom, it is but a rule for the distribution of income. Historically, this regimen emerges as a brief interlude. In fact, it already is on its way out in many parts of the world. But the social conflict will not disappear with it, precisely because only income matters in the end.

PART III: *Epistemology and Methodology*

In my statistical dissertation I developed a method for discovering the cyclic components of a movement. At that time (1930), I was reluctant to apply my method to an economic series, because I already felt that economic data cannot satisfy the conditions of the ordinary statistical theorems. Very likely, I was influenced by the characteristic skepticism of Continental schools of thought on matters such as these. Be this as it may, my first feeling grew into a firm and lasting conviction. I voiced it for the first time in my contribution to a symposium (Chapter 12). By then I had become impressed with the practice, still widespread among economists and econometricians, namely, to transform the economic data and combine them in numberless ways until a satisfactory fit is obtained. An econometrician, I said then, proceeds like an expert sculptor who would tell you that inside a given log there is a beautifully sculptured Madonna, and would prove it after carving the log with specially chosen chisels!

Even more crucial is the absence of any concern for whether the formula thus obtained will also fit other observations. It is this concern that is responsible for the success natural scientists have with their formulae. The fact that econometric models of the most refined and complex kind have generally failed to fit future data—which means that they failed to be predictive—finds a ready, yet self-defeating, excuse:

history has changed the parameters. If history is so cunning, why persist in predicting it? What quantitative economics needs above all are economists such as Simon Kuznets, who would know how to pick up a small number of relevant variables, instead of relying upon the computer to juggle with scores of variables and thus losing all mental contact with the dialectical nature of economic phenomena.

There are also sins against good practice of principles. Of the hosts of writers who compute regressions and claim success by t- or F-test, I know of no one who tried to see whether the data are normal, homoscedastic variables. Here again there is an excuse, that there are no tests for other kinds of variables. Would a respectable physician argue that, since there is no proven test yet for finding out whether some cell in a patient's body is just becoming cancerous, he can use a test for diabetes instead?

In Chapter 10, I also showed that only in quite special cases does one regression line coincide with the analytical law. Still worse, without the possibility of a controlled experiment, we can never discover the analytical law. If any one then takes the regression line (which one?) for the analytical law, one may be rightly reminded of a Josh Billings' proverb: "It's better to know nothing than to believe in what ain't so."

In the essay written in honor of P.C. Mahalanobis (Chapter 11), the reader will find an epistemological discussion of the idea that any kind of measure, if it is to be at all relevant, must reflect some physical properties, some possible actual operations. I applied this idea step by step to the construction of an axiomatic basis for cardinal measurability. On this occasion, I proved that the Archimedean Axiom (which occupies a prominent place in the theory of measurability) is not a sufficient condition for an ordinal set to be ordinally measurable. The essay also sought to clarify the role of quality in problems concerning the theory of the firm, especially the problem of optimum size and of the measurability of management. Finally, the fact that the formulae involving quantified qualities are generally nonlinear was explained as reflecting a qualitative residual.

Chapter 9 is the most recent of all essays included in the volume. Its thesis is that standard economics has concentrated on *mathematico-imaginative* and *mechanico-descriptive* types of inquiries, thus leaving out practically any *physiological analysis*, that is, leaving out what made the glory of the Classical leaders as well as of a few modern economists—Alfred Marshall and, above all, Joseph A. Schumpeter. The temper stems from the ease with which one can do mathematics on paper. Mathematics, in turn, kills the bud of any thought of qualitative analysis. For a most convincing illustration, I have considered the sacramental formula of dynamic economics: "save-invest-grow." This formula does work, but only if what one saves—which can consist only of consumer goods—can be exchanged for capital equipment from other economies. For a closed economy, such as the whole world, dynamic models cannot even represent mere growth, let alone evolutionary development—a point which I established with the aid of probably the most popular model, the Leontief dynamic system. Deep at the bottom of this system lies the assumption that either consumers eat bulldozers or yogurt is capital equipment.

PART IV: *Pure Theory*

Chapter 13 was written for the Pareto Commemorative Congress by which the Italian National Academy marked the fifty years since his death in 1923. As its title explains, it is a general discussion of Pareto's momentous contribution to economics from the viewpoint of our present knowledge of consumer behavior. Its thread is in part historical, but my main intention in writing it was to integrate Pareto's theory of ophelimity into a more general scheme of choice inspired by Paul Samuelson's famous approach known as Revealed Preference. In doing so, I was led to put the missing dots on some i's of this last theory and separate some chaff from the grain by showing which relations represent only paper-and-pencil (PAP) operations and which relations describe facts of choice.

The salient result reached in the end vindicates the position I took in a 1936 article (*Analytical Economics*, Chapter 1) concerning the integrability of the Pfaffian derived from consumer budget data. To recall, in his *Manuale* Pareto argued that the indifference varieties can be obtained by integrating that Pfaffian. To this, Vito Volterra immediately objected that (even "smooth") Pfaffians are not necessarily integrable for more than two variables—a well-known mathematical proposition. In retrospect, I feel that Pareto was intimidated by Volterra's towering mathematical prestige. Be this as it may, Pareto's attention was diverted thereby into a wrong avenue. The integrability issue was thus born.

What Volterra failed to take into account is a point established in the cited article and apparently totally ignored until 1950 when it popped up in the literature under a different garb and under the given name of "the Houthakker-Samuelson Strong Axiom of Revealed Preference." What I proved by a particular example was that even for two commodities (or for movements in a three-commodity budget plane) when a regular Pfaffian is always integrable, the integral varieties do not necessarily provide a basis for an ophelimity potential. And I concluded that integrability is "without any meaning outside the transitivity condition [of preference]." This was a rather clumsy, yet quite proper way at the time of saying that even if integral varieties exist they do not represent a preference field unless the existence of consistent binary preference is *already assumed*. In Chapter 13 below, I added an even more convincing example, that of an integrable Pfaffian derived from community demand data, which constitutes a double play—one against the sufficiency of the Revealed Preference theory, the other against Volterra's intervention. This vindicates Pareto's initial, but not explicitly outlined position, which was that integrating the Pfaffian is just another method for determining the binary preference field whose existence cannot be derived from integral varieties, but must necessarily be presumed. This method, like all behavioristic methods, uses data obtained by "spying" on the consumer. The initial method of Pareto, that of binary choice, requires "confessional-type" data. The former succeeds, only if we know that the latter works.

The last of the exceptions mentioned at the outset, a 1952 essay, has been included in this volume as Chapter 14 because of several far-reaching results. The first is that from the purely formal viewpoint there is no difference between the definition of complementarity between consumer goods first proposed by R. Auspitz and R. Lieben in 1889 (and later expanded by Vilfredo Pareto and F.Y. Edgeworth) and the modern definition introduced in 1934 by Sir John Hicks and Sir Roy Allen. Complementarity, however conceived, needs a background against which the individual's preference may be revealed. The classical definition uses the cardinally measurable utility for such a background; that is, the individual is observed while he moves on his utility surface. The number of *utils* the individual gains by obtaining two commodities separately is compared with the number of *utils* gained when those commodities are taken together. The Allen-Hicks definition does not rest on the cardinal measurability of utility, but, just like the classical one, it observes the individual as he moves on a surface, any of the indifference surfaces between Marshall-Hicks money and the two commodities in question. What is used now for comparison is money instead of cardinal utility. As Samuelson rephrased this point in his commemorative piece, the Allen-Hicks definition represents the same idea as the classical concept, but uses a "money-metric" instead of a "utility-metric."[8]

In the same essay, I introduced a new analytical tool, the *isotimetic*, the loci of constant marginal utility, which can be profitably extended to the marginal substitution of money for a consumer good in a money-metric or to the marginal productivity (when some factor prices remain constant in marginal pricing). With it, I proved the basic points on which Sir John Hicks' analysis of the multi-exchange equilibrium rests in *Value and Capital*. In particular, I showed that, whatever the metric, no three goods can be complementary with each other; complementarity between C_1 and C_2 implies high rivalry for the pair (C_1, C_3) and (C_2, C_3).

Finally, the same paper introduced a simple topological property of "independent" utilities or, in a more general context, of "separability" of variables. The property concerns the case in which the family $F(x_1, x_2, \ldots, x_n)$ = constant can be represented by the form

(1) $f(x_1, x_3, \ldots, x_n) + g(x_2, x_3, \ldots, x_n)$ = constant,

in which variables x_1 and x_2 are separated. Let a coordinate rectangle (or a "box") be construed so that three of its vertices are on any two isoquants corresponding to x_i = constant for $i > 2$. The form (1) is available if and only if the fourth vertex of the box slides at all times on

---

[8] Paul A. Samuelson, "Complementarity—An Essay on the 40th Anniversary of the Hicks-Allen Revolution in Demand Theory," *Journal of Economic Literature*, XII, No. 4 (December 1974): 1255-1286.

one and the same isoquant.[9] As Samuelson has recently shown,[10] the property comes handy in problems inviting the old fallacy that comparability of differences suffices for the construction of a weak cardinal scale—a fallacy which he exploded in 1938.[11] This is only one example showing that the analytical relevance of the property in question exceeds the domain of economic theory. It is worthwhile therefore to point out that, years later, G. Debreu also devised a topological property for separability.[12] That property, however, is applicable only for $n \geqslant 3$ and even then works only in the case in which either f is a function of x_1 alone or g a function of x_2 alone.

Whoever paid some special attention to the definitions of complementarity—the classical or the modern one—must have reached the conclusion that neither of these definitions is satisfactory—the classical one, because utility is not cardinally measurable (and in my opinion not even ordinally measurable),[13] and the modern one, because it requires an artificial experiment hard to achieve (to keep the individual equally happy while substituting money for the two commodities in question). Besides, the modern definition does not reflect the common sense notion in a transparent way. It may be well, therefore, to mention on this occasion a new and more satisfactory definition of complementarity, based only on binary choice and susceptible of simple experimental verification. Let $x = (x_1, x_2, \ldots, x_n)$ denote a basket of commodities, and let $[x, y]$ be the alternative in which the individual may consume x during a certain period and y during a subsequent period. Commodities C_1 and C_2 are complementary, independent, and rival according to whether $[x_1 + \Delta x_1, x_2 + \Delta x_2, x_3, \ldots, x_n); (x_1, x_2, \ldots, x_n)]$ is preferred, indifferent, or nonpreferred to $[(x_1 + \Delta x_1, x_2, \ldots, x_n); (x_1, x_2 + \Delta x_2, x_3, \ldots, x_n)]$, where $\Delta x_1, \Delta x_2 > 0$.

* * *

Before this book will come off the press, I will retire from Vanderbilt University after 27 years of service. Most of the essays written in my life and all those included in this volume belong to this period, which has been a fruitful one for me. Except for some very brief interludes, I had to teach and do research at the same time. All the greater is my gratitude to those members of the Vanderbilt University administration who have granted me generous (albeit modest, because of the modesty of the resources) financial support for my basic research needs. On the threshold of my retirement I can say without fear of being

---

[9] The property remains true if the "box" is constructed so that the three vertices lie on any three (instead of two) isoquants.

[10] Paul A. Samuelson, "Speeding Up Time With Age in Recognition of Life as Fleeting," in *Evolution, Welfare, and Time In Economics: Essays in Honor of Nicholas Georgescu-Roegen*, A.M. Tang, F. Westfield, and J.S. Worley, eds. Lexington, Mass.: Lexington Books, D.C. Heath, 1976.

[11] For Samuelson's 1938 paper and the conditions which in the paper just quoted he relates to a functional equation of the Norwegian mathematician, Niels Henrik Abel, see relations (3) and (4) of Chapter 13.

[12] G. Debreu, "Topological Methods in Cardinal Utility," in *Mathematical Methods in the Social Sciences*, K.J. Arrow *et al.* eds. Stanford, Cal.: Stanford University Press, 1960, pp. 16-26.

[13] Cf. Chapter 3 in *Analytical Economics*.

embarrassed that Chancellor Alexander Heard, quietly but steadily, has contributed in many ways to keep me going. This is a most appropriate occasion to thank him for his confidence in my endeavors. Some of my former students—Elton Hinshaw, Anthony M. Tang, Fred Westfield, and James S. Worley—who later became my colleagues, have been individually and together a continuous source of precious moral support for me.[14]

As always, my last and everlasting word of gratitude is for Otilia, my wife. Only those who have been fortunate to personally know her will fully understand why.

Nicholas Georgescu-Roegen

April 1976
Vanderbilt University

[14] Elton Hinshaw, who as a student in my courses set me straight on some analytical points, has graciously helped me set the style of this Foreword in better form. Stanley Keith Berry, my research assistant during this academic year, worked with me in preparing some of the material for this volume.

Acknowledgments

Every essay included in this volume is reprinted with the permission of the publisher of the original source as shown in full in the following list.

1. "Energy and Economic Myths," *The Southern Economic Journal*, XLI, No. 3 (January, 1975), pp. 347-381.
2. "Process Analysis and the Neoclassical Theory of Production," *American Journal of Agricultural Economics*, LIV, No. 2 (May, 1972), pp. 279-294. (Invited paper for the 1971 Annual Meeting of the American Association of Agricultural Economics.)
3. *The Entropy Law and the Economic Problem*, Department of Economics, The Graduate School of Business and Office for International Programs, The University of Alabama, 1971. (Distinguished Lecture Series, No. 1, delivered December 3, 1970.)
4. "The Economics of Production," *American Economic Review*, LX, No. 2 (May, 1970), pp. 1-9. (The 1969 Richard T. Ely Lecture.)
5. "Process in Farming versus Process in Manufacturing: A Problem of Balanced Development," ch. 24 in *Economic Problems of Agriculture in Industrial Societies*, (Proceedings of a Conference of the International Economic Association, Rome, September, 1965), Ugo Papi and Charles Nunn eds., London: Macmillan & Co. Ltd., and New York: St. Martin's Press, Inc., 1969, pp. 497-528.
6. "Economic Theory and Agrarian Economics," *Oxford Economic Papers*, N.S., XII, No. 1 (February, 1960), pp. 1-40.
7. "Structural Inflation-Lock and Balanced Growth," *Economies et Sociétés*, Cahiers de l'Institut de science économique appliquée, Laboratoire du Collège de France associé au C.N.R.S. et fondé en 1944 par François Perroux, IV, No. 3 (March, 1970), pp. 557-605.
8. "The Institutional Aspects of Peasant Economies: An Analytical View," ch. iv in *Subsistence Agriculture and Economic Development* (A seminar on Subsistence and Peasant Agriculture, Honolulu, March 1965), C.R. Wharton, Jr. ed., Chicago: Aldine Publishing Company, 1969, pp. 61-93.
9. "Dynamic Models and Economic Growth," in *Equilibrium and Disequilibrium in Economic Theory* (Proceedings of a Conference organized by the Institute for Advanced Studies, Vienna, July, 1974), G. Schwödiauer ed., Dordrecht, Holland: D. Reidel Publishing Company (forthcoming, 1976).
10. "Further Thoughts on Corrado Gini's *Dellusioni dell'econometria*," *Metron*, XXV, No. 104, (1966), pp. 265-279. (Paper for the International Symposium on Statistics as Methodology in the Social Sciences, A Symposium in Honor of Corrado Gini, Rome, March, 1966).

11. "Measure, Quality, and Optimum Scale," in *Essays on Econometrics and Planning Presented to Professor P.C. Mahalanobis on His 70th Birthday*, C.R. Rao ed., Oxford: Pergamon Press, 1964, pp. 231-256.
12. "Toward Partial Redirection of Econometrics, Part III," *Review of Economics and Statistics*, XXXIV, No. 3 (August, 1952), pp. 206-211.
13. "Vilfredo Pareto and His Theory of Ophelimity," in *Atti del Convegno su Vilfredo Pareto*, (Roma, October 1973), Accademia Nazionale dei Lincei, Roma (forthcoming).
14. "A Diagrammatic Analysis of Complementarity," *The Southern Economic Journal*, XIX, No. 1 (July, 1952), pp. 1-20.

Part I
Natural Resources
and the
Economics of Production

CHAPTER 1

(1972)

ENERGY AND ECONOMIC MYTHS*

> So you can now all go home and sleep peacefully in your beds tonight secure in the knowledge that in the sober and considered opinion of the latest occupant of the second oldest Chair in Political Economy in this country, although life on this Earth is very far from perfect there is no reason to think that continued economic growth will make it any worse.
>
> Wilfred Beckerman

I. INTRODUCTION

There is an appreciable grain of truth in one of Percy Bridgman's remarks that the economic profession is the most opportunistic of all. Indeed, economists' attention has continually shifted from one problem to another, the problems often being not even closely related. Search all economic periodicals of the English-speaking world before 1950, for example, and you will hardly find any mention of "economic development." It is curious, therefore, that economists have over the last hundred years remained stubbornly attached to one particular idea, the mechanistic epistemology which dominated the orientation of the founders of the Neoclassical School. By their own proud admission, the greatest ambition of these pioneers was to build an economic science after the model of mechanics—in the words of W. Stanley Jevons—as "*the mechanics of utility and self-interest*" [48, 23]. Like almost every scholar and philosopher of the first half of the nineteenth century, they were fascinated by the spectacular successes of the science of mechanics in astronomy and accepted Laplace's famous apotheosis of mechanics [53, 4] as the evangel of ultimate scientific knowledge. They thus had some attenuating circumstances, which cannot, however, be invoked by those who came long after the mechanistic dogma had been banished even from physics [23, 69–122; 5].

The latter-day economists, without a single second thought, have apparently been happy

---

* This paper represents the substance of a lecture delivered on November 8, 1972, at Yale University, School of Forestry and Environmental Studies, within the series *Limits to Growth: The Equilibrium State and Human Society,* as well as on numerous other occasions elsewhere. During July 1973 a version prepared for a planned volume of the series was distributed as a working document to the members of the Commission on Natural Resources and the Committee on Mineral Resources and the Environment (National Research Council). The present version contains a few recent amendments.

to develop their discipline on the mechanistic tracks laid out by their forefathers, fiercely fighting any suggestion that economics may be conceived otherwise than as a sister science of mechanics. The appeal of the position is obvious. At the back of the mind of almost every standard economist there is the spectacular feat of Urbain Leverrier and John Couch Adams, who discovered the planet Neptune, not by searching the real firmament, but "at the tip of a pencil on a piece of paper." What a splendid dream to be able to predict by some paper-and-pencil operations alone where a particular stock will be on the firmament of the Stock Exchange Market tomorrow or, even better, one year from now!

The consequence of this indiscriminate attachment to the mechanistic dogma, whether in an explicit or a tacit manner, is the viewing of the economic process as a mechanical analogue consisting—as all mechanical analogues do—of a principle of conservation (transformation) and a maximization rule. The economic science itself is thus reduced to a *timeless* kinematics. This approach has led to a mushrooming of paper-and-pencil exercises and increasingly complicated econometric models which often serve only to conceal from view the most fundamental economic issues. Everything now turns out to be just a pendulum movement. One business "cycle" follows another. The pillar of equilibrium theory is that, if events alter the demand and supply propensities, the economic world always returns to its previous conditions as soon as these events fade out. An inflation, a catastrophic drought, or a stock-exchange crash leaves absolutely no mark on the economy. Complete reversibility is the general rule, just as in mechanics.[1]

Nothing illustrates better the basic epistemology of standard economics than the usual graph by which almost every introductory manual portrays the economic process as a self-sustaining, circular flow between "production" and "consumption."[2] But even money does not circulate back and forth within the economic process; for both bullion and paper money ultimately become worn out and their stocks must be replenished from external sources [31]. The crucial point is that the economic process is not an isolated, self-sustaining process. This process cannot go on without a continuous exchange which alters the environment in a cumulative way and without being, in its turn, influenced by these alterations. Classical economists, Malthus in particular, insisted on the economic relevance of this fact. Yet, both standard and Marxist economists chose to ignore the problem of natural resources completely, so completely that a distinguished and versatile economist recently confessed that he had just decided that he "ought to find out what economic theory has to say" about that problem [75, 1f].

One fundamental idea dominated the orientation of both schools. A. C. Pigou stated it most explicitly: "In a stationary state factors of production are stocks, unchanging in amount, out of which emerges a continuous flow, also unchanging in amount, of real income" [68, 19]. The same idea—that a constant flow can arise from an unchanging structure—is at the basis of Marx's diagram of simple reproduction [61, II, ch. xx]. In the diagram of expanded reproduction [61, II, ch. xxi], Marx actually anticipated the modern models—such as that with which W. W. Leontief swept the profession off its feet—which ignore the problem of the primary source of the flow even in the case of a

---

[1] Some economists have insisted that, on the contrary, irreversibility characterizes the economic world [e.g., 60, 461, 808; 25], but the point, though never denied, was simply shelved away. It is in vain that some now try to claim that standard equilibrium analysis has always considered negative feedbacks [4, 334]. The only feedbacks in standard theory are those responsible for maintaining equilibrium, not for evolutionary changes.

[2] For a highly significant sample, see G. L. Bach, *Economics*, 2d ed. Englewood Cliffs, N.J.: Prentice-Hall, 1957, p. 60; Paul A. Samuelson, *Economics*, 8th ed. New York: McGraw-Hill, 1970, p. 72; Robert L. Heilbroner, *The Economic Problem*, 3rd ed. Englewood Cliffs, N.J.: Prentice-Hall, 1972, p. 177.

growing economy. The only difference is that Marx preached overtly that nature offers us everything gratis, while standard economists merely went along with this tenet tacitly. Both schools of thought shared, therefore, the Pigouvian notion of a stationary state in which a material flow emerges from an invariable source. In this idea there lies the germ of an economic myth which, as we shall see (Section VIII), is now preached by many concerned ecologists and some awakened economists. The myth is that a stationary world, a zero-growth population, will put an end to the ecological conflict of mankind. Mankind will no longer have to worry about the scarcity of resources or about pollution—another miracle-program to bring the New Jerusalem into the earthly life of man.

Myths have always occupied a prominent role in the life of man. To be sure, to act in accord with a myth is the distinctive characteristic of man among all living beings. Many myths betray man's greatest folly, his inner compulsion to believe that he is above everything else in the actual universe and that his powers know no limits. In Genesis man proclaimed that he was made in the image of God Himself. At one time, he held that the entire universe revolves around his petty abode—at another, that only the sun does so. Once, man believed that he could move things without consuming any energy, which is the myth of perpetual motion of the first kind—certainly, an essentially economic myth. The myth of perpetual motion of the second kind, which is that we may use the same energy over and over again, still lingers on in various veiled forms.

Another economic myth—that man will forever succeed in finding new sources of energy and new ways of harnessing them to his benefit—is now propounded by some scientists, but especially by economists of both standard and Marxist persuasions (Section VI). Come what may, "we will [always] think up something" [4, 338]. The idea is that, if the individual man is mortal, at least the human species is immortal. Apparently, it is below man's dignity to accept the verdict of a biological authority such as J. B. S. Haldane that the most certain fate of mankind is the same as that of any other species, namely, extinction. Only, we do not know when and why it will come. It may be sooner than the optimists believe or much later than the pessimists fear. Consequences of the accumulation of environmental deterioration may bring it about; but some persistent virus or a freak infertility gene may also cause it.

The fact is that we know little about why any species bowed out in the past, not even why some seem to become extinct before our own eyes. If we can predict approximately how long a given dog will live and also what will most probably end its life, it is only because we have had repeated occasions to observe a dog's life from birth to death. The predicament of the evolutionary biologist is that he has never observed another human species being born, aging, and dying [29, 91; 32, 208–210]. However, a species reaches the end of its existence by a process analogous to the aging of any individual organism. And even though aging is still surrounded by many mysteries [32, 205], we know that the causes which bring about the end of a species work slowly, but *persistently and cumulatively,* from the first moment of its birth. The point is that everyone of us ages with each minute, nay, with each blink, even though we are unable to realize the difference.

It is utterly inept to argue—as some economists implicitly do—that since mankind has not met with any ecological difficulty since the age of Pericles, it will never meet with one (Section VI). If we keep our eyes open, however, we will detect, as time goes by, some sufficiently apparent symptoms which may help us arrive at some general idea of the probable causes of aging and, possibly, of death. True, man's needs and the kinds of resources required for their satisfaction are far more complex than those of any other species. In exchange, our knowledge of these factors and their interrelations is, naturally, more extensive. The upshot is that even a simple analysis of the energy aspects of man's existence may help us reach at least a general picture of the ecological problems and arrive at a few, but relevant, conclusions. This, *and nothing else,* is what I have endeavored to do in this paper.

II. MECHANICS VERSUS THERMODYNAMICS

No analysis of a material process, whether in the natural sciences or in economics, can be sound without a clear and comprehensive analytical picture of such a process. The picture must first of all include the boundary—an abstract and void element which separates the process from its "environment"—as well as the duration of the process. What the process needs and what it does are then described analytically by the complete time schedule of all inputs and outputs, i.e., the precise moments at which each element involved crosses the boundary from outside or from inside. But where we draw the abstract boundary, what duration we consider, and what qualitative spectrum we use for classifying the elements of the process depend on the particular purpose of the student, and by and large on the science in point.[3]

Mechanics distinguishes only mass, speed, and position, on which it bases the concept of kinetic and potential energy. The result is that mechanics reduces any process to locomotion and a change in the distribution of energy. The constancy of total mechanical energy (kinetic plus potential) and the constancy of mass are the earliest principles of conservation to be recognized by science. A few careful economists, such as Marshall [60, 63], did observe that man can create neither matter nor energy. But in doing so, they apparently had in mind only the *mechanical* principles of conservation, for they immediately added that man can nevertheless produce utilities by moving and rearranging

[3] For a detailed discussion of the analytical representation of a process, see Georgescu-Roegen [32, ch. ix].

matter. This viewpoint ignores a most important issue: How can man do the moving? For anyone who remains at the level of mechanical phenomena, every bit of matter and every bit of mechanical energy which enter a process must come out in exactly the same *quantity* and *quality*. Locomotion cannot alter either.

To equate the economic process with a mechanical analogue implies, therefore, the myth that the economic process is a circular merry-go-round which cannot possibly affect the environment of matter and energy in any way. The obvious conclusion is that there is no need for bringing the environment into the analytical picture of that process.[4] The old tenet of Sir William Petty, that keen student of human affairs who insisted that labor is the father and nature the mother of wealth, has long since been relegated to the status of a museum piece [29, 96; 31, 280]. Even the accumulation of glaring proofs of the preponderant role played by natural resources in mankind's history failed to impress standard economists. One may think of the Great Migration of the first millenium which was the ultimate response to the exhaustion of the soil of Central Asia following a long period of persistent grazing. Remarkable civilizations—Maya is one example—crumbled away from history because their people were unable to migrate or to counteract by adequate technical progress the deterioration of their environment. Above all, there is the indisputable fact that all struggles between the Great Powers have not turned idly around ideologies or national prestige but around the control of natural resources. They still do.

Because mechanics recognizes no qualitative change but only change of place, any

[4] If "land" appears as a variable in some standard production functions, it stands only for Ricardian land, i.e., for mere space. The lack of concern for the true nature of the economic process is also responsible for the inadequacy of the standard production function from other, equally crucial, viewpoints. See Georgescu-Roegen [27; 30; 33].

mechanical process may be reversed, just as a pendulum, for instance, can. No laws of mechanics would have been violated if the earth had been set in motion in the opposite direction. There is absolutely no way for a spectator to discover whether a movie of a purely mechanical pendulum is projected in the direction in which it was taken or in the reverse. But actual phenomena in all their aspects do not follow the story of the famous Mother Goose rhyme in which the brave Duke of York kept marching his troops up the hill and down the hill without giving battle. Actual phenomena move in a definite direction and involve qualitative change. This is the lesson of thermodynamics, a peculiar branch of physics, so peculiar that purists prefer not to consider it a part of physics because of its anthropomorphic texture. Even though it is hard to see how the basic texture of any science could be otherwise than anthropomorphic, the case of thermodynamics is unique.

Thermodynamics grew out of a memoir by a French engineer, Nicolas Sadi Carnot, on the efficiency of heat engines (1824). Among the first facts it brought to light is that man can use only a particular form of energy. Energy thus came to be divided into *available* or *free* energy, which can be transformed into work, and *unavailable* or *bound* energy, which cannot be so transformed.[5] Clearly, the division of energy according to this criterion is an anthropomorphic distinction like no other in science.

The distinction is closely related to another concept specific to thermodynamics, namely, to entropy. This concept is so involved that one specialist judged that "it is not easily understood even by physicists" [40, 37].[6] But for our immediate purpose we may be satisfied with the simple definition of entropy as an *index* of the amount of unavailable energy in a given thermodynamic system at a given moment of its evolution.

Energy, regardless of quality,[7] is subject to a strict conservation law, the First Law of Thermodynamics, which is formally identical to the conservation of mechanical energy mentioned earlier. And since work is one of the multiple forms of energy, this law exposes the myth of perpetual motion of the first kind. It does not, however, take account of the distinction between available and unavailable energy; *by itself the law does not preclude the possibility that an amount of work should be transformed into heat and this heat reconverted into the initial amount of work*. The First Law of Thermodynamics thus allows any process to take place both forward and backward, so that everything is again just as it was at first, with no trace left by the happening. With only that law we are still in mechanics, not in the domain of actual phenomena, which certainly includes the economic process.

The irreducible opposition between mechanics and thermodynamics stems from the Second Law, the Entropy Law. The oldest of its multiple formulations is also the most transparent for the nonspecialist: "Heat flows by itself only from the hotter to the colder body, never in reverse." A more involved but equivalent formulation is that the entropy of a *closed* system continuously (and irrevocably) increases toward a maximum; i.e.,

---

[5] The technical definition of available (unavailable) energy does not coincide with that of free (bound) energy. But the difference is such that we may safely ignore it in the present discussion.

[6] This judgment is vindicated by the discussion of the Entropy Law in [44, 17]. Even the familiar notion of heat raises some delicate issues, with the result that some physicists may go wrong on it, too. See *Journal of Economic Literature*, X (December 1972), p. 1268.

[7] Let us also note that even energy does not lend itself to a simple, formal definition. The familiar one, that energy is the capacity of a system to perform work, clashes with the definition of unavailable energy. We must then explain that all energy can in principle be transformed into work provided that the corresponding system is brought in contact with another which is at the absolute zero of temperature. This explanation has only the value of a pure extrapolation because, according to the Third Law of Thermodynamics, this temperature can never be reached.

the available energy is continuously transformed into unavailable energy until it disappears completely.[8]

In broad lines, the story is relatively simple: *All kinds of energy are gradually transformed into heat and heat becomes so dissipated in the end that man can no longer use it.* Indeed, a point that goes back to Carnot is that no steam engine can provide work if the same temperature, however high, prevails in the boiler and the cooler.[9] To be available, energy must be distributed unevenly; energy that is completely dissipated is no longer available. The classical illustration is the immense heat dissipated into the water of the seas, which no ship can use. Although ships sail on top of it, they need available energy, the kinetic energy concentrated in the wind or the chemical and nuclear energy concentrated in some fuel. We may see why entropy came to be regarded also as an index of disorder (of dissipation) not only of energy but also of *matter* and why the Entropy Law in its present form states that *matter, too, is subject to an irrevocable dissipation.* Accordingly, the ultimate fate of the universe is not the Heat Death (as it was believed at first) but a much grimmer state—Chaos. No doubt, the thought is intellectually unsatisfactory.[10] But what interests us is that, according to all the evidence, our immediate environment, the solar system, tends toward a thermodynamic death,[11] at least as far as life-bearing structures are concerned.

III. THE ENTROPY LAW AND ECONOMICS

Perhaps no other law occupies a position in science as singular as that of the Entropy Law. It is the only natural law which recognizes that even the material universe is subject to an irreversible qualitative change, to an evolutionary process.[12] This fact led some natural scientists and philosophers to suspect an affinity between that law and life phenomena. By now, few would deny that the *economy* of any life process is governed, not by the laws of mechanics, but by the Entropy Law [32, xiii, 191–194]. The point, as we shall now see, is most transparent in the case of the economic process.

Economists have occasionally maintained that, since some scientists trespass into economics without knowing much about the subject, they, too, are justified in talking about science, notwithstanding their ignorance in that domain [4, 328f]. The thought reflects an error, which unfortunately is general with economists. But whatever the economic expertise of other scientists, economists could not fare continuously well in their own field without some solid understanding of the Entropy Law and its consequences.[13] As I argued some years ago, thermodynamics is at bottom a physics of economic value—as Carnot unwittingly set it going—and the En-

---

[8] A system is closed if it exchanges no matter and no energy with its "environment." Clearly, in such a system the amount of matter-energy is constant. However, the constancy of this amount alone does not warrant the increase of entropy. Entropy may even decrease if there is exchange.

[9] There is no truth, therefore, in Holdren's idea [44, 17] that temperature measures "the usefulness" of heat. The most we can say is that the *difference* of temperature is a rough index of the usefulness of the hotter heat.

[10] One alternative, supported by statistical thermodynamics (Section VI), is that entropy may decrease in some parts of the universe so that the universe both ages and rejuvenates. But no substantial evidence exists for this possibility. Another hypothesis, set forth by a group of British astronomers, is that the universe is an everlasting steady state in which individual galaxies are born and die continuously. But facts do not fit this hypothesis either. The issue of the true nature of the universe is far from settled [32, 201f, 210].

[11] To preclude some erring, we should emphasize the point that a reversal of this trend would be just as bad for the preservation of life on earth.

[12] Rudolf Clausius coined "entropy" from a Greek word meaning "transformation," "evolution." See [32, 130].

[13] As we shall see later on, some highly interesting examples are provided by Harry G. Johnson [49] and, in an unceremonious, assertive manner, by Robert A. Solo [73]. As for Robert M. Solow, who at first also refused to swerve a hair from the standard position [74], he recently found it opportune to concede that "it takes economics and the law of entropy" to deal with the problem of resources [75, 11]. But at bottom, he still remained attached to his old creed.

tropy Law is the most economic in nature of all natural laws [29, 92–94; 32, 276–283].

The economic process, like any other life process, is irreversible (and irrevocably so); hence, it cannot be explained in mechanical terms alone. It is thermodynamics, through the Entropy Law, that recognizes the qualitative distinction which economists should have made from the outset between the inputs of valuable resources (low entropy) and the final outputs of valueless waste (high entropy). The paradox suggested by this thought, namely, that all the economic process does is to transform valuable matter and energy into waste, is easily and instructively resolved. It compels us to recognize that the real output of the economic process (or of any life process, for that matter) is not the *material flow* of waste, but the still mysterious *immaterial flux* of the enjoyment of life.[14] Without recognizing this fact we cannot be in the domain of life phenomena.

The present laws of physics and chemistry do not explain life completely. But the thought that life may violate some natural law has no place in science. Nevertheless, as has long been observed—and more recently in an admirable exposition by Erwin Schrödinger [71, 69–72]—life seems to evade the entropic degradation to which inert matter is subject. The truth is that any living organism simply strives at all times to compensate for its own continuous entropic degradation by sucking low entropy (negentropy) and expelling high entropy. Clearly, this phenomenon is not precluded by the Entropy Law, which requires only that the entropy of the entire system (the environment *and* the organism) should increase. Everything is in order as long as the entropy of the environment increases by more than the compensated entropy of the organism.

Equally important is the fact that the Entropy Law is the only natural law that does not predict quantitatively. It does not specify how great the increase should be at a future moment or what particular entropic pattern will result. Because of this fact, there is an entropic indeterminateness in the real world which allows not only for life to acquire an endless spectrum of forms but also for most actions of a living organism to enjoy a certain amount of freedom [32, 12]. Without this freedom, we would not be able to choose between eating beans or meat, between eating now or later. Nor could we aspire to implement economic plans (at any level) of our own choosing.

It is also because of the entropic indeterminateness that life does matter in the entropic process. The point is no mystical vitalism, but a matter of brute facts. Some organisms slow down the entropic degradation. Green plants store *part* of the solar radiation which in their absence would immediately go into dissipated heat, into high entropy. That is why we can burn now the solar energy saved from degradation millions of years ago in the form of coal or a few years ago in the form of a tree. All other organisms, on the contrary, speed up the march of entropy. Man occupies the highest position on this scale, and this is all that environmental issues are about.

Most important for the student of economics is the point that the Entropy Law is the taproot of economic scarcity. Were it not for this law, we could use the energy of a piece of coal over and over again, by transforming it into heat, the heat into work, and the work back into heat. Also, engines, homes, and even living organisms (if they could exist at all) would never wear out. There would be no economic difference between material goods and Ricardian land. In such an imaginary, purely mechanical world, there would be no true scarcity of energy and materials. A population as large as the space of our globe would allow could live indeed forever. An increase in the real income per capita could be supported in part

---

[14] It seems idle therefore to ask—as Boulding [8, 10] does—whether well-being is a flow or a stock.

by a greater velocity of use (just as in the case of money circulation) and in part by additional mining. But there would be no reason for any real struggle, whether intra-species or inter-species, to arise.

Economists have been insisting that "there is no free lunch," by which they mean that the price of anything must be equal to the cost; otherwise, one would get something for nothing. To believe that this equality also prevails in terms of entropy constitutes one of the most dangerous economic myths. *In the context of entropy, every action, of man or of an organism, nay, any process in nature, must result in a deficit for the entire system.* Not only does the entropy of the environment increase by an additional amount for every gallon of gasoline in your tank, but also a substantial part of the free energy contained in that gasoline, instead of driving your car, will turn directly into an additional increase of entropy. As long as there are abundant, easily accessible resources around, we might not really care how large this additional loss is. Also, when we produce a copper sheet from some copper ore we decrease the entropy (the disorder) of the ore, but only at the cost of a much greater increase of the entropy in the rest of the universe. If there were not this entropic deficit, we would be able to convert work into heat, and, by reversing the process, to recuperate the entire initial amount of work—as in the imaginary world of the preceding paragraph. In such a world, standard economics would reign supreme precisely because the Entropy Law would not work.

IV. ACCESSIBLE ENERGY AND ACCESSIBLE MATTER

As we have seen, the distinction between available and unavailable energy (generalized by that between low and high entropy) was introduced in order that thermodynamics may take into account the fact that only one particular state of energy can be used by man. But the distinction does not mean that man can *actually* use any available energy regardless of the place and form in which it is found. If available energy is to have any value for mankind, it must also be *accessible*. Solar energy and its by-products are accessible to us with practically no effort, no consumption of additional available energy. In all other cases, we have to spend some work and materials in order to tap a store of available energy. The point is that even though we may land on Mars and find there some gas deposits, that available energy will not be accessible to us if it will take more than the equivalent energy of a cubic foot of gas *accessible on earth* to bring a cubic foot of gas from that planet. There certainly are oil shales from which we could extract one ton of oil only by using more than one ton of oil. The oil in such a shale would still represent available, but not accessible, energy. We have been reminded ad nauseam that the real reserves of fossil fuel are certainly greater than those known or estimated [e.g. 58, 331]. But it is equally certain that a substantial part of the real reserves does not constitute accessible energy.

The distinction regards efficiency in terms of energy, not in terms of economics. Economic efficiency implies energetic efficiency, but the converse is not true. The use of gas, for example, is energetically more efficient than the use of electricity, but electricity happens to be cheaper in many instances [79, 152]. Also, even though we can make gas from coal, it is cheaper to extract gas from natural deposits. Should the natural resources of gas become exhausted before those of coal, we will certainly resort to the method that is now economically inefficient. The same idea should be borne in mind when discussing the future of direct uses of solar radiation.

Economists, however, insist that "resources are properly measured in economic, not physical, terms" [51, 663; also 3, 247]. The advice reflects one of the most enduring myths of the profession (shared also by others). It is the myth that the price mechanism can offset any shortages, whether of land, energy or materials.[15] This myth will be duly examined later on, but here we need only emphasize the point that from the point of view of the longrun it is only efficiency in terms of energy that counts in establishing accessibility. To be sure, actual efficiency depends at any one time on the state of the arts. But, as we know from Carnot, in each particular situation *there is a theoretical limit independent of the state of the arts, which can never be attained in actuality*. In effect, we generally remain far below it.

Accessibility, as here defined, bears on the fact that although mankind's spaceship floats within a fantastic store of available energy, only an infinitesimal part of this store is potentially accessible to man. For even if we were to travel in space with the greatest speed, that of light, we would still be confined to a speck of cosmos. A journey just to scout the nearest sun outside the solar system for possible, yet uncertain, earth-like satellites would take nine years! If we have learned anything from the landing on the moon, it is that there is no promise of resources in interplanetary, let alone intersidereal, travel.

Still narrower limits to the accessible energy are set by our own biological nature, which is such that we cannot survive at too high or too low a temperature or when exposed to some radiations. It is for this reason that the mining of nuclear fuel and its use on a large scale has raised issues which now divide laymen as well as authorities on the subject (Section IX). There are also limits set by some purely physical obstacles. The sun cannot possibly be mined even by a robot. From the sun's immense radiating energy, only the small amount which reaches the earth counts in the main (Section IX). Nor can we harness the immense energy of the terrestrial thunders. Unique physical obstacles also stand hopelessly in the way of the peaceful use of thermonuclear energy. The fusion of deuterium requires the fantastic temperature of 0.2 billion°F, one order of magnitude hotter than the sun's interior. The difficulty concerns the material container for that reaction. As has been explained in layman's terms, the solution now sought is similar to holding water inside a mesh of rubber bands. In this connection we may recall that the chemical energy of dynamite and gunpowder, although in use for a long time, cannot be controlled so as to drive a turbine or a motor. Perhaps the use of thermonuclear energy will also remain confined to a "bomb."[16] Be this as it may, with or without thermonuclear energy, the amount of accessible energetic low entropy is finite (Section IV).

Similar considerations lead to the conclusion that the amount of accessible material low entropy is finite, too. But although in both cases only the amount of low entropy matters, it is important that the two accounts be kept separate in any discussion of the environmental problem. As we all know, available energy and ordered material structures fulfill two distinct roles in mankind's life. However, this anthropomorphic distinction would not be compelling by itself.

There is, first, the physical fact that, despite the Einstein equivalence of mass and energy, there is no reason to believe that we can convert energy into matter except at the atomic scale in a laboratory and only for some special elements.[17] We cannot produce a copper sheet, for example, from energy alone. All the copper in that sheet must exist as copper (in pure form or in some chemical

---

[15] The evidence is ample [3, 240f; 4, 337f; 49; 51, 663, 665; 74, 46f; 80; 69, 9f, 14f]. The appeal of the myth is seen in that even many on the other side of the fence share it [58; 62, 65; 6, 10, 12; and Frank Notestein, quoted in 62, 130].

[16] The technical difficulties at the present moment are surveyed in [63]. On the other hand, we should remember that in 1933 Ernest Rutherford greatly doubted that atomic energy could be controlled [82, 27].

[17] The point is that even the formation of an atom of carbon from three atoms of helium, for example, requires such a sharp timing that its probability is astronomically small, and hence the event may occur on a large scale only within astronomically huge masses.

compound) beforehand. Therefore, the statement that "energy is convertible into most of the other requirements of life" [83, 412] is, in this unqualified form, apt to mislead. Second, no material macrostructure (whether a nail or a jet) whose entropy is lower than that of its surroundings may last forever in its original form. Even the singular organizations characterized by the tendency to evade the entropic decay—the life-bearing structures—cannot so last. The artifacts which now are an essential part of our mode of life have therefore to be renewed continuously from some sources. The final point is that the earth is a thermodynamic system open only with respect to energy. The amount of meteorite matter, though not negligible, comes already dissipated.

The result is that we can count only on the mineral resources, which, however, are both irreplaceable and exhaustible. Many of a particular kind have been exhausted in one country after another [56, 120f].[18] At present, important minerals—lead, tin, zinc, mercury, precious metals—are scarce over the entire world [17, 72–77; 56]. The widespread notion that the oceans constitute an almost inexhaustible source of minerals and may even become a link in a perpetual, natural recycling system [3, 239; 69, 7f] is denounced as mere hyperbole by geological authorities [17, 85–87].[19]

The only way we can substitute energy for material low entropy is through physicochemical manipulations. By using larger and larger amounts of available energy we can sift copper out from poorer and poorer ores, located deeper and deeper in the earth. But the energy cost of mining low-content ores increases very fast [56, 122f]. We can also recycle "scrap." There are, however, some elements which, because of their nature and the mode in which they participate in the natural and man-conducted processes, are highly dissipative. Recycling, in this case, can hardly help. The situation is particularly distressing for those elements which, in addition, are found in very small supply in the environment. Phosphorus, a highly critical element in biological processes, seems to belong to this category. So does helium, another element with a strictly specific role [17, 81; 38].

An important point—apparently ignored by economists [49, 8; 69, 16, 42]—is that recycling cannot be complete.[20] Even though we can pick up all the pearls from the floor and reconstitute a broken necklace, no actual process can possibly reassemble all the molecules of a coin after it has been worn out.

This impossibility is not a straightforward consequence of the Entropy Law, as Solow believes [75, 2]. Nor is it quite exact to say, with Boulding [8, 7], that "there is, fortunately, no law of increasing material entropy." The Entropy Law does not distinguish between matter and energy. This law does not exclude (at least not in principle) a complete unshuffling of a *partial* material structure, provided that there is enough free energy to do the job. And if we have enough energy, we could even separate the cold molecules of a glass of water and assemble them into ice cubes. If, in practice, however, such operations are impossible, it is only because they would require a practically infinite time.[21]

V. DISPOSABLE WASTE

Since Malthus did not see that waste also raises some economic problems, it was normal for the schools of economic thought which ignored even the input of natural resources to pay no attention to the output of

---

[18] See the interesting story of the Mesabi Range in [14, 11f].

[19] The widespread notion that the oceans may be turned into an immense source of food also is a great delusion [13, 59f].

[20] Data on recycling are scarce and inadequate; a few are found in [12, 205; 16, 14]. For steel, see [14].

[21] All this proves that, even though the Entropy Law may sound extremely simple, its correct interpretation requires special care.

waste. As a result, waste, just like natural resources, is not represented in any manner in the standard production function. The only mention of pollution was the occasional textbook example of the laundry enterprise which suffers a loss because of a neighboring smokestack. Economists must therefore have felt some surprise when pollution started to strike everybody in the face. Yet, there was nothing to be surprised about. Given the entropic nature of the economic process, waste is an *output* just as unavoidable as the input of natural resources [27, 514f, 519, 523f]. "Bigger and better" motorcycles, automobiles, jet planes, refrigerators, etc., necessarily cause not only "bigger and better" depletion of natural resources but also "bigger and better" pollution [31; 32, 19f, 305f]. But by now, economists can no longer ignore the existence of pollution. They even have suddenly discovered that they "actually have something important to say to the world," namely, that if prices are right there is no pollution [74, 49f; also 10, 12, 17; 49, 11f; 80, 120f][22]—which is another facet of the economists' myth about prices (Sections IV and XI).

Waste is a physical phenomenon which is, generally, harmful to one or another form of life, and, directly or indirectly, harmful to human life. It constantly deteriorates the environment in many ways: chemically, as in mercury or acid pollution; nuclearly, as by radioactive garbage; physically, as in strip mining or in the accumulation of carbon dioxide in the atmosphere. There are a few instances in which a substantial part of some waste element—carbon dioxide is the salient example—is recycled by some "natural" processes of the environment. Most of the obnoxious waste—garbage, cadavers, and excrement—is also gradually reduced by natural processes. These wastes only require

---
[22] In addition, Harry Johnson finally came to see that a complete representation of a production process must necessarily include the output of waste [49, 10].

some space in which to remain isolated until their reduction is completed. There are troublesome hygienic problems involved, but the important point is that such wastes do not cause permanent, irreducible harm to our environment.

Other wastes are *disposable* only in the sense that they may be converted into less noxious ones by certain actions on our part, as when part of carbon monoxide is transformed into carbon dioxide and heat through improved combustion. A great part of sulphur dioxide pollution, another example, may be avoided through some special installations. Still other wastes cannot be so reduced. A topical example is the fact that we cannot reduce the highly dangerous radioactivity of nuclear garbage [46, 233]. This activity diminishes by itself with time, but very slowly. In the case of plutonium-239, the reduction to fifty percent takes 25,000 years! However, the harm done by radioactivity concentration to life may very well be irreparable.

Here, just as for the accumulation of any waste, from rubbish of all kinds to heat, the difficulty is created by the finitude of accessible space. Mankind is like a household which consumes the limited supply from a pantry and throws the inevitable waste into a finite trash can—the space around us. Even ordinary rubbish is a menace; in ancient times, when it could be removed only with great difficulties, some glorious cities were buried under accumulated rubbish. We have better means to remove it, but the continuous production calls for another dumping area, and another, and another... In the United States the annual amount of waste is almost two tons per capita and increasing [14, 11n.]. We should also bear in mind that for every barrel of shale oil we are saddled with more than one ton of ashes and to obtain five ounces of uranium we must crush one cubic meter of rock. What to do even with these "neutral" residuals is a problem vividly illustrated by the consequences of strip-mining.

To send the residuals into outer space would not pay on a large and continuous scale.[23]

The finitude of our space renders more dangerous wastes which persist for a long time and especially those which are completely irreducible. Typical of the last category is thermal pollution, the dangers of which are not fully appreciated. The *additional* heat into which all energy of terrestrial origin is ultimately transformed when used by man[24] is apt to upset the delicate thermodynamic balance of the globe in two ways. First, the islands of heat created by power plants not only disturb (as is well known) the local fauna and flora of rivers, lakes, and even coastal seas, but they may also alter climatic patterns. One nuclear plant alone may heat up the water in the Hudson River by as much as 7°F. Then again the sorry plight of where to build the next plant, and the next, is a formidable problem. Second, the additional global heat at the site of the plant and at the place where power is used may increase the temperature of the earth to the point at which the icecaps would melt —an event of cataclysmic consequences. Since the Entropy Law allows no way to cool a continuously heated planet, thermal pollution could prove to be a more crucial obstacle to growth than the finiteness of accessible resources [79, 160].[25]

We apparently believe that we just have to do things differently in order to dispose of pollution. The truth is that, like recycling, disposal of pollution is not costless in terms of energy. Moreover, as the percentage of pollution reduction increases, the cost increases even more steeply than for recycling [62, 134f]. We must therefore watch our step—as some have already warned us [6, 9]—so as not to substitute a greater but distant pollution for a local one. In principle at least, a dead lake may certainly be revitalized by pumping oxygen into it, as Harry Johnson suggests [49, 8f]. But it is as certain that the additional operations implied by this pumping not only require enormous amounts of additional low entropy but also create additional pollution. In practice, the reclamation efforts undertaken for lands and streams degraded by strip-mining have been less than successful [14, 12]. Linear thinking—to borrow a label used by Bormann [7, 706]—may be "in" nowadays, but precisely as economists we ought to abide by the truth that what is true for one dead lake is not true for all dead lakes if their number increases beyond a certain limit. To suggest further that man can construct at a cost a new environment tailored to his desires is to ignore completely that cost consists in essence of low entropy, not of money, and is subject to the limitations imposed by natural laws.[26]

Often, our arguments spring from the belief in an industrial activity free of pollution. It is a myth just as lulling as the belief in everlasting durability. The sober truth is that,

---

[23] The cover photograph of *Science*, 12 April 1968, and the photographs in *National Geographic*, December 1970, are highly instructive on this point. It may be true that—as Weinberg and Hammond [83, 415] argued—if we had to supply energy even for 20 billion people at an annual average of 600 million BTU per capita, we would have to crush rock only at twice the speed at which coal is now being mined. We would still face the problem of what to do with the crushed rock.

[24] Solar energy (in all its ramifications) constitutes the only (and a noteworthy) exception (Section IX).

[25] The continuous accumulation of carbon dioxide in the atmosphere has a greenhouse effect which should aggravate the heating of the globe. There are, however, other divergent effects from the increase of scattered particles in the atmosphere: agriculturally oriented changes of vegetation, interference with the normal distribution of surface and underground water, etc. [24; 57]. Even though experts cannot determine the resultant trend of this complex system in which a small disturbance may have an enormous effect, the problem is not "an old scare," as Beckerman says in dismissing it [4, 340].

[26] Solo [73, 517] also asserts that because of growth and technology, the present society could eliminate all pollution "(with the one possible exception of radiation refuse)" at a bearable cost. It is only because of some perversity of our values that we are not doing it. That we could devote more effort to pollution disposal is beyond doubt. But to believe that with nonperverse values we could defeat the natural laws reflects an indeed perverse view of reality.

our efforts notwithstanding, the accumulation of pollution might under certain circumstances beget the first serious ecological crisis [62, 126f]. What we experience today is only a clear premonition of a trend which may become even more conspicuous in the distant future.

VI. MYTHS ABOUT MANKIND'S ENTROPIC PROBLEM

Hardly anyone would nowadays openly profess a belief in the immortality of mankind. Yet many of us prefer not to exclude this possibility; to this end, we endeavor to impugn any factor that could limit mankind's life. The most natural rallying idea is that mankind's entropic dowry is virtually inexhaustible, primarily because of man's inherent power to defeat the Entropy Law in some way or another.

To begin with, there is the simple argument that, just as has happened with many natural laws, the laws on which the finiteness of accessible resources rests will be refuted in turn. The difficulty of this historical argument is that history proves with even greater force, first, that in a finite space there can be only a finite amount of low entropy and, second, that low entropy continuously and irrevocably dwindles away. The impossibility of perpetual motion (of both kinds) is as firmly anchored in history as the law of gravitation.

More sophisticated weapons have been forged by the statistical interpretation of thermodynamic phenomena—an endeavor to reestablish the supremacy of mechanics propped up this time by a *sui generis* notion of probability. According to this interpretation, the reversibility of high into low entropy is only a highly improbable, not a totally impossible event. And since the event is *possible,* we should be able by an ingenious device to cause the event to happen as often as we please, just as an adroit sharper may throw a "six" almost at will. The argument only brings to the surface the irreducible contradictions and fallacies packed into the foundations of the statistical interpretation by the worshipers of mechanics [32, ch. vi]. The hopes raised by this interpretation were so sanguine at one time that P. W. Bridgman, an authority on thermodynamics, felt it necessary to write an article just to expose the fallacy of the idea that one may fill one's pockets with money by "bootlegging entropy" [11].

Occasionally and *sotto voce* some express the hope, once fostered by a scientific authority such as John von Neumann, that man will eventually discover how to make energy a free good, "just like the unmetered air" [3, 32]. Some envision a "catalyst" by which to decompose, for example, the sea water into oxygen and hydrogen, the combustion of which will yield as much available energy as we would want. But the analogy with the small ember which sets a whole log on fire is unavailing. The entropy of the log and the oxygen used in the combustion is lower than that of the resulting ashes and smoke, whereas the entropy of water is higher than that of the oxygen and hydrogen after decomposition. Therefore, the miraculous catalyst also implies entropy bootlegging.[27]

With the notion, now propagated from one syndicated column to another, that the breeder reactor produces more energy than it consumes, the fallacy of entropy bootlegging seems to have reached its greatest currency even among the large circles of literati, including economists. Unfortunately, the illusion feeds on misconceived sales talk by some nuclear experts who extol the reactors which transform fertile but nonfissionable material into fissionable fuel as the breeders that "produce more fuel than they consume" [81, 82]. The stark truth is that the breeder is in no way different from a plant which produces hammers with the aid of some hammers. According to the deficit principle of the Entropy Law (Section III), even in breeding chickens a greater amount of low entropy is consumed than is contained in the product.[28]

---

[27] A specific suggestion implying entropy bootlegging is Harry Johnson's: it envisages the possi-

Apparently in defense of the standard vision of the economic process, economists have set forth themes of their own. We may mention first the argument that "the notion of an absolute limit to natural resource availability is untenable when the definition of resources changes drastically and unpredictably over time.... A limit may exist, but it can be neither defined nor specified in economic terms" [3, 7, 11]. We also read that there is no upper limit even for arable land because "arable is infinitely indefinable" [55, 22]. The sophistry of these arguments is flagrant. No one would deny that we cannot say *exactly* how much coal, for example, is accessible. Estimates of natural resources have constantly been shown to be too low. Also, the point that metals contained in the top mile of the earth's crust may be a million times as much as the present known reserves [4, 338; 58, 331] does not prove the inexhaustibility of resources, but, characteristically, it ignores both the issues of accessibility and disposability.[29] Whatever resources or arable land we may need at one time or another, they will consist of accessible low entropy and accessible land. *And since all kinds together are in finite amount, no taxonomic switch can do away with that finiteness.*

The favorite thesis of standard and Marxist economists alike, however, is that the power of technology is without limits [3; 4; 10; 49; 51; 74; 69]. We will always be able not only to find a substitute for a resource which has become scarce, but also to increase the *productivity* of any kind of energy and material. Should we run out of some resources, we will always think up something, just as we have continuously done since the time of Pericles [4, 332–334]. Nothing, therefore, could ever stand in the way of an increasingly happier existence of the human species. One can hardly think of a more blunt form of linear thinking. By the same logic, no healthy young human should ever become afflicted with rheumatism or any other old-age ailments; nor should he ever die. Dinosaurs, just before they disappeared from this very same planet, had behind them not less than one hundred and fifty million years of truly prosperous existence. (And they did not pollute environment with industrial waste!) But the logic to be truly savored is Solo's [73, 516]. If entropic degradation is to bring mankind to its knees sometime in the future, it should have done so sometime after A.D. 1000. The old truth of Seigneur de La Palice has never been turned around—and in such a delightful form.[30]

In support of the same thesis, there also are arguments directly pertaining to its substance. First, there is the assertion that only a few kinds of resources are "so resistant to technological advance as to be incapable of eventually yielding extractive products at constant or declining cost" [3, 10].[31] More recently, some have come out with a specific law which, in a way, is the contrary of Malthus' law concerning resources. The idea is

---

bility of reconstituting the stores of coal and oil "with enough ingenuity" [49, 8]. And if he means with enough energy as well, why should one wish to lose a great part of that energy through the transformation?

[28] How incredibly resilient is the myth of energy breeding is evidenced by the very recent statement of Roger Revelle [70, 169] that "farming can be thought of as a kind of breeder reactor in which much more energy is produced than consumed." Ignorance of the main laws governing energy is widespread indeed.

[29] Marxist economists also are part of this chorus. A Romanian review of [32], for example, objected that we have barely scratched the surface of the earth.

[30] To recall the famous old French quatrain: "Seigneur de La Palice / fell in the battle for Pavia. / A quarter of an hour before his death / he was still alive." (My translation.) See *Grand Dictionnaire Universel du XIX-e Siècle*, Vol. X, p. 179.

[31] Even some natural scientists, e.g., [1], have taken this position. Curiously, the historical fact that some civilizations were unable "to think up something" is brushed aside with the remark that they were "relatively isolated" [3, 6]. But is not mankind, too, a community completely isolated from any external cultural diffusion and one, also, which is unable to migrate?

that technology improves exponentially [4, 236; 51, 664; 74, 45]. The superficial justification is that one technological advance induces another. This is true, only it does not work cumulatively as in population growth. And it is terribly wrong to argue, as Maddox does [59, 21], that to insist on the existence of a limit to technology means to deny man's power to influence progress. Even if technology continues to progress, it will not necessarily exceed any limit; an increasing sequence may have an upper limit. In the case of technology this limit is set by the theoretical coefficient of efficiency (Section IV). If progress were indeed exponential, then the input i per unit of output would follow in time the law $i = i_0 (1 + r)^{-t}$ and would constantly approach zero. Production would ultimately become incorporeal and the earth a new Garden of Eden.

Finally, there is the thesis which may be called the fallacy of endless substitution: "Few components of the earth's crust, including farm land, are so specific as to defy economic replacement; ... nature imposes particular scarcities, not an inescapable general scarcity" [3, 10f].[32] Bray's protest notwithstanding [10, 8], this *is* "an economist's conjuring trick." True, there are only a few "vitamin" elements which play a totally specific role such as phosphorus plays in living organisms. Aluminum, on the other hand, has replaced iron and copper in many, although not in all uses.[33] However, *substitution within a finite stock of accessible low entropy* whose irrevocable degradation is speeded up through use cannot possibly go on forever.

In Solow's hands, substitution becomes the key factor that supports technological progress even as resources become increasingly scarce. There will be, first, a substitution within the spectrum of consumer goods. With prices reacting to increasing scarcity, consumers will buy "fewer resource-intensive goods and more of other things" [74, 47].[34] More recently, he extended the same idea to production, too. We may, he argues, substitute "other factors for natural resources" [75, 11]. One must have a very erroneous view of the economic process as a whole not to see that there are no material factors other than natural resources. To maintain further that "the world can, in effect, get along without natural resources" is to ignore the difference between the actual world and the Garden of Eden.

More impressive are the statistical data invoked in support of some of the foregoing theses. The data adduced by Solow [74, 44f] show that in the United States between 1950 and 1970 the consumption of a series of mineral elements per unit of GNP decreased substantially. The exceptions were attributed to substitution but were expected to get in line sooner or later. In strict logic, the data do not prove that during the same period technology necessarily progressed to a greater economy of resources. The GNP may increase more than any input of minerals even if technology remains the same, or even if it deteriorates. But we also know that during practically the same period, 1947–1967, the consumption per capita of basic materials increased in the United States. And in the world, during only one decade, 1957–1967, the consumption of steel per capita grew by 44 percent [12, 198–200]. What matters in

---

[32] Similar arguments can be found in [4, 338f; 59, 102; 74, 45]. Interestingly, Kaysen [51, 661] and Solow [74, 43], while recognizing the finitude of mankind's entropic dowry, pooh-pooh the fact because it does not "lead to any very interesting conclusions." Economists, of all students, should know that the finite, not the infinite, poses extremely interesting questions. The present paper hopes to offer proof of this.

[33] Even in this most cited case, substitution has not been as successful in every direction as we have generally believed. Recently, it has been discovered that aluminum electrical cables constitute fire hazards.

[34] The pearl on this issue, however, is supplied by Maddox [59, 104]: "Just as prosperity in countries now advanced has been accompanied by an actual decrease in the consumption of bread, so it is to be expected that affluence will make societies less dependent on metals such as steel."

the end is not only the impact of technological progress on the consumption of resources per unit of GNP, but especially the increase in the rate of resource depletion, which is a side effect of that progress.

Still more impressive—as they have actually proved to be—are the data used by Barnett and Morse to show that, from 1870 to 1957, the ratios of labor and capital costs to net output decreased appreciably in agriculture and mining, both critical sectors as concerns depletion of resources [3, 8f, 167–178]. In spite of some arithmetical incongruities,[35] the picture emerging from these data cannot be repudiated. Only its interpretation must be corrected.

For the environmental problem, it is essential to understand the typical forms in which technological progress may occur. A first group includes the *economy-innovations*, which achieve a *net* economy of low entropy —be it by a more complete combustion, by decreasing friction, by deriving a more intensive light from gas or electricity, by substituting materials costing less in energy for others costing more, and so on. Under this heading we should also include the discovery of how to use new kinds of accessible low entropy. A second group consists of *substitution-innovations*, which simply substitute physico-chemical energy for human energy. A good illustration is the innovation of gunpowder, which did away with the catapult. Such innovations generally enable us not only to do things better but also (and especially) to do things which could not be done before —to fly in airplanes, for example. Finally, there are the *spectrum-innovations*, which bring into existence new consumer goods, such as the hat, nylon stockings, etc. Most of the innovations of this group are at the same time substitution-innovations. In fact, most innovations belong to more than one category. But the classification serves analytical purposes.

Now, economic history confirms a rather elementary fact—the fact that the great strides in technological progress have generally been touched off by a discovery of how to use a new kind of accessible energy. On the other hand, a great stride in technological progress cannot materialize unless the corresponding innovation is followed by a great mineralogical expansion. Even a substantial increase in the efficiency of the use of gasoline as fuel would pale in comparison with a manifold increase of the known, rich oil fields.

This sort of expansion is what has happened during the last one hundred years. We have struck oil and discovered new coal and gas deposits in a far greater proportion than we could use during the same period (note 38, below). Still more important, all mineralogical discoveries have included a substantial proportion of *easily* accessible resources. This exceptional bonanza by itself has sufficed to lower the real cost of bringing mineral resources *in situ* to the surface. Energy of mineral source thus becoming cheaper, substitution-innovations have caused the ratio of labor to net output to decline. Capital also must have evolved toward forms which cost less but use more energy to achieve the same result. What has happened during this period is a modification of the cost structure, the flow factors being increased and the fund factors decreased.[36] By examining, therefore, only the relative variations of the fund factors during a period of exceptional mineral bonanza, we cannot prove either that the unitary total cost will always follow a declining trend or that the continuous progress of technology renders accessible resources almost inexhaustible—as Barnett and Morse claim [3, 239].

Little doubt is thus left about the fact that the theses examined in this section are

---

[35] The point refers to the addition of capital (measured in *money terms*) and labor (measured in *workers employed*) as well as the computation of net output (by subtraction) from *physical* gross output [3, 167f].

[36] For these distinctions, see [27, 512–519; 30, 4; 32, 223–225].

anchored in a deep-lying belief in mankind's immortality. Some of their defenders have even urged us to have faith in the human species: such faith will triumph over all limitations.[37] But neither faith nor assurance from some famous academic chair [4] could alter the fact that, according to the basic law of thermodynamics, mankind's dowry is finite. Even if one were inclined to believe in the possible refutation of these principles in the future, one still must not act on that faith now. We must take into account that evolution does not consist of a linear repetition, even though over short intervals it may fool us into the contrary belief.

A great deal of confusion about the environmental problem prevails not only among economists generally (as evidenced by the numerous cases already cited), but also among the highest intellectual circles simply because the sheer entropic nature of all happenings is ignored or misunderstood. Sir Macfarlane Burnet, a Nobelite, in a special lecture considered it imperative "to prevent the progressive destruction of the earth's irreplaceable resources" [quoted, 15, 1]. And a prestigious institution such as the United Nations, in its Declaration on the Human Environment (Stockholm, 1972), repeatedly urged everyone "to improve the environment." Both urgings reflect the fallacy that man can reverse the march of entropy. The truth, however unpleasant, is that the most we can do is to prevent any unnecessary depletion of resources and any unnecessary deterioration of the environment, but without claiming that we know the precise meaning of "unnecessary" in this context.

VII. GROWTH: MYTHS, POLEMICS, AND FALLACIES

A great deal of confusion stains the heated arguments about "growth" simply because the term is used in multiple senses. One confusion, against which Joseph Schumpeter insistently admonished economists, is that between *growth* and *development*. There is growth when only the production per capita of current types of commodities increases, which naturally implies a growing depletion of equally accessible resources. Development means the introduction of any of the innovations described in the foregoing section. In the past, development has ordinarily induced growth and growth has occurred only in association with development. The result has been a peculiar dialectical combination also known as "growth," but for which we may reserve another current label, namely, "economic growth." Economists measure its level by the GNP per capita at constant prices.

Economic growth, it must be emphasized, is a dynamic state, analogous to that of an automobile traveling on a curve. For such an automobile it is not possible to be inside a curve at one moment and outside it at the very next moment. The teachings of standard economics that economic growth depends only on the decision at a point in time to consume a larger or a smaller proportion of production [4, 342f; 74, 41] are largely off base. In spite of the superb mathematical models with which Arrow-Debreu-Hahn have delighted the profession and of the pragmatically oriented Leontief models, not all production factors (including goods in process) can serve *directly* as consumer goods. Only in a primitive agricultural society, employing no capital equipment, would it be true that the decision to save more corn from the current harvest will increase the next year's average crop. Other economies are growing now because they grew yesterday and will grow tomorrow because they are growing today.

The roots of economic growth lie deep in human nature. It is because of man's Veblenian instincts of workmanship and idle curiosity that one innovation fosters another —which constitutes development. Given,

---

[37] See the dialogue between Preston Cloud and Roger Revelle quoted in [66, 416]. The same refrain runs through Maddox's complaint against those who point out mankind's limitations [59, vi, 138, 280]. In relation to Maddox's chapter, "Man-made Men," see [32, 348–359].

also, man's craving for comfort and gadgets, every innovation leads to growth. To be sure, development is not an inevitable aspect of history; it depends on many factors as well as on accidents, which explains why mankind's past consists mainly of long stretches of quasi stationary states and why the present effervescent era is just a very small exception.[38]

On purely logical grounds, however, there is no necessary association between development and growth; conceivably, there could be development without growth. Because of the failure to observe the preceding distinctions systematically, it was possible for environmentalists to be accused of being against development.[39] Actually, the true environmentalist position must focus on *the total rate* of resource depletion (and the rate of the ensuing pollution). It is only because in the past economic growth has resulted not only in a higher rate of depletion but even in an increase of per capita consumption of resources that the argument drifted so as to turn around the economist's guidepost—the GNP per capita. As a result, the real issue came to be buried under the sort of sophistries mentioned in the preceding section. For even though on purely logical grounds economic growth might occur even with a decrease in the rate of resource depletion, pure growth cannot exceed a certain, albeit unknowable, limit without an increase in that rate—unless there is a substantial decrease in population.

It was natural for economists—who unflinchingly have hung on to their mechanistic framework—to remain completely indifferent when, at various times, the Conservation Movement or some isolated literati, such as Fairfield Osborn and Rachel Carson, called attention to the ecological harm of growth and the necessity of slowing down. But a few years ago the environmentalist movement gained momentum around the problem of population—*The Population Bomb,* as Paul Ehrlich epitomized it. Also, a few unorthodox economists shifted to a physiocratic position, albeit in greatly modified forms, or made a try at blending ecology into economics [e.g., 8; 9; 19; 29; 32]. Some became concerned with good, instead of affluent life [8; 65]. Moreover, a long series of incidents proved to everybody's satisfaction that pollution is not a plaything of ecologists. Although depletion of resources has also been going on with increased intensity at all times, it ordinarily is a volume phenomenon below the earth's surface, where no one can see it truly. Pollution, on the other hand, is a surface phenomenon, the existence of which cannot possibly be ignored, much less denied. Those economists who have reacted to these events have generally tried to harden further the position that economic rationality and the right kind of price mechanism can take care of all ecological problems.

But, curiously, the recent publication of *The Limits to Growth* [62], a report for the Club of Rome, caused an unusual commotion within the economics profession. In fact, criticism of the report has come mainly from economists. A manifesto of similar tenor, "A Blueprint for Survival" [6], has been rather spared this glory, apparently not because it was endorsed by a numerous group of highly respected scholars. The reason for the difference is that the *The Limits to Growth* employed analytical models of the kind used in econometrics and simulation works. From all one can judge, it was this fact that irked economists to the point of resorting to direct or veiled insults in their attack against the Trojan Horse. Even *The Economist* [55] disregarded proverbial British good form and in

---

[38] Some who do not understand how exceptional, perhaps even abnormal, the present interlude is (*Journal of Economic Literature,* June 1972, pp 459f), ignore the facts that coal mining began eight hundred years ago and that, incredible though it may seem, half of the total quantity ever mined has been extracted in the last thirty years. Also, half of the total production of crude oil has been obtained in the last ten years alone! [46, 166, 238; 56, 119f; also 32, 228]

[39] Solow also claims that to be against pollution is to be against economic growth [74, 49]. However, harmful pollution can be kept very low if appropriate measures are taken and *pure* growth is slowed down.

the editorial "Limits to Misconception" branded the report as "the highwater mark of old-fashioned nonsense." Beckerman even ignored the solemnity of an inaugural lecture and assailed the study as "a brazen, impudent piece of nonsense [by] a team of whizz-kids from MIT" [4, 327].[40]

Let us begin by recalling, first, that economists, especially during the last thirty years, have preached right and left that only mathematical models can serve the highest aims of their science. With the advent of the computer, the use of econometric models and simulation became a widespread routine. The fallacy of relying on arithmomorphic models to predict the march of history has been denounced occasionally with technical arguments.[41] But all was in vain. Now, however, economists fault *The Limits to Growth* for that very sin and for seeking "an aura of scientific authority" through the use of the computer; some have gone so far as to impugn the use of mathematics [4, 331–334; 10, 22f; 51, 660; 52; 69, 15–17]. Let us observe, secondly, that aggregation has always been regarded as a mutilating yet inevitable procedure in macroeconomics, which thus greatly ignores structure. Nevertheless, economists now denounce the report for using an aggregative model [4, 338f; 52; 69, 61f, 74]. Thirdly, one common article of economic faith, known as the acceleration principle, is that output is proportional to capital stock. Yet some economists again have indicted the authors of *The Limits* for assuming (implicitly) that the same proportionality prevails for pollution—which is an output, too! [4, 399f; 52; 69, 47f][42] Fourthly, the price complex has not prevented economists from developing and using models whose blueprints contain no prices explicitly—the static and dynamic Leontief models, the Harrod-Domar model, the Solow model, to cite some of the most famous ones. In spite of this, some critics (including Solow himself) have decried the value of *The Limits* on the sole ground that its model does not involve prices [4, 337; 51, 665; 74, 46f; 69, 14].

The final and most important point concerns the indisputable fact that, except for some isolated voices in the last few years, economists have always suffered from growthmania [65, Ch. 1]. Economic systems as well as economic plans have always been evaluated only in relation to their ability to sustain a great rate of economic growth. Economic plans, without a single exception, have been aimed at the highest possible rate of economic growth. The very theory of economic development is anchored solidly in exponential growth models. But when the authors of *The Limits* also used the assumption of exponential growth, the chorus of economists cried "foul!" [4, 332f; 10, 13; 51, 661; 52; 74, 42f; 69, 58f] This is all the more curious since some of the same critics concomitantly maintained that technology grows exponentially (Section VI). Some, while admitting at long last that economic growth cannot continue forever at the present rate, suggested, however, that it could go on at some lower rates [74, 666].

Going through this peculiar criticism, one gets the impression that the critics from the economics profession proceeded according to the Latin adage—*quod licet Jovi non licet bovi*—what is permitted to Zeus is not permitted to a bovine. Be this as it may, standard economics will recover only with difficulty

---

[40] And later he asked, "How silly do you have to be to be allowed to join [the Club of Rome]?" [4, 339]. Kaysen [51] also is caustic in places. Solow [75, 1] just says that, like everyone else, he was "suckered into reading the *Limits to Growth*," while Johnson [49, 1] disqualifies intellectually all concerned ecologists right from the outset. Outside the economists' circle, John Maddox stands out by himself for seeking to impress the reader by similar "arguments."

[41] See in particular, [26] and [28]; also [32, 339–341]. More recently, and from a different viewpoint, W. Leontief also took up the issue in his Presidential Address to the AEA [54]. Symptomatically, the frank verdict of Ragnar Frisch in his address to the First World Congress of the Econometric Society (1965) still awaits publication.

[42] Some of the foregoing objections were also voiced from outside the economics profession [1; 59, 284f].

from the exposure of its own weaknesses by these efforts at self-defense.

Outside these circles, the report has been received with sufficient appreciation, certainly not with vituperation.[43] The most apt verdict is that despite its imperfections, "it is not frivolous." [44] True, the presentation is rather half-baked, betraying the rush for early publicity [34]. But even some economists have recognized its merit in drawing attention to the ramified consequences of pollution [69, 58f]. The study has also brought to the fore the importance of duration in the actual course of events [62, 183] —a point often emphasized by natural scientists [43, 144; 56, 131] but generally overlooked by economists [32, 273f]. We need a time lead not only to reach a higher level of economic growth but also to descend to a lower one.

But the much publicized conclusion—that at most one hundred years separate mankind from an ecological catastrophe [62, 23 and *passim*]—lacks a scientifically solid basis.

There is hardly any room for quarreling about the general pattern of relations assumed in the various simulations covered by the report. However, the *quantitative* forms of these relations have not been submitted to any factual verification. Besides, by their very rigid nature, the arithmomorphic models used are incapable of predicting the evolutionary changes these relations may suffer over time. The prediction, which sounds like the famous scare that the world would come to an end in A. D. 1000, is at odds with everything we know about biological evolution. The human species, of all species, is not likely to go suddenly into a short coma. Its end is not even in distant sight; and when it comes it will be after a very long series of surreptitious, protracted crises. Yet, as Silk pointed out [72], it would be madness to ignore the study's general warnings about population growth, pollution, and resource depletion. Indeed, any of these factors may cause the world's economy to experience some shortness of breath.

Some critics have further belittled *The Limits* for merely using an analytical armamentarium in order to emphasize an uninteresting tautology, namely, that continuous exponential growth is impossible in a finite environment [4, 333f; 51, 661; 74, 42f; 69, 55]. The indictment is right, but only on the surface; for this was one of those occasions when the obvious had to be emphasized because it had been long ignored. However, the greatest sin of the authors of *The Limits* is that they have concealed the most important part of the obvious by focusing their attention exclusively on exponential growth, as Malthus and almost every other environmentalist has done.

VIII. THE STEADY STATE: A TOPICAL MIRAGE

Malthus, as we know, was criticized primarily because he assumed that population and resources grow according to some simple mathematical laws. But this criticism did not touch the real error of Malthus (which has apparently remained unnoticed). This error is the implicit assumption that population may grow beyond any limit both in number and time *provided that it does not grow too rapidly*.[45] An essentially similar error has been committed by the authors of *The Limits,* by the authors of the nonmathematical yet more articulate "Blueprint for Survival," as well as by several earlier writers. Because, like Malthus, they were set exclu-

---

[43] A notable exception is Maddox [59]. His berating review of "A Blueprint for Survival" ("The Case Against Hysteria," *Nature,* 14 January 1972, pp. 63–65) drew numerous protests: *Nature,* 21 January 1972, p. 179, 18 February 1972, pp. 405f. But given the position of economists in the controversy, it is understandable that Beckerman [4, 341f] cannot conceive why natural scientists have not assailed the report and why they seem even to accept its thesis.

[44] *Financial Times,* 3 March 1972, quoted in [4, 337n]. Denis Gabor, a Nobelite, judged that "whatever the details, the main conclusions are incontrovertible" (quoted in [4, 342]).

[45] Joseph J. Spengler, a recognized authority in this broad domain, tells me that indeed he knows of no one who may have made the observation. For some very penetrating discussions of Malthus and of the present population pressure, see [76; 77].

sively on proving the impossibility of growth, they were easily deluded by a simple, now widespread, but false syllogism: since exponential growth in a finite world leads to disasters of all kinds, ecological salvation lies in the stationary state [42; 47; 62, 156–184; 6, 3f, 8, 20].[46] H. Daly even claims that "the stationary state economy is, therefore, a necessity" [21, 5].

This vision of a blissful world in which both population and capital stock remain constant, once expounded with his usual skill by John Stuart Mill [64, Bk. 4, Ch. 6], was until recently in oblivion.[47] Because of the spectacular revival of this myth of ecological salvation, it is well to point out its various logical and factual snags. The crucial error consists in not seeing that not only growth, but also a zero-growth state, nay, even a declining state which does not converge toward annihilation, cannot exist forever in a finite environment. The error perhaps stems from some confusion between finite stock and finite flow rate, as the incongruous dimensionalities of several graphs suggest [62, 62, 64f, 124ff; 6, 6]. And contrary to what some advocates of the stationary state claim [21, 15], this state does not occupy a privileged position vis-à-vis physical laws.

To get to the core of the problem, let S denote the actual amount of accessible resources in the crust of the earth. Let P_i and s_i be the population and the amount of depleted resources per person in the year i. Let the "amount of total life," measured in years of life, be defined by $L = \Sigma P_i$, from $i = 0$ to $i = \infty$. S sets an upper limit for L through the obvious constraint $\Sigma P_i s_i \leqslant S$. For although s_i is a historical variable, it cannot be zero or even negligible (unless mankind reverts sometime to a berry-picking economy). Therefore, $P_i = 0$ for i greater than some finite n, and $P_i > 0$ otherwise. That value of n is the maximum duration of the human species [31, 12f; 32, 304].

The earth also has a so-called carrying capacity, which depends on a complex of factors, including the size of s_i.[48] This capacity sets a limit on any single P_i. But this limit does not render the other limits, of L and n, superfluous. It is therefore inexact to argue—as the Meadows group seems to do [62, 91f]—that the stationary state can go on forever as long as P_i does not exceed that capacity. The proponents of salvation through the stationary state must admit that such a state can have only a finite duration—unless they are willing to join the "No Limit" Club by maintaining that S is inexhaustible or almost so—as the Meadows group does in fact [62, 172]. Alternatively, they must explain the puzzle of how a whole economy, stationary for a long era, all of a sudden comes to an end.

Apparently, the advocates of the stationary state equate it with an open *thermodynamic* steady state. This state consists of an *open* macrosystem which maintains its entropic structure constant through material exchanges with its "environment." As one would immediately guess, the concept constitutes a highly useful tool for the study of biological organisms. We must, however, observe that the concept rests on some special conditions introduced by L. Onsager [50, 89–97]. These conditions are so delicate (they are called the principle of *detailed* balance) that in actuality they can hold only "within a deviation of a few percent" [50, 140]. For this reason, a steady state may exist in fact only in an approximated manner and over a finite duration. This impossibility of a macrosystem not in a state of chaos to be perpetually durable may one day be explicitly recognized by a new thermodynamic law just as the impossibility of perpetual motion once

---

[46] The substance of the argument of *The Limits* beyond that of Mill's is borrowed from Boulding and Daly [8; 9; 20; 21].

[47] In *International Encyclopedia of the Social Sciences*, for example, the point is mentioned only in passing.

[48] Obviously, any increase in s_i will generally result in a decrease of L and of n. Also, the carrying capacity in any year may be increased by a greater use of terrestrial resources. These elementary points should be retained for further use (Section X).

was. Specialists recognize that the present thermodynamic laws do not suffice to explain all nonreversible phenomena, including especially life processes.

Independently of these snags there are simple reasons against believing that mankind can live in a perpetual stationary state. The structure of such a state remains the same throughout; it does not contain in itself the seed of the inexorable death of all open macrosystems. On the other hand, a world with a stationary population would, on the contrary, be continually forced to change its technology as well as its mode of life in response to the inevitable decrease of resource accessibility. Even if we beg the issue of how capital may change qualitatively and still remain constant, we would have to assume that the unpredictable decrease in accessibility will be miraculously compensated by the right innovations at the right time. A stationary world may for a while be interlocked with the changing environment through a system of balancing feedbacks analogous to those of a living organism during one phase of its life. But as Bormann reminded us [7, 707], the miracle cannot last forever; sooner or later the balancing system will collapse. At that time, the stationary state will enter a crisis, which will defeat its alleged purpose and nature.

One must be cautioned against another logical pitfall, that of invoking the Prigogine principle in support of the stationary state. This principle states that the minimum of the entropy produced by an Onsager type of open thermodynamic system is reached when the system becomes steady [50, ch. xvi]. It says nothing about how this last entropy compares with that produced by other open systems.[49]

The usual arguments adduced in favor of the stationary state are, however, of a different, more direct nature. It is, for example, argued that in such a state there is more time for pollution to be reduced by natural processes and for technology to adapt itself to the decrease of resource accessibility [62, 166]. It is plainly true that we could use much more efficiently today the coal we have burned in the past. The rub is that we might not have mastered the present efficient techniques if we had not burned all that coal "inefficiently." The point that in a stationary state people will not have to work additionally to accumulate capital (which in view of what I have said in the last paragraphs is not quite accurate) is related to Mill's claim that people could devote more time to intellectual activities. "The trampling, crushing, elbowing, and treading on each other's heel" will cease [64, 754]. History, however, offers multiple examples—the Middle Ages, for one—of quasi stationary societies where arts and sciences were practically stagnant. In a stationary state, too, people may be busy in the fields and shops all day long. Whatever the state, free time for intellectual progress depends on the intensity of the pressure of population on resources. Therein lies the main weakness of Mill's vision. Witness the fact that—as Daly explicitly admits [21, 6-8]—its writ offers no basis for determining even in principle the optimum levels of population and capital. This brings to light the important, yet unnoticed point, that *the necessary conclusion*

---

[49] The point recalls Boulding's idea that the inflow from nature into the economic process, which he calls "throughput," is "something to be minimized rather than maximized" and that we should pass from an economy of flow to one of stock [8, 9f; 9, 359f]. The idea is more striking than enlightening. True, economists suffer from a flow-complex [29, 55, 88]; also, they have little realized that the proper analytical description of a process must include *both flows and funds* [30; 32, 219f, 228-234]. Entrepreneurs, as far as Boulding's idea is concerned, have at all times aimed at minimizing the flow necessary to maintain their capital funds. If the present inflow from nature is incommensurate with the safety of our species, it is only because the population is too large and part of it enjoys excessive comfort. Economic decisions will always forcibly involve both flows and stocks. Is it not true that mankind's problem is to economize S (a stock) for as large an amount of life as possible, which implies to minimize s_t (a flow) for some "good life"? (Section XI).

of the arguments in favor of that vision is that the most desirable state is not a stationary, but a declining one.

Undoubtedly, the current growth must cease, nay, be reversed. But anyone who believes that he can draw a blueprint for the ecological salvation of the human species does not understand the nature of evolution, or even of history—which is that of a permanent struggle in continuously novel forms, not that of a predictable, controllable physico-chemical process, such as boiling an egg or launching a rocket to the moon.

IX. SOME BASIC BIOECONOMICS[50]

Apart from a few insignificant exceptions, all species other than man use only *endosomatic* instruments—as Alfred Lotka proposed to call those instruments (legs, claws, wings, etc.) which belong to the individual organism *by birth*. Man alone came, in time, to use a club, which does not belong to him by birth, but which extended his endosomatic arm and increased its power. At that point in time, man's evolution transcended the biological limits to include also (and primarily) the evolution of *exosomatic* instruments, i.e., of instruments produced by man but not belonging to his body.[51] That is why man can now fly in the sky or swim under water even though his body has no wings, no fins, and no gills.

The exosomatic evolution brought down upon the human species two fundamental and irrevocable changes. The first is the irreducible social conflict which characterizes the human species [29, 98–101; 32, 306–315, 348f]. Indeed, there are other species which also live in society, but which are free from such conflict. The reason is that their "social classes" correspond to some clear-cut biological divisions. The periodic killing of a great part of the drones by the bees is a natural, biological action, not a civil war.

The second change is man's addiction to exosomatic instruments—a phenomenon analogous to that of the flying fish which became addicted to the atmosphere and mutated into birds forever. It is because of this addiction that mankind's survival presents a problem entirely different from that of all other species [31; 32, 302–305]. It is neither only biological nor only economic. It is bioeconomic. Its broad contours depend on the multiple asymmetries existing among the three sources of low entropy which together constitute mankind's dowry—the free energy received from the sun, on the one hand, and the free energy and the ordered material structures stored in the bowels of the earth, on the other.

The *first* asymmetry concerns the fact that the terrestrial component is a *stock,* whereas the solar one is a *flow.* The difference needs to be well understood [32, 226f]. Coal *in situ* is a stock because we are free to use it all today (conceivably) or over centuries. But at no time can we use any part of a future flow of solar radiation. Moreover, the flow rate of this radiation is wholly beyond our control; it is completely determined by cosmological conditions, including the size of our globe.[52] One generation, whatever it may do, cannot alter the share of solar radiation of any future generation. Because of the priority of the present over the future and the irrevocability of entropic degradation, the opposite is true for the terrestrial shares. These shares are affected by how much of the terrestrial dowry the past generations have consumed.

Second, since no practical procedure is available at human scale for transforming energy into matter (Section IV), accessible material low entropy is by far the most critical element from the bioeconomic viewpoint.

---

[50] I saw this term used for the first time in a letter from Jiří Zeman.

[51] The practice of slavery, in the past, and the possible procurement, in the future, of organs for transplant are phenomena akin to the exosomatic evolution.

[52] A fact greatly misunderstood: Ricardian land has economic value for the same reason as a fisherman's net. Ricardian land catches the most valuable energy, roughly in proportion to its total size [27, 508; 32, 232].

True, a piece of coal burned by our forefathers is gone forever, just as is part of the silver or iron, for instance, mined by them. Yet future generations will still have their inalienable share of solar energy (which, as we shall see next, is enormous). Hence, they will be able, at least, to use each year an amount of wood equivalent to the annual vegetable growth. For the silver and iron dissipated by the earlier generations there is no similar compensation. This is why in bioeconomics we must emphasize that every Cadillac or every Zim—let alone any instrument of war—means fewer plowshares for some future generations, and implicitly, fewer future human beings, too [31, 13; 32, 304].

Third, there is an astronomical difference between the amount of the flow of solar energy and the size of the stock of terrestrial free energy. At the cost of a decrease in mass of 131×10^{12} tons, the sun radiates annually 10^{13} Q—one single Q being equal to 10^{18} BTU! Of this fantastic flow, only some 5,300 Q are intercepted at the limits of the earth's atmosphere, with roughly one half of that amount being reflected back into outer space. At our own scale, however, even this amount is fantastic; for the total world consumption of energy currently amounts to no more than 0.2 Q annually. From the solar energy that reaches the ground level, photosynthesis absorbs only 1.2 Q. From waterfalls we could obtain at most 0.08 Q, but we are now using only one tenth of that potential. Think also of the additional fact that the sun will continue to shine with practically the same intensity for another five billion years (before becoming a red giant which will raise the earth's temperature to 1,000°F). Undoubtedly, the human species will not survive to benefit from all this abundance.

Passing to the terrestrial dowry, we find that, according to the best estimates, the initial dowry of fossil fuel amounted to only 215 Q. The outstanding recoverable reserves (known and probable) amount to about 200 Q. These reserves, therefore, could produce only two weeks of sunlight on the globe.[53] If their depletion continues to increase at the current pace, these reserves may support man's industrial activity for just a few more decades. Even the reserves of uranium-235 will not last for a longer period if used in the ordinary reactors. Hopes are now set on the breeder reactor, which, with the aid of uranium-235, may "extract" the energy of the fertile but not fissionable elements, uranium-238 and thorium-232. Some experts claim that this source of energy is "essentially inexhaustible" [83, 412]. In the United States alone, it is believed, there are large areas covered with black shale and granite which contain 60 grams of natural uranium or thorium per metric ton [46, 226f]. On this basis, Weinberg and Hammond [83, 415f] have come out with a grand plan. By strip-mining and crushing all these rocks, we could obtain enough nuclear fuel for some 32,000 breeder reactors distributed in 4,000 offshore parks and capable of supplying a population of twenty billion for millions of years with twice as much energy per capita as the current consumption rate in the USA. The grand plan is a typical example of linear thinking, according to which all that is needed for the existence of a population, even "considerably larger than twenty billion," is to increase all supplies proportionally.[54] Not that the authors deny that there also are non-technical issues; only, they play them down with noticeable zeal [83, 417f]. The most important issue, of whether a social organization compatible with the density of population and the nuclear manipulation at the

---

[53] The figures used in this section have been calculated from the data of Daniels [22] and Hubbert [46]. Such data, especially those about reserves, vary from author to author but not to the extent that really matters. However, the assertion that "the vast oil shales which are to be found all over the world [would last] for no less than 40,000 years" [59, 99] is sheer fantasy.

[54] In an answer to critics (*American Scientist,* LVIII, No. 6, p. 619), the same authors prove, again linearly, that the agro-industrial complexes of the grand plan could easily feed such a population.

grand level can be achieved, is brushed aside by Weinberg as "transscientific" [82].[55] Technicians are prone to forget that due to their own successes, nowadays it may be easier to move the mountain to Mohammed than to induce Mohammed to go to the mountain. For the time being, the snag is far more palpable. As responsible forums openly admit, even one breeder still presents substantial risks of nuclear catastrophes, and the problem of safe transportation of nuclear fuels and especially that of safe storage of the radioactive garbage still await a solution even for a moderate scale of operations [35; 36; especially 39 and 67].

There remains the physicist's greatest dream, controlled thermonuclear reaction. To constitute a real breakthrough, it must be the deuterium-deuterium reaction, the only one that could open up a formidable source of terrestrial energy for a long era.[56] However, because of the difficulties alluded to earlier (Section IV), even the experts working at it do not find reasons for being too hopeful.

For completion, we should also mention the tidal and geothermal energies, which, although not negligible (in all 0.1 Q per year), can be harnessed only in very limited situations.

The general picture is now clear. The terrestrial energies on which we can rely effectively exist in very small amounts, whereas the use of those which exist in ampler amounts is surrounded by great risks and formidable technical obstacles. On the other hand, there is the immense energy from the sun which reaches us without fail. Its direct use is not yet practiced on a significant scale, the main reason being that the alternative industries are now much more efficient economically. But promising results are coming from various directions [37; 41]. What counts from the bioeconomic viewpoint is that the feasibility of using the sun's energy directly is not surrounded by risks or big question marks; it is a proven fact.

The conclusion is that mankind's entropic dowry presents another important differential scarcity. From the viewpoint of the extreme longrun, the terrestrial free energy is far scarcer than that received from the sun. The point exposes the foolishness of the victory cry that we can finally obtain protein from fossil fuels! Sane reason tells us to move in the opposite direction, to convert vegetable stuff into hydrocarbon fuel—an obviously natural line already pursued by several researchers [22, 311–313].[57]

Fourth, from the viewpoint of industrial utilization, solar energy has an immense drawback in comparison with energy of terrestrial origin. The latter is available in a concentrated form, in some cases, in a too concentrated form. As a result, it enables us to obtain almost instantaneously enormous amounts of work, most of which could not even be obtained otherwise. By great contrast, the flow of solar energy comes to us with an extremely low intensity, like a very fine rain, almost a microscopic mist. The important difference from true rain is that this radiation rain is not collected naturally into streamlets, then into creeks and rivers, and finally into lakes from where we could use it in a concentrated form, as is the case with waterfalls. Imagine the difficulty one would face if one tried to use *directly* the kinetic energy of some microscopic rain drops as

---

[55] For a recent discussion of the social impact of industrial growth, in general, and of the social problems growing out of a large scale use of nuclear energy, in particular, see [78], a monograph by Harold and Margaret Sprout, pioneers in this field.

[56] One percent only of the deuterium in the oceans would provide 10^9 Q through that reaction, an amount amply sufficient for some hundred millions of years of very high industrial comfort. The reaction deuterium-tritium stands a better chance of success because it requires a lower temperature. But since it involves lithium-6, which exists in small supply, it would yield only about 200 Q in all.

[57] It should be of interest to know that during World War II in Sweden, for one, automobiles were driven with the poor gas obtained by heating charcoal with kindlings in a container serving as a tank!

they fall. The same difficulty presents itself in using solar energy directly (i.e., not through the chemical energy of green plants, or the kinetic energy of the wind and waterfalls). But as was emphasized a while ago, the difficulty does not amount to impossibility.

Fifth, solar energy, on the other hand, has a unique and incommensurable advantage. The use of any terrestrial energy produces some noxious pollution, which, moreover, is irreducible and hence cumulative, be it in the form of thermal pollution alone. By contrast, any use of solar energy is *pollution-free*. For, whether this energy is used or not, its ultimate fate is the same, namely, to become the dissipated heat that maintains the thermodynamic equilibrium between the globe and outer space at a propitious temperature.[58]

The *sixth* asymmetry involves the elementary fact that the survival of every species on earth depends, directly or indirectly, on solar radiation (in addition to some elements of a superficial environmental layer). Man alone, because of his exosomatic addiction, depends on mineral resources as well. For the use of these resources man competes with no other species; yet his use of them usually endangers many forms of life, including his own. Some species have in fact been brought to the brink of extinction merely because of man's exosomatic needs or his craving for the extravagant. But nothing in nature compares in fierceness with man's competition for solar energy (in its primary or its by-product forms). Man has not deviated one bit from the law of the jungle; if anything, he has made it even more merciless by his sophisticated exosomatic instruments. Man has openly sought to exterminate any species that robs him of his food or feeds on him— wolves, rabbits, weeds, insects, microbes, etc.

But this struggle of man with other species for food (in ultimate analysis, for solar energy) has some unobtrusive aspects as well.

And, curiously, it is one of these aspects that has some far-reaching consequences in addition to supplying a most instructive refutation of the common belief that every technological innovation constitutes a move in the right direction as concerns the economy of resources. The case pertains to the economy of modern agricultural techniques.

X. MODERN AGRICULTURE: AN ENERGY SQUANDERER

Given the extant spectrum of green plants and their geographical distribution at any one time, the biological carrying capacity of the earth is determined, even though we could compute it only with difficulty and only approximately. It is within this capacity that man struggles with other life-bearing structures for food. But man is unique among all species in that he can influence, within limits, not only his share of food but also the efficiency of the transformation of solar energy into food. With time, man learned to plow deeper, to rotate the use of land, to fertilize the soil with manure, and so on. In his farming activity, man also came to derive an immense benefit from the use of domesticated draft animals.

Two evolutionary factors have influenced farming technology over the years. The oldest one is the continuous pressure of population on the extant land under cultivation. Village swarming, at first, and later migration, were able to relieve the pressure. Means of increasing the yield of land also helped ease the tension. The main source of release, however, remained the clearing of vast tracts of land. The second factor, a by-product of the Industrial Revolution, was the extension to agriculture of the process by which low entropy from mineral sources was substituted for that of biological nature. The process is even more conspicuous in agriculture. Tractors and other agricultural machines have taken the place of man and draft animals, and chemical fertilizers, that of manuring and fallowing.

However, mechanized agriculture does not fit small family farms which have at their

---

[58] One necessary qualification: even the use of solar energy may disturb the climate if the energy is released in another place than where collected. The same is true for a difference in time, but this case is unlikely to have any practical importance.

disposal a large supply of free hands. Yet even in this case it had to come. The peasant who practices organic agriculture, who uses animals for power and manure as fertilizer, must grow not only food for his family but also fodder for his helpers. The increasing pressure of population thus forced even the small farmer, practically everywhere, to do away with the beasts of burden so as to use his entire land for food [27, 526; 31, 11f; 32, 302f].

The point beyond any possible doubt is that, given the pressure of population in the greater part of the globe, there is no other salvation from the calamities of undernutrition and starvation than to force the yield on the land under cultivation by an increasingly mechanized agriculture, an increasing use of chemical fertilizers and pesticides, and an increasing cultivation of the new high-yield varieties of cereal grains. However, contrary to the generally and indiscriminately shared notion, this modern agricultural technique is in the longrun a move against the most elementary bioeconomic interest of the human species.

First, the replacement of the water buffalo by the tractor, of fodder by motor fuels, of manure and fallowing by chemical fertilizers substitutes scarcer elements for the most abundant one—solar radiation. Secondly, this substitution also represents a squandering of terrestrial low entropy because of its strongly decreasing returns.[59] What modern agricultural technique does is to increase the amount of photosynthesis on the same piece of cultivated land. But this increase is achieved by a more than proportional increase in the depletion of the low entropy of terrestrial origin, which is the only critically scarce resource. (We should note that decreasing returns in substituting solar for terrestrial energy would, on the contrary, constitute a good energetic deal.) This means that, if half of the input of terrestrial energy (counted from the mining operation) required by modern agriculture for one acre—cultivated, say, with wheat—is used each year, in two years the less industrialized agriculture would produce more than twice as much wheat from the same piece of land. This diseconomy—surprising as it may seem to the worshipers of machinery—is especially heavy in the case of the high-yield varieties which earned their developer, Norman E. Borlaug, a Nobel Prize.

A highly mechanized and heavily fertilized cultivation does allow a very large population, P_i, to survive, but the price is an increase of the per capita depletion of terrestrial resources s_i, which *ceteris paribus* means a proportionally greater reduction of the future amount of life (Section VIII). In addition, if growing food by "agro-industrial complexes" becomes the general rule, many species associated with old-fashioned, organic agriculture may gradually disappear, a result which may drive mankind into an ecological cul-de-sac from which there would be no return [31, 12].

The above observations bear upon the perennial question of how many people the earth could support. Some population experts claim that there would be enough food even for some forty billion people at a diet of some 4,500 kilocalories provided that the best farming methods were used on every acre of potentially arable land.[60] The logic rests on multiplying the amount of potentially arable land by the current average yield in Iowa. The calculations may be as "careful" as boasted—they represent, nonetheless, linear thinking. Clearly, neither these authors nor those less optimistic have thought of the crucial question of *how long* a population of forty billion—nay, even one of only one million for that matter—can last [31, 11; 32, 20, 301f]. It is this question which, more

---

[59] Between 1951 and 1966, the number of tractors increased by 63 percent, phosphate fertilizers by 75 percent, nitrate fertilizers by 146 percent, and pesticides by 300 percent. Yet the crops, which may be taken as a good index of yield, increased by only 34 percent! [6, 40]

[60] This position has been advanced, for example, by Colin Clark in 1963 [see 31, 11; 32, 20], and very recently by Revelle [70].

than most others, lays bare the most stubborn residual of the mechanistic view of the world, which is the myth of the optimum population "as one that can be sustained indefinitely" [6, 14; also 62, 172f; 74, 48].

XI. A MINIMAL BIOECONOMIC PROGRAM

In "A Blueprint for Survival" [6, 13], the hope is expressed that economics and ecology will one day merge. The same possibility has already been considered for biology and physics, with most opinions agreeing that in the merger biology would swallow up physics [32, 42]. For essentially the same reason—that the phenomenal domain covered by ecology is broader than that covered by economics—economics will have to merge into ecology, if the merger ever occurs. For, as we have seen in the preceding two sections, the economic activity of any generation has some influence on that of the future generations—terrestrial resources of energy and materials are irrevocably used up and the harmful effects of pollution on the environment accumulate. One of the most important ecological problems for mankind, therefore, is the relationship of the quality of life of one generation with another—more specifically, the distribution of mankind's dowry *among all generations*. Economics cannot even dream of handling this problem. The object of economics, as has often been explained, is the administration of scarce resources; but to be exact, we should add that *this administration regards only one generation*. It could not be otherwise.

There is an elementary principle of economics according to which the only way to attribute a relevant price to an irreproducible object, say, to Leonardo's Mona Lisa, is to have absolutely everyone bid on it. Otherwise, if only you and I were to bid, one of us could get it for just a few dollars. That bid, i.e., that price, would clearly be parochial.[61]

This is exactly what happens for the irreproducible resources. Each generation can use as many terrestrial resources and produce as much pollution as its own bidding alone decides. Future generations are not, simply because they cannot be, present on today's market.

To be sure, the demand of the present generation reflects also the interest to *protect* the children and perhaps the grandchildren. Supply may also reflect expected future prices over a few decades. But neither the current demand nor the current supply can include even in a very slight form the situation of more remote generations, say, those of A.D. 3,000, let alone those that might exist a hundred thousand years from now.

Not all the details, but certainly the most important consequences of allocation of resources among generations by the market mechanism may be brought to the fore by a very simple, actually a highly simplified diagram. We shall assume that demand for some mineral resource already mined (say, coal-on-the-ground) is the same for each successive generation and that each generation must consume at least one "ton" of coal. The demand schedule is also assumed to include the preference for protecting the interests of a few future generations. In Figure 1, D_1, D_2, ... D_{15} represent the aggregate demands of successive generations, beginning with the present one. The interrupted line *abcdef* represents the average cost of mining the deposits of various accessibilities. Total reserves amount to 15 tons. Now, if we ignore for a moment the effect of the interest rate on the supply of the coal *in situ* by the owners of the mines, then the first generation will mine the amount $a'b'$, the shaded area representing the differential rent of the better mines. We may safely regard aa' as the price of the coal contained in these mines. The second generation will mine the amount $b'c'$. But

---

[61] Yet the economist's myth that prices reflect values in some generally relevant sense is now shared by other professions as well. The Meadows group, for example, speaks of the cost of resource depletion [62, 181], and Barry Commoner, of the cost of environmental deterioration [18, 253f and *passim*]. These are purely verbal expressions, for there is no such thing as the cost of irreplaceable resources or of irreducible pollution.

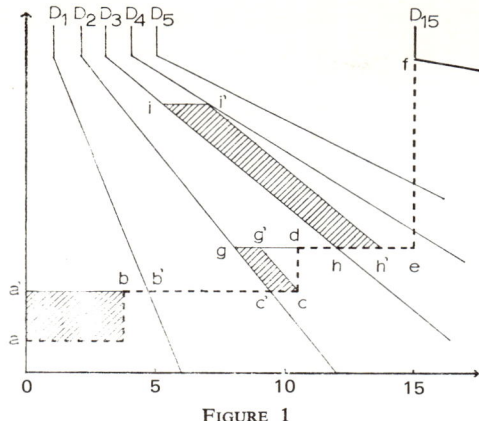

FIGURE 1

since no mine will earn a differential rent, the price of the coal *in situ* will be zero. During the third generation, the marginal cost of mining will be at the level of h; the quantity mined will be gh, with the quantity $c'c = gg'$ earning the rent shown by the shaded area. Finally, the fourth generation is left with the amount hh' (determined by the condition that $g'd = h'e$), which will earn a pure scarcity rent, represented by the shaded area $hh'i'i$. Nothing will be left for the following generations.

Several things are now obvious. First, the market mechanism *by itself* results in resources being consumed in higher amounts by the earlier generations, that is, faster than they should be. Indeed, $a'b' \geqslant b'c' > gh > hh'$, which confirms the dictatorship of the present over the future. Should all the generations bid from the outset for the total deposit of coal, the price of coal *in situ* will be driven up to infinity, a situation which would lead nowhere and only explode the entropic predicament of mankind. Only an omniscient planner could avoid this situation by simply allocating one ton of coal *in situ* to each of the first fifteen generations, each ton consisting of the same qualitative composition.[62]

Bringing in the interest rate modifies the

[62] In a pioneering work [45], Hotelling demonstrated once for all that one cannot speak of optimum allocation of resources unless the demand over the entire future is known.

picture somewhat and allows us to see even more clearly the impotence of the market to prevent the excessive depletion of resources by the earlier generations. Let us consider the case which I earlier called a bonanza era. Specifically, it is the situation in which the best quality of coal mine suffices to satisfy the present demand as well as that of the future generations *as far as the* present economic time horizon goes. Within this horizon, then, there is no rent at any time and hence no inducement to save coal *in situ* for future generations. Coal *in situ* can thus have no price during the present generation.

The question ignored by the few economists who have recently tackled some market aspects of natural resources [e.g., 75] is why resources *in situ* may, after all, have a positive price even if there are no self-imposed restrictions by the mine owners. The answer is that if present resources have a price, it is not ordinarily because of present scarcity, but because of some expected differential scarcity within the present time horizon. To illustrate the rationale of this process, let C_1, C_2, C_3 be coal mines of different qualities, the costs of mining one unit of coal being $k_1 < k_2 < k_3$, respectively. Let us further assume that C_1 is expected to be exhausted during the third generation after the present one, when C_2 will become economically efficient. Let us also assume that C_2, in turn, will be exhausted during the second generation thereafter, and that C_3 will then suffice for the remainder of the time horizon. During the third future generation, C_1 will prove to enjoy a differential rent $r_1 = k_2 - k_1$ with respect to C_2, and after two more generations the differential rent of C_2 over C_3, $r_2 = k_3 - k_2$, will become manifest. Only C_3 has no differential rent, and hence, as we have seen in the previous paragraph, its price is zero throughout. On the other hand, because C_2 necessarily earns a rent in the fifth generation from now, it must have a present positive price, namely, $p_2^0 = r_2/(1+i)^5$, where i is the interest rate (assumed constant throughout the time horizon). In the j-th generation from now, the price will be $p_2{}^j = r_2/(1 + i)^{5-j}$. A similar logic determines the present price of C_1. Only, we must observe that during the generation when the differential rent of C_1 becomes manifest, the price of C_2 is $p_2^3 = r_2/(1 + i)^2$. The rent must therefore be added to this price. Hence, the present price of the coal of C_1 is $p_1{}^0 = (r_1 + p_2^3)/(1 + i)$.[3]

The formulae just established show that the effect of the interest rate in the presence of a qualitative spectrum of mines is to extend the use of coal mined from more accessible sources (in comparison to the quantities determined by Figure 1). In some rather idle way, we may say that the existence of the interest rate helps the economy of resources. But let us not ignore the far more important conclusion of the foregoing analysis, which is especially striking in the case of an era of bonanza. Serious scarcities may become effective (as will certainly happen) beyond the present time horizon. That future fact can in no way influence our present market decisions; it is virtually inexistent as far as these decisions are concerned.

Nothing need be added to convince ourselves that the market mechanism cannot protect mankind from ecological crises in the future (let alone to allocate resources optimally among generations) even if we would try to set the prices "right."[63] The only way to protect the future generations, at least from the excessive consumption of resources during the present bonanza, is by reeducating ourselves so as to feel some sympathy for our *future* fellow humans in the same way in which we have come to be interested in the well-being of our *contemporary* "neighbors." This parallel does not mean that the new ethical orientation is an easy matter. Charity for one's contemporaries rests on some objective basis, namely, the individual self-interest. The difficult question one has to face in spreading the new gospel is not "what has posterity done for me?"—as Boulding wittily put it—but, rather, "why should I do anything for posterity?" What makes you think, many will ask, that there will be any posterity ten thousand years from now? And indeed, it would certainly be poor economics to sacrifice anything for a nonexistent beneficiary. These questions, which pertain to the new ethics, are not susceptible of easy, convincing answers.

Moreover, there is the other side of the coin, also ethical and even more urgent, on which Kaysen [51] and Silk [72], in particular, have rightly insisted. The nature of Mohammed-men being what it is, if we stop economic growth everywhere, we freeze the present status and thus eliminate the chance of the poor nations to improve their lot. This is why one wing of the environmentalist movement maintains that the issue of population growth is only a bogy used by the rich nations in order to divert attention from their own abuse of the environment. For this group, there is only one evil—inequality of development. We must proceed, they say, toward a radical redistribution of productive capacity among all nations. Another view argues that, on the contrary, population growth is the most menacing evil of mankind and must be dealt with urgently and independently of any other action. As expected, the two polarized views have never ceased clashing in useless and even violent controversies—as happened especially at the Stockholm Conferences in 1972, and, quite recently, at the Bucharest Conference on Population.[64] The difficulty is again seated in human nature: it is mutual, deep-rooted

---

[63] The economist's characteristic confidence in the omnipotence of the price mechanism (Section IV, note 15) led many of my auditors to counter that the choice between satisfying present or future needs, with the usual reward for postponing consumption, will set the prices right for optimal use of resources. The argument fails to take into account precisely the limitation of our time horizon, which does not extend beyond a couple of decades [10, 10]. Even Solow, in an illustration defending the standard position [74, 427], assumes a horizon of thirty years only.

[64] For a highly interesting account of the crosscurrents at the Stockholm Conference, see [2].

mistrust—of the rich that the poor will not cease growing in numbers and of the poor that the rich will not stop getting richer. Sane reason, however, invites us to recognize that the differential gradient between the poor and the rich nations is an evil in itself, and although closely connected with continuous population growth, it must be dealt with directly as well.

Because pollution is a surface phenomenon which also strikes the generation which produces it, we may rest assured that it will receive much more official attention than its inseparable companion, resource depletion. But since in both cases there is no such thing as the cost of undoing an irreparable harm or reversing an irrevocable depletion, and since no relevant price can be set on avoiding the inconvenience if future generations cannot bid on the choice, we must insist that the measures taken for either purpose should consist of quantitative regulations, notwithstanding the advice of most economists to increase the allocation efficiency of the market through taxes and subsidies. The economists' plank will only protect the wealthy or the political protégés. Let no one, economist or not, forget that the irresponsible deforestation of numerous mountains took place because "the price was right" and that it was brought to an end only after quantitative restrictions were introduced. But the difficult nature of the choice should also be made clear to the public—that slower depletion means less exosomatic comfort and that greater control of pollution requires proportionately greater consumption of resources. Otherwise, only confusion and controversies at cross-purposes will result.

Nor should any reasonable ecological platform ignore the basic fact that, from all we know about the struggle for life in general, man will probably not let himself down, when pressed for his needs, natural or acquired, by sparing his competitors (including future humans). There is no law in biology stating that a species must defend the existence of others at the cost of its own existence. The most we can reasonably hope is that we may educate ourselves to refrain from "unnecessary" harm and to protect, even at some cost, the future of our species by protecting the species beneficial to us. Complete protection and absolute reduction of pollution are dangerous myths which must be exposed as such (Section V).

Justus von Liebig observed that "civilization is the economy of power" [32, 304]. At the present hour, the economy of power in all its aspects calls for a turning point. Instead of continuing to be opportunistic in the highest degree and concentrating our research toward finding more economically efficient ways of tapping mineral energies—all in finite supply and all heavy pollutants—we should direct all our efforts toward improving the direct uses of solar energy—the only clean and essentially unlimited source. Already known techniques should without delay be diffused among all people so that we all may learn from practice and develop the corresponding trade.

An economy based primarily on the flow of solar energy will also do away, though not completely, with the monopoly of the present over future generations, for even such an economy will still need to tap the terrestrial dowry, especially for materials. The depletion of these critical resources must therefore be rendered as small as feasible. Technological innovations will certainly have a role in this direction. But it is high time for us to stop emphasizing exclusively—as all platforms have apparently done so far—the increase of supply. Demand can also play a role, an even greater and more efficient one in the ultimate analysis.

It would be foolish to propose a complete renunciation of the industrial comfort of the exosomatic evolution. Mankind will not return to the cave or, rather, to the tree. But there are a few points that may be included in a minimal bioeconomic program.

First, the production of all instruments of war, *not only of war itself,* should be prohibited completely. It is utterly absurd (and also hypocritical) to continue growing tobacco if, avowedly, no one intends to smoke. The nations which are so developed as to be the main producers of armaments should be able to reach a consensus over this prohibition without any difficulty if, as they claim, they also possess the wisdom to lead mankind. Discontinuing the production of all instruments of war will not only do away at least with the mass killings by ingenious weapons but will also release some tremendous productive forces for international aid without lowering the standard of living in the corresponding countries.

Second, through the use of these productive forces as well as by additional well-planned and sincerely intended measures, the underdeveloped nations must be aided to arrive as quickly as possible at a good (not luxurious) life. Both ends of the spectrum must effectively participate in the efforts required by this transformation and accept the necessity of a radical change in their polarized outlooks on life.[65]

Third, mankind should gradually lower its population to a level that could be adequately fed only by organic agriculture.[66] Naturally, the nations now experiencing a very high demographic growth will have to strive hard for the most rapid possible results in that direction.

Fourth, until either the direct use of solar energy becomes a general convenience or controlled fusion is achieved, all waste of energy—by overheating, overcooling, overspeeding, overlighting, etc.—should be carefully avoided, and if necessary, strictly regulated.

Fifth, we must cure ourselves of the morbid craving for extravagant gadgetry, splendidly illustrated by such a contradictory item as the golf cart, and for such mammoth splendors as *two-garage* cars. Once we do so, manufacturers will have to stop manufacturing such "commodities."

Sixth, we must also get rid of fashion, of "that disease of the human mind," as Abbot Fernando Galliani characterized it in his celebrated *Della moneta* (1750). It is indeed a disease of the mind to throw away a coat or a piece of furniture while it can still perform its specific service. To get a "new" car every year and to refashion the house every other is a bioeconomic crime. Other writers have already proposed that goods be manufactured in such a way as to be more durable [e.g. 43, 146]. But it is even more important that consumers should reeducate themselves to despise fashion. Manufacturers will then have to focus on durability.

Seventh, and closely related to the preceding point, is the necessity that durable goods be made still more durable by being designed so as to be repairable. (To put it in a plastic analogy, in many cases nowadays, we have to throw away a pair of shoes merely because one lace has broken.)

Eighth, in a compelling harmony with all the above thoughts we should cure ourselves of what I have been calling "the circumdrome of the shaving machine," which is to shave oneself faster so as to have more time to work on a machine that shaves faster so as to have more time to work on a machine that shaves still faster, and so on *ad infinitum*. This change will call for a great deal of recanting on the part of all those professions which have lured man into this empty infinite regress. We must come to realize that an important prerequisite for a good life is a substantial amount of leisure spent in an intelligent manner.

Considered on paper, in the abstract, the foregoing recommendations would on the whole seem reasonable to anyone willing to examine the logic on which they rest. But one thought has persisted in my mind ever since

---

[65] At the Dai Dong Conference (Stockholm, 1972), I suggested the adoption of a measure, which seems to me to be applicable with much less difficulty than dealing with installations of all sorts. My suggestion, instead, was to allow people to move freely from any country to any other country whatsoever. Its reception was less than lukewarm. See [2, 72].

[66] To avoid any misinterpretation, I should add that the present fad for organic foods has nothing to do with this proposal, which is based only on the reasons expounded in Section X.

I became interested in the entropic nature of the economic process. Will mankind listen to any program that implies a constriction of its addiction to exosomatic comfort? Perhaps, the destiny of man is to have a short, but fiery, exciting and extravagant life rather than a long, uneventful and vegetative existence. Let other species—the amoebas, for example—which have no spiritual ambitions inherit an earth still bathed in plenty of sunshine.

REFERENCES

1. Abelson, Philip H., "Limits to Growth." *Science,* 17 March 1972, 1197.
2. Artin, Tom. *Earth Talk: Independent Voices on the Environment.* New York: Grossman Publishers, 1973.
3. Barnett, Harold J. and Chandler Morse. *Scarcity and Growth.* Baltimore: Johns Hopkins Press, 1963.
4. Beckerman, Wilfred, "Economists, Scientists, and Environmental Catastrophe." *Oxford Economic Papers,* November 1972, 327–344.
5. Blin-Stoyle, R. J., "The End of Mechanistic Philosophy and the Rise of Field Physics," in *Turning Points in Physics,* edited by R. J. Blin-Stoyle, et al. Amsterdam: North-Holland, 1959, pp. 5–29.
6. "A Blueprint for Survival." *The Ecologist,* January 1972, 1–43.
7. Bormann, F. H.: "Unlimited Growth: Growing, Growing, Gone?" *BioScience,* December 1972, 706–709.
8. Boulding, Kenneth, "The Economics of the Coming Spaceship Earth," in *Environmental Quality in a Growing Economy,* edited by Henry Jarrett. Baltimore: Johns Hopkins Press, pp. 3–14.
9. ———, "Environment and Economics," in [66], pp. 359–367.
10. Bray, Jeremy. *The Politics of the Environment,* Fabian Tract 412. London: Fabian Society, 1972.
11. Bridgman, P. W., "Statistical Mechanics and the Second Law of Thermodynamics," in *Reflections of a Physicist,* 2d ed. New York: Philosophical Library, 1955, pp. 236–268.
12. Brown, Harrison, "Human Materials Production as a Process in the Biosphere." *Scientific American,* September 1970, 195–208.
13. Brown, Lester R. and Gail Finsterbusch, "Man, Food and Environment," in [66], pp. 53–69.
14. Cannon, James, "Steel: The Recyclable Material." *Environment,* November 1973, 11–20.
15. Cloud, Preston, ed. *Resources and Man.* San Francisco: W. H. Freeman, 1969.
16. ———, "Resources, Population, and Quality of Life," in *Is There an Optimum Level of Population?,* edited by S. F. Singer. New York: McGraw Hill, 1971, pp. 8–31.
17. ———, "Mineral Resources in Fact and Fancy," in [66], pp. 71–88.
18. Commoner, Barry. *The Closing Circle.* New York: Knopf, 1971.
19. Culbertson, John M. *Economic Development: An Ecological Approach.* New York: Knopf, 1971.
20. Daly, Herman E., "Toward a Stationary-State Economy," in *Patient Earth,* edited by J. Hart and R. Socolow. New York: Holt, Rinehart & Winston, pp. 226–244.
21. ———. *The Stationary-State Economy.* Distinguished Lecture Series No. 2, Department of Economics, University of Alabama, 1971.
22. Daniels, Farrington. *Direct Use of the Sun's Energy.* New Haven: Yale University Press, 1964.
23. Einstein, Albert and Leopold Infeld. *The Evolution of Physics.* New York: Simon and Schuster, 1938.
24. "The Fragile Climate of Spaceship Earth." *Intellectual Digest,* March 1972, 78–80.
25. Georgescu-Roegen, Nicholas, "The Theory of Choice and the Constancy of Economic Laws." *Quarterly Journal of Economics,* February 1950, 125–138. Reprinted in [29], pp. 171–183.
26. ———, Chapter 12 in this volume.
27. ———, Chapter 5 in this volume.
28. ———, Chapter 10 in this volume.
29. ———. *Analytical Economics: Issues and Problems.* Cambridge, Mass.: Harvard University Press, 1966.
30. ———, Chapter 4 in this volume.
31. ———. Chapter 3 in this volume.
32. ———. *The Entropy Law and the Economic Process.* Cambridge, Mass.: Harvard University Press, 1971.
33. ———, Chapter 2 in this volume.
34. Gillette, Robert, "The Limits to Growth: Hard Sell for a Computer View of Doomsday." *Science,* 10 March 1972, 1088–1092.
35. ———, "Nuclear Safety: Damaged Fuel Ignites a New Debate in AEC." *Science,* 28 July 1972, 330–331.
36. ———, "Reactor Safety: AEC Concedes Some Points to Its Critics." *Science,* 3 November 1972, 482–484.
37. Glaser, Peter E., "Power from the Sun: Its Future." *Science,* 22 November 1968, 857–861.
38. Goeller, H. E., "The Ultimate Mineral Resource Situation." *Proceedings of the National Academy of Science, USA,* October 1972, 2991–2992.
39. Gofman, John W., "Time for a Moratorium." *Environmental Action,* November 1972, 11–15.
40. Haar, D. ter, "The Quantum Nature of Matter

and Radiation," in *Turning Points in Physics,* [5], pp. 30–44.
41. Hammond, Allen L., "Solar Energy: A Feasible Source of Power?" *Science,* 14 May 1971, 660.
42. Hardin, Garrett, "The Tragedy of the Commons." *Science,* 13 December 1968, 1234–1248.
43. Hibbard, Walter R., Jr., "Mineral Resources: Challenge or Threat?" *Science,* 12 April 1968, 143–145.
44. Holdren, John and Philip Herera. *Energy.* San Francisco: Sierra Club, 1971.
45. Hotelling, Harold, "The Economics of Exhaustible Resources." *Journal of Political Economy,* March–April 1931, 137–175.
46. Hubbert, M. King, "Energy Resources," in [15], pp. 157–242.
47. Istock, Conrad A., "Modern Environmental Deterioration as a Natural Process." *International Journal of Environmental Studies,* 1971, 151–155.
48. Jevons, W. Stanley. *The Theory of Political Economy,* 2d ed. London: Macmillan, 1879.
49. Johnson, Harry G. *Man and His Environment.* London: The British-North American Committee, 1973.
50. Katchalsky, A. and Peter F. Curran. *Nonequilibrium Thermodynamics in Biophysics.* Cambridge, Mass.: Harvard University Press, 1965.
51. Kaysen, Carl, "The Computer that Printed Out W*O*L*F*." *Foreign Affairs,* July 1972, 660–668.
52. Kneese, Allen and Ronald Ridker, "Predicament of Mankind." *Washington Post,* 2 March 1972.
53. Laplace, Pierre Simon de. *A Philosophical Essay on Probability.* New York: John Wiley, 1902.
54. Leontief, Wassily, "Theoretical Assumptions and Nonobservable Facts." *American Economic Review,* March 1971, 1–7.
55. "Limits to Misconception." *The Economist,* 11 March 1972, 20–22.
56. Lovering, Thomas S., "Mineral Resources from the Land," in [15], pp. 109–134.
57. MacDonald, Gordon J. F., "Pollution, Weather and Climate," in [66], pp. 326–336.
58. Maddox, John, "Raw Materials and the Price Mechanism." *Nature,* 14 April 1972, 331–334.
59. ———. *The Doomsday Syndrome.* New York: MacGraw Hill, 1972.
60. Marshall, Alfred. *Principles of Economics,* 8th ed. London: Macmillan, 1920.
61. Marx, Karl. *Capital.* 3 vols. Chicago: Charles H. Kerr, 1906–1933.
62. Meadows, Donella H., et al. *The Limits to Growth.* New York: Universe Books, 1972.
63. Metz, William D., "Fusion: Princeton Tokamak Proves a Principle." *Science,* 22 December 1972, 1274B.
64. Mill, John Stuart. *Principles of Political Economy,* in *Collected Works,* vols. II–III. Edited by J. M. Robson. Toronto: University of Toronto Press, 1965.
65. Mishan, E. J. *Technology and Growth: The Price We Pay.* New York: Praeger, 1970.
66. Murdoch, William W., ed. *Environment: Resources, Pollution and Society.* Stamford, Conn.: Sinauer, 1971.
67. Novick, Sheldon, "Nuclear Breeders." *Environment,* July–August 1974, 6–15.
68. Pigou, A. C. *The Economics of Stationary States.* London: Macmillan, 1935.
69. *Report on Limits to Growth.* Mimeographed. A Study of the Staff of the International Bank for Reconstruction and Development, Washington, D. C., 1972.
70. Revelle, Roger, "Food and Population." *Scientific American,* September 1974, 161–170.
71. Schrödinger, Erwin. *What is Life?* Cambridge, England: The University Press, 1944.
72. Silk, Leonard, "On the Imminence of Disaster." *New York Times,* 14 March 1972.
73. Solo, Robert A., "Arithmomorphism and Entropy." *Economic Development and Cultural Change,* April 1974, 510–517.
74. Solow, Robert M., "Is the End of the World at Hand?" *Challenge,* March–April 1973, 39–50.
75. ———, "The Economics of Resources or the Resources of Economics," Richard T. Ely Lecture. *American Economic Review,* May 1974, 1–14.
76. Spengler, Joseph J., "Was Malthus Right?" *Southern Economic Journal,* July 1966, 17–34.
77. ———, "Homosphere, Seen and Unseen; Retreat from Atomism." *Proceedings of the Nineteenth Southern Water Resources and Pollution Control Conference,* 1970, pp. 7–16.
78. Sprout, Harold and Margaret Sprout. *Multiple Vulnerabilities.* Mimeographed. Research Monograph No. 40, Center of International Studies, Princeton University, 1974.
79. Summers, Claude M., "The Conversion of Energy." *Scientific American,* September 1971, 149–160.
80. Wallich, Henry C., "How to Live with Economic Growth." *Fortune,* October 1972, 115–122.
81. Weinberg, Alvin M., "Breeder Reactors." *Scientific American,* January 1960, 82–94.
82. ———, "Social Institutions and Nuclear Energy." *Science,* 7 July 1972, 27–34.
83. ——— and R. Philip Hammond, "Limits to the Use of Energy." *American Scientist,* July–August 1970, 412–418.

CHAPTER 2

(1971)

Process Analysis and the Neoclassical Theory of Production

To ABUSE a term is to use it without any attempt at explaining its meaning. In this sense, "process" has been abused in all sciences, but in none as much as in social sciences. Most curiously, in economics the greatest abuse has taken place where one would least expect it to happen, namely, in production theory. Neoclassical economists as well as the standard economists of latter days have never paused to describe the process of production in some operational manner so that you and I may know what they meant by the term. In comparison with our classical forefathers—who went to great pains to describe and analyze some processes of production, as Adam Smith, for example, did in his famous illustration of the pin factory—modern economists have found intellectual comfort in pure symbolism, so that they have gradually stopped considering even the traditional classification of the production factors.

Glaring evidence of the modern economist's craving for hollow symbolism is the fact that to this day Philip H. Wicksteed's presentation of the concept of production function constitutes the standard approach to the topic. "*The product being a function of the factors of production we have $P = f(a, b, c, \cdots)$*," Wicksteed [17, p. 4] said, and economists, generally, still define this fundamental concept in the same cavalier fashion.[1] If we have changed anything, we have replaced "product" and "factors of production" by the vapid terms "output" and "inputs," a substitution which only increases the reader's illusion that he is offered a cogent analytical definition. Now everyone can rest satisfied with the simple etymological translation: "input" is what we put in, and "output" is what is put out.

To be sure, symbolism has been the soul of science ever since man began to organize his knowledge about actuality. Yet symbolism, if not supported by an operational interpretation of each symbol (or at least of each primary symbol), silently but unfailingly leads the student away from the most arduous and most important task of any special science, that of bringing the human mind in closer contact with actuality. The neglect of clarifying even partially the concept of production function is all the more puzzling in view of the "practical" nature of the economic science as attested, in particular, by the immense number of works which only compute one gigantic "concrete" production function after another.[2] In any case, the omission is not a matter of purely academic interest only. On the contrary, as I have argued in a series of essays [6, 8, 9, 10, Ch. ix], it falsifies our understanding of the production process, a fact responsible for several important blank spots in neoclassical theory of production. One such blank spot concerns the fundamental difference between productive processes in agriculture (or other strongly seasonal activities) and productive processes in manufacturing. An-

---

[1] This summary presentation of the concept of production function does not only characterize most textbooks—some widely used, e.g., Stonier and Hague [16, p. 219], Leftwich [11, p. 109], Samuelson [15, pp. 515ff]—but it also appears in the writings of some consecrated pundits of our profession, e.g., Frisch [3, p. 41] and Samuelson [14, p. 57].

[2] It is not only because of this neglect that the relevance of these production functions must be questioned. The other reason pertains to the current econometric practices which also involve a chasm between the nature of statistical observations and the stochastic axioms of multivariate analysis. For this last point see Georgescu-Roegen [7].

other covers the difference between the economy of the productive processes in primary activities, mining and agriculture, and the productive processes in secondary activities. Still another concerns one of the most important factors of economic development, namely, the economy of capital utilization. In the present paper, I propose to review and expand these points and to present some additional results concerning the analysis of cost of production and of factor allocation.

Contradictions, Omissions, and Surmises

If our theory of production is to be an adequate logical representation of actuality, it is absolutely necessary that before making even one step further we insist on knowing what corresponds in actuality to every symbol—including F—in the popular formula

(1) $\qquad Q = F(X, Y, \cdots, Z),$

by which standard economics describes any "static" production process. The symbol F must also be included because "function" by itself is an ambiguous term even in mathematics. However, the point is that in the standard theory not even the other symbols are connected in some definite operational manner with data observable in actuality. A few careful authors do go beyond Wicksteed's hollow definition. But they tell two different stories. Some conceive the production function (1) as a relation between the *quantity* of product Q and the *quantities* of factors $X, Y, \cdots, Z$. Others conceive the production function as a relation between the output *per unit of time* q and the inputs *per unit of time* $x, y, \cdots, z$, briefly, as a relation between *rates* of flow:[3]

(2) $\qquad q = f(x, y, \cdots, z).$

A clear symptom of the standard economist's lack of respect for epistemological problems is the fact that no such economist seems to have thought of the possibility that the two viewpoints may not be equivalent. Even authorities use both definitions interchangeably.[4] Having been exposed by a long rote to this equivalence, we have become so firmly convinced of its validity that some of my fellow econometricians, on hearing for the first time its denunciation, immediately protested. As Patinkin admitted subsequently, the novelty so astounded him that he could but be certain that the argument misinterpreted the property of homogeneous functions.[5]

The argument, however, is extremely elementary. By definition we have $Q = tq$, $X = tx$, $Y = ty, \cdots, Z = tz$, for *any* interval of time t. If the two definitions are equivalent, (1) and (2) yield straightforwardly

(3) $\quad tf(x, y, \cdots, z) \equiv F(tx, ty, \cdots, tz).$

And from this identity we deduce, first, that

(4) $\qquad f(x, y, \cdots, z) \equiv F(x, y, \cdots, z),$

and, second, that this common function is homogeneous of the first degree. The presupposed equivalence between (1) and (2) implies, therefore, that absolutely every production process is indifferent to scale, a position which, I hope, is no longer defended by anybody.[6]

An ad hoc attempt to save the day against the foregoing argument led to the suggestion that the quantities are accumulated variable flows:

(5) $$Q(t) = \int_0^t q(t)dt,$$
$$X(t) = \int_0^t x(t)dt, \cdots, Z(t) = \int_0^t z(t)dt.$$

This idea makes matters worse. Instead of (3), we have

(6) $\quad f[x(t), y(t), \cdots, z(t)] = \sum x(t) \dfrac{\partial F}{\partial X},$

for any $x(t), y(t), \cdots, z(t)$, which requires that the production function be a simple linear expression of the inputs [8, p. 43]

(7) $\quad f(x, y, \cdots, z) = ax + by + \cdots + cz.$

The standard economist's lack of interest in arriving at a clear idea of the process of production is reflected also in his silence on the measurability of the "output" and "inputs" involved in (1) or (2). True, from the confusing arguments concerning the measurability of utility there emerged the idea that cardinal measurability is a bogey. A carpenter, we are told,

---

[3] See the sample list references given in [9], notes 4 and 5.
[4] Ragnar Frisch [3, p. 43] uses the two definitions on one and the same page.
[5] See "Discussion" in [6, p. 528].
[6] In fairness to Wicksteed, one should note that the modern incongruity cannot be laid at his door; he explicitly (although without any explanation) assumed that all production functions are homogeneous of the first degree [17, p. 33].

may very well lay down his measuring rod once, four times, eight times, $\cdots$ and count "one," "two," "three," $\cdots$. Perhaps this idea prompted recent theorists of production to be content with any form of measurability for the factors and the product. As far as pure symbolism is concerned, such a position can raise no objection. But the situation changes if the production function is to reflect a part of actuality. In actuality, practically everything we buy and sell is cardinally measurable, for otherwise we would not be able to speak of uniform prices, i.e., of prices per unit. Even land and labor, which exist in a broad spectrum of qualities, are sold on the basis of a cardinal measure—acres or hours.[7]

Actually, of all students, economists should be the last to accept the doctrine that only ordinal measure counts. For if this doctrine were true, our most powerful tools of analysis would go overboard. If commodities were not cardinally measurable by one method or another, then the principle of decreasing marginal substitution, of increasing or decreasing returns, for instance, would lose the basis on which they are defined. One could, in that case, cause the isoquants to have practically any shape one pleases [5, p. 234; 8, pp. 38ff].

The fact that ordinal scales allow us a tremendous freedom of manipulation has led some to surmise that in each case the scales of the factors can be chosen so that the production function be homogeneous of the first degree (an idea that aims at denying the existence of optimum plants). Joan Robinson [13, pp. 109, 332ff] was first to argue that such a feat could be achieved if factors are measured in "efficiency" units. While she retracted her error at the first opportunity, others persisted in it.[8] That the surmise is not true as a general proposition may be shown by a simple example. Let $q = xy + y$ be the production function in case the product and the factors are measured in usual units. Let us assume that with the "efficiency" scales $x = h(u)$, $y = k(v)$, the production function $q(u, v) = h(u)k(v) + k(v)$ is homogeneous of the first degree. The supposition entails $h(0) = 0$, $k(0) = 0$. If we make $u = 0$ in the defining identity

(8) $h(\lambda u)k(\lambda v) + k(\lambda u) = \lambda[h(u)k(v) + k(v)],$

we obtain $k(\lambda v) = \lambda k(v)$. The introduction of this relation in (8) yields $h(\lambda u) = h(u)$, which is absurd.[9]

Finally, there is another thesis that aims at denying the existence of optimum scale of any process. The thesis is that if all pertinent factors are taken into account all natural laws (and hence any process of production as well) are expressed by homogeneous functions of the first degree. As Samuelson [14, p. 84] pointed out, the thesis is operationally idle. It can be proved that it is, in addition, downright incongruous [10, p. 107]. For if in any relation $y = f(x_1, x_2, \cdots, x_n)$, assumed to be the complete expression of a natural law, m factors are left out, the observations of those retained will generally fill a subspace of $n - m$ dimensions. Consequently, the observations of the $n - m$ retained factors could not possibly reveal to us the existence of any law.

Process: An Analytical Tangle

There is, however, some reason why the meaning of "process," a term so frequently used, is hardly clarified in the special scientific literature. Process implies Change, and Change is the most baffling concept in philosophy. Perhaps Change is only illusion—as many philosophers maintain. Perhaps everything is permanently there, just as the paintings of a museum which gradually emerge into our apprehension as we walk from one room to another. Be this as it may, to explain Change is the highest aim of any special science, even though we usually proclaim that science can study only what does not change. This proclamation seems intended to conceal the true difficulty of science in general, which is that our Understanding cannot conceive an action without an agent. The verb "becomes," like all verbs, requires a subject (in grammar as well as in our comprehension). And the rub is that "Jim became tired" is a very unsatisfactory thought. How can we conceive Jim as the *same* person before and after the event of his becoming tired? The fact that there is something to which we refer as "Jim" both before and after that event does not suffice in the least to establish sameness. Witness the

---

[7] On the issue of cardinality, see Georgescu-Roegen [5].
[8] E.g., McLeod and Hahn [12, pp. 132ff]. The stronger and hence the more inept position that factors can be measured in efficient units even when they differ qualitatively is basic in many econometric works on technological change.

[9] In [8, p. 39 n] I stated without proof that also the assertion by Dorfman [2, p. 82] that one can define the inputs and outputs "in such a way that all production functions are linear" is not true. This refutation is a little more involved because of the additional degree of freedom supplied by the scale of the product.

fact that there is an entity to which we refer as "the President of the United States" both before and after the Inauguration. Sameness raises formidable problems of epistemology. Yet, curiously, most of them seem hardly touched by philosophers. For good measure, think of a cosmic event perceived by one observer as a flash, and as a slight wave of warmth by another who travels at a greatly different relative speed. Unsuspected though the fact may be, the issue of sameness is, as we shall see presently, a most crucial one in arriving at an analytical representation of a production process.

One may circumvent all these difficulties by subscribing to the basic tenet of Dialectics, which is that Being is Becoming. However, science cannot be erected on this foundation. For science must distinguish between "object" and "event," that is, between Being and Becoming. It is this distinction that draws the line between Dialectics and Analysis. Analysis offers science the great advantages of precision and easier description of actuality. But these advantages have a price: the endless paradoxes and logical contradictions that emerge from almost every analytical framework to remind us of the dialectical nature of actuality as well as of our thought. Actuality—we must stress the point—is seamless. Hence, violence is done to it as Analysis slices it into discretely distinct pieces in order to facilitate our Understanding.[10]

The upshot is that any analytical science can study only partial processes, i.e., only slices of actuality. Every *analytical* process, therefore, can be but a partial process. To determine such a process, we need an *analytical boundary*. This boundary must include the *duration* of the process and the "geographical" *frontier*, which separates the process from the rest of actuality at all times. The conclusion is clear: no analytical boundary, no analytical process.

Several points now deserve pointed emphasis. Since actuality is seamless we can, in principle, draw a boundary wherever we please. However, a student is always guided by some purpose, proper to the domain of his inquiry. Second, the analytical boundary is a void, for otherwise we would need boundaries of boundaries of boundaries . . . , and would be engulfed into an infinite regress. Third, what an analyt-

ical process does can be described only by listing everything that crosses its frontier and at what time. If we are interested in studying also what happens inside an analytical process, we must divide it into other processes by drawing new boundaries.[11] Lastly, in drafting the list of the elements that cross the frontier, one must bear in mind that, a partial process being an artificial slice of actuality, such a process is inexistent both before the origin of its duration, $t=0$, and after the end of that duration, $t=T$.

Within this framework "input" and "output" acquire an operational as well as analytical definite meaning. However, Analysis must make now another heroic step, which is to assume that the input and output elements, C_k, exist in a finite number of discretely distinct and measurable qualities. On this basis, the analytical coordinates of a process form a point in a functional space,

$$(9) \quad [I_1{}_0^T(t), I_2{}_0^T(t), \cdots, I_n{}_0^T(t); O_1{}_0^T(t), \cdots, O_n{}_0^T(t)],$$

where $I_k(t)$ is the cumulative input and $O_k(t)$ the cumulative output of C_k up to time t (inclusive). The same element may appear both as input and as output. Think of a process consisting only of locomotion (which involves no qualitative change) or one in which electrical motors are used to produce electrical motors.

The analytical picture (9) may suit many a special science, but not economics. In economics, commodities play the same fundamental role as that played by cells in biology, by molecules in chemistry, by elementary particles in physics. Commodity fetishism, notwithstanding its loud denunciation by Karl Marx, is the indispensable foundation of economics. Witness the fact that if we reexamine the entire literature on economic production, Marx's included, in the light of the argument of this section, we see that every economist has drawn economic boundaries only in relation to some commodity. An engineer, for example, may draw a boundary between the furnace with melted glass and the rolling machines of a plate glass factory, but not so an economist in the past or at present. For melted glass is not and has never been a commodity. The conclusion is that if any ele-

---

[10] For a detailed discussion of these issues, see Georgescu-Roegen [10].

[11] A point that illustrates the ways in which Analysis may baffle us: if we push this process to the limit, the whole happening slips through our analytical mesh.

ment *that is not a commodity* is an input or an output of a production process represented by (9), a thorny problem confronts the economist. For, in that case, how can he set up that indispensable tool in any science, namely, the balance between inputs and outputs, which in economics means a value balance? And the fact is that numerous elements of any production process are not commodities proper—tired workers, worn-out tools, and waste are normal outputs, while free goods are normal inputs.

The solution of the impasse lies in the analytical fiction of the static process or, as Marx more properly described it, the process that reproduces itself. Obviously, for a process to be reproducible, some of its elements must remain intact despite the change caused by the process. This is the inevitable consequence of the epistemological position specific to Analysis—that there is both Being and Becoming. It is the common conception that in any production process a number of agents act upon some materials so as to transform them into products. In a static process, the distinction necessarily leads to the division of the analytical coordinates into two categories—the *flow* and the *fund* (not stock!) elements. A flow is an element that either is only "consumed" or only "produced" by the process; it is either only an input or only an output. A fund is both an input and an output; more precisely, it is a factor whose economic efficiency is maintained by the very process in which it participates. The distinction, obviously, breaks down in the case of locomotion. However, locomotion as such presents no interest whatsoever for the economist. Even when locomotion is apparently the case—as it is in an oil pipeline—the economist has all the reasons in the world for treating the oil at the source and the oil at the destination as two different commodities.

The distinction may, however, raise other, more delicate issues which must be handled with what Alfred Marshall so rightly called "delicacy and sensitiveness of touch." For we must not forget the fact that Analysis has to cut some slits in a Whole that has no joints or seams. Ricardian land provides the clearest illustration of the concept of fund, but a machine that is continuously maintained and repaired also fits the definition. Accordingly, a machine coming out of a process is a fund even though it may have no part whatsoever in common with the "same" machine that went into the process at some time in the past. Still more important is another point, which hardly needs any elaboration: the clover seed is a flow in producing clover fodder, but a fund in producing clover seed.

A host of economists have assailed the notion of reproducible (static) process on the ground that it is a far remove from reality. But this objection is as unfounded as that against the fiction of uniform motion. What do most factories do other than reproduce today the process of yesterday? The main difficulty of the notion of reproducible process is of an analytical nature. First, in order to maintain the efficiency of a machine intact, we need other machines. Machines and tools to maintain other machines and tools lead to a regress which may stop only if the reproducible process includes practically every production process in the world. But even this difficulty may be circumvented by introducing the necessary services among inputs. One cannot, however, dispose as easily the fact that workers, other than the self-sufficient farmers, are not maintained within the production process. This analytical blemish, however, does not deprive the notion of reproducible process of its usefulness.[12] Yet the notion carries with it some danger. Its analytical transparency has led and it still leads some economists to ignore the entropic nature of the economic process (which continuously and irrevocably degrades man's environmental dowry) and, as a result, to maintain that the solution to mankind's economic problem lies in the stationary state. The truth, sad though it is, is that this problem is the heart of the evolution of the human species and will end only with the end of that species.

The Elementary Process: A Fundamental Concept

We are now in the position to introduce the basic analytical element of production theory. By an elementary process we shall understand the reproducible process by which one unit (natural or appropriately chosen) of the particular output called product is produced from certain other specified elements. The process by which a table is made by a cabinetmaker from dressed lumber, prepared coating materials, ready-made hardware, is a good illustration in

---

[13] On the issues mentioned in the last paragraphs, the reader may consult Chapter ix in Georgescu-Roegen [10].

point. But the elementary process is the basis of any production, whether of an automobile, of a pound of steel, or of a transatlantic ship.[13] Given the distinction we can make between flow and fund coordinates, we shall measure the former in ordinary cumulative amounts (flows) and the latter in cumulative services. This idea may seem simple enough, but one point, often neglected, should retain our attention. There is a dimensional difference between flows and services as well as between rates of flow and rates of services. Curiously enough, the rate of service is independent of the time dimension. The rate of service of two workers is just two workers; only their service may be one hundred or eight man-hours depending on its duration. A further point is that "the flow of services" is a license which throws an opaque blanket over an important problem of value. For only the woolen fabric flows into the coat, not so the services of the needle. If you find a needle or a part of it in your new coat, it is only because of a regrettable accident.

In a generic fashion, let us denote the services of Ricardian land by $L(t)$, those of capital proper by $K(t)$, and those of labor by $H(t)$. Similarly, we may denote by $Q(t)$, $R(t)$, $I(t)$, $M(t)$, and $W(t)$ the flows of products, of natural resources, of manufactured articles, of maintenance supplies, and of waste. The complete description of any elementary process is

(10)
$$[\overset{T}{\underset{0}{Q}}(t), \overset{T}{\underset{0}{R}}(t), \overset{T}{\underset{0}{I}}(t), \overset{T}{\underset{0}{M}}(t), \overset{T}{\underset{0}{W}}(t); \overset{T}{\underset{0}{L}}(t), \overset{T}{\underset{0}{K}}(t), \overset{T}{\underset{0}{H}}(t)],$$

which can also be represented by a series of graphs, one for each element [6, p. 515; 9, p. 3].

This analytical setup brings to light two points of exceptional economic importance. The first concerns the long-standing denial of the value of Ricardian land and natural resources. But there can be no doubt that Ricardian land must be included in a complete description of a productive process. In agriculture, in particular, Ricardian land plays a role wholly analogous to that of a fisherman's net; only, instead of fish, it catches solar energy and its by-products [6, p. 508]. As to the natural resources, we may recall that Karl Marx, of all the economists, recognized that no one can fish from a lake where there are no fish. The omission of these two factors in standard analytical economics as well as in most applied works may also be imputed to the neglect of standard economists for any epistemological clarification of the production function. The consequence has been that the economic process is now viewed as a simple circular affair, as a mechanical analogue, which, like a pendulum, just beats time, but makes no history. The actual economic process, on the contrary, is making its own history, through the continuous tapping of natural resources and the search for a more efficient use of these scarce economic factors.

The second point is that in absolutely any elementary process numerous factors are inevitably idle during parts of the duration. This *technical* idleness is the worst form of economic waste and a great hamper to economic progress. Unfortunately, natural as well as economic factors prevent its complete elimination. The natural factors pertain to season. Some activities, especially farming, have a temporal rhythm over which, for all practical purposes, man has hardly any control. One can begin at any time of the day, of the week, or of the year, an elementary process by which a table or an automobile is produced. But a corn grower must sow corn in the fields within a specific, very short period of the year if he wants to have a crop. The difference is of paramount importance. In the case of manufactured or mined products, we can arrange the elementary processes *in line* in such a manner that each fund shifts to another process as soon as it has finished its task in the previous one. This is the way any factory operates, like an assembly line even though one is not in direct view.[14] There are inexorable physical factors, therefore, that oblige the farmers all over the world (with only a few, but highly instructive exceptions[15]) to grow practically all their products by elementary process *in parallel* (or almost so). In this arrangement, processes begin and end at the same time, with the result that the idleness coefficient of most factors is reduced very little, if at all. Think how long a plow remains idle on a farm compared with the continuous use of a furnace in steel works, for example.

Patinkin, in his argument against the im-

---

[13] The complications inherent to joint products must be left out at this stage.

[14] The point may be clarified by a diagram as in Georgescu-Roegen [6, p. 517].

[15] The exceptions consist of a few spots around the equator. For their object-lessons, see Georgescu-Roegen [6, pp. 524ff; 10, p. 253].

portance and the relevance of this difference, maintains that in the agriculture of the world viewed as a single process funds are continuously employed. The point is obviously inept: the tractors in the Northern Hemisphere *are* idle during the time those in the Southern Hemisphere *are* active and both *are* often idle over long stretches of time.[16] On the other hand, the economic importance of the difference between the factory and the farming system is splendidly dramatized by the "chicken war" in Europe, a direct consequence of the transformation of the American chicken farms into "chicken factories." It is not excluded that with time we may find the means for growing pigs or even cattle in a factory system. But in spite of many headline claims progress in that direction does not seem at this time too encouraging. In any case, the old Marxist dream of "factories in the open air" is not going to become a general fact for the farmer, not until man conquers the cosmic power necessary for rearranging the position of the globe on the ecliptic.

For the economic historian, too, the analytical concept developed in this section is highly instructive in that it casts a great deal of light on the intimate connection between the factory system and the intensity of demand, on the one hand, and on the assumed correlation between that system and modern technology. Certainly, if only one table is demanded during the period T, production can be carried on only by elementary processes *in series*, with one process beginning when the previous one ends. Nowadays, this is the case for the production of large ships, of space rockets, for example. At one time, this was the general situation in every artisan shop. If elementary processes must be arranged in series, there is hardly any possibility for reducing the technical idleness. Nor is there any incentive or occasion for some division of labor and even of machinery. A cabinet-maker of old times would not have acted economically if he had divided his tasks between him and another worker or if he had used different saws for different cuts. On the other hand, there is absolutely no reason why a factory system should be incompatible with the technology of the ancient times. Indeed, nothing militates against the idea that in the busy shops work proceeded according to that system even in antiquity. In fact, all these remarks justify the view that with the gradual increase in demand the factory system came to be practiced first in some shops of the eighteenth century and that this evolution spurred the division of labor and induced the fever that caused the Industrial Revolution. The relation of cause to effect may after all be the reverse of that of the accepted doctrine in economic history. And from all one can judge, Adam Smith seems to have had in mind this causal order in his argument relating the division of labor to demand.

The Production Functions

We may next assume—a quite strong assumption, if one stops to think about it—that, for a given product, the representations by (10) of all possible elementary processes form some "surface" in the corresponding functional space. The equation of this surface expressed in terms of the product coordinate is the production function of the elementary processes in point,[17]

$$(11) \quad \overset{T}{\underset{0}{Q}}(t) = \mathcal{F}[\overset{T}{\underset{0}{R}}(t), \overset{T}{\underset{0}{I}}(t), \overset{T}{\underset{0}{M}}(t), \overset{T}{\underset{0}{W}}(t);$$

$$\overset{T}{\underset{0}{L}}(t), \overset{T}{\underset{0}{K}}(t), \overset{T}{\underset{0}{H}}(t)].$$

This "function" is what mathematicians call *functional*. It relates functions to functions, not numbers to numbers as (1) or (2) do.

We may now consider complex processes, beginning with those in which elementary processes are arranged in parallel. One category is illustrated by the growing of corn from seed on several farms in the same geographic area. If the elementary process represents what one farm does, then the corresponding production function is immediately derived from (11):

$$(12) \quad n\overset{T}{\underset{0}{Q}}(t) = \mathcal{F}[n\overset{T}{\underset{0}{R}}(t), n\overset{T}{\underset{0}{I}}(t), \ldots, n\overset{T}{\underset{0}{H}}(t)].$$

This is a situation in which the familiar yet misleading tenet "doubling the inputs doubles the output" applies. But there are other cases of processes in parallel to which the tenet does not apply. If (10) is a process by which one bread loaf is baked in a bakery, the baking of a batch of loaves does not require the amplification of all coordinates. The *same* mixing ma-

---

[16] See "Discussion" in [6, pp. 528ff].

[17] Two remarks are in order here. First, it is not necessary that the production surface should cover the entire factor space. Second, normally $Q(T)=1$ and $Q(t)=0$ for $t \neq T$.

chine, oven, and building can take care of the multiple task.

Difficulties of the kind just mentioned render very tedious the derivation of the factory production function from (11). But this can be achieved in a direct fashion by considering the general representation (10) and observing that in a factory system all coordinates, flows and services, are proportional to the time the factory remains open. Hence, we have for *any* T

$$(13) \quad (\overset{T}{\underset{0}{qt}}) = \mathcal{G}[(\overset{T}{\underset{0}{rt}}), (\overset{T}{\underset{0}{it}}), (\overset{T}{\underset{0}{mt}}), (\overset{T}{\underset{0}{wt}}); (\overset{T}{\underset{0}{Pt}}), (\overset{T}{\underset{0}{Ht}})],$$

where P stands for all funds other than labor. As is easily seen, (13) is a *degenerate* functional which can be reduced in two ways to a point function. The first form is

$$(14) \quad q = \Theta(r, i, m, w; P, H),$$

where every symbol represents either a *rate* of flow or a *rate* of service. The second form is

$$(15) \quad Q = \psi(R, I, M, W; \mathcal{P}, \mathcal{H}; t)$$

where every symbol—except t—is either the *amount* of some flow or the *amount* of some service over the period t:

$$(16) \quad \begin{array}{c} R = rt, \quad I = it, \quad M = mt, \quad W = wt; \\ \mathcal{P} = Pt, \quad \mathcal{H} = Ht. \end{array}$$

These two forms uncover the root of the confusion created by the two different interpretations (1) and (2) and exposed in the second section, above. It is seen that if the production function refers to a factory process (or to a process consisting of identical factories in parallel), all variables must be *rates of flows or of services*, as in (14). If one insists on describing the factory process in terms of *quantities of flows and services*, the corresponding production must include also the *duration* over which these quantities are accumulated, as (15) shows. Furthermore, we also find that there is some homogeneity of the first degree here, namely, of ψ. It is this form that lends another justification to the tenet mentioned earlier, "doubling the factors doubles the output," provided we specify that the doubling includes the time of production t as well. We also obtain a relation analogous to (3). Only, its correct form is

$$(17) \quad t\Theta(r, i, m, w; P, H) = \psi(R, I, M, W; \mathcal{P}, \mathcal{H}; t),$$

which yields

$$(18) \quad \Theta = [\psi]_{t=1}.$$

The existence of two distinct forms by which a factory process may be represented reflects a normal duality, even though we may not be fully conscious of it. Form (14) describes only what the process is capable of doing and what it needs for being carried on. It is the same way in which we describe an electric bulb by "100 watts, 110 volts." By contrast, (15) tells us what the factory process has done, is doing, or will do during any period t of full activity. The point that in order to describe what a factory actually does we have to refer either to ψ, which includes the time of production t, or to multiply Θ by this factor vindicates Marx's insistence on the importance of the time of production and of the working day for the theory of value. By comparison, we can see how far removed from the basic economic facts standard theory of production is because of the superficial treatment of such a fundamental concept as the production process. In the standard theory of production there is no room for the time of production or for the working day. The result is that planners everywhere seem to ignore completely one of the most effective and readily available levers of economic development—which is the minimizing of plant idleness. Witness the fact that even in the least developed countries plants lie idle the greatest part of the time either because of an anachronic regimen of a 40-hour week or (and) because they are used only by one shift.

The Components of the Factory Production Function

If we try now to project in greater detail the production function (14) against actuality, we immediately see that any such production function is limitational, that is, the nonwasteful combinations of factors satisfy not one but several relations.[18] It is unquestionable that what a factory can do is what its physical plant is capable of doing. Indeed, a qualified engineer can determine the value of q by merely study-

---

[18] For an introduction to the problem of limitationality and its impact on marginal pricing, see Georgescu-Roegen [4].

ing the blueprints of the factory. This means that the production capacity of a plant q^* is determined only by L and K,

(19) $$q^* = G^*(L, K),$$

and that the same is true of the labor power H^* necessary for utilizing that capacity,

(20) $$H^* = H^*(L, K).$$

A factory, however, may operate below its capacity, in which case what it can do depends also on the labor power $H \leq H^*$ actually employed. A more general relation must therefore replace (19):

(21) $$q = G(L, K, H) \leq q^*.$$

To express what a factory does, we need an additional coordinate, namely, the proportion of the day during which the factory works or, in other words, the working time of the factory. This coordinate is, of course, closely associated with the worker's working day. If we denote it by τ, $0 \leq \tau \leq 1$, and if we assume that H does not vary during τ—which is a reasonable assumption—the amount of *daily* production is given by[19]

(22) $$Q = \tau G(L, K, H).$$

The complete analytical picture of a factory process involves several other categories of factors. The element that comes first to mind represents the inventories, nonspeculative, to be sure. Lacking some detailed inquiries into the matter, we can only surmise that the level of the technical inventories is determined by the size of the other funds considered so far:

(23) $$S = S(L, K, H).$$

The next element is a fund of a peculiar nature. It consists of what we generally call "goods in process." The term is certainly misleading, for half-tanned hides and melted glass, for instance, can hardly constitute goods in the proper sense of the term. The surprising point is that in spite of all our heroic assumptions, it is just not possible to arrive at a picture of a static factory process that involves only commodities proper (besides free goods and waste). But the most peculiar thing about this fund is that in effect it depicts a Becoming, namely, the Becom-

ing represented by all transformation stages through which the product is created from certain material inputs. For this reason we may refer to it as the *process-fund* of a factory. With the same caveat as for the inventory fund, we may assume that the process-fund is determined by

(24) $$\Gamma = \Gamma(L, K, H).$$

There remain the flow coordinates. It stands to reason that for the maintenance flow we have

(25) $$m = m(K, H), \quad m = w_1,$$

where w_1 is the flow rate of waste correlated to the ingredients and parts of which m consists. The last equality merely expresses the well-known principle of conservation of matter-energy: no process whatsoever can yield an excess or a deficit in these terms. Next, there are the input flows of natural resources and manufactured goods that are transformed by the agents into the output flows of product and correlated waste. Since this transformation, too, must obey the principle of conservation of matter-energy, all these last flows must be related by a homogeneous function of the first order, $qt = g(rt, it, w_2 t)$, which yields

(26) $$q = g(r, i, w_2).$$

Finally, we must take into account the fact that what differentiates one factory process from another is the size of the waste flow, i.e., the amounts of the input flows that go to waste. But the size of the waste flow is determined by the structure of the physical plant as well as by the size and the kind of manpower employed. This means that

(27) $$w_2 = w_2(L, K, H),$$

so that (26) must be replaced by

(28) $$q = g[r, i, w_2(L, K, H)].$$

It is thus seen that the factory production function (14) actually consists of seven main components—relations (21), (23)–(25), and (27)–(28)—and one auxiliary relation, (20). The purely economic coordinate, τ, determines then the daily production Q, the daily maintenance flow $M = \tau m$, and the proper daily flows $R = \tau r$, $I = \tau i$.

Nothing need be added to convince ourselves that this analytical picture of a factory process is a far cry from the crude, colorless formula to which standard economics reduces the descrip-

---

[19] If we wish to get closer to the real facts in applied endeavors, we must replace τ by a function of the worker's working day and of the number of shifts.

tion of the same process. Some aspects, however, deserve to be emphasized. First, in contrast with the standard formula, the new picture provides a clear analytical basis to some elements that are ignored by that formula. The elements are the working time τ, the production capacity q^*, and the employment H^* for this capacity. As a result, we can now supply specific definitions for several notions that are used but only loosely defined by standard economics. The ratio q/q^* measures the capacity utilization at *a point in time*, while τ measures the capacity utilization *over time*. The overall capacity utilization is measured by the product $\tau q/q^* = Q/Q^*$, where Q^* is the maximum daily production (obtainable for $\tau = 1$). We can also define in a specific way the frequently used notion of labor-capital ratio. If this coordinate is to reflect the organic composition of capital, it must necessarily refer not to the labor power actually employed (which often may be nill) but to the labor power called for by the existing capital. Hence, the labor-capital ratio must be expressed by H^*/K.

In all probability, the fact that τ, q^*, H^* are not parts of the standard analytical framework (which means the standard production function) is responsible for the absence of the corresponding data from current statistical information. The closest element revealed is the number of man-hours supplied during the year or, what comes to almost the same thing, the total wage bill for each industry. What we generally know, therefore, is only $\sum \tau_i H_i$ over the year. Such data have clearly no connection with H^*. They do not even provide some indirect index of the average employment during the year. Certainly, 1600 man-hours may be the result of 400 men each working four hours as well as 200 men each working eight hours.[20] The results of those applied works in which the capital intensity is measured by the ratio between the total wage bill and the value of fixed capital must, therefore, be viewed with great reservations.

The new analytical representation of a factory process enables us to see clearly through the arguments repeatedly advanced in support

---

[20] This is the elementary reason why t must appear explicitly in (14). The point bears also upon the fact that the Leontief system is generally described in terms of ws and the total "flow of labor services," with the one can never know the actual scale of operation. is last point, see Georgescu-Roegen [10, pp.

of the thesis that there are no optimum units, that doubling the inputs doubles the output. As already mentioned, the thesis is confirmed if the doubling is achieved by doubling the time during which the factory works—a point clearly expressed by (22). The homogeneity of the first degree of the function (26) also bears out the thesis provided that we formulate it correctly: if all inputs *and the waste output* are doubled, then the product output must necessarily double. However, the issue of the existence of optimum units does not hinge upon these two almost tautological aspects. Instead, it hinges upon whether doubling the funds, L, K, H in (21) doubles the potential flow rate q. That numerous defenders of the position that scale does not matter in production have ignored the last point may again be due to the fact that the standard production function makes no distinction between flow and fund elements.

Analysis of Cost and Allocation in a Factory System

Much of the standard picture found in every textbook which aims at showing how cost is derived from the production function and how the optimum allocation of liquid capital is achieved on the same basis must now be changed. But before proceeding, I should stress the fact that, in contrast with the complete silence of the standard theory on this restriction, the analysis presented here applies only to the factory. This case, we should note, is the simplest of all. Other types of processes present difficulties which must be left for another occasion.

The total cost T of a daily output Q always consists of two parts: τT_V, the cost that varies proportionately with τ, and T_F, the cost that is independent of τ. (The distinction is analogous but not identical to the standard distinction between variable and fixed cost.) We thus have the identities

(29) $$T = \tau T_V + T_F,$$
$$T_V = \sum H p^H + \sum r p^r + \sum i p^i + \sum m p^m,$$
$$T_H = \sum L p^L + \sum K p^K + \sum S p^S + \Gamma p^{\Gamma},$$

where the p are prices.

If the management can vary only τ, then the cost of producing a daily output Q, Q varying from $Q = 0$ to $Q = q \times 1$, is $T = T_F + (Q/q)T_V$. In this situation, the optimum τ is determined by

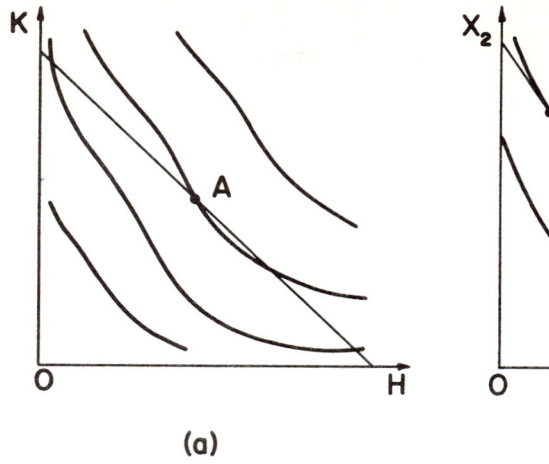
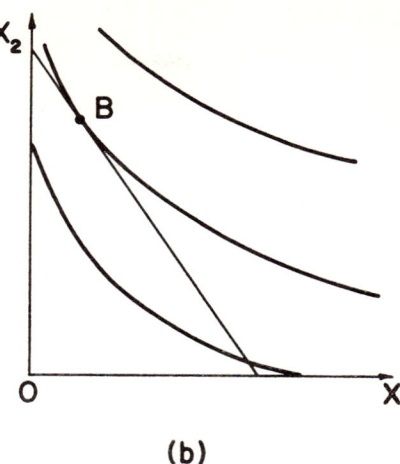

Figure 1

the maximization of profit $R-T$, where $R(Q)$ is the daily total receipts for the offer Q. If $R=pQ$, then the optimum is either $\tau=0$ or $\tau=1$ according to whether $p<$ or $>T_V/q$ and $0\leq\tau\leq 1$ if $p=T_V/q$. If R has the usual parabolic shape, then optimum τ is given by $dR/dQ = T_V/q$.

If, on the other hand, τ is constant and all other elements may vary, then the total cost function must be determined in relation to the usual optimality criterion—the greatest daily output Q for every given total cost T. However, the assumption that τ does not vary is inept, even in the situations in which the worker's working day is limited by law. And since the analysis developed from this assumption is as complicated as the general case, to avoid duplication we may pass directly to the general case in which both τ and all production factors are free to vary—a problem that the standard theory of production could not, per force, even entertain. The main relevant results reached in this direction can be safely brought home by a simplified, albeit unrealistic, structure that leaves out the Ricardian land, the inventories, and the process-fund as well as the maintenance flow. If we also use a uniform notation for all remaining input flows, the production function is reduced to only two components,

(30) $\quad Q = \tau G(H, K), \quad Q = \tau g[x, w_2(H, K)]$.

In the still more simplified structure in which there are only one kind of labor, one kind of capital, and two kinds of input flows, these relations may be represented graphically by two families of isoquants. These isoquants remain the same as τ varies; only the corresponding product values increase in the same proportion with τ. Figure 1a shows the isoquants of G, Figure 1b those of g.[21] We must, however, make a special note of the fact that, in general, the last map changes as the fund structure (H, K) changes. The graphical representation will enable us to pinpoint some of the results to which we now turn.

The optimization problem in this case may be stated as follows: to find the maximum of $Q=\tau g$ for a given

(31) $\quad T = \tau \left(\sum_a p_a^H H_a + \sum_c p_c^x x_c \right) + \sum_b p_b^K K_b,$

and the additional constraint

(32) $\quad G(H, K) = g[x, w_2(H, K)]$.

In (31), p_a^H, p_b^K are the prices of the services of various kinds of labor and capital *per day*, and p_c^x are the prices *per customary unit of measurement* of the various input factors. The necessary condition for the maximum τg is $d(\tau g) = 0$, to which we must add the constraints $dT=0$, $dG=dg$. Because of $d(\tau g)=0$ and of (32), the last relation may be replaced by the simpler one, $d(\tau G)=0$. We thus must have:

[21] The reason why the isoquants in Figure 1a are not necessarily convex toward the origin—as the standard theory assumes them to be—is revealed below.

$$dT = \tau \left(\sum_a p_a{}^H dH_a + \sum_c p_c{}^x dx_c \right)$$
$$+ \sum_b p_b{}^K dK_b + T_\nu d\nu,$$

(33) $$d(\tau G) = \tau \left(\sum_a G_a{}^H dH_a + \sum_b G_b{}^K dK_b \right)$$
$$+ G d\tau$$

$$d(\tau g) = \tau \left(\sum_a g_a{}^H dH_a + \sum_b g_b{}^K dK_b \right.$$
$$\left. + \sum_c g_c{}^x dx_c \right) + g d\tau,$$

where the partial derivatives are denoted according to the pattern $G_a{}^H = \partial G/\partial H_a$.

To avoid the complications associated with a corner solution (which are irrelevant for the immediate argument), we shall assume that these linear equations may be simultaneously satisfied for differential elements that otherwise vary freely. This means that the corresponding linear forms are not independent. Hence:

(34) $$\begin{aligned} p_a{}^H - \lambda G_a{}^H - \mu g_a{}^H &= 0, \\ p_b{}^K - \tau\lambda G_b{}^K - \tau\mu g_b{}^K &= 0, \\ p_c{}^x - \mu g_c{}^x &= 0, \\ T_V - \lambda G - \mu g &= 0. \end{aligned}$$

If τ is given, then the last relation must be eliminated.

In either case, from (34) we obtain

(35) $$p_c{}^x/g_c{}^x = \mu,$$

(36) $$\frac{p_a{}^H - \mu g_a{}^H}{G_a{}^H} = \frac{p_b{}^K - \tau\mu g_b{}^K}{\tau G_b{}^K} = \lambda,$$

for all a, b, and c.

Relations (35) show that for the optimal factor combination the marginal rates of substitution of the flow factors are equal to the price ratios. In Figure 1b, the corresponding budget line is, therefore, tangent to the isoquant at the optimal combination $B(x_1, x_2)$. This result does not differ from the standard theory of factor allocation. For comparison, we may recall that, if (1) is the production function and if $T = \sum p_i X_i$ is the total cost, instead of (33) we have

(37) $$dT = \sum p_i dX_i = 0, \quad dF = \sum F_i dX_i = 0,$$

where $F_i = \partial F/\partial X_i$. As above, we obtain

(38) $$p_i - \mu F_i = 0,$$

and further

(39) $$\frac{p_i}{F_i} = \mu = \frac{dT}{dQ} = \frac{T}{F^*},$$

where

(40) $$F^* = \sum X_i F_i.$$

But (36) offers the surprise that the ratios of the fund prices are no longer equal to the corresponding marginal rates of substitution derived from either relation (30). The result is that in Figure 1a the budget line of the fund factors, $\tau p^H H + p^K K = T_0$, is not necessarily tangent to the isoquant of the optimal combination $A(H, K)$. From (36), it is seen that the tangency obtains if and only if

(41) $$\frac{g_a{}^H}{G_a{}^H} = \frac{g_b{}^K}{G_b{}^K}$$

for all a and b. In view of the definition of g by (30), the last relations become

(42) $$\left(\frac{\partial w_2}{\partial H_a}\right)\bigg/ G_a{}^H = \left(\frac{\partial w_2}{\partial K_b}\right)\bigg/ G_b{}^K.$$

Recalling a main theorem of the Jacobian, we easily recognize that (42) represents the necessary and sufficient conditions for w_2 to be a function of G, in effect, of q. Hence, the second equation (30) may be solved for q and replaced by

(43) $$q = h(x).$$

If this is the case, the isoquant map of the flow factors no longer shifts with the fund structure. The isoquants of the two families are now paired, $G = q^0$ is paired with $g = q^0$. Without some knowledge of the actual structures of factory production functions, it is hard to judge whether this case is realistic in any way. A reasonable guess would be that it is not. If waste depends only on the size of the product flow, it should be the same regardless of the extent to which capital is substituted for labor. Be this as it may, even if (43) is true, (36) shows that the optimal factory depends not only on prices but *also on* τ (if τ is fixed). And in either case, the marginal rates of substitution between funds are not equal to the price ratios unless $\tau = 1$.

The point that marginal productivity makes

no sense in the case of limitational factors is elementary. We must therefore be careful not to equate any of the partial derivatives contained in the foregoing calculations with the marginal productivity of the corresponding factor. An increase in the input flow rate x, for example, will not cause any increase in the product flow unless the fund factors are appropriately adjusted to handle that increase. However, the partial derivatives still enter into the basic formulas. These formulas may be derived either by blind calculus, as we have done above, or directly by determining the increments that match each other in the optimum manner [4]. For example, any increase dx must be accompanied by an adjustment of at least one fund factor, which means that the corresponding increments must satisfy the relations

(44) $\quad g_a^H dH_a + g_c^x dx_c = dq, \quad G_a^H dH_a = dq.$

Since the corresponding additional cost is $dT = \tau(p_c^x dx_c + p_a^H dH_a)$, (44) yields

(45) $\quad \dfrac{dT}{dQ} = \dfrac{p_c^x}{g_c^x}\left(1 - \dfrac{g_a^H}{G_a^H}\right) + \dfrac{p_a^H}{G_a^H}.$

If the adjustment is made by a change in K, the additional cost is $dT = \tau p_c^x dx_c + p_b^K dK_b$, and instead of (45) we have

(46) $\quad \dfrac{dT}{dQ} = \dfrac{p_c^x}{g_c^x}\left(1 - \dfrac{g_b^K}{G_b^K}\right) + \dfrac{p_b^K}{\tau G_b^K}.$

Now, since dT/dQ must be the same for any possible combination of increments, (45) and (46) yield (35) and (36). In fact, (45) and (46) are the correlative relations of $dT/dQ = p_i/F_i$ in (39). Finally, because of (34) we have straightforwardly $dT = \lambda d(\tau G) + \mu d(\tau g)$ or

(47) $\quad \dfrac{dT}{dQ} = \lambda + \mu,$

a relation that mirrors $dT/dQ = \lambda$ of (39).

In order that the solutions supplied by (34) should correspond to a maximum of $Q = \tau g$, we must have $d^2(\tau g) < 0$ subject to the first two constraints of (33). The simple structure of Figure 1 will suffice to pinpoint the relevant results of this new problem. According to a classic theorem of quadratic forms, the condition that $d^2(\tau g) < 0$ subject to $dT = 0$, $d(\tau G) = 0$, is equivalent to a condition pertaining to the signs of the progressive sequence of the principal minors of the matrix[22]

(48) $\begin{bmatrix} 0 & 0 & 0 & 0 & \tau G^H & \tau G^K & G \\ 0 & 0 & \tau p_1 & \tau p_2 & \tau p^H & p^K & T_V \\ 0 & \tau p_1 & \tau g^{11} & \tau g^{12} & \tau g^{1H} & \tau g^{1K} & g^1 \\ 0 & \tau p_2 & \tau g^{21} & \tau g^{22} & \tau g^{2H} & \tau g^{2K} & g^2 \\ \tau G^H & \tau p^H & \tau g^{H1} & \tau g^{H2} & \tau g^{HH} & \tau g^{HK} & g^H \\ \tau G^K & p^K & \tau g^{K1} & \tau g^{K2} & \tau g^{KH} & \tau g^{KK} & g^K \\ G & T_V & g^1 & g^2 & g^H & g^K & 0 \end{bmatrix}$

These signs must be alternatively negative and positive beginning with the first minor marked.[23] We may recall that in the standard case the same condition establishes the convexity toward the origin of the isoquants. In the present case, the same convenient general geometry does not obtain. Only the isoquants of Figure 1b must have this property, a result that follows from the sign of the first minor after p_1 and p_2 are replaced by μg^1, μg^2 on the basis of (34). The existence of a maximum output does not set any simple restriction on the shape of the isoquants $G = \text{const.}$ [4].

The introduction of the working day uncovers some unsuspected problems. To begin with, we may distinguish now two planning cost curves.[24] The first, $T(Q; \tau)$ represents the minimum total cost for any given Q when τ is fixed. This function is determined by (30), (31) and the system consisting only of the first three relations of (34). The second planning cost curve $T(Q)$ represents the minimum cost for every Q when τ is free to vary. In this case, the complete system (34) determines also the optimal value of τ for every Q. Naturally, $T(Q) \leq T(Q; \tau)$, the equality prevailing only for those values of Q for which the optimal value of τ coincides with the fixed τ.[25]

In case τ is not fixed, by (30) the last relation

---

[22] The new notations in the matrix are self-explanatory; the double superscript denotes a partial derivative of the second order. For the theorem, see, for example, Samuelson [14, p. 378].
[23] If τ is fixed, the same conditions prevail for the reduced matrix obtained by eliminating the last row and the last column from (48).
[24] The term "planning" describes better than "long-run" the situation in which absolutely no factor is fixed.
[25] The dichotomy goes further. There are also two types of *plant* cost curves—one in which τ is free to vary, $T(Q; K)$, and one in which τ is fixed, $T(Q; K, \tau)$. They are derived from (34) by making all $dK = 0$.

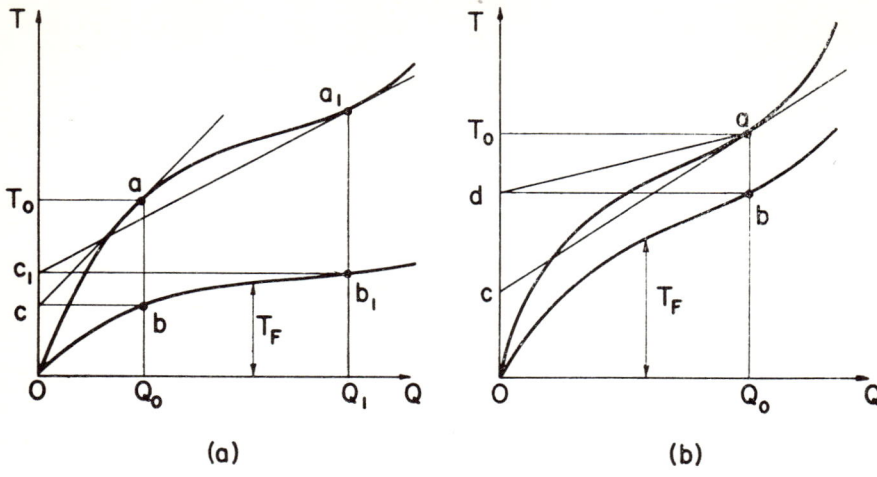

(a)   (b)

Figure 2

of (34) becomes $T_V = (\lambda+\mu)q$ which further yields

(49) $$T = (\lambda + \mu)Q + T_F.$$

Let now $T(Q)$ and $T_F(Q)$ be represented graphically in Figure 2a and let ac be the tangent at a to the total cost curve. By (47), the slope of this tangent is $\lambda+\mu$; hence $T_0 c = (\lambda+\mu)Q_0 = ab$ is the variable, and $Oc = Q_0 b$ is the fixed cost $T_F(Q_0)$. If all the variables of g are now kept constant and only τ is allowed to vary, the new total cost is represented by a segment of the straight line ca. All is in order if, as in the situation shown, the marginal cost is decreasing—i.e., if $d^2T/dQ^2 < 0$. For the resulting cost outside $Q = Q_0$ is greater than the minimum minimorum cost, $T(Q)$. However, if the marginal cost is decreasing—as for the scale Q_1—by varying only τ we obtain a total cost represented by a segment of the tangent $a_1 c_1$ smaller than the minimum minimorum cost. We thus reach the conclusion that marginal cost dT/dQ must *always* be decreasing. The conclusion seems to make sense also from a different angle. From (49) we get

(50) $$\frac{dT}{dQ} = \mu + \lambda + Q\frac{d(\mu+\lambda)}{dQ} + \frac{dT_F}{dQ},$$

and if we take into account (47), this relation yields

(51) $$Q\frac{d^2 T}{dQ^2} + \frac{dT_F}{dQ} = 0.$$

Since it stands to reason that $dT_F/dQ > 0$, we must have $d^2T/dQ^2 < 0$.

However, this conclusion is not necessarily true. A point which is all too often ignored is that calculus conditions for extrema under constraint fool us whenever the extremum is a corner solution. In case of τ, only $\tau = 1$ may be a corner solution. If this is the case, the last equation of (34) is no longer valid. Instead, it is replaced by the corner inequality $T_V < (\lambda+\mu)q$. Also, (49), (51) as well as the geometry of Figure 2a must be discarded. However, (47) still follows from the system (33) modified so as to take into account that $\tau = 1$.[26] Therefore, the intercept of the tangent ac still gives $T_0 c = (\lambda+\mu)Q$. Only, the variable cost ab is now smaller than $T_0 c$ and hence the fixed cost $Q_0 b$ is greater than Oc, as shown in Figure 2b. The cost function obtainable by varying only τ is in this case represented by the segment da.[27] Since this cost is greater than the minimum minimorum, all is again in order. The point is that the planning marginal cost does not have to be always decreasing.

Let us also note that $T_V < (\lambda+\mu)q$ may obtain even for a scale for which $d^2T/dQ^2 < 0$, such as Q_0 in Figure 2a. But in that case, the optimal arrangement necessarily implies $\tau = 1$. For if

---

[26] In the modified system $\tau = 1$ and $d\tau = 0$.
[27] The corresponding optimal value being $\tau = 1$, the daily production cannot exceed Q_0.

$\tau < 1$, by increasing τ we could again obtain a cost for Q lower than $T(Q)$.

Concluding Remarks

Although the results presented in this paper represent only a first attempt at providing the concept of production process with an adequate analytical representation and at examining the bearings of such a representation on the theory of production, they teach a few general yet vital object lessons. The first such object lesson is that the analytical description of a concept which necessarily involves Change cannot possibly be reduced to the shallow, facile formula of the standard production function. The nature of Change would not have tormented past and present philosophical minds if the problem were that simple. Another object lesson is that time—not in the sense of a dimensional measure as in "sixty miles per hour," but in the full sense of duration—is one of the most important coordinates of any process, hence of any production process as well. Even in the case of a factory—a process which appears to start producing the instant the factory opens—the working time is a valuable economic coordinate which raises some intriguing questions of the sort examined in the last part of the preceding section. The analysis of the process of production brought to light also the fact that the economics of production must take into account not only the working time but also the time of idleness, an issue that distinguishes in a palpable way the economy of farming, especially, from that of manufacturing.

We have also seen that a fitting representation of the simplest production process (that of a factory), even if stripped of many essential but cumbersome components, does not lend itself to an analysis as simple as the seducing geometry of the familiar isoquant map and of the standard cost curves picture. Relations such as (45) and (46), although referring to a highly simplified structure, defeat any attempt at a simple, transparent translation in terms of specific factor productivities, to which the standard theory has accustomed us. The results reached for the total cost function, $T(Q)$, are somewhat simpler and more definite. However, the actual shape of this function is still an open question. The basis from which $T(Q)$ is derived being at least as complicated as the systems (35) and (36) are, it is hard to see how the structure of that function could be directly related to the shape properties of the functions that describe the factory process. In fact, because of the computational facilities offered by the Cobb-Douglas type function, no effort worthy of mention seems to have been directed toward determining the shape properties of the basic function (21) on which most standard works often rely.

The results presented in this paper, it must be emphasized, do not provide us with a set of engineering instructions to be applied in each particular circumstance. But they help us become aware of a series of new problems, some of a paramount importance for policy. It is vain to hope that the production theory may reach the stage when its general analysis will yield practical recipes. There is one crucial fact which we all recognize under examination but otherwise ignore completely. The passage from one fund structure to another in the production of the "same" product does not mean (outside some truly exceptional instances) that some capital funds are increased and some labor funds are decreased (or vice versa). Such a substitution involves above all a qualitative change. True, nothing stands in the way of introducing as many different qualities as we may wish in our production functions *on paper*.[28] Such a feat may indeed clarify some issues, but it does not provide us with an operational basis for cost analysis. In actuality cost analysis is not derived from some gigantic production functions involving all possible qualities of factors. Instead, the decisions are made by comparing the costs of a few possible processes. It would seem, therefore, that the cost analysis of the type propagated by Jacob Viner and amended by E. H. Chamberlin [1, App. B], although not as high in our esteem as that based on production functions, may after all be the more appropriate for the task. As far as factor allocation is concerned, there is only one practical counsel: every marginal dollar must bring the same physical return, which is all that we can really extract from (45) and (46). From all we know, only cost is a fact; the production functions are analytical fictions in a broader sense of the term than the formulas of the natural sciences are. The latter are calculating devices, the former are analytical similes which only help our Understanding to deal with a complex actuality pervaded by qualitative change. All the more necessary it is that these similes should be as faithful as Analysis can allow them to be.

---

[28] Cf. Georgescu-Roegen [5].

References

[1] CHAMBERLIN, EDWARD H., *The Theory of Monopolistic Competition*, 6th ed., Cambridge, Harvard University Press, 1948.

[2] DORFMAN, ROBERT, *Application of Linear Programming to the Theory of the Firm*, Berkeley, University of California Press, 1951.

[3] FRISCH, RAGNAR, *Theory of Production*, Chicago, Rand McNally, 1965.

[4] GEORGESCU-ROEGEN, NICHOLAS, "Fixed Coefficients of Production and Marginal Productivity Theory," *Rev. Econ. Stud.* 3:40–49, Oct. 1935. Reprinted with a 1964 Postscript in Nicholas Georgescu-Roegen, *Analytical Economics: Issues and Problems*, Cambridge, Harvard University Press, 1966, pp. 279–299.

[5] ———, Chapter 11 in this volume.

[6] ———, Chapter 5 in this volume.

[7] ———, Chapter 10 in this volume

[8] ———, "Chamberlin's New Economics and the Unit of Production," Ch. 2 in *Monopolistic Competition Theory: Studies in Impact*, ed., R. E. Kuenne, New York, John Wiley & Sons, Inc., 1967, pp. 31–62.

[9] ———, Chapter 4 in this volume.

[10] ———, *The Entropy Law and the Economic Process*, Cambridge, Harvard University Press, 1971.

[11] LEFTWICH, RICHARD H., *The Price System and Resource Allocation.*, rev. ed., New York, Holt, Rinehart and Winston, 1960.

[12] McLEOD, A. N., AND F. H. HAHN, "Proportionality, Divisibility, and Economics of Scale: Two Comments, I and II," *Quart. J. Econ.* 63:128–137, Feb. 1949.

[13] ROBINSON, JOAN, *Economics of Imperfect Competition*, London, The Macmillan Company, 1933.

[14] SAMUELSON, PAUL A., *Foundations of Economic Analysis*, Cambridge, Harvard University Press, 1948.

[15] ———, *Economics*, 8th ed., New York, McGraw-Hill Book Company, 1970.

[16] STONIER, ALFRED W., AND DOUGLAS C. HAGUE, *A Textbook of Economic Theory*, 2nd ed., New York, Longmans, Green and Co., 1958.

[17] WICKSTEED, PHILIP, *An Essay on the Co-ordination of the Laws of Distribution*, (London, 1894), Reprints of Scarce Tracts, No. 12, London School of Economics and Political Science, London, 1932.

CHAPTER 3

(1970)

The Entropy Law and the Economic Problem

I

A curious event in the history of economic thought is that, years after the mechanistic dogma had lost its supremacy in physics and its grip on the philosophical world, the founders of the Neoclassical school set out to erect an economic science after the pattern of mechanics—in the words of Jevons, as *"the mechanics of utility and self-interest."*[1] And while economics has made great strides since, nothing has happened to deviate economic thought from the mechanistic epistemology of the forefathers of standard economics. A glaring proof is the standard textbook representation of the economic process by a circular diagram, a pendulum movement between production and consumption within a completely closed system.[2] The situation is not different with the analytical pieces that adorn the standard economic literature; they, too, reduce the economic process to a self-sustained mechanical analogue. The patent fact that between the economic process and the material environment there exists a continuous mutual influence which is history-making carries no weight with the standard economist. And the same is true of Marxist economists, who swear by Marx's dogma that everything nature offers man is a spontaneous gift.[3] In Marx's famous diagram of reproduction, too, the economic process is represented as a completely circular and self-sustaining affair.[4]

Earlier writers, however, pointed in another direction, as did Sir William Petty in arguing that labor is the father and nature is the mother of wealth.[5] The entire economic history of mankind proves beyond question that nature, too, plays an important role in the economic process as well as in the formation of economic value. It is high time, I believe, that we should accept this fact and consider its consequences for the economic problem of mankind. For, as I shall endeavor to show in this paper, some of these consequences have an exceptional importance for the understanding of the nature and the evolution of man's economy.

II

Some economists have alluded to the fact that man can neither create nor destroy matter or energy[6]—a truth which follows from the Principle of Conservation of Matter-Energy, alias the First Law of Thermodynamics. Yet no one seems to have been struck by the question—so puzzling in the light of this law—"what then does the economic process do?" All that we find in the cardinal literature is an occasional remark that man can produce only utilities, a remark which actually accentuates the puzzle. How is it possible for man to produce something material, given the fact that he cannot produce either matter or energy?

To answer this question, let us consider the economic process as a whole and view it only from the purely physical viewpoint. What we must note first of all is that this process is a partial process which, like all partial processes, is circumscribed by a boundary across which matter and energy are exchanged with the rest of the material universe.[7] The answer to the question of what this *material* process does is simple: it neither produces nor consumes matter-energy; it only absorbs matter-energy and throws it out continuously. This is what pure physics teaches us. However, economics—let us say it high and loud—is not pure physics, not even physics in some other form. We may trust that even the fiercest partisan of the position that natural resources have nothing to do with value will admit in the end that there is a difference between what goes into the economic process and what comes out of it. To be sure, this difference can be only qualitative.

An unorthodox economist—such as myself—would say that what goes into the economic process represents *valuable natural resources* and what is thrown out of it is *valueless waste*. But this qualitative difference is confirmed, albeit in different terms, by a particular (and peculiar) branch of physics known as thermodynamics.

53

From the viewpoint of thermodynamics, matter-energy enters the economic process in a state of *low entropy* and comes out of it in a state of *high entropy*.[8]

To explain in detail what entropy means is not a simple task. The notion is so involved that, to trust an authority on thermodynamics, it is "not easily understood even by physicists."[9] To make matters worse not only for the layman, but for everyone else as well, the term now circulates with several meanings, not all associated with a physical coordinate.[10] A recent edition of *Webster's Collegiate Dictionary* (1965) has three entries under "entropy." Moreover, the definition pertaining to the meaning relevant for the economic process is likely to confuse rather than enlighten the reader: "a measure of unavailable energy in a closed thermodynamic system so related to the state of the system that a change in the measure varies with change in the ratio of the increment of heat taken in the absolute temperature at which it is absorbed." But (as if intended to prove that not all progress is for the better) some older editions supply a more intelligible definition. "A measure of the unavailable energy in a thermodynamic system"—as we read in the 1948 edition—cannot satisfy the specialist but would do for general purposes. To explain (again in broad lines) what unavailable energy means is now a relatively simple task.

Energy exists in two qualitative states—*available* or *free* energy, over which man has almost complete command, and *unavailable* or *bound* energy, which man cannot possibly use. The chemical energy contained in a piece of coal is free energy because man can transform it into heat or, if he wants, into mechanical work. But the fantastic amount of heat-energy contained in the waters of the seas, for example, is bound energy. Ships sail on top of this energy, but to do so they need the free energy of some fuel or of the wind.

When a piece of coal is burned, its chemical energy is neither decreased nor increased. But the initial free energy has become so dissipated in the form of heat, smoke and ashes that man can no longer use it. It has been degraded into bound energy. Free energy means energy that displays a differential level, as exemplified most simply by the difference of temperatures between the inside and the outside of a boiler. Bound energy is, on the contrary, chaotically dissipated energy. This difference may be expressed in yet another way. Free energy implies some ordered structure, comparable with that of a store in which all meat is on one counter, vegetables on another, and so on. Bound energy is energy dissipated in disorder, like the same store after being struck by a tornado. This is why entropy is also defined as a measure of disorder. It fits the fact that a copper sheet represents a lower entropy than the copper ore from which it was produced.

The distinction between free and bound energy is certainly an anthropomorphic one. But this fact need not trouble a student of man, nay, even a student of matter in its simple form. Every element by which man seeks to get in mental contact with actuality can be but anthropomorphic. Only, the case of thermodynamics happens to be more striking. The point is that it was the economic distinction between things having an economic value and waste which prompted the thermodynamic distinction, not conversely. Indeed, the discipline of thermodynamics grew out of a memoir in which the French engineer Sadi Carnot (1824) studied for the first time the *economy* of heat engines. Thermodynamics thus began as a physics of economic value and has remained so in spite of the numerous subsequent contributions of a more abstract nature.

III

Thanks to Carnot's memoir, the elementary fact that heat moves by itself only from the hotter to the colder body acquired a place among the truths recognized by physics. Still more important was the consequent recognition of the additional truth that once the heat of a closed system has diffused itself so that the temperature has become uniform throughout the system, the movement of the heat cannot be reversed without external intervention. The ice cubes in a glass of water, once melted, will not form again by themselves. In general, the free heat-energy of a closed system continuously and irrevocably degrades itself into bound energy. The extension of this property from heat-energy to all other kinds of energy led to the Second Law of Thermodynamics, alias the Entropy Law. This law states that the entropy (i.e., the amount of bound energy) of a closed system continuously increases or that the order of such a system steadily turns into disorder.

The reference to a closed system is crucial. Let us visualize a closed system, a room with an electric stove and a pail of water that has just been boiled. What the Entropy Law tells us is, first, that the heat of the boiled water will continuously dissipate into the system. Ultimately, the system will attain thermodynamic equilibrium—a state in which the temperature is uniform throughout (and all energy is bound). This applies to every kind of energy in a closed system. The free chemical energy of a piece of coal, for instance, will ultimately become degraded into bound energy even if the coal is left in the ground. Free energy will do so in any case.

The law also tells us that once thermodynamic equilibrium is reached, the water will not start boiling by itself.[11] But, as everyone knows, we can make it boil again by turning on the stove. This does not mean, however, that we have defeated the Entropy Law. If the entropy of the room has been decreased as the result of the temperature differential created by boiling the water, it is only because some low entropy (free energy) was brought into the system from the outside. And if we include the electric plant in the system, the entropy of this new system must have decreased, as the Entropy Law states. This means that the decrease in the entropy of the room has been obtained only at the cost of a greater increase in entropy elsewhere.

Some writers, impressd by the fact that living organisms remain almost unchanged over short periods of time, have set forth the idea that life eludes the Entropy Law. Now, life may have properties that cannot be accounted for by the natural laws, but the mere thought that it may violate some law of matter (which is an entirely different thing) is sheer nonsense. The truth is that every living organism strives only to maintain its own entropy constant. To the extent to which it achieves this, it does so by sucking low entropy from the environment to compensate for the increase in entropy to which, like every material structure, the organism is continuously subject. But the entropy of the entire system—consisting of the organism and its environment—must increase. Actually, the entropy of a system must increase faster if life is present than if it is absent. The fact that any living organism fights the entropic degradation of its own material structure may be a characteristic property of life, not accountable by material laws, but it does not constitute a violation of these laws.

Practically all organisms live on low entropy in the form found immediately in the environment. Man is the most striking exception: he cooks most of his food and also transforms natural resources into mechanical work or into various objects of utility. Here again, we should not let ourselves be misled. The entropy of copper metal is lower than the entropy of the ore from which it was refined, but this does not mean that man's *economic* activity eludes the Entropy Law. The refining of the ore causes a more than compensating increase in the entropy of the surroundings. Economists are fond of saying that we cannot get something for nothing. The Entropy Law teaches us that the rule of biological life and, in man's case, of its economic continuation is far harsher. In entropy terms, the cost of any biological or economic enterprise is always greater than the product. In entropy terms, any such activity necessarily results in a deficit.

IV

The statement made earlier—that, from a purely physical viewpoint, the economic process only transforms valuable natural resources (low entropy) into waste (high entropy)—is thus completely vindicated. But the puzzle of why such a process should go on is still with us. And it will remain a puzzle as long as we do not see that the true economic output of the economic process is not a material flow of waste, but an immaterial flux: the enjoyment of life. If we do not recognize the existence of this flux, we are not in the economic world. Nor do we have a complete picture of the economic process if we ignore the fact that this flux—which, as an entropic feeling, must characterize life at all levels—exists only as long as it can continuously feed itself on environmental low entropy. And if we go one step further, we discover that every object of economic value—be it a fruit just picked from a tree, or a piece of clothing, or furniture, etc.—has a highly ordered structure, hence, a low entropy.[12]

There are several lessons to be derived from this analysis. The first lesson is that man's economic struggle centers on environmental low entropy. Second, environmental low entropy is scarce in a different sense than Ricardian land. Both Ricardian land and the coal deposits are available in limited amounts. The difference is that a piece of coal can be used only once. And, in fact, the Entropy Law is the reason why an engine (even a biological organism) ultimately wears out and must be replaced by a *new* one, which means an additional tapping of environmental low entropy.

Man's continuous tapping of natural resources is not an activity that makes no history. On the contrary, it is the most important long-run element of mankind's fate. It is because of the irrevocability of the entropic degradation of matter-energy that, for instance, the peoples from the Asian steppes, whose economy was based on sheep-raising, began their Great Migration over the entire European continent at the beginning of the first millenium. The same element—the pressure on natural resources—had, no doubt, a role in other migrations, including that from Europe to the New World. The fantastic efforts made for reaching the moon may also reflect some vaguely felt hope of obtaining access to additional sources of low entropy. It is also because of the particular scarcity of environmental low entropy that ever since the dawn of history man has continuously sought to invent means for sifting low entropy better. In most (though not in all) of man's inventions one can definitely see a progressively better economy of low entropy.

Nothing could, therefore, be further from the truth than the notion that the economic process is an isolated, circular affair—as Marxist and standard analysis represent it. The economic process is solidly anchored to a material base which is subject to definite constraints. It is because of these constraints that the economic process has a unidirectional irrevocable evolution. In the economic world only money circulates back and forth between one economic sector and another (although, in truth, even the bullion slowly wears out and its stock must be continuously replenished from the mineral deposits). In retrospect it appears that the economists of both persuasions have succumbed to the worst economic fetishism —money fetishism.

V

Economic thought has always been influenced by the economic issues of the day. It also has reflected—with some lag—the trend of ideas in the natural sciences. A salient illustration of this correlation is the very fact that, when economists began ignoring the natural environment in representing the economic process, the event reflected a turning point in the temper of the entire scholarly world. The unprecedented achievements of the Industrial Revolution so amazed everyone with what man might do with the aid of machines that the general attention became confined to the factory. The landslide of spectacular scientific discoveries triggered by the new technical facilities strengthened this general awe for the power of technology. It also induced the literati to overestimate and, ultimately, to over- sell to their audiences the powers of science. Naturally, from such a pedestal one could not even conceive that there is any real obstacle inherent in the human condition.

The sober truth is different. Even the lifespan of the human species represents just a blink when compared with that of a galaxy. So, even with progress in space travel, mankind will remain confined to a speck of space. Man's biological nature sets other limitations as to what he can do. Too high or too low a temperature is incompatible with his existence. And so are many radiations. It is not only that he cannot reach up to the stars, but he cannot even reach down to an individual elementary particle, nay, to an individual atom.

Precisely because man has felt, however unsophisticatedly, that his life depends on scarce, irretrievable low entropy, man has all along nourished the hope that he may eventually discover a self-perpetuating force. The discovery of electricity enticed many to believe that the hope was actually fulfilled. Following the strange marriage of thermodynamics with mechanics, some began seriously thinking about schemes to unbind bound energy.[18] The discovery of atomic energy spread another wave of sanguine hopes that, this time, we have truly gotten hold of a self-perpetuating power. The shortage of electricity which plagues New York and is gradually extending to other cities should suffice to sober us up. Both the nuclear theorists and the operators of atomic plants vouch that it all boils down to a problem of cost, which in the perspective of this paper means a problem of a balance sheet in entropy terms.

With natural scientists preaching that science can do away with all limitations felt by man and with the economists following suit in not relating the analysis of the economic process to the limitations of man's material environment, no wonder that no one realized that we cannot produce "better and bigger" refrigerators, automobiles, or jet planes, without producing also "better and bigger" waste. So, when everyone (in the countries with "better and bigger" industrial production) was, literally, hit in the face by pollution, scientists as well as economists were taken by surprise. But even now no one seems to see that the cause of all this is that we have failed to acknowledge the entropic nature of the economic process. A convincing proof is that the various authorities on pollution now try to sell us, on the one hand, the idea of machines and chemical reactions that produce no waste, and, on the other, salvation through a perpetual recycling of waste. There is no denial that, in principle at least, we can recycle even the gold dispersed in the sand of the seas just as we can recycle the boiling water in my earlier example. But in both cases we must use an additional amount of low entropy much greater than the decrease in the entropy of what is recycled. There is no free recycling just as there is no wasteless industry.

VI

The globe to which the human species is bound floats, as it were, within the cosmic store of free energy, which may be even infinite. But for the reasons mentioned in the preceding section, man cannot have access to all this fantastic amount, nor to all possible forms of free energy. Man cannot, for example, tap directly the immense thermonuclear energy of the sun. The most important impediment (valid also for the industrial use of the "hydrogen bomb") is that no material container can resist the temperature of massive thermonuclear reactions. Such reactions can occur only in free space.

The free energy to which man can have access comes from two distinct sources. The first source is a *stock,* the stock of free energy of the mineral deposits in the bowels of the earth. The second source is a *flow,* the flow of solar radiation intercepted by the earth. Several differences between these two sources should be well marked. Man has almost complete command over the terrestrial dowry; conceivably, we may use it all within a single year. But, for all practical purposes, man has no control over the flow of solar radiation.

Neither can he use the flow of the future *now*. Another asymmetry between the two sources pertains to their specific roles. Only the terrestrial source provides us with the low entropy materials from which we manufacture our most important implements. On the other hand, solar radiation is the primary source of all life on earth, which begins with chlorophyll photosynthesis. Finally, the terrestrial stock is a paltry source in comparison with that of the sun. In all probability, the active life of the sun—during which the earth will receive a flow of solar energy of significant intensity—will last another five billion years.[14] But hard to believe though it may be, the entire terrestrial stock could yield only a few days of sunlight.[15]

All this casts a new light on the population problem, which is so topical today. Some students are alarmed at the possibility that the world population will reach seven billion by A.D. 2000 —the level predicted by United Nations demographers. On the other side of the fence, there are those who, like Colin Clark, claim that with a proper administration of resources the earth may feed as many as forty-five billion people.[16] Yet no population expert seems to have raised the far more vital question for mankind's future: How long can a given world population—be it of one billion or of forty-five billion—be maintained? Only if we raise this question can we see how complicated the population problem is. Even the analytical concept of optimum population, on which many population studies have been erected, emerges as an inept fiction.

What has happened to man's entropic struggle over the last two hundred years is a telling story in this respect. On the one hand, thanks to the spectacular progress of science man has achieved an almost miraculous level of economic development. On the other hand, this development has forced man to push his tapping of terrestrial resources to a staggering degree (witness off-shore oil-drilling). It has also sustained a population growth which has accentuated the struggle for food and, in some areas, brought this pressure to critical levels. The solution, advocated unanimously, is an increased mechanization of agriculture. But let us see what this solution means in terms of entropy.

In the first place, by eliminating the traditional partner of the farmer—the draft animal— the mechanization of agriculture allows the entire land area to be allocated to the production of food (and to fodder only to the extent of the need for meat). But the ultimate and the most important result is a shift of the low entropy input from the solar to the terrestrial source. The ox or the water buffalo—which derive their mechanical power from the solar radiation caught by chlorophyll photosynthesis—is replaced by the tractor—which is produced and operated with the aid of terrestrial low entropy. And the same goes for the shift from manure to artificial fertilizers. The upshot is that the mechanization of agriculture is a solution which, though inevitable in the present impasse, is anti-economical in the long run. Man's biological existence is made to depend in the future more and more upon the scarcer of the two sources of low entropy. There is also the risk that mechanized agriculture may trap the human species in a cul-de-sac because of the possibility that some of the biological species involved in the other method of farming will be forced into extinction.

Actually, the problem of the economic use of the terrestrial stock of low entropy is not limited to the mechanization of agriculture only: it is the main problem for the fate of the human species. To see this, let S denote the present stock of terrestrial low entropy and let r be some average annual amount of depletion. If we abstract (as we can safely do here) from the slow degradation of S, the *theoretical* maximum number of years until the complete exhaustion of that stock is S/r. This is also the number of years until the *industrial* phase in the evolution of mankind will forcibly come to its end. Given the fantastic disproortion between S and the flow of solar energy that reaches the globe annually, it is beyond question that, even with a very parsimonious use of S, the industrial phase of man's evolution will end long before the sun will cease to shine. What will happen then (if the extinction of the human species is not brought about earlier by some totally resistant bug or some insidious chemical) is hard to say. Man could continue to live by reverting to the stage of a berry-picking species—as he once was. But, in the light of what we know about evolution, such an evolutionary reversal does not seem probable. Be that as it may, the fact remains that the higher the degree of economic development, the greater must be the annual depletion r and, hence, the shorter becomes the expected life of the human species.

VII

The upshot is clear. Every time we produce a Cadillac, we irrevocably destroy an amount of low entropy that could otherwise be used for producing a plow or a spade. In other words, every time we produce a Cadillac, we do it at the cost of decreasing the number of human lives in the future. Economic development through industrial abundance may be a blessing for us now and for those who will be able to enjoy it in the near future, but it is definitely against the interest of the human species as a whole, if its interest is to have a lifespan as long as is compatible with its dowry of low entropy. In this paradox of economic development we can see the price man has to pay for the unique privilege of being able to go beyond the biological limits in his struggle for life.

Biologists are fond of repeating that natural selection is a series of fantastic blunders since future conditions are not taken into account. The remark, which implies that man is wiser than nature and should take over her job, proves that man's vanity and the scholar's self-confidence will never know their limits. For the race of economic development that is the hallmark of modern civilization leaves no doubt about man's lack of foresight. It is only because of his biological nature (his inherited instincts) that man cares for the fate of only some of his immediate descendants, generally not beyond his great-grandchildren. And there is neither cynicism nor pessimism in believing that, even if made aware of the entropic problem of the human species, mankind would not be willing to give up its present luxuries in order to ease the life of those humans who will live ten thousand or even one thousand years from now. Once man expanded his biological powers by means of industrial artifacts, he became *ipso facto* not only dependent on a very scarce source of life support but also addicted to industrial luxuries. It is as if the human species were determined to have a short but exciting life. Let the less ambitious species have a long but uneventful existence.

Issues such as those discussed in this lecture pertain to long-run forces. Because these forces act extremely slowly we are apt to ignore their existence or, if we recognize them, to belittle their importance. Man's nature is such that he is always interested in what will happen until tomorrow, not in thousands of years from now. Yet it is the slow-acting forces that are the more fateful in general. Most people die not because of some quickly acting force—such as a pneumonia or an automobile accident—but because of the slow-acting forces that cause aging. As a Jain philosopher remarked, man begins to die at birth. The point is that it would not be hazardous to venture some thoughts about the distant future of man's economy any more than it would be to predict in broad lines the life of a newly born child. One such thought is that the increased pressure on the stock of mineral resources created by the modern fever of industrial development, together with the mounting problem of making pollution less noxious (which places additional demands on the same stock), will necessarily concentrate man's attention on ways to make greater use of solar radiation, the more abundant source of free energy.

Some scientists now proudly claim that the food problem is on the verge of being completely solved by the imminent conversion on an industrial scale of mineral oil into food protein—an inept thought in view of what we know about the entropic problem. The logic of this problem justifies instead the prediction that, under the pressure of necessity, man will ultimately turn to the contrary conversion, of vegetable products into gasoline (if he will still have any use for it).[17] We may also be quasi-certain that, under the same pressure, man will discover means by which to transform solar radiation into motor power directly. Certainly, such a discovery will represent the greatest possible breakthrough for man's entropic problem, for it will bring under his command also the more abundant source of life support. Recycling and pollution purification would still consume low entropy, but not from the rapidly exhaustible stock of our globe.

Footnotes

1. W. Stanley Jevons, *The Theory of Political Economy* (4th edn., London, 1924), p. 21.
2. E. g., R. T. Bye, *Principles of Economics* (5th edn., New York, 1956), p. 253; G. L. Bach, *Economics* (2nd edn., Englewood Cliffs, N. J., 1957), p. 60; J. H. Dodd, C. W. Hasek, T. J. Hailstones, *Economics* (Cincinnati, 1957), p. 125; R. M. Havens, J. S. Henderson, D. L. Cramer, *Economics* (New York, 1966), p. 49; Paul A. Samuelson, *Economics* (8th edn., New York, 1970), p. 42.
3. Karl Marx, *Capital* (3 vols., Chicago, 1906-1933), I, 94, 199, 230, and *passim*.
4. *Ibid*., II, ch. XX.
5. *The Economic Writings of Sir William Petty*, ed. C. H. Hull (2 vols., Cambridge, Eng., 1899), II, 377. Curiously, Marx went along wth Petty's idea; but he claimed that nature only "helps to create use-value without contributing to the formation of exchange value." Marx, *Capital*, I, 227. See also *ibid*., p. 94.
6. E. g., Alfred Marshall, *Principles of Economics* (8th edn., New York, 1924), p. 63.
7. On the problem of the analytical representation of a process, see my *The Entropy Law and the Economic Process* (Cambridge, Mass., 1971), pp. 211-231.
8. This distinction together with the fact that no one would exchange some natural resources for waste disposes of Marx's assertion that "no chemist has ever discovered exchange value in a pearl or a diamond." *Capital*, I, 95.
9. D. ter Haar, "The Quantum Nature of Matter and Radiation," in *Turning Points in Physics*, ed. R. J. Blin-Stoyle *et al.* (Amsterdam, 1959), p. 37.
10. One meaning that has recently made the term extremely popular is "the amount of information." For an argument that this term is misleading and for a critique of the alleged connection between information and physical entropy, see *The Entropy Law and the Economic Process*, Appendix B.
11. This position calls for some technical elaboration. The opposition between the Entropy Law—with its unidirectional qualitative change—and mechanics—where everything can move either forward or backward while remaining self-identical—is accepted without reservation by every physicist and philosopher of science. However, the mechanistic dogma retained (as it still does) its grip on scientific activity even after physics recanted it. The result was that mechanics was soon brought into thermodynamics in the company of randomness. This is the strangest possible company, for randomness is the very antithesis of the deterministic nature of the laws of mechanics. To be sure, the new edifice (known as statistical mechanics) could not include mechanics under its roof and, at the same time, exclude reversibility. So, statistical mechanics must teach that a pail of water may start boiling by itself, a thought which is slipped under the rug by the argument that the miracle has not been observed because of its extremely small probability. This position has fostered the belief in the possibility of converting bound into free energy or, as P. W. Bridgman wittily put it, of bootlegging entropy. For a critique of the logical fallacies of statistical mechanics and of the various attempts to patch them, see *The Entropy Law and the Economic Process*, ch. VI.
12. This does not mean that everything of low entropy necessarily has economic value. Poisonous mushrooms, too, have a low entropy. The relation between low entropy and economic value is similar to that between economic value and price. An object can have a price only if it has economic value, and it can have economic value only if its entropy is low. But the converse is not true.
13. See note 11, above.
14. George Gamow, *Matter, Earth, and Sky* (Englewood Cliffs, N. J., 1958), pp. 493 f.
15. Four days, according to Eugene Ayres, "Power from the Sun," *Scientific American*, August 1950, p. 16. The situation is not changed even if we admit that the calculations might be in error by as much as one thousand times.
16. Colin Clark, "Agricultural Productivity in Relation to Population," in *Man and His Future*, ed. G. Wolstenholme (Boston, 1963), p. 35.
17. That the idea is not far-fetched is proved by the fact that in Sweden, during World War II, automobiles were driven by the poor gas obtained by heating wood with wood.

CHAPTER 4

(1969)

THE ECONOMICS OF PRODUCTION

RICHARD T. ELY LECTURE

For the last twenty years or so I have singled myself out among my fellow econometricians for arguing with all the means at my disposal that not every element of the economic process can be related to a number and, consequently, that this process cannot be represented in its entirety by an arithmomorphic model. At the same time, I have insisted that in our haste to mathematize economics we have often been carried away by mathematical formalism to the point of disregarding a basic requirement of science; namely, to have as clear an idea as possible about what corresponds in actuality to every piece of our symbolism. Curiously, in the home of quantity, in the natural sciences, this position does not constitute a singularity. On the contrary, essentially the same words of caution have come from many a high authority in physics—such as Max Planck or Percy William Bridgman, for example.[1] But even some engineers have raised their voices against blind symbolism. The recent remarks by a well-known British engineer are worth quoting at length:

> Contrary to common belief it is sometimes easier to talk in mathematics than to talk in English; this is the reason why many scientific papers contain more mathematics than is either necessary or desirable. Contrary to common belief it is also often less precise to do so. For mathematical symbols have a tendency to conceal the physical meaning that they are intended to represent; they sometimes serve as a substitute for the arduous task of deciding what is and what is not relevant; .... It is true that mathematics cannot lie. But it can mislead.
>
> However, the dangers of over-indulgence in formula spinning are avoided if mathematics is treated, wherever possible, as a language into which *thoughts may only be translated after they have first been [clearly] expressed in the language of words*. The use of mathematics in this way is indeed disciplinary, helpful, and sometimes indispensable.[2]

The topic of this lecture—the economics of production—presents, I believe, sufficient interest by itself. But in choosing it, I have been guided also by the fact that it may serve as a substantial illustration of the harm caused by the blind symbolism that generally characterizes a hasty mathematization.

I

What has come to be known as "the production function" is quite an old item in the economist's analytical paraphernalia. As we may recall, it was introduced in 1894 by Wicksteed with one simple remark: *"the product being a function of the factors of production we have $P = f(a, b, c, \ldots)$."*[3] This paradigm of imprecision apparently sufficed to make us accept Wicksteed's simple symbolism as an adequate analytical representation of any production process and use it indiscriminately in every kind of situation. And as the usage of the vapid terms "input" and "output" became widespread, popular manuals came to treat the subject in an even more cavalier manner than Wicksteed's. A typical presentation is that the production function expresses symbolically the fact that "the output of the firm depends on its inputs."

But even consummate economists have accepted Wicksteed's formula without any ado. They only felt that the meaning of the variables involved ought to be explained. The greater number of such authors adopt the position that the formula shows the quantities of inputs (or of factors) necessary to produce a certain quantity of output (or of product). Accordingly, all symbols in a production function,

(1) $$Q = F(X, Y, Z, \cdots),$$

stand for quantities.[4] Others conceive the same function as a relation between the inputs per unit

---

[1] Max Planck, *The New Science* (New York, 1959), pp. 43, 158–59; P. W. Bridgman, *The Logic of Modern Physics* (New York, 1949), p. 50.

[2] Reginald O. Kapp, *Towards a Unified Cosmology* (New York, 1960), p. 111. My italics.

[3] Philip H. Wicksteed, *The Co-ordination of the Laws of Distribution* (London, 1894), p. 4.

[4] For a small yet representative sample, see A. L. Bowley, *The Mathematical Groundwork of Economics* (Oxford, 1924), pp. 28–29; J. R. Hicks, *The Theory of Wages* (London, 1932), p. 237; E. Schneider, *Theorie der Produktion* (Vienna, 1934), p. 1; A. C. Pigou, *The Economics of Stationary States* (London, 1935), p. 142; P. A. Samuelson, *Foundations of Economic Analysis* (Cambridge, Mass., 1948), pp. 57–58; K. E. Boulding, *Economic Analysis* (3rd ed., New York, 1955), p. 585; Sune Carlson, *A Study on the Pure Theory of Production* (New York, 1956), p. 12; Ragnar Frisch, *Theory of Production* (Chicago, 1965), p. 41.

of time and the output per unit of time; i.e., as a relation

(2) $$q = f(x, y, z, \cdots),$$

in which all symbols stand for rates of flow.[5]

Curiously, no one seems to have been intrigued by the existence of these entirely distinct interpretations. Instead, many economists—including some analytical authorities—have used both definitions indifferently, sometimes even on the same page.[6] The undeniable inference is that the economic profession considers relations (1) and (2) as two completely equivalent ways of representing any production process whatsoever. Yet behind this belief there lies an analytical imbroglio which is easily brought to light.

We need only recall that the production function is a tool associated with a static process or, to use a more explicit expression, with a steady-going process. For such a process, the following relations

(3) $$Q = tq, \quad X = tx, \quad Y = ty, \cdots$$

hold for any time interval t and for the quantities of product and of factors corresponding to that interval. With the aid of these relations and (2), relation (1) becomes

(4) $$tf(x, y, z, \cdots) = F(tx, ty, tz, \cdots).$$

And since this relation must be true for any t, it follows, first, that f and F are one and the same function,

(5) $$f(x, y, z, \cdots) \equiv F(x, y, z, \cdots),$$

and, second, that this function is homogeneous of the first degree. Therefore, the tacit presumption that the forms (1) and (2) are equivalent implies that the returns to scale must be constant in absolutely every production process.

Nothing, I believe, need be added to convince ourselves that this imbroglio is the direct consequence of our acceptance of Wicksteed's symbolism without first probing its validity as an analytical mirror of actuality. This conclusion raises a new and troublesome issue. Does either of the forms, (1) or (2), constitute an adequate representation of a process of production and, if so, what kind of process may be represented by it? For a start, let us try to examine it in its broad lines.

[5] G. Stigler, *The Theory of Competitive Price* (New York, 1942), p. 109; T. C. Koopmans, "Analysis of Production as an Efficient Combination of Activities," in *Activity Analysis of Production and Allocation*, ed. T. C. Koopmans (New York, 1951), p. 35.
[6] E.g., Frisch, *op. cit.*, p. 43.

II

Before anything else, we should note that for no other branch of economics is the concept of process as essential as for the economics of production. But, widely used though the word "process" is in sciences and philosophy, the literature seems to offer no specific definition of it. Now and then, the concept is merely associated with change. However, change is a notoriously intricate notion which has kept philosophers divided into two opposing camps: one maintaining that all is only being; the other, that all is only becoming. Obviously, science can follow neither of these teachings. Nor can it, unfortunately, embrace Hegel's dialectical synthesis that being is becoming. Analytical science must distinguish between object and event. Consequently, it must embrace the so-called "vulgar" philosophy—according to which there are both being and becoming—and cling to it to the very end. The upshot is that science must find a way to represent a process analytically.

It is obvious that, for this purpose, we must retain one point of dialectics; namely, that change and, hence, process cannot be conceived otherwise than as a relation between some entity and its complement in the absolute whole. In viewing a living tree as a process we oppose that tree to everything else—to "its other," in Hegel's terminology. Only for the absolute whole—the universe in its eternity—has change no meaning: such a totality has no complement. The notion of partial process, therefore, implies some slits cut into the absolute whole. But as a long series of thinkers, from the ancient Anaxagoras to the modern Niels Bohr, have taught us, this whole is seamless.[7] However, in this case as in all others, analysis must proceed by some heroic simplifications and totally ignore their ultimate consequences.

The first heroic step is to divide actuality into two parts—one representing the partial process in point; the other, its environment (so to speak)—separated by a boundary consisting of an arithmomorphic void. For if the boundary would not be such a void, we would get three parts instead of two and, as is easily seen, we would be drawn into a dialectical infinite regress. So, all that exists in actuality at any moment must belong either to a process or to its environment. The basic element of the analytical picture of a process is, therefore, the boundary. No analytical boundary, no analytical process.

[7] See Fragment 8 in J. Burnet, *Early Greek Philosophy* (4th ed., London, 1930), p. 47; Niels Bohr, *Atomic Physics and Human Knowledge* (New York, 1958), p. 10.

Now, precisely because actuality is a seamless whole we can slice it wherever we may please. And, Plato to the contrary,[8] actuality has no joints to guide a carver. As economists we know only too well the unsettled issue of where the natural boundary of the economic process lies. Only our particular purpose in each case can guide us in drawing the boundary of a process. So, every scientist slices actuality in the way that suits best his own objective—an operation that cannot be performed without some intimate knowledge of the corresponding phenomenal domain.

An analytical boundary, as conceived here, must consist of two components. Like a frontier, one component separates the process at any time from the rest of actuality, although we must not think that this frontier is necessarily geographical in nature or rigidly determined. Witness the process of thought itself or that of an acorn growing into an oak. The second component is the duration of the process, determined by the time moments at which the process we have in mind begins and ends. Naturally, these moments must be at a finite distance; otherwise, we would not know all that has gone into the process or all that the process does. Nor must we allow them to coincide. For, to recall Whitehead's admonition,[9] a durationless process, an event at an instant of time as a primary fact of nature, is nonsense.[10] We can then choose the time scale so that the process begins at $t = 0$ and ends at $t = T$, with $T > 0$. T is the duration of the process, but for a production process we may prefer, instead, Marx's term: the time of production.

The next point is truly crucial: in saying that a given analytical process begins at $t = 0$ and ends at $t = T$ we must take the underscored words in their strictest sense. Before $t = 0$ and after $t = T$ the analytical process is out of existence. That is, in conceiving such a process we must totally abstract from it what happens outside the duration we have assigned to the process. The mental operation is clear: an analytical process must be viewed as a hyphen between one *tabula rasa* and another.

Our next problem is to arrive at an analytical description of the happening, associated with a given process. Because of the principle, "No analytical boundary, no analytical process," analysis

[8] Plato, *Phaedrus*, p. 265.
[9] Alfred North Whitehead, *An Enquiry Concerning the Principles of Natural Knowledge* (2nd ed., Cambridge, England, 1925), p. 2; also his *The Concept of Nature* (Cambridge, England, 1930), p. 57.
[10] All this does not mean that, in the next stage of our inquiry, we cannot arrive at the excluded cases by a passage to the limit.

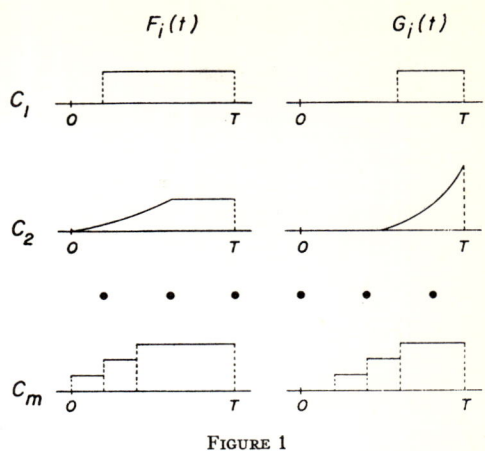

FIGURE 1

must renounce any hope of including in this description the happenings inside an analytical process. Indeed, in order to describe analytically what happens inside a process, we must divide it by a new boundary into two new processes to which the same rule will apply. And so on ad infinitum. The analytical description of a process, therefore, reduces to recording everything that crosses the boundary in either direction. In connection with this picture we can endow the terms input and output with quite precise meanings.

Analysis now needs to take another heroic step —to assume that the number of elements involved is finite and that every element is cardinally measurable (which implies that every element is a homogeneous entity). If $C_1, C_2, \ldots, C_m$ denote the distinct elements, the analytical description is complete if for every C_i we have determined two nondecreasing functions $F_i(t)$ and $G_i(t)$ over the closed interval $[0, T]$, the first function showing the cumulative input, the second the cumulative output up to time t (inclusive). Any analytical process—whether in economics or any other domain—may therefore be represented graphically by a series of curves, as in Figure 1.

In a plastic image, the coordinates of an analytical process may be likened to continuously reported data of import and export, with one important detail. Since in describing a process analytically we must begin and end with a *tabula rasa*, this hall (in which we are now gathered) must be listed both as input and as output in the process consisting of the delivery of this lecture. In the analytical approach we are not interested in how this hall came into being or in its use before or after this lecture. However, we must recognize that, as the result of every use, the hall suffers some wear and tear, imperceptible though this may be. Similarly, in any production process the same person must be listed as a rested worker among inputs and as a tired worker among outputs. A tool, too, may go in new and come out used. But even though we recognize the rested and the tired worker as being the same man, we must treat the former as a different element from the latter. Each element of an analytical process—as we have decided—must be completely homogeneous, a condition that does not always cover sameness.

These cases, of the worker and of the tool, raise a troublesome problem for the economist. The reason is that our material of study is the commodity. We slice the economic domain into units of production and units of consumption because at the boundaries thus drawn we can catch every commodity. Drawing a boundary in a glass plant between the melted glass furnace and the rolling machines would serve none of the economist's purposes: at this time, melted glass is not a commodity.[11] Briefly, the economist cannot afford to abandon his commodity fetishism any more than the chemist, for example, can renounce his fetishism of the molecule.

The difficulty, which at bottom is related to qualitative change, is that even though we cannot avoid including "tired worker" and "used tool" in the list of outputs of any production process, neither category fits the usual notion of commodity. Our entire analytical edifice would collapse if we were to accept the alternative position that the aim of economic production is to produce not only the usual products but also tired workers and used tools.

A new heroic step is needed to eliminate this difficulty. It consists of the familiar, old fiction of a process in which capital is maintained constant. The fiction does raise some analytical issues, for if all tools and all workers are to be maintained at a constant level of efficiency, any production process will have to include most of the enterprises and households in the world. Factually, however, the fiction is not more, not less reasonable than that of frictionless movement in mechanics. A simple glance at the activity inside a plant or a household suffices to convince us that efforts are constantly directed not toward keeping durable goods physically self-identical (which is quite impossible), but toward maintaining them in good working condition. And this is all that counts in production. The only factor we need neglect is the daily wear and tear of the worker. This is not too much to demand, since the worker is daily restored in the household.

The elements may now be divided into two relevant categories. In the first category we shall place those elements that appear as input and as output and are related by reason of sameness or of equality of quantity. A piece of Ricardian land, a motor, the amount of clover used as seed in growing clover seed (not clover fodder!), or a worker, illustrate this category. To elements such as these I propose to refer as funds so as to emphasize their economic invariableness. The other elements, which appear only as input or only as output, constitute the category of flows. Obviously, since the fund elements are maintained, the process may be activated again provided that the necessary inflows are still forthcoming. Labeled variously as a static or as a stationary process, or, still, as a diagram of simple reproduction, this fiction constitutes the fundamental element in the analysis of production from the classical to the hypermodern school. Reproducible, however, seems to describe the process better. The analytical picture we have thus reached is worth stressing: in a reproducible process, the fund elements are the immutable agents that transform some input flows into output flows. No picture of a process—whether static or dynamic—is complete if it does not include both categories of elements.[12] And the essential difference between these categories calls for a different representation of the fund coordinates. A flow coordinate will continue to be represented, according to the case, by the cumulative input or the cumulative output; i.e., by a quantity of some substance. Because in case of a fund the input and the output are economically the same substance, the coordinate of a fund may be represented by the difference $F_i(t) - G_i(t)$. But to maintain a convenient symmetry with the flow coordinates, we may use instead the cumulative amount of that intangible entity usually called the service of the fund.

III

In the case of a production process, the elements may be classified into some fruitful categories. The inflows that are transformed by the agents may come either from nature or from other

---

[11] That technological innovations may change this situation is evidenced by ready-mix cement and brown-and-serve rolls, for example, which only recently have become commodities.

[12] A point on which I insisted long ago: cf. my article "Aggregate Linear Production Function and Its Applications to von Neumann's Model" in *Activity Analysis of Production and Allocation*, ed. T. C. Koopmans (New York, 1951), pp. 100–01.

production processes; we shall denote them generically by (R) and (I). There also are inflows, (M), earmarked for maintenance. The output flows consist of products, (Q), and waste, (W). Finally, the funds include Ricardian land (L), capital equipment (K), and—to use Marx's very appropriate term—labor power, (H). With these notations, the analytical representation of a reproducible process is

(6) $[R_0^T(t), I_0^T(t), M_0^T(t), Q_0^T(t), W_0^T(t); L_0^T(t), K_0^T(t), H_0^T(t)];$

that is, a set of functions, which defines a point in an abstract (functional) space.

This is a far cry from the notion inherited from Wicksteed, according to which a process is represented by a point in the ordinary (Euclidean) space. The superiority of (6) over the point representation needs no elaborate argument. In (6) we have a complete set of instructions on how to set up the corresponding process. The form also reminds us continuously that a process has a duration, a time of production. Nothing is missing from it.

The difference yields an entirely new form for the production function. Since by a production function we must still understand the set of all processes that transform the same input flows into an outflow of the same product, from (6) it follows that the production function must be a relation among functions, instead of numbers. This relation, which may be written after the old pattern as

(7) $Q_0^T(t) = \mathcal{F}[R_0^T(t), I_0^T(t), M_0^T(t), W_0^T(t); L_0^T(t), K_0^T(t), H_0^T(t)],$

is what the mathematicians call a functional.

The results just reached call for numerous observations. Here, I can take up only a few and touch upon them briefly.[13]

First, the reason why I have excluded no element from the categories listed in (6), is that the scholar must never prejudge. Even an economist must first arrive at a complete description of a process and only then decide which elements may be left out because they are economically irrelevant. Nature does not indeed have a cashier's window where we may pay her for the elements (R); yet it would be utterly inept to ignore in the economics of production the fact that natural resources are neither inexhaustible nor uniformly distributed over the globe. The type of economic model now in vogue, which assumes that growth normally proceeds at a constant rate, simply blots out the most numerous and most critical cases—such as Somaliland or our own Appalachia, for instance. One may feel even more tempted to leave out the waste category, on the ground that waste by definition has no value. But again, as we have come to recognize recently on an increasing scale, the existence of waste is not an innocuous aspect of the economic process.

Second, we should not fail to observe that, since a fund enters a reproducible process and comes out without any impairment of its economic efficiency, service is the only way by which it can participate in the production of the product. While it is true that the cloth—an inflow element—effectively passes into the coat, the same cannot apply to the needle—a fund element. And if one finds the needle in the coat just bought, it certainly is a regrettable accident. The point is that the problem of how the contribution of a fund affects the value of the product is not as directly simple as in the case of a flow factor.

Third, both the value of a fund's service and that of its maintenance flow must, in principle, be imputed to the value of the product. Contrary to Marx's teachings—which have gradually infiltrated the thinking of many a standard economist—there is no economic double counting in this. No worker, no lecturer, can discharge his duties by sending to the shop or to the classroom only that "definite quantity of muscle, nerve, brain, etc., [which] is wasted" during work—as Marx claimed.[14] When one works, one must be present with his entire fund of mental and physical capabilities. The same is true for all other funds: the bridge must be there in its full material existence before we can cross the river. If it were true that we could cross a river on the maintenance flow of a bridge or drive the maintenance flow of an automobile on the maintenance flow of a highway, there would be little financial difficulty in saturating the whole world with all the river crossings and automotive facilities. Economic development could be brought about everywhere with practically no waiting.

IV

As with almost everything else, among the various processes we may envisage in production there is one process that fits the epithet "elementary." It is the process by which every unit of the product—a single piece of furniture or a molecule of gasoline—is produced. The process is

[13] For greater details see Chapter IX in my forthcoming volume, *Entropy and the Economic Process* (Harvard Univ. Press).

[14] Karl Marx, *Capital* (Chicago, 1932), Vol. I, p. 190.

directly observable in the shop of a cabinetmaker, but it can be easily determined even in a large-scale enterprise. Whatever the product, one thing is certain about the elementary process. In relation to it, most of the funds are idle over large periods of time. The plow is needed only a few days during the whole production time of growing a corn plant; the same is true for the saw or the plane in the production of a table. There is no exception to this rule. And, I contend, one of the most important aspects of the economics of production is how to minimize these periods of fund idleness, whether we are thinking of man, capital equipment, or land.[15]

Now, if only one table is demanded during the time of production, T, then obviously we need operate only one elementary process after another in succession. But if n tables, with $n > 1$, are demanded during T, then two alternatives are open to us. We may start n processes at the same time and repeat the operation when they are ended. This is the arrangement in parallel. The second arrangement is the arrangement in line, in which equal batches of processes are begun one after another at intervals equal to an aliquot part of T.

It is obvious that the production function of a system in which the elementary processes are arranged in parallel is obtained from (7) by multiplying every coordinate by n:

(8) $\quad [n\overset{T}{\underset{0}{Q}}(t)] = \mathcal{F}\{[n\overset{T}{\underset{0}{R}}(t)], \cdots, [n\overset{T}{\underset{0}{W}}(t)]; [n\overset{T}{\underset{0}{L}}(t)], \cdots\}.$

The point that deserves to be stressed is that the arrangement in parallel offers little or no economic gain. Most kinds of fund factors are now needed in an amount n times as great as in the elementary process. In addition, the idleness of each such fund factor is *ipso facto* amplified by n. The only exceptions are the fund factors that—like a large bread oven, for instance—may accommodate several elementary processes simultaneously. But even though the capacity of such a fund factor would be more fully utilized, its idleness period would remain the same.

The situation completely changes for the arrangement in line. If we assume away any incommensurabilities among the time periods involved in the schedule of the elementary process—an inevitable assumption in practice—and if n is sufficiently large, then a number of processes can be arranged in line so that no fund is idle at any time.[16] The situation is vividly exemplified by an assembly line, in which every tool and every worker shift without interruption from one elementary process to the next. The arrangement in line, however, describes any factory. In a factory, therefore, the economy of time reaches its maximum. This conclusion opens an avenue of utmost importance. To explore it, we may begin by determining the analytical representation of a factory process.

In a first approach we may consider the entire physical plant as one monolithic fund, P. Over an arbitrary interval $[0, t]$, during which the factory process is in operation, the coordinate of this fund is the function Pt. Similarly, the coordinate of labor power is Ht. And if for the convenience of diction we assume that all flow elements are continuous, their coordinates, too, are represented by linear homogeneous functions. Thus (7) becomes:

(9) $\quad (\overset{t}{\underset{0}{q t}}) = \mathcal{G}[(\overset{t}{\underset{0}{r t}}), (\overset{t}{\underset{0}{i t}}), (\overset{t}{\underset{0}{m t}}), (\overset{t}{\underset{0}{w t}}); (\overset{t}{\underset{0}{P t}}), (\overset{t}{\underset{0}{H t}})].$

Let us note that this is a very special functional: first, every function involved in it depends upon a single parameter and, second, the value of t is entirely arbitrary. For these reasons, the functional degenerates into a point function.

There are two degenerate forms. The first is

(10) $\quad q = \Theta(r, i, m, w; P, H).$

This formula reminds us of one of the current interpretations mentioned in Section I; namely, that of relation (2). We should note, however, that θ involves two dimensionally different categories of variables. The lower case symbols represent flow rates of some substances. The upper case symbols stand for the rates of service per unit of time. Strangely, however, these last rates do not involve the time element at all: P stands for the plant, and H, for the total labor power—briefly, for quantities of some substances. The second degenerate form is

(11) $\quad Q = \Psi(R, I, M, W; \mathcal{P}, \mathcal{H}; t).$

Here, the symbols in roman capital letters are again quantities of some substances; those in script letters are services, and t is the period with which these quantities and services are associated. The form (11), in turn, reminds us of relation (1); i.e., of the quantity interpretation of Wicksteed's formula. The most important difference is that Ψ includes time as an explicit variable.

---

[15] A period of idleness is characterized by the constancy of the corresponding fund coordinate.

[16] The number of elementary processes that should be started each time is the smallest common multiple of the numbers of such processes that can be accommodated at the same time by each unit of the various funds. Batches should be started at intervals of T/d, d being the greatest common divisor of T and of the intervals during which the various funds are needed in an elementary process.

This difference bears upon the earlier argument that Wicksteed's production function is homogeneous of the first degree. Actually, Ψ is such a function—as easily follows from the identities

(12) $\quad Q = qt, \cdots, W = wt, \mathcal{P} = Pt, \mathcal{K} = Ht,$

analogous to (3). There is then an intimate relation between (10) and (11); namely,

(13) $\quad t\Theta(r, i, m, w; P, H) = \Psi(R, I, M, W; \mathcal{P}, \mathcal{K}; t).$

Hence,

(14) $\quad\quad\quad\quad\quad \Theta = [\Psi]_{t=1}.$

The imbroglio created by (5) is thus resolved. Of course, this does not mean that the factory process operates with constant returns to scale. The homogeneity of Ψ corresponds to the tautology that if we double the time during which a factory works, then the quantity of every flow element and the service of every fund will also double. The issue of returns to scale pertains, instead, to what happens if the fund elements are doubled. The point may be made still clearer.

A superficial inspection of any operating plant suffices to reveal that P consists of some Ricardian land, R, some capital equipment, K, some technical inventories, S, and a special fund, Γ, usually called "goods in process." The last term is definitely a misnomer: half-tanned hides or partly wired radio sets, for example, are not goods. Process-fund seems a more exact term because Γ is in effect a becoming frozen in its various phases. If a photograph of Γ would be projected part by part, as if it were a movie, we would witness the actual change of some input flows into product and waste flows. In spite of this varied composition of any plant, what a given plant can do depends on its blueprint alone, which in turn involves only L and K. And since what a plant can do is shown by the flow rate of its product, we have a first relation

(15) $\quad\quad\quad\quad q^* = G^*(L, K).$

A second relation expresses the fact that, given the plant, we require a certain labor power, H^*, if we want to obtain the flow rate q^*. Hence,

(16) $\quad\quad\quad\quad H^* = H^*(L, K).$

Should we man the plant with less labor power than H^*, the product flow rate would also become smaller than q^*. To account for this rather common situation, we need to put

(17) $\quad\quad\quad\quad q = G(L, K, H) \leq q^*.$

But the fact that this relation looks very familiar should not mislead us: as (17) is defined here, if $q < q^*$, q does not necessarily decrease (and ordinarily does not) when L and K are decreased while H is kept constant. Actually, the ratio q/q^* measures the percentage of capacity utilized if H is the labor power employed.[17]

The next relations are self-explanatory:

(18) $\quad\quad S = S(L, K, H), \quad\quad \Gamma = \Gamma(L, K, H).$

There remain the relations binding the other flow elements. The case of the maintenance flow, m, is simple: its size must depend on the amount of equipment to be maintained and the labor fund employed. In addition, by virtue of the conservation law of matter and energy, m must be equal to w_1, the flow rate of wear-and-tear waste—burned or discarded lubricating oil, broken saw bands, and the like. Hence,

(19) $\quad\quad m = m(K, H), \quad\quad m = w_1.$

The same conservation law applies to all other flows. For example, the wood contained in a piece of furniture together with the scrap and the sawdust must *exactly* account for the wood introduced into the production of that furniture. In the case of a factory system, this relationship yields

(20) $\quad\quad\quad qt = g(rt, it, w_2 t),$

where w_2 is the flow rate of waste arising from transformation alone. Since (20) must be true for any t, the function g must be homogeneous of the first degree. To this function we may indeed apply the old-time tautology that "doubling the inputs doubles the output." The basic error in some arguments about the returns to scale is to apply this tautology to (17) instead of (20). If L, K, and H are doubled, q does not necessarily double even if we double at the same time all flow inputs. The new factory may be more efficient or more wasteful in using the input materials, which leads us to put

(21) $\quad\quad\quad\quad w_2 = w_2(L, K, H).$

Relation (10) is thus decomposed into seven basic relations, listed from (17) to (21), which together constitute the general representation of a factory process.[18]

We should now note that the picture at which we have arrived is analogous to the inscription "60 watts" on an electric bulb. That is, it tells us what the factory can do, not what it has done, is doing, or, above all, will do. Like the inscription

---

[17] As an ordinal measure of the utilized capacity we may use H/H^*. On this point, see note 21, below.

[18] Obviously, this analytical description will have to be completed with additional relations if the particular factory process happens to involve other limitationalities.

on the bulb, relation (17), for example, is true regardless of whether the factory works or is idle. To show what the factory does, we need an additional coordinate, which, under its various aspects, has deeply preoccupied Marx, but which, perhaps for easily understood reasons, is not found in the analytical tool box of the neoclassical economist. The coordinate is the time, δ, during which the factory works daily. The amount of the daily production, Q, follows immediately from (17):

(22) $\qquad Q = \delta G(L, K, H),$

a relation which vindicates Marx's dear tenet that labor time measures value even though it has no value itself.[19]

V

So much for grounded-in-actuality symbolism. Let me devote my closing remarks to some of the object lessons of this symbolism.

I have stressed the fact that in any elementary process every agent is idle over some definite periods that depend not on our choice or whim but on the state of the arts. I have also argued that we can nonetheless eliminate this kind of idleness completely and that there is only one way to achieve this: to arrange the elementary processes in a factory system. Because of this extraordinary property, the factory system deserves to be placed side by side with money as the two most fateful economic innovations for mankind. I say "economic" and not "technical" because the economy of time achieved by the factory system is independent of technology. Nothing prevents us from using the most primitive technique of cloth weaving in a factory system.

To be sure, there is a second kind of idleness, which depends entirely on our decision: it is the idleness of the factory itself if δ is shorter than a full day. In view of these two kinds of idleness, the economics of production reduces to two commandments: first, produce by the factory system and, second, let the factory operate around the clock.

The first commandment calls for two observations. Even though we can draw the blueprint of a factory for any elementary process whatsoever, not every such factory is economically advantageous. For example, we do not build transoceanic "Queens" by processes in line. The reason is that we can build a "Queen" more quickly than it is demanded in relation to time. The much extolled progress of the industrial revolution may not after all be due only to technological innovations. For these innovations as well as the in-

[19] Marx, *Capital*, Vol. I, pp. 45, 588.

creased specialization of labor could not have come about unless an increased demand had already induced most craft shops to introduce the system in line. There can be little doubt about it: the factory system was born in an artisan's workshop, not in a factory.

The second observation is that to operate an arrangement of elementary processes in line it is absolutely necessary that we have the freedom to start a process at any time of the day, of the week, and of the year. Unfortunately, we do not always have this freedom. Seasonal variations—which result from the position of our planet relative to our main source of free energy, the sun—prevent us from adapting the factory system to a series of important productive activities. The most important instance is husbandry. For the overwhelming majority of localities, there is a very short and definite period of the year during which a corn plant, for instance, can be grown in the open space from seed. This is why farmers have to work their fields in parallel; that is, in a system of production that yields practically no economy of time. The global analytical representation of that system is (8), not (9). The upshot is that the open-air factories, about which socialist writers in particular have been continuously raving, will remain a utopian dream as long as we are unable to alter the orbit of the earth.

The association between agriculture and the idleness of all agents involved is by now a commonplace. Still, not much is known or even suspected about the importance of the related loss. Two simple illustrations may bring out this point. For the first, let us consider one of the exceptional localities—such as the Island of Bali—where, because of an almost constant climate throughout the year, rice could be grown in an open-air factory. In this case, every day the same number of hands would move over the fields with the same funds of plows, buffaloes, sickles, and flails to plow, sow, harvest, and thresh. Every day the villagers would eat the rice sown that very day, as it were, and they would no longer have to bear the burden of the debts specific to agriculture. But most important of all, we would also discover that, without diminishing the old production at all, there would remain a substantial number of superfluous workers as well as a substantial stock of superfluous equipment—a palpable measure of the overcapitalization of farming in comparison with manufacturing. The second illustration pertains to the current technique by which chickens are raised in the United States. In fact, in this country there are no longer any chicken farms—even though the term continues to be used. Instead, there are chicken factories, with elementary processes arranged in line. The "chicken war" of yesteryear would not have come about if the difference between the old and the new techniques had not been so great as to exceed the shipping cost over the Atlantic plus the wage differential between this country and Europe.

But if not every production activity can be turned into a factory, we should at least try to render the idleness of the agents as small as possible in each particular case. In other words, to bring even a whole economy as near as possible to the functioning of a factory system should be the guiding thought of any planner at any level. In the activity of the countryside, the cottage industry propounded by the agrarians was one answer to this idea. In Romania (so I was told) tractors and drivers shuttle between the plain regions—where two crops are grown each year— and the hilly regions—where only one crop can be grown because of a shorter vegetation period. The necessary funds of tractors and drivers are thus substantially reduced at the cost of some gasoline, oil, and spare parts flows. Less costly solutions would be obtained by mixing several crops within the same locality, the crops being chosen so as to minimize idleness (and hence capital cost). Formally, the problem boils down to splicing graphic patterns with a minimizing condition—a problem of a special type of combinatorial analysis which, I am sure, will prove highly rewarding.

The second commandment is particularly relevant for the underdeveloped economies. In a rich country, it makes perfect sense to operate every factory with one shift, even if the shift be of six or four hours only. In a rich country, there also is no need for night shifts, except whenever technology imposes around-the-clock production. Briefly, in a rich country leisure is a commodity which people may prefer to higher income. Things are different in almost every underdeveloped country where—as every government pronouncement urges—the order of the day is not only development but rapid development. In such countries, the regimen of the eight-hour working day and the reluctance to use night shifts are anachronistic factors that work against the avowed aims.[20] There may be many reasons why planners as well as our planning theory have overlooked the simplest and the most direct lever of economic development; namely, the length of the working day. But one possible reason is that this element of the problem has been left out of the neoclassical representation of a production process. The same omission—we should note—vitiates also the familiar comparisons of the capital-output and capital-labor ratios computed from current statistical data. Since the theoretical apparatus ignores the working time, δ, the most sophisticated statistical agencies, too, have felt no need to include it in their usual collections. Thus, we are unable to obtain valid statistical estimates of K/q^* and K/H^*, the basic technical and theoretical elements.[21]

Another omission of the neoclassical representation is that, as a rule, only the funds (variously defined) are included in the production function. The fact that after a factory is built, production cannot go on unless the input flow factors are forthcoming, has thus been pushed away from the focus of attention. None other than an authority such as A. C. Pigou preached that "in a stationary state factors of production are stocks, unchanging in amount, out of which emerges a continuous flow, also unchanging in amount, of real income."[22] The omission of the input flow factors is not unrelated to the present race of all underdeveloped countries to build one factory after another without a thorough examination of the availability of the necessary flow inputs. I am confident that if the prospective economic plan of every country were realized by miracle overnight, we would discover that we have long since been planning for a world with an immense excess capacity of industrial production.

The thoughts I shared with you here may seem simple. Perhaps they are simple, once we have untangled the imbroglio hatched by blind symbolism. The economics of production, its elementary nature notwithstanding, is not a domain where one runs no risk of committing some respectable errors. In fact, the history of every science, including that of economics, teaches us that the elementary is the hotbed of the errors that count most.

---

[20] I may hasten to admit that (22) is only a first approximation formula: a factory working with one shift of ten hours will not produce 25 percent more than with a shift of eight hours. To take better account of facts, we should replace δ by a function of the number of shifts and the number of working hours of each shift. But this amendment does not affect in the least the validity of the statements just made.

[21] The difficulty is especially serious in the case of comparisons between two different industries. Even if we know that each industry has always used its full capacity, i.e., has worked with the corresponding H^*, the values of capital-labor ratios derived from the usual statistical data are neither comparable nor strictly relevant—unless we also know that both industries employed the same number of shifts. In fact, the Census of Manufactures provides no information on the number of shifts and on the percentage of utilized capacity.

[22] Pigou, *op. cit.*, p. 19.

CHAPTER 5

(1965)

PROCESS IN FARMING VERSUS PROCESS IN MANUFACTURING: A PROBLEM OF BALANCED DEVELOPMENT

I. INTRODUCTION: THE ASYMMETRY OF A SYMMETRY

1. It is Wicksteed who first pointed out the 'fascinating' analogy between the laws of satisfaction and those of production.[1] A trivial idea by now, the formal identity between consumption and production theory comes from the fact that in both cases the main problem is one of maximizing an ordinary function of several independent variables subject to a budget constraint. In consumption, it is the utility,

$$u = \varphi(\xi, \eta, \zeta, \ldots), \qquad (1)$$

that must be maximized within a given income; in production, it is the maximization of the output,

$$q = f(x, y, z, \ldots), \qquad (2)$$

for any given outlay. And there seems to be no greater delight for the author of a text-book than to comb every formal symmetry between indifference curves and isoquants on the one hand, and between the marginal utility and the marginal productivity theorems on the other. Yet, should one compare the march of ideas in the micro-theory of consumption and that of production, one would not fail to be surprised by the marked contrast between the two.

An unending series of contributions have taken us from the Gossen–Jevons conception of utility as a cardinal magnitude depending separately on the amount of each commodity, over an ever-changing panorama of consumer's theory. There came in succession

[1] Philip H. Wicksteed, *The Co-ordination of the Laws of Distribution* (Reprinted as No. 12 of the Scarce Tracts in Economic and Political Science, London School of Economics and Political Science, London, 1894), pp. 8 f., 48.

Edgeworth's complementarity, Pareto's ordinal index of ophelimity, Samuelson's revealed preference, then a new cardinal utility, and in more recent time the view that the preference field cannot even be completely represented by a Dirichlet function such as (1). Certainly, that is not the end of this most elusive problem.

On the other hand, ever since the essential idea behind Walras's coefficients of production found a more adequate expression in (2), this form of representing a productive process has constituted the most solid corner-stone of production theory. To be sure, our present notion of a production function differs from that of Wicksteed when he first introduced the concept in the work cited above. But the difference is rather superficial. As a result of the contributions of Pareto and others after him, we have become aware of the fact that (2) does not cover all situations: joint products and limitational factors require amended models. But these models too involve only Dirichlet functions. And though careful writers discriminate now between the production function of a unit of production and that of an aggregate of such units, the distinction actually tends to strengthen the position that (2) is the basic analytical representation of a productive process.

2. But no student of production theory, it seems, felt the need for raising the same kind of *epistemological* questions about the production function as those that have continuously tormented the students of consumer's behaviour. This is the contrast to which I have referred above. It has, I submit, a very natural explanation. The servants of quantitative economic analysis, especially the mathematical economists, have bent their efforts to defending that sector of their theoretical citadel where the opposite camp has always concentrated its attack; for the suggestion that human actions can be analysed into laws of mathematical rigidity had been decried long before economics came out with a mathematical theory of utility. On the other hand, even the fiercest enemies of mathematical economics have never challenged the use of the mathematical tool for studying production: after all, production being a physico-chemical process must run according to the rigid laws of nature. And from what one can infer, economists too have concurred in that production theory, for one, cannot possibly upset the apple cart of mathematical economics. There is no need to search further for an explanation of their lack of interest in a critique of formula (2) as an analytical representation of a productive process.

It is precisely such a critique that led the author to the results presented in this paper. These results include, first, the conclusion that there are two analytically distinct types of productive processes, neither of which is described completely by the traditional production function. Second, they will expose the fallacies of the rather widespread practice of conceiving the economic process as a circular flow. As an immediate and perhaps a most eloquent application of these results, they will be used in analysing, from a yet untried viewpoint, the balance between agriculture and industry in economic development.

II. THE PRODUCTION FUNCTION: A CRITIQUE

1. One difficulty of a student approaching for the first time the subject of consumer's behaviour is the vast literature covering the various meanings of the function (1) as well as the experiments by which one could conceivably construct the particular 'utility' function of a given person. On the other hand, the discussion of the production function often amounts to no more than a repetition of Wicksteed's initial theme: 'The Product being a function of the factors of production we have $P = f(a, b, c, ...)$.'[1] In some of the not too recent works one finds a more or less comprehensive discussion of the nature of the production factors, that is, of the elements supposed to be represented by the symbols $x, y, z, ...$. But ever since 'output' and 'input' displaced the classical terminology, most economists, it seems, have felt that the immediately obvious etymology of the new terms renders superfluous any further explanation of what corresponds in actuality to these symbols. Indeed, search as one may, even in the works of the most respected modern authorities one finds hardly any discussion of the empirical scaffold of the production function. All seem satisfied with some purely formal definition, some variation on Wicksteed's theme.

For a representative sample, we may cite, first, Boulding's definition:

> the basic transformation function of an enterprise is its *production function*, which shows what quantities of inputs (factors) can be transformed into what quantities of output (product);[2]

---

[1] Wicksteed, *op. cit.* p. 4. This cavalier treatment is characteristic of most textbooks nowadays. Cf. A. W. Stonier and D. C. Hague, *A Textbook of Economics* (2nd edn., London, 1958), p. 119, or Richard H. Leftwich, *The Price System and Resource Allocation* (rev. edn., New York, 1960), p. 108.

[2] Kenneth E. Boulding, *Economic Analysis* (3rd edn., New York, 1955), p. 585. This definition represents the most popular variant. Cf. A. L. Bowley, *The*

second, Stigler's:

> A production function may be defined as the relationship between inputs of productive services *per unit of time* and outputs of products *per unit of time*.[1]

and, finally, Carlson's, according to which the production function is

> the relationship between the quantity of input bought at the beginning of the production period and the quantity of output turned out by the process at the period's end.[2]

We find no attempt to supplement such formal definitions with some operational instructions about how to determine a production function at least under hypothetically ideal conditions. Boulding likens a production process to a recipe from a cook-book. Just as one such recipe says that 'if we put 4 eggs, 2 cups of milk . . . in a waffle mixture and obey the instructions, we should expect to get ten waffles out of it, no more and no less', so the cook-book of the iron manufacturer tells him that if he 'mixes so much ore, so much lime, so much coke, and so much heat for so many hours', he will get 'so much iron'.[3] But the widely shared position that it is not the business of the economist to know more about the relationship between output and inputs is more directly expressed by Stigler, who says that production functions are 'taken from disciplines such as engineering and industrial chemistry: to the economic theorist they are data of analysis'.[4]

2. But is it not strange to think that the economist can analyse these data without knowing more about their nature than that q is 'output' and $x, y, z, \ldots$ are 'inputs'? In fact no sooner have we proclaimed that all we need to know is that a relation such as (2) exists, that we

---

Mathematical Groundwork of Economics (Oxford, 1924), pp. 28 f.; J. R. Hicks, *The Theory of Wages* (London, 1932), p. 237; Erich Schneider, *Theorie de Produktion* (Vienna, 1934), p. 1; A. C. Pigou, *The Economics of Stationary States* (London, 1935), p. 142; Paul A. Samuelson, *Foundations of Economic Analysis* (Cambridge, Mass., 1948), p. 57.

In view of the argument to be developed in the following section, we should note that some of these authors — Boulding (p. 201), most explicitly — identify the quantities of inputs with the *services* of the production factors.

[1] George J. Stigler, *The Theory of Price* (New York, 1949), p. 109 (italics added). The treatment of the production function in the 1952 revised edition is lowered to the level of the ordinary text-book and thus is quite uninteresting for the present discussion.

[2] Sune Carlson, *A Study on the Pure Theory of Production* (New York, 1956), p. 12. Carlson uses 'input' as a plural noun.

[3] Boulding, *op. cit.*, pp. 585 f.

[4] Stigler, *op. cit.*, pp. 109 f. For the general agreement on this point see Hicks, *op. cit.*, p. 237; Pigou, *op. cit.*, p. 142; Samuelson, *op. cit.*, p. 57; Carlson, *op. cit.*, p. 10.

turn to discuss such specific structural properties of the production function as the convexity of isoquants or the laws of returns to factors and scale. The space occupied in economic literature by the problem of whether or not there is an optimum size of the unit of production plainly proves that economists have never been consistent as regards what they need to know about the production function. All the harder is it then to understand their manifest indifference toward the foundations of this concept.

A rather piquant consequence of this indifference is the fact that no one seems to have noticed that the Boulding and the Stigler definitions cannot be equivalent unless all production functions are homogeneous of the first degree. Yet, the point is most elementary: Let $q, x, y, \ldots$ denote the flow rates of output and inputs so that (2) should correspond to Stigler's definition. Clearly, then, during any time interval t the quantity of product is tq and the quantities of inputs are $tx, ty, \ldots$ According to Boulding's definition, we must also have $tq = F(tx, ty, \ldots)$. Hence, $f \equiv F$ is a homogeneous function of the first degree, as said.[1]

3. A series of issues — some immediately obvious and familiar to all of us, others brought to light as a surprise by a highly interesting study of Chenery[2] — show how foolhardy is the belief that the production function which the economist needs for his special job comes ready-made from the technical 'cook-books'.

3. 1. The first issue, a very familiar one, hinges upon the qualitative variations of some inputs, especially of the human inputs. Technical blue-prints hardly ever include the administrative, often not even the technical personnel. No chemical formula includes the chemist himself, not to mention the bursar from whom he gets his salary cheque. Engineering blue-prints, therefore, can throw no light upon the problem of the influence that the quality of the entrepreneur, of plant managers, or even of foremen, has upon the amount of output obtainable from the same set of material inputs. The engineer's concern is the efficiency of machines, not the efficiency of 'human inputs'. But in the actual process it is man who operates the machines, who responds actively to their signals and, above all, who keeps the chain of the process going. And at least the economic theorist should know better than to be satisfied with the answer that technical blue-prints pertain to the results obtainable with the

[1] We shall see later on (*IV.* 4) the reason why a great confusion, closely related to the point just made, reigns over mathematical economics.
[2] Hollis B. Chenery, 'Engineering Production Function', *Quarterly Journal of Economics*, lxiii (1949), 507–531.

'average' quality of human inputs.[1] What this average is in each activity sector is one important outcome of the general economic process; hence, for the economic theorist it is an item of explanation, not a datum.

But one may interject that non-quantifiable qualities — such as those, especially, of the persons who do not directly operate machines — cannot in any case be included in the function (2) because all variables must first of all be measurable.[2] This is indeed correct, but if and only if we insist that (2) should be a Dirichlet function. But why should the economist bide by this restriction — legitimate though it may be in engineering — if the economic problem involves also qualitative variables?

For a suggestion of how the restriction may be avoided, let $q = F(x, y, z, \ldots)$ represent the engineering production function of a plant and let us assume that the relation pertains to an *ideal* process. (An illustration of such an engineering production function is the theoretical formula $2nH + nO = n$ molecules of water.) Let now $q = E(m_i;\ x, y, z, \ldots)$ be the corresponding *economic* production function of the same plant. Here, m_i, $i \epsilon H$, denotes a particular qualitative unit of 'plant manager'. There is no need for H, the set of all such qualities, to be a linear continuum and, hence, for i to be a number. But since there is no *ideal* manager, the difference $F - E$ is necessarily positive. The smaller this difference, the better is the managerial quality of m_i.

True, one could no longer speak of the partial derivative of E with respect to m_i, but the differences between the values of E for m_i and m_j, $i \neq j$, would still have a meaning useful in analysing the pricing process. Equally true is the fact that from the operational viewpoint E has an important shortcoming: there is no way to determine exactly and *ex ante* what is the result of substituting m_j for m_i. But the construct can serve the purposes of economic analysis in the same way in which the system of arbitrary (unspecified) functions do, or the circular concept of 'the fittest' benefits biology.[3] We should not

---

[1] More often than not, however, the technical blue-prints pertain to what happens in the 'test tube' or, at most, under laboratory-like conditions. That is why, if statistical data are sufficiently reliable, the input coefficients computed from them describe more adequately the actual situation than those supplied by the blue-prints.

[2] This opinion is emphatically expressed by Samuelson, *op. cit.*, p. 84, and R. G. D. Allen, 'The Mathematical Foundations of Economic Theory', *Quarterly Journal of Economics*, lxiii (1949), 116, n. 6. Pigou (*op. cit.*, p. 137), however, argues that leaving out of account the qualitative elements of the problem does not affect the relevance of economic analysis.

[3] One example of the analytical service of the economic production function

overlook the fact that the theorem that maximization of output with constant outlay requires that the marginal product of the 'dollar' be equal for all inputs has been proved without implying that inputs are cardinally measurable. And if the issue of the indivisibility of m_i is critically raised, we should observe that the same issue calls for amending the above theorem even in case all factors were measurable.

3. 2. The other issues, though less familiar, carry more weight because they do not depend upon how one views the role of economics. To begin with, an engineering production formula — as Chenery reminds us [1] — describes an isolated process (say, a refrigeration system or an electrical transformer), not the full process of a plant. Secondly, in such a formula most outputs and inputs are not commodities, that is, entities with which economics is concerned. Instead, they represent physico-chemical properties: pressure, temperature, stress, specific gravity, pH, viscosity, etc. etc.[2] Moreover, an engineering formula may not be available for every individual process within a plant. But even if there is such a formula for every process in a given plant, to obtain the engineering production function for the whole plant may be well-nigh impossible. To transform an engineering formula into a functional relation between commodities is not an easy task. Nor are we certain that it is always feasible. There is also the staggering problem of covering all the innumerable combinations of engineering processes by which a given product may be produced.

More important still is the point that the choice of the processes studied by engineering — and for which alone there exist engineering formulae — is narrowly determined by the price constellation of the production factors. As a result, engineering formulae are available mostly for processes that can be used economically only by economies rather advanced in development. Consequently, even if one could derive an engineering production function for every product, the information would not cover all possible factor combinations.

4. The preceding critique is not meant at all to deny that the physico-chemical laws constitute the backbone of any material production. Nor should one interpret it as a belittling of the scientific value of engineering economics — a field which is now open to econometricians thanks to the pioneering work of Tjalling C.

---

suggested above is offered in the author's paper, 'Measure, Quality, and Optimum Scale', Chapter 11 in this volume.

[1] Chenery, *op. cit.*, pp. 529 f.   [2] *Ibid.*, pp. 510 f.

Koopmans — or of all special works concerned with determining an economic production function of the type (2) for some particular industry with the aid of statistical data. The purpose of the critique is to point out that in actuality there is also a great deal of purely economic flesh around the engineering backbone. The economic theorist cannot ignore this economic flesh. There are also good reasons why the offerings of engineering are not of much help to him. Algebraically simple though most of engineering formulae are, because of the difficulties mentioned in the preceding section they can provide hardly any basis for those principles without which the edifice of economic analysis would completely collapse : the principle of decreasing marginal rate of substitution and the laws of returns to factor and scale. As a matter of fact, from the gas pipeline formula [1] we should conclude that output increases more than proportionally with the size, and hence with the quantity, of pipe. And though no one, it is true, can tell what output corresponds to every factor combination yet untried by engineering, economic analysis must arrive at some idea, at least, about how output varies in general. The economic theorist himself, therefore, has to construct the proper tool for the analysis of economic production, and he must tend this delicate job with all the sophistication required by it.

III. THE PARTIAL PROCESS AND ITS ANALYTICAL CO-ORDINATES

1. If the few remarks scattered in literature are pieced together, they reveal that what we want in the very first place to represent by the concept of production function is a catalogue listing every optimal process by which a certain product can be obtained from each possible factor combination.[2] But then it follows with impeccable logic that before we can say anything about the structure of such a catalogue we must have a clear picture of the individual category to which the catalogue pertains. That is, we must first arrive at an analytical representation of a productive process.

No other word is used in economics with such careless ease as 'process'. Yet if one pauses to think about it, no other concept is as full of epistemological thorns as that of process. The reason is that we cannot talk about process without getting entangled in the

[1] Chenery, op. cit., p. 515.
[2] 'This catalog of possibilities is the production function', Samuelson, op. cit., p. 57.

difficult problem of Change. Ever since Herakleitos intrigued his contemporaries by teaching that 'you cannot step twice into the same rivers' — because the second time neither you nor the rivers are the *same* — the problem of the opposition between Being and Becoming has tormented the mind of every great philosopher. Science, however, has long since decided to embrace the viewpoint of 'vulgar philosophy', which viewpoint is that there is both Being and Becoming. From all we know, to abandon this dualism is to renounce *analysis* ; and to renounce analysis is to do away with *theoretical sciences*. However, we must not expect that analysis can remain entirely immune to the epistemological ills inherent in any dualism. It does not take much to see this truth.

2. Formidable though the notion of the whole universe in eternity certainly is, at least it does not raise one particularly difficult issue specific to the concept of *partial* process. For the universal process we need not specify what is included in it : the Whole is self-sufficient. That is not true for a partial process. To speak of such a process we must first of all determine its boundary with respect to both time and substance : *no boundary, no partial process*.

But as the entity from which we wish to carve a slice is a *seamless* whole, it contains no lines already traced to guide the carver. For the economist nothing could illustrate this point more convincingly than the still unsettled controversy concerning where the *natural* boundary of the economic process lies. And this situation is not the consequence of the complexity — as one may say — of the economic world itself. There is no natural boundary either between physics and chemistry, or between chemistry and biology. The issue is present everywhere we turn.

For a very elementary example from economics, let us take the case of a truck carrying over a highway a load of lumber from a mill to a furniture factory. Is there any objective criterion for deciding whether the travelling truck is part of the mill process or of the factory process? Perhaps one could answer that it depends upon which enterprise operates the truck. This, indeed, is the practice of book-keepers or cost analysts. But general economic analysis can hardly accept a criterion which becomes idle in case the mill and the factory belong to the same enterprise. And only harm can result from restricting that analysis to processes delimited by the money boundaries of enterprises.

3. Precisely because reality is a seamless whole one can draw a process boundary anywhere one pleases. However, an arbitrary slice

of reality does not correspond to our conception of process. On the other hand, it is hard to say exactly what sort of boundaries determine processes. To wit, every special science draws process boundaries where it suits its own *purpose* best. Without an intimate knowledge of this particular purpose, therefore, one could not say what a process means from the viewpoint of thermo-nuclear physics, for instance. In other words, any analytical process presupposes a purpose, and consequently is essentially a *primary* notion, that is, it may be clarified by discussion, but never reduced to other notions. Almost every important problem in social sciences supplies a striking illustration of these points. So, let us turn to economics.

The purpose of economics, to recall Say's definition, is to study the production, the distribution and the consumption of riches. Nothing is more natural, therefore, than that all partial processes in economics should be delimited in strict reference to riches, viz. commodities. The analytical definition of industry is referred back to the notion of commodity; a market process too is defined in a similar manner. Still more telling is the fact that the economist is not interested, for instance, in separating the process of plate-glass manufacturing at the point where the melted glass pours into the rolling machines: melted glass, under present technology, is not a commodity. From this comes the principle, heard now and then, that what happens inside a productive process is the engineer's, not the economist's, concern. But if strictly followed, the principle becomes a fetter. Without a broad understanding of the internal articulations of a productive process, it seems difficult to arrive at an analytical picture of such a process adequate for the economist's general task.

4. In economic literature we find two entirely distinct modes of representing a productive process; each one embodies one of the two traditional conceptions of such a process.

In one model we easily recognize the idea behind Boulding's and, more directly so, Stigler's definition (*II*. 1): process is represented by a set containing only *flow co-ordinates*. This mode of representing the process of an industry made its conspicuous entrance into the literature through the Leontief 'static system'.[1] But we owe to Koopmans its general formalization: a process P is represented by a vector

$$P(a_1, a_2, ..., a_n), \qquad (3)$$

where each a_i indicates 'the rate of flow per unit of time of each of the

---

[1] Wassily W. Leontief, *The Structure of American Economy, 1919–1929* (2nd edn., New York, 1951), pp. 12 f., 37.

n commodities involved' in the process, negative values corresponding to inputs, and positive to outputs.[1]

The second model, which corresponds to Carlson's definition — and to a certain extent to Boulding's as well — was explicitly formulated for the first time by John von Neumann.[2] In this model, a process P is represented by a two-row matrix

$$P \begin{bmatrix} B_1, B_2, ..., B_n \\ A_1, A_2, ..., A_n \end{bmatrix}, \qquad (4)$$

where the vectors (A) and (B) represent the quantities of commodities existing at the beginning and at the end of the period during which the process is completed.

It is puzzling though that the same concept, that of a process, should be represented by two models which are not wholly equivalent. Indeed, (3) contains only import-export type of data, *i.e.* data reported by an observer concerned with *what crosses the boundary of the process within an infinitesimal period of time*. On the other hand, the data in (4) are census type data : both (A) and (B) represent *what exists within the boundary of the process at an instant of time*. Or, to put it in a still more instructive way, (3) ignores what crosses the *time* sector of the process boundary and takes into account only what crosses the *physical* sector of the same boundary ; (4) proceeds the other way round. Clearly, each model — as I observed in an earlier paper [3] — tells only one-half of the whole story. For one thing, if $(A)=(B)$ we have no means of knowing whether (4) represents a uniformly going process or a totally frozen state.

5. In order to avoid irrelevant technical aspects and long notations, let us consider the *scientific report* of a very simple process, that of growing maize by a method requiring only a spade for preparing the ground, all other operations being done by hand. A complete list of the elements involved in the process is easily drawn :

1. Land space.
2. Solar radiant energy.
3. Chemicals from the air.
4. Soil with chemical nutrients.
5. Labourers.
6. Spades.
7. Maize grain.
8. Stubble, cobs, etc.

[1] Tjalling C. Koopmans, 'Analysis of Production as an Efficient Combination of Activities', *Activity Analysis of Production and Allocation*, ed. Tjalling C. Koopmans (New York, 1951), p. 36.
[2] John von Newmann, 'A Model of General Economic Equilibrium', *Review of Economic Studies*, xiii (1945), 2. (The original article, in German, appeared in 1935).
[3] 'Aggregate Linear Production Function and Its Applications to von Neumann's Economic Model', *Activity Analysis of Production and Allocation* (*op. cit.*), pp. 100 f.

Let us choose the time boundary so that the process *begins* with the first act of spading and *ends* with the shelling of the last kernel. Let this interval — or the time of production, as Marx called it [1] — be denoted by $(0, T)$, where T is measured in full days. Let us also use the terms 'input' and 'output' with their basic meaning, that is, to denote elements crossing the boundary *into* or *out of* the process, respectively. This is a most natural thing to do. It also has the merit of bringing two important issues into plain view.

First, most of the categories listed above appear both as inputs and as outputs. Second, the spade that goes into the process is *qualitatively different* from that coming out of it. And, as alluded to above, qualitative change does not fit into ananalytical representation. To circumvent this incompatibility at least in part, we may follow the common practice of identifying each spade by the total time it has been in actual use since it was 'new'. The report then may look as follows:

Category	*Inputs*	*Time*	*Outputs*	*Time*
Land-space	One acre	0	One acre	T
Solar energy		$E(t)$		
Air chemicals		$C(t)$		$C_1(t)$
Soil chemicals		$S(t)$		$S_1(t)$
Labourers	One rested man	$0, 1, T-\tau$	One tired man	$\tau, 1+\tau, T$
Spades	One new	0	One used	τ
Maize grain	One bag	1	Eleven bags	T
Stubble, etc.			X	T

The time co-ordinates represent either the instant a finite entity crosses the boundary or, in the case of a 'continuous' exchange, the total amount of an input (output) up to the instant t. For the sole purpose of avoiding notational complications the length of the working day is assumed constant and is represented by τ.

6. Complete though it is, such a report can hardly serve the purpose of the economic theorist: it includes many items that are not commodities in the usual sense of the term.

The case of solar energy and air chemicals is simple. Land-space not only provides a spatial base for the process at hand, but it also constitutes a *catching net*, as it were, for solar energy and chemicals from the atmosphere. Within a homogeneous climatic region, the average amount of solar energy and exchange of oxygen or carbon dioxide between the atmosphere and some given crop is proportional with the area of the land-space. And since even over a very long

[1] Karl Marx, *Capital* (Chicago, 1933), ii, 272 ff.

time period the two elements just mentioned vary little in intensity, they can be regarded as the 'original and indestructible' attributes of land-space. Hence, solar energy and air chemicals can be lumped together with land-space as 'Ricardian land'.

The problem of the tired labourer or of the used spade, however, is far more complex. By definition a used tool must be an output of some productive process. Yet in no sense can we say that it is the aim of economic production to produce used tools. Consequently, they have no cost of production. Moreover, with the exception of used automobiles and used dwellings, no used capital equipment has a regular market and, hence, a price in the same sense in which new equipment has. Used equipment, therefore, is not a commodity proper, and yet no report of a productive process can be complete without reference to it.

To be sure, in practice one may adopt one of the numerous conventions used in computing depreciation. But such a solution, besides involving some arbitrariness, is logically circuitous: it presupposes prices and interest rate to be already given. Economic theory has endeavoured to avoid the Gordian knot altogether by building its foundation only upon a process in which *all capital equipment is continuously maintained in its original efficiency*. The idea underlies Marx's diagram of simple reproduction as well as the neoclassical concept of static process. But not all its analytical snags have been completely elucidated.

7. To transform our illustrative process into a static one, we should include in it the activities by which the spade is kept sharp and its rivets, handle or blade are replaced when necessary. These activities imply additional labour power, additional inputs and, above all, additional tools. And since the efficiency of these tools must be, in turn, kept constant, we are drawn into a regress which may perhaps stop only after the whole production sector has been included in the process at hand. Moreover, if strictly interpreted, a static process must also maintain its labourers — *i.e.* the variable capital of Marx [1] or the personal (human) capital of Walras.[2] Thus, in the end, we have to include the consumption sector as well — a glaring illustration of the observations made earlier concerning the analytical difficulties of the concept of partial process. To avoid the regress to the whole economic process, we may assume without fear of being unrealistic

---

[1] Karl Marx, *Capital* (Chicago, 1932), i, 232 f.
[2] Léon Walras, *Elements of Pure Economics*, tr. W. Jaffé (London, 1954), pp. 214 ff.

that in every partial process part of the capital equipment and all human capital are maintained by outside processes, each one in turn to be analysed separately. After all, analysis cannot proceed without some heroic abstractions at one stage or another.

8. Once we have removed the qualitative difference from the picture, the analytical representation of a partial process no longer presents any difficulty. The elements participating in a process now fall into two simple and distinct categories. The first includes the elements that figure in the scientific report only as inputs or as outputs; the second comprises the elements that appear both as inputs and as outputs, more exactly, the elements that enter and come out of the process in an economically, if not also physically, identical form and in the same amount. It is appropriate to refer to the elements of the first category as *flow* elements and to the others as *fund* elements.[1]

The participation of any factor in the process is easily described by a function of t. In case of a flow item, the function, say, $A(t)$, shows *the total amount of the corresponding flow during the time interval* $(0, t)$, this interval being open or closed to the right according to whether the item is an input or an output. By convention, $A(t) \leq 0$ for an input flow and $A(t) \geq 0$ for an output flow.[2] For a fund element, the function, say, $B(t)$ shows the *amount of the corresponding fund effectively participating in the process at the instant t.*[3]

To determine the analytical co-ordinates of a static partial process according to the method just outlined is a very simple matter, except for one particular category. In our example this category is represented by maize. Maize figures both as input and as output, but the output is greater than the input. The case therefore presents a difficulty which does not exist for Ricardian land, labour power or equipment. The solution, however, is straightforward and, moreover, inescapable: *one bag of maize is a fund element, whereas ten bags of maize constitute a flow element.*

One point may now be stressed: the analytical co-ordinates of any

---

[1] The reason why here 'fund' is to be preferred to 'stock' should be obvious; but see p. 524, n. 1 below.

[2] Apart from being non-decreasing in absolute value, the function $A(t)$ may have any form whatever. However, for 'continuous' items, we may safely assume that $A(t)$ has a derivative function $a(t)$, which then represents the instantaneous flow rate at the instant t. In the case of a discontinuous input, for example — $A(t)$ is ordinarily such that $A(t) = 0$ for $t < t_0$ and $A(t) = A_0$ for $t \geq t_0$, A_0 being the global input at t_0.

[3] As a mere matter of consistent formulation, the function $B(t)$ for a fund B_0 going in at t_0 and coming out at t_1, should be defined so that $B(t) = 0$ for $t < t_0$ and $t \geq t_1$, and $B(t) = B_0$ for $t_0 < t \leq t_1$.

process are strictly determined by its boundary. Therefore, it would be a mistake to believe that commodities can be divided into two distinct classes, funds and flows : the same commodity may be a flow in one process and a fund in another.[1] Thus, the clover seed in a process the *purpose* of which is to produce clover seed is a fund, but in a process aimed at producing clover fodder, it is a flow.[2]

9. There are certain implications of the preceding argument on which one would perhaps welcome additional discussion.

9. (*a*) The first such implication is that gross output is never an analytical co-ordinate of a partial process. This does not imply, however, that gross output data are meaningless. After all, the farmer of our illustration brings home eleven bags of maize, an amount which corresponds to gross output. As a purely descriptive co-ordinate of non-analysed facts gross output raises no problem. But facts must ultimately be analysed and analytical categories, once established, must always be kept separate. And the point is that two analytically distinct categories, a fund and a flow, are added arithmetically to arrive at gross output.

One is likely though to ask why in a farming process the fertilizer input, for instance, should be treated differently from the seed input : both undergo a qualitative change inside the process. The issue is easily clarified if we refer to a more stringent illustration, namely, a process in which one hammer is used to hammer (shape) an additional hammer. According to all classifications ever proposed in economics, the first hammer is classified *as capital* (fixed or constant), *not as output*, whereas the reverse is true for the second hammer. The physical identity of the two hammers is no reason against recognizing their distinct roles in the productive process.[3] Nor is the lack of

---

[1] One may note in passing that this relativity bespeaks the impossibility of analysis to fare well and indefinitely with the dualist separation between Being and Becoming.

[2] The point that the analytical co-ordinates of a process depend strictly upon its boundary has some bearing upon such discussions of the consolidation problem in input-output *analysis* as offered by Leontief, *op. cit.*, pp. 14 ff., and R. Dorfman, P. A. Samuelson and R. M. Solow, *Linear Programming and Economic Analysis* (New York, 1958), Chapter 9 and 10.

[3] Nothing illustrates better the imbroglio produced by the failure to keep fund and flow co-ordinates separate than the celebrated problem of the transformation of values into prices in Marx's economic theory. It all comes from this : in his diagram of simple reproduction Marx fails to distinguish analytically between the hammering hammer, c_1, and the hammered hammer, $v_1 + s_1$. The error is made elementarily obvious by the cacophony used by Paul M. Sweezy, *The Theory of Capitalist Development* (New York, 1942), pp. 76 f., to explain that diagram : gross output (total value) is obtained by adding 'the constant *capital engaged* [in production with] the income of the capitalist [and] the *income* of the worker'. (My emphasis.)

physical continuous identity of the maize seed a reason against treating it as a fund in our analysis.

9. (*b*) The flow-complex that seems to dominate the current economic thought is the only reason why we should also discuss the point that the analytical representation of a partial process necessarily includes both flow and fund co-ordinates. The necessity, however, should be immediately obvious: the *fund* co-ordinates represent the material base of the process, the *flow* co-ordinates describe the change (transformation) achieved with the aid of this base. A framework based upon both Being and Becoming must necessarily include one analytical category for each, fund elements to represent *the unchangeable agents* and flow elements to represent *the object changed by the agents*.

There is no need nowadays to insist upon the analytical difference between *stock* and *flow*.[1] But a great deal of uncertainty still exists concerning the difference between *stock* and *fund* and, especially, between *flow* and *service*. To recall, it was Walras who incorporated in production analysis J. B. Say's old idea that services constitute the fundamental element in a productive process. But while he carefully distinguished between capital funds and their services, he failed to note the difference between these concepts on the one hand and those of stock and flow on the other. Somehow the lacuna has perpetuated itself through all subsequent writings. For a most convincing illustration in point we may choose such an authority as Pigou:

> In a stationary state factors of production are stocks, unchanging in amount, out of which emerges a continuing flow, also unchanging in amount, of real income.[2]

The physical picture becomes even more perplexing as we are further instructed, first, that the services of factors 'become embodied in other goods', and, second, that 'what is directly demanded is the flow of [such] services'.[3] Clearly, one root of the confusion is the improper use of the term 'flow'. Once this term is assigned to convey the idea of *some material substance crossing the process boundary*, it is utterly misleading to apply it also to services. The inevitable trap is that, because a flow can be stored up, we come to speak of services being 'embodied in the product'. The fact that, under certain

---

[1] On which we have been instructed first by S. Newcomb, *Principles of Political Economy* (New York, 1886), and then by Irving Fisher in a series of contributions from 'What Is Capital?' *Economic Journal*, vi (1896), 509–534, to *The Nature of Capital and Income* (New York, 1919), Chapter iv.

[2] Pigou, *op. cit.*, p. 19. Cf. Walras, *op. cit.*, Lesson 17, especially para. 169.

[3] Pigou, *op. cit.*, pp. 20, 117.

circumstances, the value of services is 'embodied' in the value of the product cannot alter the physical side of a process : only the elements that flow in the process can be *physically* embodied in the outflowing product.

But perhaps we fear that by overtly recognizing that an inflow can be transformed into an outflow only with the aid of services we would implicitly commit ourselves to the view that Ricardian land and capital equipment *must* earn an income. Be this as it may, the fact is that these fund co-ordinates are ordinarily deleted from the analytical picture of a static process.[1] The procedure, in addition to prejudicing the analysis of income distribution, constitutes an analytical distortion full of pitfalls — as we shall see later on.

The familiar leitmotiv that there would be double counting if both the maintenance flow of capital and its service are included in the analytical picture of a partial process is grounded on presumptions as regards value distribution. Marx alone endeavoured to justify the principle that even in the case of labour maintenance flows take care of services. And although Marx painstakingly avoided mentioning service by using instead such expressions as the work performed by a machine or the life-activity of the labourer,[2] there can be no doubt that he treated the labourer as a fund in our own sense.[3] Now Marx does start out by admitting that labour power is

> the aggregate of those mental and physical capabilities existing in a human being, which he exercises whenever he produces a use-value of any description.[4]

But in the end he reduces the participation of the labourer in a production process to 'a definite quantity of muscle, nerve, brain, etc. [which] is wasted' during work.[5] By this volte-face Marx set the pattern for ignoring the brute fact that any productive process requires the participation of the worker's entire fund of muscle, nerve, brain, etc. Why, nature is so constructed that no professor can discharge his duties by sending to class only that part of his nervous or muscular energy which ordinarily he spends during a lecture.

---

[1] The most illustrious example is the Leontief static system. Koopmans, *op. cit.*, pp. 39 f., does include land in his model ; but he does not distinguish between flow and service co-ordinates.
[2] Cf. *Capital*, i, 589, or his 'Wage Labour and Capital' reprinted in K. Marx and F. Engels, *Selected Works* (Moscow, 1958), i, 82.
[3] *Capital*, i, 189 f., especially p. 622. But the most incontrovertible statement is that of F. Engels, 'Marx's *Capital*' in *Selected Works, op. cit.*, i, 464 : 'Labour power exists in the form of the living worker who requires a definite amount of means of subsistence for his existence as well as for the maintenance of his family.'
[4] *Capital*, i, 186.   [5] *Ibid.*, p. 190.

And it is absurd to think that one can cross a river on the maintenance flow of a bridge: a non-existent bridge cannot possibly render a service; nor can it be maintained.

10. To sum up, the analytical picture of a process by which some product — whether a bushel of maize, a ton of coal or a piece of furniture — is obtained, consists of the following broad categories of co-ordinates:

FLOW CO-ORDINATES
Inputs: From nature $R(t)$
From other processes:
  (a) current inputs $I(t)$
  (b) maintenance $M(t)$
Outputs: Product $O(t)$
Waste $W(t)$

FUND CO-ORDINATES
Ricardian land $L(t)$
Capital $K(t)$
Labour $H(t)$

Over the production time interval $(0, T)$, each of these co-ordinates is defined as explained in section *III*. 8 above. Their most likely forms are shown in Fig. 1. To retain Marx's convenient terminology, the total time $T_H = \overline{Ot_1} + \overline{t_2 t_3} + \overline{t_4 t_5} + \overline{t_6 T}$ will be referred to as the working period. Needless to add, in most cases $T_H < T$, either because some phases of the process require no labour services whatsoever, or because the process is intentionally interrupted for some reason or other.

A catalogue of all possible partial processes pertaining to the same product is thus contained in the following general formula

$$O(t) = \mathscr{G}\,[R(t),\ I(t),\ W(t),\ M(t),\ L(t),\ K(t),\ H(t)], \tag{5}$$

which being a functional mapping is a far cry from the Dirichlet function (2). To be sure, among the arguments of $\mathscr{G}$ there normally exist additional relations corresponding to some technical restrictions. They permit the elimination of the flow co-ordinates from (5), in which case we have

$$O(t) = \mathscr{F}\,[L(t),\ K(t),\ H(t)]. \tag{5a}$$

This formula, we should note, shows how misleading is the familiar proposition that output is a function of the *services* of the

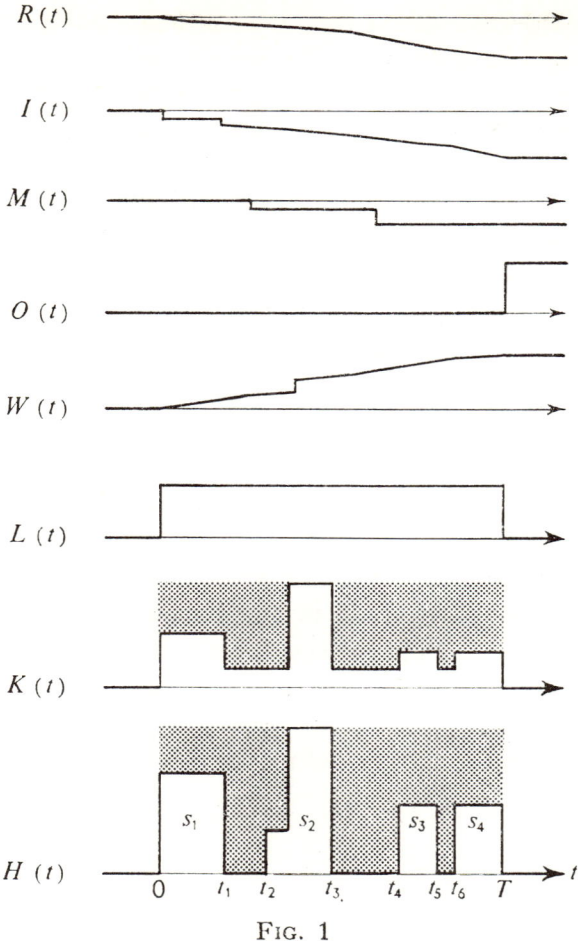

Fig. 1

production factors. Indeed, according to the consecrated meaning of the term, the services of labour, for example, are represented in Fig. 1 by the sum of the areas s_i. Clearly, this sum, though determined by $L(t)$, does not determine this co-ordinate.

IV. THE FACTORY PROCESS

1. A factory is such a familiar object nowadays — especially, to those who are the progeny of an industrialized society — that we are apt to lose sight of two essential facts : first, that the factory system of production represents one of the greatest *economic* innovations in history, and second, that the system is not (and, most likely, will never be) applicable to all production sectors.

Now, any factory represents a partial process in the broad sense of this term. However, a factory process is not a partial process in the sense adopted by the analysis of the preceding section. True, in both processes some material flows (inputs) are transformed into other material flows (outputs); otherwise they would not be processes at all. But whereas a partial process consists of a *temporal sequence* of operations, each requiring the services of different factors and for different periods of time, in a factory process all these operations are performed *simultaneously and in a special arrangement*. To the analysis of this difference, which has far-reaching economic consequences, we must now direct our attention.

One point, already hinted, needs to be stressed at the outset: *the partial process constitutes the basis of all production, whether in agriculture, mining or manufacturing*. As an analytical device, the diagram of Fig. 1 has therefore a universal applicability. Among other things, it makes perfectly clear a fact of special importance for the present argument: a farmer's plough, a miner's pick, or a carpenter's plane, do not *continuously* participate in the partial process in which each happens to be used. This is true for most fund factors, including labour, and for all processes. There is then an inherent reason why fund factors may have to remain idle for varying periods of time. Whether this idleness, which definitely represents the most relevant form of economic waste, can in fact be avoided depends upon several conditions.

One such condition is the possibility of using the same fund factor in another partial process. And whether this is economically feasible depends in turn upon demand, joint or simple. For a clear illustration of the role of simple demand — which alone is relevant for the topic of this section — let us refer to the production of tables. If only one table is all that is demanded during a time interval greater or equal to the corresponding period of production T, then obviously production must be carried out by partial processes *in series*, *i.e.* such that none overlaps in time. In this case, there is no way of avoiding a rather long idleness of the plane, of the carpenter himself, etc., unless these same factor funds can be employed in the production of other goods for which there is sufficient demand. But if during an interval equal to the same production period more than one table is demanded, then there are two alternatives: (1) a sufficient number of partial processes are started *in parallel*, *i.e.* at the same time, and (2) a sufficient number of processes are started *in line*, *i.e.* a different process is started at each time instant of a chosen sequence.

Clearly, the analytical picture of the processes run in parallel leads to a diagram identical to Fig. 1, except for the fact that all ordinates are amplified by the number of processes. Consequently, idleness (represented by the dotted areas) also is amplified. The diagrammatical representation of processes run in line, however, is entirely different and depends upon the way their sequence is arranged in time. We need not bother about the general case. Nor is it necessary here to go into the mathematical niceties by which the following two theorems are proved:

A. *Given the number of necessary partial processes, they can be arranged in line so that the idleness of some arbitrarily chosen fund factor is minimum.*

B. *If the number of necessarily partial processes is sufficiently large and end points of all periods during which each fund factor renders a service are commensurable with T, then there is a minimum number of processes that can be arranged in line so that no fund factor used in this arrangement is idle.*[1]

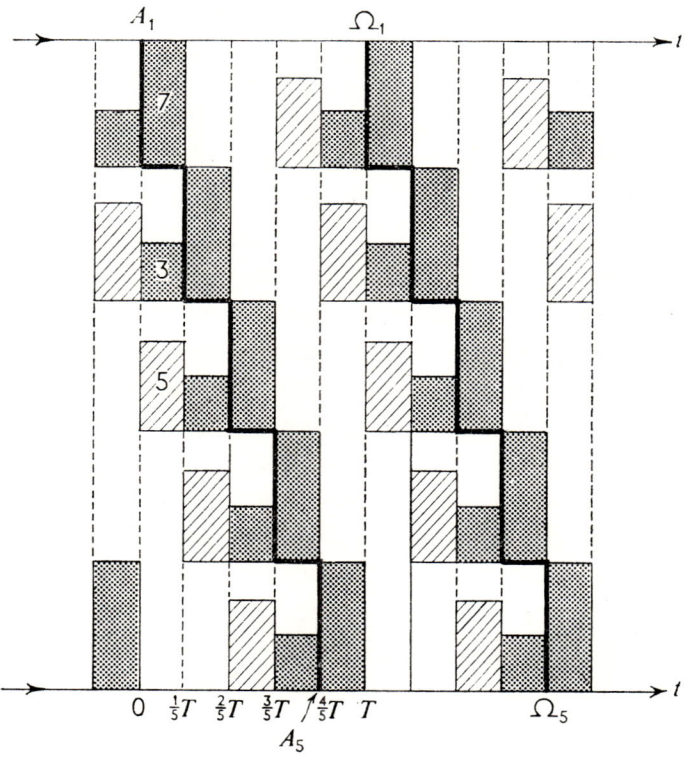

FIG. 2

[1] If m is the smallest integer such that the ratio between the service period of each fund factor and $t' = T/m$, is an integer, then the arrangement consists of m processes spaced in line at an interval equal to t'.

The diagram of Fig. 2 illustrates the last theorem for a partial process simple enough to avoid unnecessary complications. This process involves only two funds, say, two types of labour. The services of one of these factors are marked by shaded areas. Within the interval $(0, T)$, there are five processes starting in line at 0, $T/5$, $2T/5$, $3T/5$, $4T/5$, respectively. They are delimited by the step lines $A_1 A_5$ and $\Omega_1 \Omega_5$. Ten labourers of one category and five of the second category are periodically shifted from one process to the next. Needless to add, it is this sort of arrangement that characterizes the factory system.[1] The view expressed at the beginning of this section concerning the emergence of the factory system is thus vindicated.

2. There is some advantage in considering first a factory process that goes on uninterruptedly, as is actually the case in steel or glass production. For the analytical description of such a process, it is wholly irrelevant that the *same service* is rendered during one shift by Mr. A and during another by Mr. B. The corresponding labour fund, like any other physical fund, never leaves the process. But this is no justification for the flow-complex which would have us ignore the existence of factor funds in describing a steady-going process. After all, a factory is not a phantom without material basis any more than a bridge or a working human being is.

No analytical description of a factory process, therefore, could be complete without the fund co-ordinates. But in the case of an uninterrupted process, we can no longer determine these co-ordinates by following the same procedure as for a partial process. One could determine, it is true, the labour funds from the line of employees passing through the factory gate at the beginning of a shift. But this simple solution does not work for other funds, for they never pass through the gate. The only solution is to take a *census* of what exists inside the process. And since the process is steady-going, it does not matter at what moment this census is taken.

Such a census will reveal the existence of several broad categories of funds. There are first the familiar ones: Ricardian land, L, capital equipment of all sorts, K, and labour of various kinds, H. But two new categories now show up. The first covers the stores (inventories) of commodities which appear in some flow co-ordinates. Let it be denoted by S. The second new category consists of 'goods in process', though 'goods' is here a misnomer: melted glass or half-tanned hides, for example, can hardly fit the term. It is none the

[1] Incidentally, the aggregate of the minimum number of processes arranged in line so as to occupy continuously all factor funds may provide a basis for defining a unit-plant.

less true that *at any time* there exists inside the boundary of a factory a *process fund* consisting of the successive phases through which the input flow passes to become ultimately the product flow. From the analytical viewpoint it is highly instructive to note that this fund is a picture of the qualitative change achieved by the factory process, just as a movie film contains a whole drama at once. The co-ordinate corresponding to this Becoming frozen into a Being, is therefore a complex entity. Let us denote it by $\mathscr{C}$.[1]

The flow co-ordinates are the same as in the case of a partial process. However, a few additional observations seem in order. First, the factory process offers further support to the view that gross output is not a proper analytical co-ordinate of a process: the number of hammers used in producing hammers will never appear as a flow co-ordinate of the hammer factory. *Instead, there will be a fund co-ordinate of hammers.*[2] Second, all input flows can be lumped together as a general maintenance flow: the role of the current input and of part of the direct input from nature is to maintain $\mathscr{C}$. There is nevertheless some advantage in abiding by the old classification.

To simplify the analytical picture, let us make the customary assumption of absolute continuity. In this case, every flow factor can be represented by an instantaneous flow rate, and the analytical representation of a factory process assumes the following general pattern:

FLOW CO-ORDINATES
    Inputs: From nature r
                From other processes:
                    (a) current input i
                    (b) maintenance m
    Outputs: Product q
                  Waste w

FUND (SERVICE) CO-ORDINATES
        Ricardian land L
        Capital proper K
        Labour H
        Process fund $\mathscr{C}$
        Stores S

---

[1] For practical purposes, one can divide $\mathscr{C}$ into a finite number of sections such that within each the qualitative variations could be ignored in the first approximation. Then $\mathscr{C}$ can be approximately represented by a vector (C). However, one must not overlook the fact that C_1 is not necessarily a commodity.

[2] Surprising though it may seem, the only way of finding out how many hammers are used during a given interval is from the waste flows. However, 'worn out hammers' may not necessarily appear as an item of waste: the heads might be remelted.

3. A few points about this table need special emphasis. First, *every co-ordinate is a point-coordinate measuring some intensity*. The flow co-ordinates measure the intensity of inputs and outputs; their dimension is (substance)/(time). The fund co-ordinates too measure intensities, of services; their dimension is (service)/(time). The fact that this last measure coincides with the measure of the corresponding fund — which has the dimension (substance) — should not obscure the dimensional difference between the two concepts. The upshot of these remarks is that there is absolutely no sense in which one can speak of price in connection with any co-ordinate in the table. A water bill is for the water consumed, not for the gauge of the connecting pipe.

Second, let us observe that we would arrive at exactly the same co-ordinates even if the factory would not work around the clock.[1] The table thus does not describe *what the factory actually does, but only what it is capable of doing while working*. The information it provides is analogous to knowing only that Mr. C is a 'civil engineer'; it describes the potentiality of the factory, whether or not it is working. And like Mr. C, most factories can interrupt and resume work at will. Moreover, because $\wp$ is continuously maintained by the very act of production, the product starts to flow the instant input flows move in. A factory is like a music box which starts playing as soon as it is opened regardless of when this happens. In a factory process, therefore, *there is no time-lag between input and output flows*. And this, we should note, is not an analytical simplification of actuality.

4. To know what a factory actually does, one needs an additional co-ordinate, not included in our table. This co-ordinate is the time length, $\delta \leqslant 1$, the factory works each day. The quantities of material inputs and services consumed and of the product produced *during the period δ*, are then immediately obtained if every co-ordinate of our table is multiplied by δ. However, some fund factors by necessity provide services around the clock even if $\delta < 1$. In the new picture, the corresponding co-ordinates must therefore be multiplied by *one unit of time*. Only these new co-ordinates have a price, whether positive, null, or negative.[2]

---

[1] The matter of user's cost may safely be ignored for our immediate purpose.

[2] The point is of particular importance in cost analysis. In principle, only the daily *variable* cost is proportional to δ, the prices of land and capital services being customarily set for a full day. In actuality, however, there is also a cost of changing shifts and possibly a differential wage rate between shifts. Hence, total cost $TC = A_j + B_j \delta$, j indicating the number of shifts used. Pure competition then determines δ by the condition $TC = (pq + p_w w)\delta$. The reader may find it instructive

A catalogue of all factory processes must nevertheless be based upon the original table. And since all analytical co-ordinates are now *real numbers*, it can be represented by a Dirichlet function:

$$q = G(r, i, m, w\,;\, L, K, H, \mathcal{C}, S). \tag{6}$$

It is normal to expect that once the individual items covered by L, K, H, are determined quantitatively and qualitatively, the other elements are also determined through the usual technical constraints represented by limitationality relations. Hence, (6) can be reduced to

$$q = F(L, K, H). \tag{6a}$$

But this reduction should not induce us to ignore — as is frequently done — the other elements of the process not represented in (6a).

5. There is one consequence of the preceding analysis which I have repeatedly found hard to bring home, elementary though the point is. If $Q_0 = \delta q$ is the daily output, from (6a) it immediately follows that

$$Q_0 = \delta F(L, K, H). \tag{7}$$

But this is a far cry from the neo-classical definition of the production function which leads to

$$Q_0 = \emptyset\,(L^\circ, K^\circ, H^\circ), \tag{8}$$

where L°, K°, H° stand for the amounts of services.[1] There seem to be only two fair interpretations of this formula. The first is to read $L^\circ = \delta L$, $K^\circ = \delta K$, $H^\circ = \delta H$. As already explained (*II.* 2), this implies that F is homogeneous of the first degree, a condition met neither by a production unit (factory), nor by every industrial aggregate. The second interpretation is to read $L^\circ = L \times 1$, $K^\circ = K \times 1$, $H^\circ = H \times \delta$. But the basic analytical sin of (8) is still untouched. Let $H^1 = H' \times \delta'$; according to (8), $\emptyset\,(L^\circ, K^\circ, H^1) = \emptyset\,(L^\circ, K^\circ, H^\circ)$, which obviously cannot be true in general. The work of one worker for six days is not necessarily equivalent to the work of six workers for one day — as Wicksell once observed — even though the wage bill is the same in both cases. In many mathematical models, the confusion is further increased by treating K° as a stock

---

to compare these points with Marx's attack on Senior's 'last hour', *Capital*, i, 248 ff.

[1] I take it that the neo-classical definition is that of Boulding and of any other author cited in note 2, p. 499 who cared to describe the variables entering into the function. Actually, it seems difficult to find an author not endorsing the same definition. Stigler, Koopmans and perhaps Pigou are among the very few who view the matter in the same sense as (6a).

capable of being increased by the amount of new capital accumulation ΔK.

Of course, the *numerical*, though not the *dimensional*, discrepancy between the generally used formula $Q_0 = F(L, K, H)$ and (7) disappears if δ is taken as time unit. Q_0 would still represent 'the daily output'. But δ must be constant in order to serve as unit. Perhaps neo-classical economists have proceeded on the assumption that δ is an institutionally determined constant; but if so, they failed to state it explicitly. Be this as it may, treating an important variable as a constant has seriously impaired the value of neo-classical analysis of production — as will be seen presently (V. 6).

One has to turn to Marxian economics to find a more adequate approach. In the light of what precedes, Marx was right in allocating a most prominent place to the working day in his economic analysis. And in a certain sense, (7) lends some support to Marx's dearest tenet that labour time, though it has no value, is a measure of value.[1] Unfortunately, for reasons easy to guess, Marx's analytical findings never travelled beyond Marxian economics. All the greater then is the merit of W. H. Nicholls for having broken the neo-classical tradition by introducing the length of the working day as an independent variable in the production function.[2]

V. BALANCE IN ECONOMIC DEVELOPMENT

1. From Marx, however, neo-classical economics has accepted, it seems, two dogmas pertaining to economic development: first that over-population is a bogey, and second, that the same economic laws govern agricultural and industrial production.[3] Yet Engels in 1884,

---

[1] *Capital*, i, 45, 588.
[2] W. H. Nicholls, *Labor Productivity Functions in Meat Packing* (Chicago, 1948). In essence, his analysis is based on the formula $Q_0 = \psi(\tau, L, K, H)$, which for L and K constant leads to $Q_0 = f(\tau, H)$, the notations being ours. None of the simple formulae tried out by Nicholls (pp. 98 ff.) comes near the form $Q_0 = \tau g(H)$, as should follow from (7). There are many reasons why one should not expect this form to fit actual data: the worker's fatigue (cf. Nicholls, p. 25, n. 2), the opening- and closing-effect, etc. Yet none of these reasons work against the separation of the variables as in $Q_0 = h(\tau)g(H)$, with h and g having an S-shape. This form, however, involves some computational difficulties because the simplest S-shape curve is a third degree parabola.
[3] For an analysis of over-population see my essay 'Economic Theory and Agrarian Economics', Chapter 6 in this volume. To straighten out some misinterpretations of that analysis I wish to emphasize that, as ought to be clear, my definition of over-population is a short-run concept. Consequently, even though it might be hard to find in the *changing* actuality a situation where labour marginal productivity is *mathematically* equal to zero, this does not affect the validity of my analysis or

while quoting approvingly L. H. Morgan's statement that 'Mankind are the only beings who may be said to have gained an absolute control over the production of food', finds it necessary to insert 'almost' to tone down 'absolute'.[1] Were Engels still living, the disturbing evidence accumulated by recent history would perhaps cause him to have many second thoughts on this controversial issue. As it happens, the results obtained in the preceding sections of this paper can throw a great deal of light on it.

2. Let us first draw an analytical picture of the whole economic process, including both production and consumption, at any instant of time. Since by consolidation all flows between production and consumption units must cancel out, the global picture includes only two flow co-ordinates: an input flow from nature and an output flow of waste. As far as the material elements are concerned, the economic process simply transforms natural resources into economically valueless waste.[2]

As it has been repeatedly recognized, man can neither create nor destroy matter and energy. But this is only one-half of the story — the half told by mechanics, that cherished model of most social scientists. However, natural resources do not consist of mere matter and mere energy, but of *matter arranged in some definite structures* and of *free energy*. Mere matter, such as the gold scattered over the bottom of the oceans, has no value for us: we need gold ore where gold is arranged so that we can extract it in useful time. Nor has the immense heat energy contained in the ocean waters any value for us: a sailing ship needs fuel, *i.e.* energy in the free state. All the carbon, oxygen, hydrogen, etc., in the world could not support a human life if they were not arranged in a molecule of sugar, starch or protein.

In the story of nature told by thermodynamics, the laws of which are as inexorable as those of mechanics, the energy-matter constituting natural resources is *qualitatively* different from that forming waste. The energy-matter of natural resources is arranged in some orderly patterns, or as the physicists say, it has a low entropy. In

---

policy recommendations for those numerous economies where the *actual* labour productivity is negligible, or only smaller than the prevailing minimum of subsistence. The whole edifice of economic science would collapse if we confuse analytical concepts with evolutionary facts.

[1] F. Engels, *The Origin of the Family, Private Property and the State* (4th edn., New York, 1942), p. 19.

[2] Since this fact might come as a shock to some, I should add that, on the contrary, it reveals a much neglected truth: the actual 'product' of the economic process is not a material flow, but a psychological (or vital) flux: the mere enjoyment of life, as Irving Fisher tried hard to teach us. Real income and labour or, alternatively, real income and leisure, are only the material 'measure' of that flux.

waste we find only disorder, that is, *high entropy*. That is not all. The Second Law of thermodynamics tells us also that the whole universe is subject to a continuous qualitative degradation : entropy increases and the increase is irrevocable. Consequently, *natural resources can pass through the economic process only once : waste is irrevocably waste.* Man cannot defeat this law, any more than he can stop the law of gravitation from working. The economic process, like biological life itself, is *unidirectional*. Money alone moves in a circular flow, because no one throws it away though it is only an artificial token.

3. Natural resources fall into two distinct categories. Some exist as *stocks* in the crust of our planet.[1] These can be used with a speed and rhythm which, in principle, depends only on man's choice. Conceivably, we could exhaust all the known stocks of oil within one year if we wanted to do so and made our plans accordingly. The direct point is that any mine can be operated as a factory process around the clock. *Mining, therefore, does not compel us to keep idle any of the factor funds involved in the process.* The same is true for all manufacturing industries. We should further observe that it is precisely this freedom which man has in using almost at will any mineral deposit — once he has discovered how to use it advantageously — that is responsible for the spectacular progress of technology.

If one would like to brag about man's feats, one should rather choose man's control over inert matter, not over food. The taproot of all food is in photosynthesis ; and photosynthesis requires first of all solar energy. But in contrast with most other stocks of energy, that of the sun is — and may for ever remain — beyond man's control. Solar energy comes to each place on the earth at various epochs of the year in a definite flow rate.

The consequence is that, with a very few exceptions, partial processes in agriculture cannot be arranged *in line*, as in a factory process. They can be arranged only *in parallel*, all beginning at the appropriate phase of the climatic annual cycle in each place. And to pinpoint further the only reason for this necessity : in the island of Bali, for instance, where the climate varies little throughout the year, nothing stands in the way of growing rice *in line*. The same number of buffaloes, ploughs, sickles, flails and villagers could move over the entire field of the village, ploughing, seeding, planting, weeding,

[1] Since entropy continuously increases, no natural resource, not even the solar free energy, can be a fund. At the same time, it is now obvious that only energy-matter represents a fund in the strict sense of the term.

harvesting and threshing without interruption. Moreover, the people could then eat each day the rice sown that very day, as it were (*IV*. 3). They would no longer need to wait the long days between ploughing and threshing and, especially, to bear the burden of agricultural loans. Unfortunately, man's condition is such that there are very few spots where this Bali formula can work and, hence, where 'the open-air factory' can become a reality. Elsewhere every kind of agricultural production *inevitably* imposes some idleness on *both* capital and labour over the production period and complete idleness on every fund factor during the rest of the year (as shown in Fig. 1).[1] Although this alone would suffice as proof that industry and agriculture are governed by different laws, the difference is sharper still.

4. An expert on thermodynamics could, no doubt, compute the maximum amount of photosynthesis a certain crop could achieve during one average year on an acre of land in a given location. The size of our catching net — the earth's surface — thus sets a limit to the amount of total photosynthesis each year. To think that one could get a greater yield than this *theoretical* maximum would be absurd. Equally absurd would be to ignore that solar energy is not the only necessary input.

The input that illustrates most conspicuously the fateful work of the Entropy Law is the soil nutrients. The long history of peasant societies may be summarized in a few words: a continuous struggle with the effects of the Entropy Law. Under its pressure, the village economy passed from swidden agriculture, to shifting cultivation, and finally to crop rotation. It also invented, in succession, the caschrom, the scratch-plough, and the ordinary plough; it also discovered rotation and manuring. These are momentous achievements which contrast with the economic inertia of the contemporary peasant economies. In the contemporary era, however, the peasant economy has come to a crisis that the village alone can no longer solve. The Entropy Law makes the crisis inevitable: the population explosion has only speeded its coming. But leaving aside the population explosion — which is a biological rather than an economic phenomenon — we can easily see that the crisis stems from the scarcity of land — about which we can do rather little — and from the qualitative deterioration of agricultural land through millenary use with manuring only. The tables have turned: it is the turn of the town

---

[1] Clearly, this does not cover the total idleness of the labour fund in excess of the maximum service intensity. But this idleness, if present, is not an immediate consequence of material laws.

now to support the economy of the countryside: the 'manure' must now come from the industrial sector in the form of fertilizers.

5. Undoubtedly, there are many improvements that a sound economic policy can and ought to pursue inside the agricultural sector. We must however decide upon them with the analytical picture of Fig. 1 in mind. Let us ignore the problem of employing villagers outside agriculture; important though this employment is from the viewpoint of the peasants' income, *it clearly cannot solve the food crisis*. Agronomists, I am sure, can offer many valuable suggestions how to lengthen the service period of land by staggering different crops. A co-operative use of capital equipment could in this case reduce the amount of capital equipment per acre: for, if the whole village cultivates only one crop, everyone needs one plough with its team during the few days of ploughing.

'The horse eats people' is an old peasant saying, which attests the economic awareness of the villagers. And, indeed, the maintenance of draft animals greatly cuts into the net output of a peasant's enterprise. Surprising though it may seem, it is in the agricultural over-populated countries — such as Pakistan, or India, or Indonesia, to name only some — that a replacement of draft animals by mechanical power is most urgent. From this, however, we need not jump to the conclusion that in such — ordinarily low income — countries, the buffaloes must be replaced by heavy tractors like those of the U.S.A. farms. A heavy tractor, in comparison with a small (garden) tractor, certainly saves *labour fund* and *labour time* as well; but it does not appreciably increase the yield if *ceteris paribus*. If labour is plentiful, too much mechanization is anti-economical; and if income is low, as is the rule in over-populated countries, heavy machinery is a luxury comparable to that of a splendid villa on the Riviera used for a couple of weeks each year.[1]

We can then say, for the second time, that it is the turn of the town to raise 'buffaloes' and grow 'cattle fodder'.

6. The problem then is how to transfer the modern 'manure', 'buffaloes', and 'fodder' from the town to the country-side. Clearly, in any economic system whatsoever, they must be paid for: the industrial worker must have his income. But the country-side being as desperately poor as we know it is in all over-populated, underdeveloped countries, we seem to be confronted with the old vicious

---

[1] It is interesting that Gerald K. Boon, *Economic Choice of Human and Physical Factors in Production* (Amsterdam, 1964), pp. 162, 259, arrives at an equivalent conclusion by a careful analysis of cost functions. Boon's study, I may add, is highly valuable in many other respects.

circle (and a vicious thought, as well) that the poor cannot but stay poor. Many may also argue that allocating industrial resources to the production of things needed by agriculture is a fundamentally wrong move: because it slows down industrialization and because salvation can come only from industrialization. But the double dilemma is only apparent, and it is caused by the fact that neo-classical analysis of production has ignored an important variable.

One of the main secrets by which the Western industrialized societies have achieved their spectacular development is — as correctly assessed by Marx and confirmed by (7) — a long working day. This secret solves our dilemma also. For an illustration that would be familiar to all planners, let us take the current industrial bill of goods of a Leontief input-output model. Clearly, if no industry works around the clock, then this bill can be increased by, say, ten per cent immediately with the existent capacity: all we have to do is to lengthen the working day. Industrial development can then go on at the same speed, the real income of the industrial worker can remain untouched, and, at the same time, there would be a surplus available for the industrial needs of agriculture. And to make sure that we have fully grasped the difference between an agricultural and an industrial process, we should note that a lengthening of the working day in agriculture cannot possibly increase output, not even in a Bali formula. It can only release labour power for other possible uses.[1]

I am fully aware of the practical difficulties of all sorts involved in implementing an economic policy based upon these conclusions. But I wish to submit that, in view of our proclaimed economic aims, the

[1] A disregard for dimensions is responsible for the fact that the numerous contributions on linear processes have missed the above conclusions. Thus Leontief, *op. cit.*, p. 173, defines X_i, $1 \leqslant i < n$, as 'total net outputs of all various branches of the national economy [during a particular year]'. Clearly, every X_i is a flow, equivalent to our $Q_0 = \delta Q$. But X_n is there defined as 'total employment' measured in number of persons (pp. 173, 179), and in another place (pp. 42, 160) as 'the output of services (by the household industry)'. That is, in one case X_n is the intensity rate of labour services, equivalent to our H, in the other it is the service δH. Then, x_{ik} is also defined as an input flow, equivalent to our δi or δm. The important point, however, is that the coefficients $a_{ik} = x_{ik}/X_i$ have no longer any relation to the time factor, $a_{ik} \simeq i/Q$. The linearity assumption, correctly expressed is that a_{ik} is constant for *all efficient processes* when these are described as in the table of section IV. 2. Indeed, a_{ik} is constant for *any non-agricultural* process if x_{ik}, X_i vary only because δ varies.

Not to depart from Leontief's analytical assumption, let $(1, -a_{21}; H_1)$, $(-a_{12}, 1; H_2)$, represent the analytical co-ordinates of the industrial and agricultural processes, respectively. Let B_1, B_2, be the desired daily (or annual) net output. The standard system then becomes

$$\delta X_1 - a_{12} X_2 = B_1, \quad -\delta a_{21} X_1 + X_2 = B_2,$$

where X_1, X_2, represent the physical scales of the two processes. There are three unknowns, with $\delta \leqslant 1$.

legal rule of a forty- or even forty-eight-hour week constitutes an anachronism for the under-developed countries that possess some industrial potential of a non-parasitary nature. By this remark, a natural conclusion of the analysis presented in this paper, I may have touched a sore spot : the conflict of interest between the town and the country-side. About this conflict Marx said, in passing though, that it constitutes the pivot of all history.[1] Yet, this conflict seems even more important than that upon which he built his doctrine, for no other reason than the fact that its roots lie in an evolutionary law of nature, the Entropy Law.[2]

---

[1] *Capital*, i, 387.
[2] For additional details on the economic importance of the Entropy Law, see the author's volume *Analytical Economics: Issues and Problems* (Cambridge, Mass., 1965), Part I, Chapter v, sect. 1.

CHAPTER 6

(1960)

Economic Theory and Agrarian Economics

According to some recent studies, more than 1.3 billion people still live in a self-subsistence economy, that is, as peasants. Most of these also live on the verge of starvation. Asia and Africa, which together represent more than 60 per cent of the world's population, produce only a little more than 30 per cent of the world's agricultural output. Conservative estimates show that if basic nutritional needs for the entire population of the world are to be met, it is necessary that the food production be increased by at least 30 per cent.[1] Neither the overwhelming numerical importance of peasant economy nor the scarcity of food is a new economic development peculiar to our own time.

In spite of all this, agrarian economics — by which I mean the economics of an overpopulated agricultural economy and not merely agricultural economics — has had a very unfortunate history. Noncapitalist economies simply presented no interest for Classical economists. Marxists, on the other side, tackled the problem with their characteristic impetuosity, but proceeded from preconceived ideas about the laws of a peasant economy. A less-known school of thought — Agrarianism — aimed at studying a peasant economy and only this. An overt scorn for quantitative theoretical analysis, however, prevented the Agrarians from constructing a proper theory of their particular object of study, and consequently from making themselves understood outside their own circle. There remain the Standard economists (as a recent practice calls the members of the modern economic school for which neither Neoclassical nor General Equilibrium suffices as a single label). Of late, as economic development has become tied up with precarious international politics, Standard economists have been almost compelled to come to grips with the problem of underdevel-

---

1. The above data are found in W. S. and E. S. Woytinsky, *World Population and Production* (New York, 1953), pp. 307, 435, and *passim*.

oped economies, and hence with noncapitalist economics. But in their approach they have generally committed the same type of error as Marxists.

Thus, the agrarian economy has to this day remained a reality without a theory. And the topical interest of a sound economic policy in countries with a peasant overpopulation calls for such a theory as at no other time in history. But one cannot aspire to present a theory of a reality as complex as the peasant economy within the space of an article. My far more modest aim is to point out the basic features that differentiate an overpopulated agricultural economy from an advanced economy. I have endeavored to present the argument in terms of the familiar analytical tools of Standard theory or others akin to these. The brief historical critique which prefaces the theoretical analysis is intended to place the latter in a better perspective, particularly as concerns policy implications.

I. THEORY, REALITY, AND POLICY

1. *Theory and reality.* Theory is in the first and last place a logical file of our factual knowledge pertaining to a certain phenomenological domain.[2] Only mathematics is concerned with the properties of "any object whatever," for which reason since Aristotle's time it has been generally placed in a special category by itself. To each theory, therefore, there must correspond a specific domain of reality. In any science, the problem of precisely circumscribing this domain faces well-known difficulties. Where physics ends and chemistry begins, and where economics ends and ethics begins, are certainly thorny questions, although not equally so. Here, however, I want to discuss a quite pedestrian query pertaining to the problem of the proper domain of a theory. And this query is: Can an economic theory which successfully describes the capitalistic system, for instance, be used to analyze another economic system, say feudalism, successfully?

Let us observe that a similar question hardly ever comes up in the physical sciences, for no evidence exists to make physicists believe that matter behaves differently today from yesterday. In contrast, we find that human societies vary with both time and locality. To be sure, one school of thought still argues that these variations are only different instances of a unique archetype and that consequently all social phenomena can be encompassed by a single theory. This is not the place to show in detail where the weakness of the various attempts in this direction lies. Suffice it to mention here that when the theories constructed by these attempts

---

2. That is not to deny that theory may serve other purposes, but these are by-products of its essential nature.

do not fail in other respects, they are nothing but a collection of generalities of no operational value whatever. As Kautsky once judiciously remarked, "Marx designed to investigate in his 'Capital' the capitalistic mode of production [and not] the forms of production which are common to all people, as such an investigation could, for the most part, only result in commonplaces." [3] For an economic theory to be operational at all, i.e. to be capable of serving as a guide for policy, it must concern itself with a specific type of economy, not with several types at the same time.

What particular reality is described by a given theory can be ascertained only from that theory's axiomatic foundation. Thus, Standard theory describes the economic process of a society in which the individual behaves *strictly* hedonistically, where the entrepreneur seeks to maximize his cash-profit, and where any commodity can be exchanged on the market at uniform prices and none exchanged otherwise. On the other hand Marxist theory refers to an economy characterized by class monopoly of the means of production, money-making entrepreneurs, markets with uniform prices for all commodities, and complete independence of economic from demographic factors.[4] Taken as abstractions of varying degree, both these axiomatic bases undoubtedly represent the most characteristic traits of the capitalist system.[5] Moreover, far from being absolutely contradictory, they are complementary, in the sense of Bohr's Principle of Complementarity.[6] This is precisely why one may speak of Marx as "the flower of Classical economics." [7]

A far more important observation is that the theoretical foundations of both Standard and Marxist theories consist of cultural or, if you wish, institutional traits. Actually, the same must be true of any economic theory. For what characterizes an economic system is its institutions, not the technology it uses. Were this not so, we would have no basis for distinguishing between Communism and Capitalism, while, on the other hand, we should regard capitalism of today and capitalism of, say, fifty years ago as essentially different systems.

As soon as we realize that for economic theory an economic system is

3. Karl Kautsky, *The Economic Doctrines of Karl Marx* (New York, 1936), p. 1.
4. I refer to the fact that the assumption of a permanent reserve army simply means that at the subsistence wage rate the supply of labor is "unlimited" both in the short and in the long run, whereas Classical economics held that this was true only in the long run. *Infra*, note 51.
5. I have left the surplus value proposition out of the Marxist axioms because this proposition — as I shall argue later — belongs to feudalism, not to capitalism.
6. This principle by which Bohr overcame the impasse created by the modern discoveries in physics states that reality "cannot be comprehended in a single picture" and that "only the totality of the phenomena exhausts the possible information about objects." Niels Bohr, *Atomic Physics and Human Knowledge* (New York, 1938), pp. 40 and *passim*.
7. Terence McCarthy in the preface to K. Marx, *A History of Economic Theories* (New York, 1952), p. xi.

characterized exclusively by institutional traits, it becomes obvious that neither Marxist nor Standard theory is valid as a whole for the analysis of a noncapitalistic economy, i.e. of the economy of a society in which part or all of the capitalist institutions are absent. A proposition of either theory may eventually be valid for a noncapitalistic economy, but its validity must be established *de novo* in each case, either by factual evidence or by logical derivation from the corresponding axiomatic foundation. Even the analytical concepts developed by these theories cannot be used indiscriminately in the description of other economies. Among the few that are of general applicability there is the concept of a production function together with all its derived notions. But this is due to the purely physical nature of that concept. Most economic concepts, on the contrary, are hard to transplant. "Social class" seems the only exception, obviously because it is inseparable from "society" itself (save the society of Robinson Crusoe and probably that of the dawn of the human species). This is not to say that Marxist and Standard theories do not provide us with useful patterns for asking the right kind of questions and for seeking the relevant constituents of any economic reality. They are, after all, the only elaborate economic theories ever developed.

All this may seem exceedingly elementary. Yet this is not what Standard and (especially) Marxist theorists have generally done when confronted with the problem of formulating policies for the agrarian overpopulated countries. And, as the saying goes, "economics is what economists do."

2. *A reality without theory.* As has often been remarked, economists of all epochs have been compelled by the social environment to be far more opportunistic than their colleagues in other scientific fields, with the result that their attention has been concentrated upon the economic problems of their own time.[8] And as the transition of economic science from the purely descriptive (i.e. taxonomic) to the theoretical stage coincided with the period during which in Western Europe feudalism was rapidly yielding to capitalism, it was only natural that capitalism should become the object of study of the first theoretical economists. That may explain only why most Western economists have been interested in developing the theory of the capitalist system, but not why none attempted a theory of a noncapitalist economy. The only explanation of this omission is the insuperable difficulty in getting at the cultural roots of a society other than that to which one actually belongs. And, as we have hinted, an in-

8. The point finds an eloquent illustration in the vogue that the problem of economic development has recently acquired among Western economists: we have reached the point where the development of underdeveloped nations is as much an economic problem of the West as of these other nations.

tuitive knowledge of the basic cultural traits of a community is indispensable for laying out the basis of its economic theory.

By its very nature, a peasant village is the milieu least fit for modern scientific activity. The modern scientist had therefore to make the town his headquarters. But, from there, he could not possibly observe the life of a peasant community. London, for instance, offers indeed "a favorable view . . . for the observation of bourgeois society" — a circumstance immensely appreciated by Marx[9] — but not even a pinhole through which to look at a peasant economy. Even if, unlike Marx, an economist was born in a village, he had to come to town for his education. He thus became a true townee himself, in the process losing most, if not all, *Verstehen* of the peasant society. It was natural, therefore, that to Marx as well as to other Western economists (to those coming from a peasantless country, especially) the peasant should seem "a mysterious, strange, often even disquieting creature." [10] Yet none showed Marx's unlimited contempt for the peasantry. For him, the peasantry was just a bag of potatoes, not a social class. In the *Communist Manifesto* he denounced "the idiocy of rural life" to the four corners of the world. But these Marxist hyperboles apart, there is, as we shall presently see, a spotless rationale behind Marx's attitude towards the peasant.

The difference between the philosophy of the industrial town and of the agricultural countryside has often attracted the attention of sociologists and poets alike.[11] But few have realized that this difference is not like going to another church, and that it involves every concrete act concerning production and distribution as well as social justice. Undoubtedly the basis of this difference is the fact that the living Nature imposes a different type of restriction upon *homo agricola* than the inert matter upon *homo faber*.

To begin with, no parallelism exists between the law of the scale of production in agriculture and that in industry. One may grow wheat in a pot or raise chickens in a tiny backyard, but no hobbyist can build an automobile with only the tools of his workshop. Why then should the optimum scale for agriculture be that of a giant open-air factory? In the second place, the role of the time factor is entirely different in the two activities. By mechanical devices we can shorten the time for weaving an ell of cloth, but we have as yet been unable to shorten the gestation period in animal husbandry or (to any significant degree) the period for maturity

---

9. K. Marx, *A Contribution to the Critique of Political Economy* (Chicago, 1904), preface, p. 14.

10. Karl Kautsky, *La question agraire* (Paris, 1900), p. 3.

11. In the Western literature, Oswald Spengler is probably the best-known author for placing a great historical value upon this difference. See especially his *The Decline of the West* (New York, 1928), vol. II, chap. iv.

in plants. Moreover, agricultural activity is bound to an unflinching rhythm, while in manufacture we can very well do tomorrow what we have chosen not to do today. Finally, there is a difference between the two sectors which touches the root of the much discussed law of decreasing returns (in the evolutionary sense). For industrial uses man has been able to harness one source of energy after another, from the wind to the atom, but for the type of energy that is needed by life itself he is still wholly dependent on the most "primitive" source, the animals and plants around him. These brief observations are sufficient to pinpoint not only why the philosophy of the man engaged in agriculture differs from that of the townee but also why agriculture and industry still cannot be subsumed under the same law. Whether future scientific discoveries may bring life to the denominator of inert matter is, for the time being, a highly controversial — and no less speculative — topic.

Probably the greatest error of Marx was his failure to recognize the simple fact that agriculture and industry obey different laws; as a result, he proclaimed that the law of concentration applies equally well to industry and agriculture.[12] To repeat, Marx had no opportunity to observe a peasant economy. Nor is there anything in his vast literary activity to indicate that he ever studied a noncapitalist agriculture.[13] The analysis of rent in *Capital* is based entirely on capitalist production even during Marx's brief excursion into peasant agriculture.[14]

Probably no other theoretical aberration has been refuted by historical developments as promptly and as categorically as the Marxist law of concentration in agriculture. During the second half of the nineteenth century one census after another revealed that in agriculture concentration was continuously decreasing while the peasants instead of being proletarianized became landowners in increasing numbers. In Kautsky's own lamenting words, "the capitalists were on the increase, not the proletarians." The indictment was all the more unappealable since this phenomenon was taking place in capitalist countries without any planned intervention. That convinced everyone save the ultra-orthodox Marxists that the concentration law is false.

3. *Policy and factitious theory*. The aftermath of "the sorest experience of Marxist doctrine" — as Veblen labeled the refutation of the concentration law[15] — can be best appraised in the light of the Hegelian tenet which is the cornerstone of the Marxist doctrine. To recall, according to that tenet it is beyond man's power to change the course of history. This

12. K. Marx, *Capital*, vol. I (Chicago, 1906), chap. xiv, sec. 10.
13. Kautsky, *La question agraire*, p. xii. Also F. Engels in the preface to the third volume of *Capital* (Chicago, 1909), p. 16.
14. Marx, *Capital*, vol. III (Chicago, 1909), chap. xlvii, sec. 5.
15. Thorstein Veblen, *The Place of Science in Modern Civilization* (New York, 1919), pp. 450 ff.

is why Marx argued that socialism is to come as the natural product of the evolution of the relations of production, not because the interests of the working class would in any sense be superior to or more important than those of capitalists. Marx even scoffed at those who wanted to base a socialist platform on such "unscientific" arguments as greater social justice. But, always according to the Marxist Hegelianism, man can speed up the historical process so as to shorten the periods of growing pains. A right policy must be based on the acceptance of the inexorable outcome. Because of the belief in the concentration law in agriculture, Socialists were advised to more than welcome any measure that would tend to proletarianize the peasants so that the advent of socialism would be hastened. But since the peasant did not want to hear of proletarianization, Socialist parties found themselves rejected everywhere by the peasant masses. Failures on the electoral front, combined with the mounting evidence against the Marxist theory, brought about the internal crisis known as the Agrarian Question. At the Frankfurt (1894) and Breslau (1895) congresses, the Question almost wrecked the unity of the party.[16] Even though officially this unity was then saved, the Question continued to make life difficult for Marxism. In the end Marx himself was obviously disturbed by the overwhelming evidence and the mounting criticism, for in the last two years of his life he painfully sought to amend his theory, but not so as to jeopardize the political movement which he had set in motion and to which he was attached from first to last.[17] But his desire was unrealizable, because contradictory. After Marx's death the party made great efforts to cover up the Agrarian Question. They vacillated between Leninist opportunism, according to which they proclaimed loudly that no one intends to destroy the peasant, and various dialectic circumvolutions aimed at proving that there *is* concentration although in an

---

16. For the Agrarian Question one may consult Kautsky, *La question agraire*, and G. Gatti, *Le socialisme et l'agriculture* (Paris, 1902). The Kautsky work is important because it appeared (German edition, *Die Agrarfrage*, Stuttgart, 1899) only a few years after the Breslau congress, where Kautsky had a decisive role in defeating the "deviationist" motion. Gatti, on the other hand, was a prominent Socialist who ultimately embraced the non-Marxist view on agriculture.

17. Marx's public concession, though somewhat veiled, is found in the preface to the 1882 Russian edition of the *Communist Manifesto* — see K. Marx and F. Engels, *Correspondence, 1846-1895*, ed. Dona Torr (New York, 1935), p. 355. A clearer expression of the deviation from "the Marxist line" came in a letter Marx wrote in 1881 to Vera Zasulich in answer to a definite question regarding the necessity of speeding up the proletarianization of the Russian peasant. The letter, however, was published by the Marx-Engels Institute only in 1924, when the struggle between Russian Marxists and their adversaries was long since over. (D. Mitrany, *Marx against the Peasant*, University of North Carolina Press, 1951, pp. 31–33, was the first to draw the attention of the English-speaking reader to this letter.) We know also that in his last years Marx decided to learn Russian (and apparently even Turkish) to have access to the original sources concerning the agrarian problems of Eastern Europe (*Correspondence*, p. 353). For more than one reason, it was too late.

entirely new sense.[18] The Agrarian Question was thus kept on a low flame until Stalin decided to solve it by proclaiming a holy war against the peasants, a war with which neo-Marxism has since become almost synonymous.

It is hard not to see in this momentous decision the ultimate product of Marx's scorn for the peasant. Indeed, this scorn constituted a lasting ferment for the thinking of Marxist leaders. Quite early, none other than Engels spoke of the necessity for the proletariat "to crush a general peasant uprising." [19]

Be this as it may, the Stalinist war, which by its number of victims surpasses all other wars known to history, could not have found sufficient momentum in the cultural opposition between the urban and the rural sectors. Nor could this war feed on "sacking the rich," for precisely in the regions where Stalinism has till now spread, the capitalist-bourgeois class was paper-thin, and the rich peasant quite a rarity. The war must have had other springs.

That the interests of the town conflict with those of the countryside is by now a well-established fact. However, it is not always realized that the price-scissors do not tell the whole story. For this story, we must observe that food is indispensable, whereas the need for industrial products is secondary, if not superfluous. To obtain foodstuff from the agricultural sector, and moreover to obtain it *cheaply*, constitutes a real problem for the industrial community. In the ultimate analysis, "cheap bread" is a cry directed against the tiller of the soil rather than against the capitalist partner of the industrial worker. In some circumstances this conflict may become very spiny. And it is permanently spiny in the overpopulated countries where the income of the masses allows only the satisfaction of the most elementary needs and where the population of the town is unduly swollen by a rural exodus. That has been the situation in all countries — with one or two exceptions — where Stalinism has come to power. And it is in this situation that the war against the peasant found its needed spring.[20]

18. An epitome of these endeavors is offered by Kautsky, *La question agraire*. He argued that, although the concentration law is not true as to the size of the holdings, it is true as to the global ownership, with more landowners having important outside sources of income. Then he threw everything overboard by arguing that peasant agriculture must disappear in any case because the optimum scale of production is that of latifundia.

19. Quoted in Mitrany, *Marx against the Peasant*, p. 219.

20. The conflict between the interests of the agricultural and industrial sectors exists also in the advanced economies, including the United States. Cf. J. D. Black, "Discussion," *Proceedings of the Fifth International Conference of Agricultural Economists* (London, 1939), pp. 86–87. The only difference is that in these economies the conflict is attenuated by the high income, and therefore it can be resolved by such methods as the Agricultural Price Support Program. Overpopulation is the necessary condition for the conflict to become a social *vis viva*.

Clearly, the Stalinist formula constitutes a solution (at least a temporary one) of the conflict between the industrial and agricultural sectors. But the solution is based on the primacy of the interests of the industrial and bureaucratic sections of the society, not on some evolutionary law regarding the inexorable proletarianization of the peasants.[21] Consequently according to the very essence of Marxism the Stalinist formula cannot claim to be "scientific."[22]

Marx was, however, aware of the conflict between the industrial and agricultural divisions of society. He once remarked quite *en passant* that "the whole economic history of society is summed up in the movement of this antithesis [the division between the city and the countryside]." [23] This remark is extremely important. It shows that Marx, for once, recognized the existence of an antithesis which — as we argued in the preceding section — seems rooted in the permanent conditions of the human species, and which should therefore outweigh any antithesis peculiar to a particular economic system. Unfortunately, Marx did not explore this point further to explain how he would have envisaged the *scientific* (in the Hegelian sense) solution of that antithesis.

4. *Policy without theory.* In the first half of the nineteenth century, while the West grew intensively preoccupied with the lot of the industrial masses, Russia witnessed the rise of a social movement concerned solely with the peasant. Time and again, essentially different economic conditions imposed entirely different preoccupations. It was not, therefore, because of the much discussed intellectual isolation of Russia that the founders of this new ideology borrowed nothing from Western economic theories. They simply drew the logical consequences from the fact that these theories were molded on a different economic reality. But as their intellectual inheritance contained nothing regarding the economics of a peasant community, the new social reformers had to start from scratch. They soon discovered that their personal social background could not help them in grasping the problems in which they were interested, and as a consequence decided to go "to the people." This slogan earned them the

---

21. A London tailor, J. G. Eccarius, *Eines Arbeiters Widerlegung der nationalökonomischen Lehren John Stuart Mills* (Zürich, 1868), was the first to argue that to guarantee "cheap bread" to the industrial worker the peasant must be placed under the dictatorship of the proletariat. The book, it is said, enjoyed great prestige among Marxists during the 1870's (see Mitrany, *Marx against the Peasant*, p. 15). That Eccarius's view has become the basis of Communist agrarian policy is beyond question: "general collectivization of the peasants is indeed a means of . . . securing the supply of food [for the towns]" (V. Lenin, *Selected Works*, Moscow, 1934–1939, XII, 13). The fact is that Narodnikism is an older school of thought than Marxism.

22. As we shall see, neither can it be justified on positive welfare grounds; *infra*, footnotes 83 and 85.

23. Marx, *Capital*, vol. I, chap. xiv, sec. 4, p. 387.

Russian name of *Narodniki*, but outside Russia they became generally known as Populists.[24]

As Marxism began to acquire a basis of its own in Russia, the incompatibility between Marxist theory and the Russian reality gave rise to a fierce and more lasting conflict between *Narodniki* and Marxists than that between the orthodox Marxists and the Agrarian Socialists in the West. Some *Narodniki* did become attracted by Marxism, primarily because its programmatic implications and social dialectics appealed to their revolutionary spirit. But as it was impossible to fit the peculiarities of an agrarian economy into the Marxist frame, most of these succumbed as hetero-Marxists. The great majority of the *Narodniki*, however, refused to be lured into denying the specific traits of that economy. And thus the Agrarian ideology came to be identified with a double negation: not Capitalism, not Socialism. It is precisely this double negation that has been called into question by Western economists, whether Marxist or not.

David Mitrany observes that Marx's view on peasant agriculture combines "the townsman's contempt for all things rural and the economist's disapproval of small scale production." [25] But this is true for most Western social scientists. Add to this, especially, their usual disdain for any idea that is not presented through a mathematical model, and you have the explanation for the misunderstanding of the Agrarians by the West.[26] Indeed the *Narodniki*, like the Agrarians of latter days, have not only failed to construct a theory of the peasant economy — as the others have done for capitalism — but they have distinguished themselves by a lack of interest in, almost a spurn for, analytical preoccupations. They relied exclusively on the intuitive approach, on the *Verstehen* of the peasant's *Weltanschauung*, much as the German historical school advocated (although there was hardly any direct contact between the two schools). Populism, like Marxism, represented not only an economic doctrine but a faith as well. And this faith "fed on a strong sentimental undercurrent,

---

24. Alexander Herzen, who in 1847 went into exile because of his political activity, is generally regarded as "the founder of Russian 'Socialism,' or 'Narodnikism,'" as Lenin put it. Quoted in Marx and Engels, *Correspondence*, p. 285.

25. Mitrany, *Marx against the Peasant*, p. 6.

26. In this respect, it is highly instructive to compare, for instance, the analysis of Populism by L. H. Roberts, *Rumania* (New Haven, 1951), pp. 142 ff, with that by Rosa Luxemburg, *The Accumulation of Capital* (London, 1959), pp. 271–291. Although Rosa Luxemburg was "a more genuine Marxist than any other member of the German movement" (Paul M. Sweezy, *The Theory of Capitalist Development*, New York, 1942, p. 207), her analysis is far more objective than Roberts's.

On the *Narodniki*, one may fruitfully consult also J. Delewski, "Les idées des 'narodniki' russes," *Revue d'économie politique*, XXXV (1921), 432–462, and above all Mitrany, *Marx against the Peasant*, chap. iv. The memoirs of the "grandmother" of the Russian revolution, Katerina Breshkovskaia, *Hidden Springs of the Russian Revolution* (Stanford, 1931), are interesting as personal history.

on the emotional piety and rustic ties" of its believers.[27] All this laid Populism open to the accusation of romanticism.

The particular circumstances in which *Narodnikism* began its career may account for much of its peculiar spirit. But the lack of any true theorizing in the Populist doctrine was due more to the unusual difficulty of casting the peasant's economic conduct into a schema than to anything else. For this we have the testimony of one of the most praiseworthy Russian Agrarians, Alexander Tschajanov, who gave to one of his works the symptomatic title: *Die Lehre von der bäuerlichen Wirtschaft: Versuch einer Theorie der Familienwirtschaft im Landbau* (Berlin, 1932). In the concluding remarks of this book, in which he submits only the various activities of agricultural production to quantitative analysis, Tschajanov confesses his dissatisfaction over the fact that we still do not possess a theory of the economic behavior of the peasant. He significantly observes that the relation between Classical economics and an economic theory of a peasant community seems to be similar to that between Euclidian and non-Euclidian geometry. Yet he ends with the admission that an abstract theory of agrarian economics cannot easily be constructed.[28]

Whatever the explanation for the outlook of the Agrarians, there is no more dramatic example of the disaster that awaits him who in formulating an economic policy disregards theoretical analysis, than the well-known fate of the Agrarian parties of Eastern Europe.

II. OVERPOPULATION: A RE-EXAMINATION

1. *The facts analyzed.* The Agrarians have at all times sensed that the plague of most underdeveloped agrarian economies is overpopulation and that consequently the problem of a peasant economy is to a large extent a population problem.[29] It is natural, therefore, for us to see whether an analysis of overpopulation would not offer a lead to the solution of the Agrarian riddle.

Whoever speaks of "excess" is naturally expected to define it in terms of a point of reference which in some way must represent a normal if not an optimum situation. But to define "normal" or "optimum" is not easy, especially if one faces a quibbling relativist. Such a relativist may argue, for instance, that the excess capacity of a monopolistic industry is a fiction because all capacity could be used if monopoly were removed and a new

27. Mitrany, *Marx against the Peasant*, p. 40.
28. Tschajanov, *Die Lehre* (as cited in text), p. 130.
29. Cf. Tschajanov, p. 131, for instance.

system of distribution were introduced. A wholly analogous position is adopted by Marx in arguing that overpopulation exists only relative to "the average needs of the self-expansion of capital." [30] Be this as it may, we must recognize that the concept of overpopulation presents unusual difficulties. Normal (or optimum) population implies the concept of normal (or optimum) life. And even if the latter were not such an elusive concept, we would still find it impossible to choose a "normal" valid for all times and localities. To avoid the trivial conclusion that every population is normal for the time and place in which it lives, it is necessary to adopt some criterion of normality. This criterion may be dynamic or static, depending on the problem at hand.[31]

Ever since statistical data have been used for comparative purposes, it has become obvious that some agricultural countries presented symptoms suggesting the existence of some sort of overpopulation. It has been remarked that, given the following data for two prominently agricultural economies,

	Denmark	*Yugoslavia*
Inhabitants per square kilometer of arable land	36.6	157.4
Wheat yield in quintals per hectare	22.9	11.0

even if Yugoslavia could raise her agricultural yield to the Danish level, the average Yugoslav would still have only one quarter as much food as the average Dane. This observation has supplied the basis for a crude concept of relative overpopulation upon which are based the measures of overpopulation in terms of some crop basket as a standard.[32] As has generally been admitted, the concept of relative overpopulation thus defined is ambiguous and the procedure for its measure debatable.[33] The chief drawback of this approach, however, is that it sidetracked the analysis from the right direction. Indeed, a difference in the *per capita* national product (or a sector of it) may be a *symptom* of the difference between two economic systems, but by no means an *intrinsic coordinate* of that difference. Otherwise we should regard the economic system of Belgium as different from that of the United States. But the belief that the difference between an agrarian and a capitalist economy is a matter of degree only, not of essence, is still very frequent.

30. Marx, *Capital*, vol. I, chap. xxv, sec. 3, p. 695.
31. Marx, for instance, argued that the developed means of communication in the United States at the middle of the last century made that country more densely populated than India. *Ibid.*, vol. I, chap. xiv, sec. 4, p. 387.
32. Cf. W. E. Moore, *Economic Demography of Eastern and Southern Europe* (Geneva, 1945), chap. iii.
33. *Ibid.*, pp. 55 ff.

And yet the elements for the solution of the problem were not out of reach. In the 1930's, studies originating in several countries with large peasantries revealed the astounding fact that a substantial proportion of the population could disappear without the slightest decrease in the national product.[34] The closeness of the independent estimates of the *superfluous* population for each case shows that we are confronted with a real quantitative phenomenon.[35] If additional proof of this is needed, one may invoke some relevant "experiments" history carried out *in vivo*. For two years after the beginning of hostilities in 1914, agricultural production in Russia was maintained at the prewar level, although no less than 40 per cent of the *able-bodied* male peasants were in the army.[36] The same phenomenon occurred in Rumania during World War II. Whenever agricultural production collapsed in Eastern Europe during the World Wars it was solely because of the extreme requisition of draft animals, the difficulty of replacing worn-out implements, and, of course, the disturbances caused by the movement of armies. Even the disappearance of some ten million Ukrainian peasants during the so-called liquidation of kulaks, although accompanied also by a radical disturbance of the entire economy, had only an ephemeral influence on agricultural output.[37]

Now, to say simply that part of the population could disappear without causing any decrease in output is not sufficient for a theoretical characterization of overpopulation. The national product of the United States *could* easily be maintained at the same level even though a large proportion of the population were to disappear. The *differentia specifica* between the two situations is that in the latter the national product could be increased if people simply chose to have less leisure, while in the former, not.[38] This difference reveals that the situation where the marginal pro-

---

34. References to the earliest studies for Poland and Bulgaria are in Doreen Warriner, *Economics of Peasant Farming* (London, 1939), pp. 68–69.

35. For Rumania, one study (*Enciclopedia României*, Bucharest, 1939, III, 60) estimated the percentage of superfluous peasant population at 48, another, at 45 (V. Madgearu, *Evoluția economiei românești după războiul mondial*, Bucharest, 1940, p. 49). The first estimate was derived from national statistical data, but the second was checked by *direct observation* in extensive field work covering sixty villages chosen at random. Moore, *Economic Demography*, pp. 63–64, using national data, arrived at a percentage of 51.4.

36. Leonard E. Hubbard, *The Economics of Soviet Agriculture* (London, 1939), pp. 59, 65.

37. *Ibid.*, p. 117. Harvey Leibenstein, in a recent paper, "The Theory of Underemployment in Backward Economies," *Journal of Political Economy*, LXV (1957), 103, alludes to some experiences in the Soviet orbit when industrialization would have caused a shortage of labor in the agricultural sector, but fails to say precisely which events he refers to. My guess is that they exhibited only the familiar kind of *spurious* shortage caused by wholesale dislocations of persons, if they did not reflect either peasant resistance or administrative inefficiency.

38. Marx, in *Capital*, vol. I, chap. xxv, sec. 3, p. 698, asserts that if the population of England would be reduced in the same proportion for all categories, the remaining population "would be absolutely insufficient" to maintain the same level of output

ductivity of labor equals zero is the starting point in searching for a definition of overpopulation. And the existence of countries where the actual marginal productivity of labor is zero for all practical purposes has been admitted by all keen observers of peasant economies.[39]

All this clearly conflicts with Professor Schultz's categorical statement that there is "no evidence for any poor country *anywhere* that would even suggest that a transfer of some small fraction, say 5 per cent, of the existing labor force out of agriculture, with other things equal, could be made without reducing its [agricultural] production." [40] Nothing is farther from my thought than to challenge the fact that the concrete cases cited by him prove that in several Latin American countries the agricultural production did fall off after some labor had been transferred to other activities.[41] But that is not sufficient to justify his well-known position, namely that the overpopulation theory of underdevelopment "as a 'theory' . . . fails in that the expected consequences are not those that one observes." [42] The situation of most Latin American countries is not identical with that of the East European or Asiatic countries, although they all have this in common: they are underdeveloped. Although overpopulation is always accompanied by underdevelopment, it is neither a necessary nor the only cause of it. The underdevelopment of Latin American countries may have other bases than overpopulation.[43] Overpopulation, therefore, cannot provide the basis for a *general* theory of underdeveloped economies, but only of those economic realities beset by it. This is concrete illustration of the point which one of the preceding sections sought to bring home.

To regard the notion of overpopulation as a myth is undoubtedly a Marxist residual. And precisely because that notion still meets with opposition in some circles, a few further remarks seem in order. If in the "so-called" overpopulated economies the marginal productivity is zero — a critic may ask — how can we explain the fact that in such economies there is a greater need for skilled labor than in the other countries? Certainly — he may continue — you are not going to say that the marginal productivity of an engineer in India or Egypt is zero. But this way of looking at the

---

in spite of England's "colossal" means for saving labor. Clearly, this implies that no skilled labor has *free* leisure, a characteristic assumption of Marxist economics.

39. E.g., Warriner, *Economics of Peasant Farming*, p. 65.

40. Theodore W. Schultz, "The Role of Government in Promoting Economic Growth," in *The State of the Social Sciences*, ed. L. D. White (Chicago, 1956), p. 375. (My italics.)

41. *Ibid.*, pp. 375–376.

42. Theodore W. Schultz, *The Economic Test in Latin America*, New York State School of Industrial and Labor Relations, Bulletin 35, August 1956, p. 15.

43. Although my knowledge of the factual situation in those countries is very superficial, I would venture to suggest that some are "underpopulated" relative to the available land resources. Professor Schultz's evidence may even corroborate such a view.

problem is to intermingle evolutionary factors with static concepts and, above all, to confuse labor with capital. An evolutionary change is bound to bring about shortages of some types of skilled labor (and surpluses of some others) in *any* economy. Thus, Italy certainly feels today a shortage of technicians for her newly discovered oilfields. This, however, represents a quasi bottleneck, to borrow an expression coined in the Marshallian spirit by Professor Lewis.[44] If no further evolutionary changes occur, the quasi bottleneck will disappear just as any quasi rent will do. But, once the new equilibrium is reached, will the marginal productivity of a petroleum technican become zero? Not at all. For the equilibrium marginal productivity of such a technician represents not only the marginal productivity of his labor but also that of the capital invested in his training.[45] Obviously, this line of reasoning regards labor as a uniform, plastic quality of all human beings, and is — I believe — in the tradition of Classical, and somewhat in that of Marxist, economics. But I fail to see a better way to analyze the problems raised by *population* in its *purely quantitative* aspect. Actually this view of labor is even more necessary in the analysis of economic growth than of a stationary state, where population may very well be regarded as a frozen distribution of *qualities*.

Therefore, the statement that the marginal productivity of labor is zero means that the marginal productivity of skilled labor consists only of the marginal productivity of the capital invested in the production of skill. It is a most rational expectation that an overpopulated economy should feel a greater shortage of skilled labor than a nonoverpopulated one. Everything points to the fact that a shortage of skilled labor reflects a shortage of capital, not necessarily of labor. It is a peculiar feature of overpopulated economies that the skilled laborer is overburdened with work while the unskilled is loafing most of the time. Further still, the real economic aspect of spreading knowledge in an underdeveloped country now appears in full light: the need for additional education competes with the need for additional physical means of production, a fact which we are apt to overlook at times and underestimate often. Where resources are very scarce, free education for *all* types of skills is as *uneconomic* as haphazard production of capital equipment. Some countries, like Soviet Russia, seem to have grasped this truth; others, such as Italy, apparently have not.

It is a simple matter of definition to observe now that in any economy, whether overpopulated or not, there is only one way to measure the marginal productivity of labor: at the margin, i.e., where labor appears un-

---

44. W. Arthur Lewis, "Economic Development with Unlimited Supply of Labor," *The Manchester School*, XXII (1954), 145.
45. Another part may reflect the "rent" of his personal talents, but that is a side aspect of the problem.

adulterated by capital. The marginal productivity of labor in any economy then is the marginal productivity of its unskilled labor. It is a mere factual coincidence that agricultural labor generally is unskilled labor. But this fact throws a new light upon the constant correlation of overpopulation with the agricultural conditions in the economic literature. For clearly, if the marginal productivity of labor in a country is zero, so must be the marginal productivity of the peasant.

2. *A theoretical schema.*[46] To make the argument as simple as possible, let us assume that the national *product*, represented aggregatively by x, is produced by an atomistic "industry" (an assumption fully justified in overpopulated agrarian economies). This means that the production function of the entire economy is homogeneous of the first degree,

(1) $$x = F(L, T) = TG(L/T)$$

where L stands for labor, and T for a composite variable of land and capital. To obtain this function, for each proportion of the factors of production Op (Fig. 11-1), we determine the optimum size, U, of the

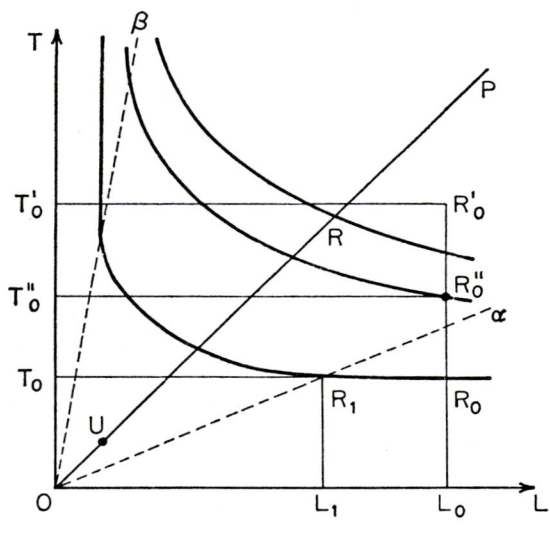

FIGURE 11-1

production unit, from the system

(2) $$T = pL, \qquad L\frac{\partial f}{\partial L} + T\frac{\partial f}{\partial T} = f(T, L),$$

where $f(T, L)$ is the production function of the "firm." The maximum product obtainable from a given combination of factors, R, is equal to

---

46. I prefer the term "schema" to the commonly used "model," for I wish thereby to emphasize the essential difference between the *blueprint-model* of natural sciences and the *simile-schema* of social sciences.

OR/OU times the output of U. That determines (1) for every R. It is important, however, to remember that to obtain the output computed from (1), the resources $R(L, T)$ must be equally divided among OR/OU identical production units.[47] It is also seen that given the amounts of the factors of production, the optimum scale of the "firm" is *uniquely* determined for every technological horizon. Hence, in the case where the geo-historical conditions of an economy are such that *all available resources must be used in production as long as they increase output*, the argument regarding the superiority of the large-scale production is poor economics.[48] In the causal relationship between the efficiency of large-scale production units and a high level of economic development (i.e., a high T/L) the latter is the cause of the former, not vice versa. There is no need to go over the reasons why the isoquants of any production function sooner or later become parallel to the axes. Thus, in the region $LO\alpha$, for instance, the output can be increased if and only if there is an increase in the factor land-capital. We shall refer to such a factor as limitative.[49] Clearly, in the region where a factor is limitative its marginal productivity is constant, while that of all other factors is zero.

A few definitions. Let the population, P, of a given economy be divided into P_w, the working class, and P_g, the "government" class. In the latter class we include all members of the economy who are not dependent on wages or salaries received from the "industry" producing x.[50] Let s and s' be the individual *minimum* standard of living of the working and government class respectively. The position taken in this paper is that both these variables are historically determined, and consequently susceptible of being changed by economic policy. They condition the *minimum* public need for x (public roads, armaments, capital accumulation, etc.). If this minimum is denoted by E, the minimum net product of the community is

(3) $$\chi = P_w s + P_g s' + E.$$

Let us also put

(4) $$F = \varphi P_w, \qquad L_0 = \delta F$$

where F is the size of the *potential* labor force, and δ represents the labor

---

47. Strictly speaking, OR/OU is not necessarily an integer, but for an atomistic industry this does not matter. We should also point out that U is placed out of scale in Fig. 11-1; otherwise it could hardly be distinguished from O on the drawing.

48. That is the fault of the argument advanced by Kautsky *et al.* against peasant holdings. *Supra*, note 18.

49. Not to be confused with *limitational*. A factor is limitational when its increase is a necessary but not a sufficient condition for an increase in output. See Georgescu-Roegen, "Limitationality, Limitativeness, and Economic Equilibrium," in *Proceedings of the Second Symposium in Linear Programming* (Washington, D.C., 1955), 301.

50. This class corresponds to what Veblen called "the kept class." It naturally includes all kinds of servants, public and personal.

time a worker can supply *above the biologically necessary minimum of sleep and rest*. For symmetry, the time necessary for sleep and rest will not be included in *leisure*.

The primordial economic problem of any community is to find a mode by which a national product equal at least to χ can be obtained with the available resources. One coordinate of the problem is the labor supply. Since *ex hypothesi* the working class cannot maintain itself on a smaller real income than sP_w, the supply curve of labor must start discontinuously from an end point M of coordinates L_0, sP_w/L_0 (Fig. 11-2). Since men

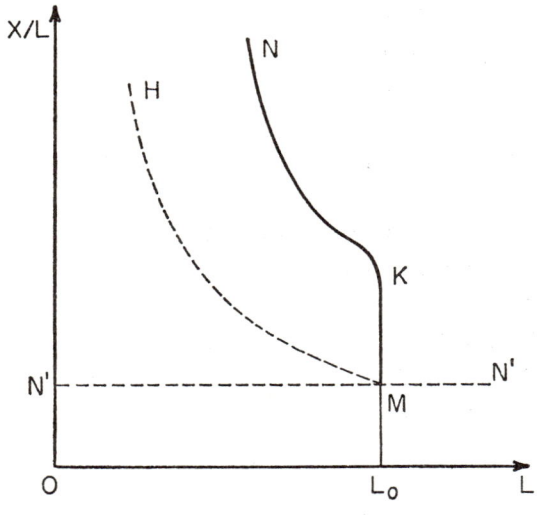

FIGURE 11-2

generally live in families, the labor supply of one individual depends upon the income of his family. To get around the difficulty of circular dependence, we may assume that the labor supply of the community is obtained by summing up the supplies of all families. With this remark, let us consider first the situation in which people can freely sell *leisure* at the market wage rate, and let MKN represent the amount of labor supplied at various wage rates. Clearly, MKN is the short-run supply of labor in a wage economy.[51] Its relation with the preference-field is shown more clearly if we refer to a familiar map of indifference curves. In Fig. 11-3, $OE = \delta$ and $OF = s$. The economic problem of the worker is in fact a discontinuous jump: from E, where his *natural* income places him, at any point in

---

51. Classical economics argued that the long-run equilibrium wage rate is constant and equal to ON', so that $N'N'$ represents the long-run supply curve of labor. For Marxist economics, however, $N'N'$ represents both the short- *and* the long-run supply of labor, a direct consequence of its assumption of a permanent reserve army. This is the analytical expression of the distinctive feature of Marxist economics in refusing to accept any relation whatever between economic and demographic factors. On this point, see a letter of Engels in Marx and Engels, *Correspondence, 1846–1895*, p. 199.

the area $XFF'E'$ in Fig. 11–3(a). Where he will finally land depends upon the type of economic system in which he lives. If this is a free wage market, his labor supply is given by a Hicksian "price-consumption" curve FF'', and all is well. A very frequent pattern of behavior is that for which the supply of labor is inelastic for wage rates just above the minimum possible (i.e. the slope of FE). Be this as it may, we can safely assume that the supply curve of labor has everywhere an elasticity smaller than unity. The literature on underdeveloped countries, however, often mentions a very curious pattern of behavior, that of the individual who after earning the minimum of subsistence becomes interested exclusively in leisure. Such behavior, understandably, would make any policy maker despair: the individual seems to refuse to be developed. The behavior leads to the indifference map shown by Fig. 11–3(b), and to a supply curve represented by a branch of an equilateral hyperbola, MH (Fig. 11–2).[52] Whatever the pattern of behavior, the curve MH constitutes a relevant element for the analysis of distribution: it represents the minimum average share per unit of labor time for each amount of employment.

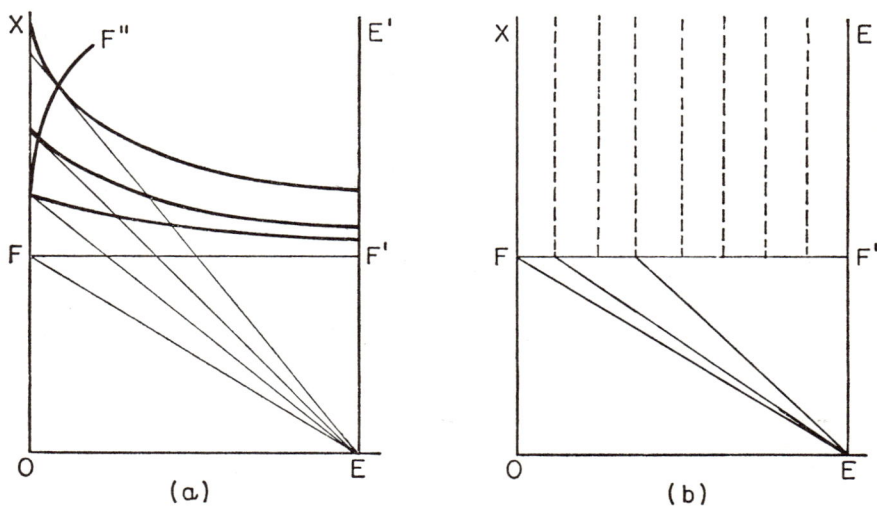

FIGURE 11–3

The second coordinate of the economic problem is the productivity of labor under the assumption that all land-capital resources, T_0, are used in the production of x. The curve, μA, representing the average productivity of labor obviously varies with T_0, but its shape presents some constant features (Fig. 11–4). All these curves begin by a horizontal segment

[52]. I feel that for this type of behavior we must assume a zero marginal rate of substitution between real income and leisure. But if I am wrong, policy makers confronted with this hopeless reaction to a wage scheme should be able to get around the difficulty by imposing a *corvée* simultaneously with a very high wage rate for freely contracted work. A simple diagram will show that in this way they can induce the individual to move inside the area $XFF'E'$.

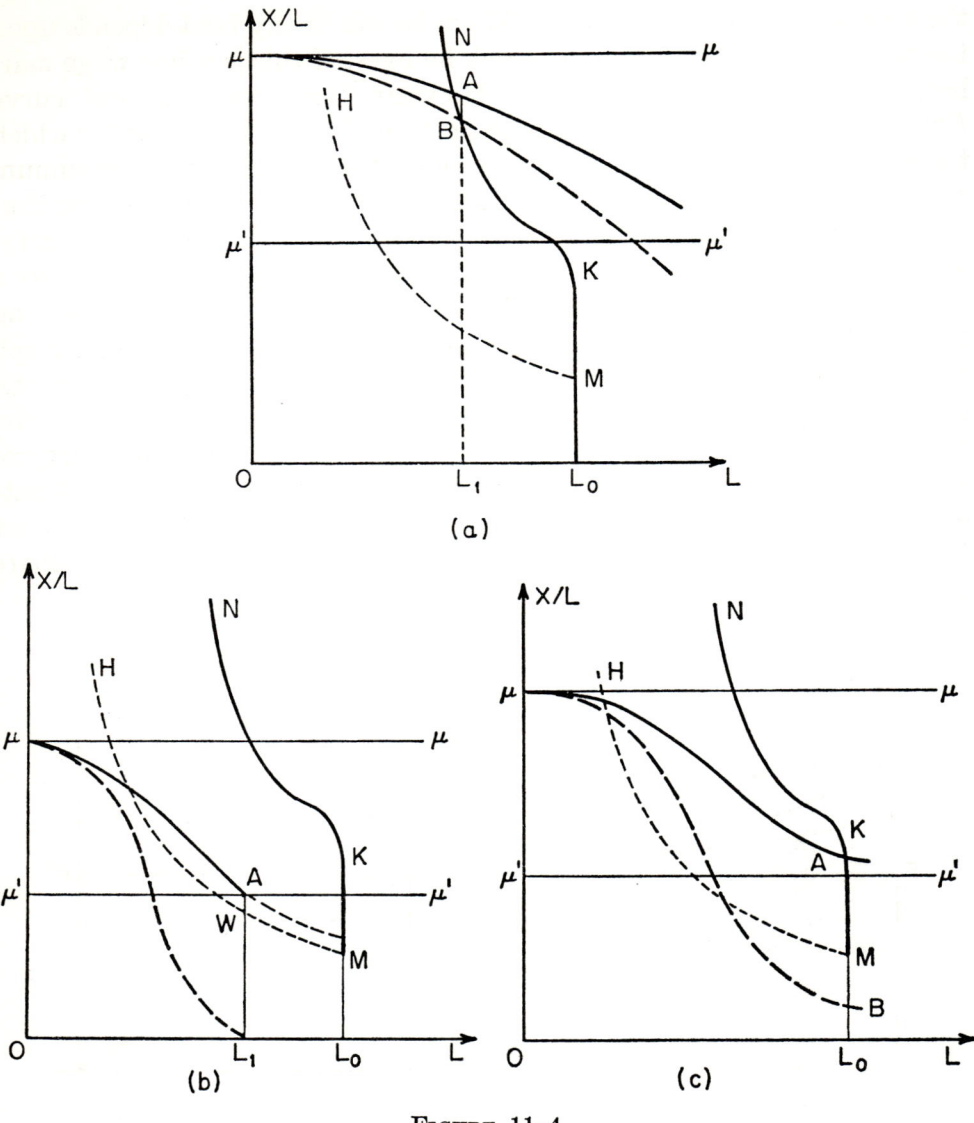

FIGURE 11-4

at the same level $O\mu$, because x/L remains constant in the domain $TO\beta$. They also coincide with equilateral hyperbolas below the level $O\mu'$ equal to the value of x/L on $O\alpha$.[53]

If T_0 is sufficiently large, the curve of the marginal productivity of labor μB intersects MN at B, as in Fig. 11-4(a). In such a case the stage is set for the solution of the community's economic problem according to the marginal productivity principle (provided $s'P_g + E$ is not too large, as has been the case in many countries during the World Wars). The economy may even allocate part of T to the direct use by consumers, so that although

53. $TO\beta$ and $O\alpha$ refer to Fig. 11-1.

the available resources are represented by R'_0, only the amounts represented by R are used in production (Fig. 11–1). At the other extreme, for low values of T_0, we find the case where the average productivity curve of labor μA lies below $\mu'\mu'$ for $L = L_0$ as in Fig. 11–4(b). That is the case of *strict overpopulation*.[54] (This corresponds to R_0 in Fig. 11–1.) Obviously, in such a situation there can be no economic advantage in using labor beyond L_1, where its marginal productivity becomes zero. To continue existing without being beset by Malthusian forces, the economy cannot have an $s'P_g + E$ larger than $AW \times L_1$. In practice, however, this maximum is always attained, so that ordinarily in overpopulated countries $\chi = X_0$, X_0 being the maximum national product obtainable with the available resources. It is obvious, however, that the economy cannot possibly function according to the principles of marginal productivity theory. This is true also for an economy where the marginal productivity of labor is positive for $L = L_0$ without being greater than $L_0 M$; see Fig. 11–4(c). For this and other good reasons, such an economy should be regarded as overpopulated, without the qualification "strictly." [55]

The important conclusion is that overpopulation is correlated with a low T_0, more precisely with a low T_0/L_0. Overpopulation, therefore, is tantamount to poverty of nonlabor resources; the opposite situation, to "land-of-plenty." In the real world, however, most underdeveloped *agrarian* economies are poor not only because of the insufficiency of land, but also because of a chronic dearth of capital. For these countries the difference between T_0 and the size of usable land, A_0, is negligible. That justifies the use of L_0/A_0 as an ordinal index of agricultural overpopulation instead of L_0/T_0. Finally, by replacing L_0/A_0 by P/A_0, we obtain the most commonly used but crude form of that index.

3. *Further remarks.* In approaching issues concerning the economy as a whole the economist has but two choices: to use either a general equilibrium apparatus or an aggregative schema. In the first case, he must resign himself to being rather sterile on practical matters; in the second, he must accept the theoretical calamities of aggregation. For more than one reason, I have chosen the latter procedure. It is, however, possible to illustrate the conditions of overpopulation by a schema in which the national product is not completely aggregated. Let us assume that the economy produces an agricultural product, X_1, and an industrial product, X_2. Bearing in mind that in an overpopulated economy the standard of living barely covers the most elementary needs, and that these needs are highly rigid,

---

54. We consider only the alternative where μA intersects MN. The opposite case involves Malthusian aspects which though interesting lie beyond the scope of this paper.

55. In Fig. 11–1 this case corresponds to R''.

we may proceed on the assumption that the two products are not substitutable. If the minimum necessary product is represented by X_1^0, X_2^0, the case of overpopulation is illustrated by Fig. 11–5(a).[56] The only solution is M, where the marginal productivity of labor is zero in both productive sectors. Hence, in this case too, no advantage can be derived from using labor beyond L_1.

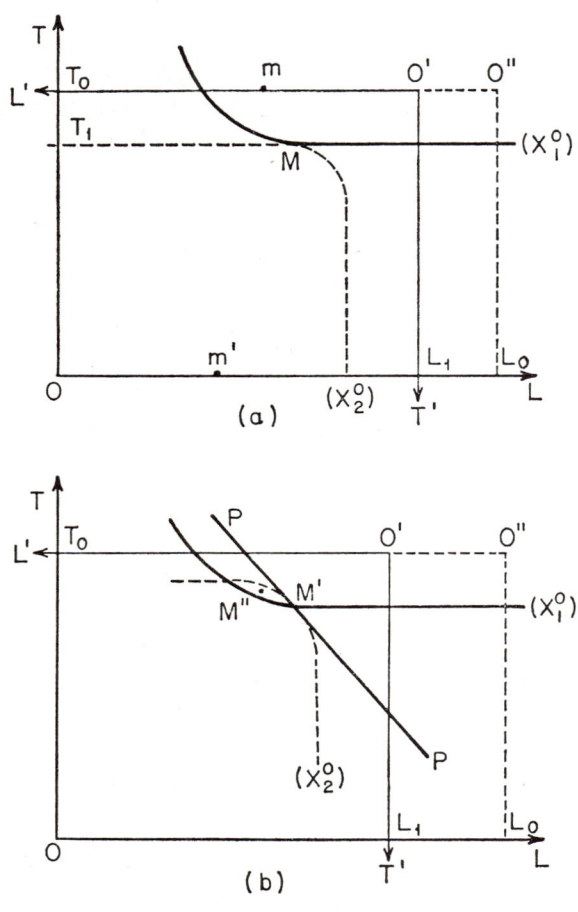

FIGURE 11–5

The schema of Fig. 11–5 forces upon us a series of highly interesting problems. We shall mention only one. A nonisolated economy has the choice between producing one kind of product at home and obtaining it through foreign trade. The question then is whether the resulting national income (X_1', X_2') could allow an overpopulated economy to move its industrial structure from M to m (or m') where the marginal productivity of labor is positive; see Fig. 11–5(a). We cannot pursue here this intricate

56. In Fig. 11–5 an "Edgeworth-box" is used with O and O' representing the origin of each system of coordinates.

problem, but we may at least remark that no agrarian country seems to have been able to escape the conditions of overpopulation by mere trading. Most probably, overpopulation will remain a local problem calling for local remedies as long as people in general neither wish nor are allowed to leave their own lands.

III. AN ANALYSIS OF THEORETICAL ISSUES

1. *Profit versus tithe*. The question of whether the Walrasian system has a mathematical solution has always been considered a crucial one for Standard theory. But no Standard economist seems to have realized that the Walrasian system raises a still more vital question: Is its mathematical solution also an *economically* valid one? That it is valid has been taken for granted by all who tackled the unusually difficult problem of the existence of the mathematical solution. To recall, Abraham Wald was content with proving that in a (simplified) Walrasian system the "equilibrium" prices are non-negative.[57] Wald was hardly an economist, but after the publication of his original paper in German (1934), no economist observed that unless we also know that the "equilibrium" price of labor is at least equal to the minimum of biological subsistence, the theorem has only meager economic relevance. The truly economic aspect of the problem is most clearly set aside in the more recent work of Arrow and Debreu. These authors start out with the assumption that *every* member of the community is endowed *ab initio* with a sufficient real income for his entire life span.[58] What we do know, however, is that man is endowed with labor of limited efficiency and can use resources of limited quantities. These limits in some cases may be such that although the economy can produce a sufficient real income for all, this *economic* solution cannot be reached by the mechanism of marginal productivity which is part and parcel of the Walrasian system.[59] We have seen that overpopulated economies are in this particular situation. The problem now is to see how production and distribution may be regulated in such an economy.

A lead for the solution of this problem is offered by the observation that agricultural overpopulation has usually been manifest in countries where

57. Abraham Wald, "On the System of Equations of Mathematical Economics," *Econometrica*, XIX (1951), 368–403.
58. Kenneth J. Arrow and Gerard Debreu, "Existence of an Equilibrium for a Competitive Economy," *Econometrica*, XXII (1954), pp. 266, 270.
59. On the surface, this statement may appear to contradict the theorem proved by Arrow and Debreu. That is not the case. Indeed, their proof assumes the economic problem already solved: the individual has already jumped from E into $XFF'E'$ (see Fig. 11–3a). In their approach, the Walrasian solution may consist of everyone conserving his initial position without any alteration.

feudalism was late in being supplanted by capitalism. To see the difference between the distribution under feudalism and capitalism, let us draw the familiar curve of the marginal productivity of labor on the available T_0, $ABCL_1$ in Fig. 11-6. Let also $abcL_1$ be drawn so that the ratio between the

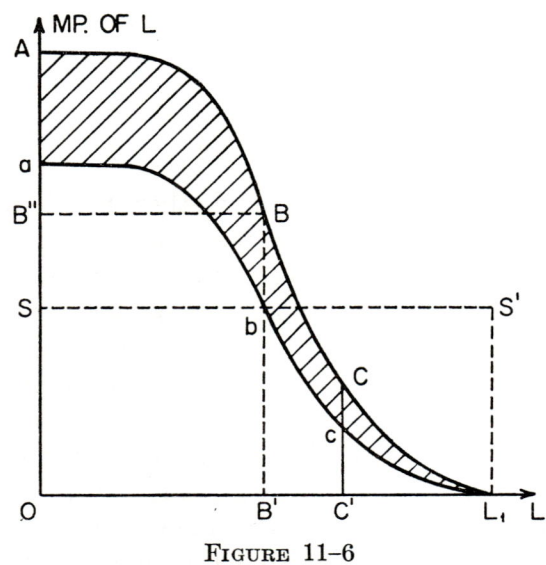

FIGURE 11-6

ordinate of the shaded area and that of $ABCL_1$, say between cC and $C'C$, be equal to the tithe ratio, ρ. It is elementary that if OB' labor is used in production, the share of the entrepreneur-landlord is $AabB$ in the feudal system, whereas the same share under capitalism is $AB''B$. The difference between the two systems is thus clear. But there is also an analogy: to the interest of the government class to maximize profit-rent in the capitalist system, in feudalism there corresponds the interest of the same class to maximize tithe.[60] Obviously, for any given value of ρ, tithe is maximum at L_1. In a strictly overpopulated country this maximum is bound to be reached as the result of workers' own necessity of securing a share at least equal to sP_w. We should keep in mind also that in the feudal system of distribution *there are workers who earn more than their specific contribution to output*. This is seen immediately if $OSS'L_1$ is drawn so that its area is equivalent to that of $OabcL_1$; all labor between B' and L_1 receives more than its *net* contribution to output. Striking historical evidence of this

---

60. Economic relations under feudalism presented almost an infinite gamut. However, the most frequent features of feudal economy were the *corvée* and the crop-sharing. The *corvée* usually consisted of work performed at the manor, on the land under special cultures appurtenant to the demesne (vineyards, orchards, gardens), and for public works. Both these institutions survived the legal abolition of the feudal system, as we have had occasion to observe during our own time everywhere in Eastern Europe. The use of the term "tithe" for the share of the landlord is, admittedly, improper, but convenient.

aspect of feudalism is provided by the gleaners, who received a share greater than the quantity of corn gleaned. In contrast with this, capitalism has no place for any "gleaners."

It goes without saying that a feudal government class would seek a *maximum maximorum* of tithe by maximizing also ρ. Obviously, the maximum of ρ is given by the ratio AW/AL_1 of Fig. 11–4(b), or alternatively by the relation

(5) $$X_0(1 - \rho) = sP_w.$$

This means that workers receive only their minimum of subsistence. Yet, this is not a *sine qua non* condition of the system, for at least in a strictly overpopulated economy the workers may still work up to L_1 even if ρ were equal to zero — unless their behavior conforms to Fig. 11–3(b).

The formula "*the laborer must be poor to be industrious*" came to be a key for the feudal system only as the conditions of strict overpopulation ceased to exist owing to land reclamation, to the increase in accumulated capital, and finally to the increase of labor productivity as the result of technical progress. These factors caused the productivity curves of labor to shift upwards and to the right. Ultimately the marginal productivity of labor for $L = L_0$ became positive, but still less than ML_0; see Fig. 11–4(c). Clearly, the economy in question was not yet ready for the capitalist formula. Hence, feudalism, which could still provide a solution to the problem of distribution, continued its existence although plagued by a new conflict. If the marginal productivity of labor is positive for $L = L_0$, maximization of tithe requires that workers should have no leisure. Whether they would work to that limit of their own accord in a (not strictly) overpopulated country depends upon the value of $X_0(1 - \rho)$, i.e. on ρ. To render the laborers willing to work up to L_0, the average share of labor must not exceed L_0K. This is how the above-mentioned formula came to be an undisputed economic dogma for late feudalism. Even Quesnay argued that the laborer who can buy his bread cheaply becomes "lazy and arrogant." [61] How deep-rooted this formula must have been (at least in late feudalism) is shown by its echoing in economic literature as late as the nineteenth century.[62]

---

61. *Oeuvres économiques et philosophiques de F. Quesnay*, ed. Auguste Oncken (Paris, 1888), p. 248. The only objection raised by Quesnay is that if abused to the extreme the formula may bring the laborer so close to the status of an animal that he will ultimately behave like one, i.e., respond only to the most elementary needs of the moment and thus lose any interest in his own economic progress (*ibid.*, p. 354). Such a behavior corresponds to that of Fig. 11–3(b). The fact that it has frequently been reported in the poorest countries of long-standing exploitation confirms Quesnay's remarks on the point.

62. Anonymous, *An Inquiry Into Those Principles Respecting the Nature of Demand as Advocated by Mr. Malthus* (London, 1821), p. 67, cited by Marx, *Capital*, vol. I, chap. xxiv, sec. 3, p. 653.

With a ratio T_0/P still too low for the curve of marginal productivity of labor to intersect the supply curve of labor, it was only natural that the feudal formula should have been used also by the nonagricultural sector during the first phase of the industrial revolution, and that it should have survived by inertia well beyond the beginning of capitalism. Only because of this circumstance was it possible for Marx to mistake the basic formula of feudalism for an essential feature of capitalism and thus to formulate a theory of surplus value which is only an elaboration of condition (5).

2. *A discussion of conduct*. Very often, some features of a system become noticeable only under the light cast by a contrasting structure. That — as we have seen — is the case of some aspects of marginal productivity theory when confronted with a feudal mechanism in an overpopulated country. The same is true of the Edgeworth-Pareto pattern of individual behavior when contrasted with the behavior of an agrarian (peasant) community.

The founders of modern utility theory all agreed that "*each individual acts as he desires*," [63] a statement that has ever since been repeated in one form or another. But its truth is so tautological that it has no value for a student of human conduct; this student wants to know precisely what pleases the individual. In answering this question, Standard theory assumes that what pleases an individual can be expressed as a function $\psi(Y)$ involving only the quantities of commodities *in his own possession* (represented by the vector Y). That is what I meant earlier by *strictly hedonistic* behavior.

Undoubtedly, such a description is rigorously true for Robinson Crusoe, but it can hardly fit most individuals living in societies. For even *homo capitalisticus* — which Standard theory is deemed to describe — often varies his tips according to his impression of the attendant's neediness, or patronizes a shop only because its owner is hard pressed. Whether or not this conduct reflects the faculty of interpersonal comparison of needs, the fact remains that many an individual responds to changes in the income of the others. A more realistic view of the matter, therefore, leads us to regard the ophelimity

$$(6) \qquad \Omega = \psi(Y; Y_s)$$

as a function not only of Y but also of Y_s, which stands for the particular criteria by which the individual views the welfare of his community. The

---

63. Irving Fisher, *Mathematical Investigations in the Theory of Value and Prices* (New Haven, 1925), p. 11. Also Vilfredo Pareto, *Manuel d'économie politique* (Paris, 1927), p. 62.

individual described by (6) still reacts *hedonistically* — that is, as he desires — but not *strictly hedonistically*.[64]

The problem of individual distribution in a small group — as we know — has no determinate equilibrium in a purely mechanistic schema, whether this refers to an oligopolistic industry or to a small exchange market. In practice a solution is, however, reached only because the group follows some institutional patterns grown out of its particular historical conditions. And we should not be mistaken about it: price leadership, cartel quotas, product competition, and what not, are just as much cultural patterns as "ploughing the land of the widow" or marital dowry, for instance. Without institutional patterns concerning distribution, even the first human societies, small by necessity, could not have arrived at that modicum of stability which is the *sine qua non* condition of organic existence. Whether or not the faculty of "sympathy" for his neighbor is part of man's original nature, that faculty must have evolved before the first viable communities could be formed. Even today, only a type of conduct conforming to (6) can account for the stability of small communities. That, of course, applies first of all to the peasant communities.

A complete description of how individual distribution in a peasant community is regulated must include the institutional patterns prevailing in that particular community. That is why a mechanistic schema of peasant behavior, like the schema of Standard theory, proved to be an impossible project for all who thought of it.[65] And that is not all. If one finally decides to study the peasant institutions so as to construct a *homo oeconomicus* to represent the peasant, one soon discovers that these institutions are almost infinitely variable, a fact that precludes any relevant classification. It is natural that such a baffling and elusive problem as the conduct of the peasant should have attracted few students and resisted being caught within a simple formula.

The small community not only needs a type of conduct oriented also by Y_s, but it also provides the necessary conditions for the operationality of formula (6). In such a community the individual is bound to realize that his own actions influence his own ophelimity also indirectly, via the coordinate Y_s. In addition, everyone naturally arrives at a fairly accurate idea of the situation of everyone else. If these two conditions are not fulfilled, the coordinate Y_s cannot be an *active* agent for the conduct of the individual, even if Y_s is an element of the individual's ophelimity. The most relevant example is provided by the urban agglomerations of

---

64. Clearly, hedonistic behavior does not necessarily imply "altruistic" behavior; all depends upon the sign of $\partial \psi / \partial Y_s$. The pattern (6) accounts very well for the individual who wants to "keep up with the Joneses," and who consequently feels a "pain" if his neighbor's income increases.

65. Tschajanov, *Die Lehre* (note 28, above), p. 131.

an industrial society. In such large communities, the individual can no longer know the situation of all his fellows. On the other hand, he is bound to realize that by his isolated actions he can exert only an infinitesimal influence upon the variable Y_s. He is thus naturally compelled to conduct himself as if Y_s did not enter into his ophelimity function. We cannot avoid noticing the great similarity of this situation with that of the individual producer in an atomistic industry, who also is compelled by his particular situation to act as if his own offer had no influence upon the market price. There are thus two reasons that account for the success Standard theory has had with the assumption of a strictly hedonistic *homo oeconomicus*. The *theoretical* success stems from the simplicity of the atomistic structure, a reason that finds confirmation also in the completeness of the theory of competitive industry.[66] The *practical* success, on the other hand, is due to the fact that the applications of the theory have always concerned a capitalist economy.

Yet, as J. M. Clark remarked long ago, the demand (i.e., the demand derived from purely hedonistic conduct) cannot reflect any of the social purposes of the community.[67] For the fact that *homo capitalisticus* in general behaves as if his ophelimity were independent of Y_s does not mean that he is basically an egoist in comparison with *homo agricola* — as has often been argued. Yet a prolonged eclipse of the social variable, Y_s, as a coordinate of conduct is apt to cause the disappearance of that variable also from the ophelimity function (i.e. from the individual's awareness). This actually happened at the height of bourgeois liberalism in the West as economic goods in the strict sense became the only coordinate of "rational" conduct. As a Populist wrote in the 1870's, in this conduct there was no room for the principles of justice and solidarity of the village life, but only for success by "shrewdness and tricks."[68] That a society could not last long on this basis is proved by the gradual emergence of the welfare state. And, to continue our parallel, let us observe that the welfare state is a genuine cartel, the cartel of an atomistic society with the purpose of dealing with a problem for which isolated action is impotent. That is the *raison d'être* of all cartels. However, the cartel of the welfare state is as much an institution as the social patterns of peasant communities.

66. The completeness of monopoly theory also has its counterpoint in the theory of the consumer. As already remarked, in a mono-society such as Crusoe's, ψ cannot possibly contain Y_s.

67. J. M. Clark, "Economic Theory in an Era of Social Readjustment," *American Economic Review*, IX (1919), Suppl., 288–289.

68. Quoted in Mitrany, *Marx against the Peasant* (note 17, above), p. 40.

IV. CONCLUDING REMARKS

Because some conclusions of the preceding arguments have a direct bearing on practical issues, it seems appropriate to present them together in this last section.

1. The undifferentiating schema of general equilibrium can only divert our attention from the unique role that leisure plays in economics. For instance, man has always endeavored to discover labor-saving devices because in the long run leisure is an economic *summum bonum* (and for no other reason). In the short run, on the other hand, leisure may be economically *unwanted*. An advanced economy, such as that represented by Fig. 11–4(a), may very well have less leisure than a strictly overpopulated economy, Fig. 11–4(b). And indeed, visitors from the lands-of-plenty often point out reprovingly that the people of poor countries indulge in greater leisure than themselves. They seem to ignore the fact that in strictly overpopulated countries people have no choice: in those countries leisure is imposed upon them by geo-historical conditions, and is not the result of an opportunity choice between greater leisure and greater real income, as is the case in advanced economies. In a strictly overpopulated economy, leisure is not properly speaking an economic good, for it has no use but as leisure. Its value then can be but zero.[69] The peculiar characteristic of the strictly overpopulated economy, namely that leisure has no value although labor has a positive "price," bears on the definition of national income.

Walras seems to have been the first economist to include leisure in national income.[70] If x is taken as *numéraire*, and if w and r represent the prices of L and T respectively, Walras's definition of the national income amounts to

$$(7) \qquad \varphi = x + wl + rt = wL_0 + rT_0,$$

where l, t are the amounts of L, T *directly used* by the consumers. The economic relevance of (7) was further revealed by Barone's famous proposition of welfare economics. To recall, this proposition states that for given prices of L and T, optimum welfare requires that the Walrasian

---

69. Further proof that Marx attributed feudal features to capitalism is the fact that in his economic theory he assumes that labor power has no use-value *to its owner*. Cf. Karl Kautsky, *The Economic Doctrines of Karl Marx*, p. 60.

70. Léon Walras, *Elements of Pure Economics* (Homewood, Ill., 1954), pp. 215, 379.

national income be maximum.[71] It follows that *if national income is to be used as an index of economic progress*, its only rational definition must be that of (7). But Barone's argument is valid only for an advanced economy, where leisure is time allocated by an opportunity choice and where its price obviously is identical with that of labor. In a strictly overpopulated economy, however, leisure has a zero value. It is natural to think of eliminating it from the national income and hence to define the latter by

(7 *bis*) $\qquad \varphi_1 = x + w(l - l_1) + rt = w(L_0 - l_1) + rT_0,$

where l_1 represents the amount of leisure, and $l - l_1$ the amount of personal services. Yet, even the labor corresponding to personal services has no alternative use. Moreover, in overpopulated economies t is usually negligible, for such economies cannot afford hunting grounds, national parks, and the like. The conclusion is that in a strictly overpopulated economy, the most rational index of progress is the national product *stricto sensu*. We should then put

(7 *ter*) $\qquad\qquad\qquad \varphi_2 = x.$

In the fact that in an overpopulated economy the feudal formula leads to the maximum national product, we have an equivalent of Barone's proposition: in such an economy the feudal formula warrants maximum welfare.

It is curious that in spite of Barone's proposition the current definition of national income in advanced economies is that of (7 *bis*). Only recently, Simon Kuznets proposed to return to the Walrasian formula on grounds that recall the theoretical implications of Barone's theorem.[72] He rightly points out that by excluding leisure from national income we may obscure an important effect of technological progress. The preceding analysis yields an even stronger conclusion: in comparing the rate of economic development of one advanced and one overpopulated country, we should use in each case the appropriate definition of national income.[73] Indeed, only for the latter economies is it appropriate to define economic progress as the increase per capita net product.

71. Enrico Barone, "Il ministro della produzione nello stato collettivista," *Giornale degli Economisti*, XX (1908), 267–293, 391–414. References will be made here to the English translation in *Collectivist Economic Planning*, ed. F. A. von Hayek (London, 1935), pp. 245–290. The above-mentioned proposition is found on pp. 253–257.
72. Simon Kuznets, "Long-term Changes in the National Income of the United States of America since 1870," in *Income and Wealth, Series II*, ed. Simon Kuznets (Cambridge, 1952), pp. 63 ff.
73. This is especially important for the topical comparisons between East European economies and the advanced economies of the West. For I greatly doubt that in any of the Eastern European countries the increase in T_0 has as yet been sufficient in face of population growth to eliminate the conditions of overpopulation.

2. Ordinary statistical data may be highly misleading for a measure of leisure in the case of overpopulated countries. As surprising as this may seem, the overpopulated countries furnish the highest occupation ratios (F/P).[74] On the whole, it appears that in those countries it is hard to find someone unemployed, yet almost everyone is loafing. The paradox is easily cleared up. With an excess of labor, everyone fights to establish a solid claim to a share of the national product. This leads to a social pattern which may be labeled "splitting the job." Several persons are on a job that technically requires only one person, but each one insists on being considered a full-time employee for fear of seeing his claim challenged.

This practice has been frequently denounced as the hallmark of inefficiency if not of remissness. The scientific critic has justified this verdict on the principle that an efficient economy should pay no factor more than its marginal productivity. It is clear, however, that this argument is an unwarranted extrapolation of a law valid only in advanced economies. Indeed, as we have shown, an overpopulated economy does not operate efficiently unless some laborers earn more than their own contribution to output.

The question of the oversized bureaucracy — an unfailing characteristic of overpopulation — also has been approached from a wrong angle. Few students have realized that in overpopulated countries (and only in these) an oversized bureaucracy is a normal economic phenomenon. With labor used up to its technical limit, nothing can be gained from a reduction in the number of public or personal servants; such a reduction may create only social turmoil. Many an overpopulated country deserves to be censured not because it has a large bureaucracy but because its entire government class has a high standard of living amidst poverty. Undoubtedly, too high a standard of living for the government class is a deterrent to economic development, for it greatly reduces the already meager power of capital accumulation of the economy. If in an advanced economy equality of standard of living answers to an ethical principle, in an overpopulated country it represents an economic imperative.

3. Glossing over academic refinements, we may regard economic development as an upward shifting of the labor productivity curves (Fig. 11-4). A fabulous amount of foreign aid apart, no economy can *jump* from the situation of Fig. 11-4(b) to that of Fig. 11-4(a). In other words, it is quasi certain that in its development a strictly overpopulated economy has to pass through a phase like that depicted by Fig. 11-4(c), i.e. through a phase where the working class has no leisure at all. This situation with

---

74. For example, the occupation ratio in Rumania before World War II was one of the highest in the world, *Enciclopedia României*, I, 154.

its sixteen-hour day and seven-day week is well known owing to its detailed description by the socialist literature of the last century.[75] As already pointed out, Marx erroneously took it for a basic feature of capitalism. The period in the economic history of the West that served him as a model for depicting "the calvary of capitalism" corresponds rather to the growing-pains of capitalism. For capitalism, understood as an economic system regulated by profit maximization, could really exist only after the marginal productivity of labor had reached a sufficiently high level so that it could be equated with the wage rate. Capitalist development proper began only after this phase had been consummated. Increasing leisure (not *unwanted* leisure) for the working class constitutes its most distinctive feature. To wit, the forty-hour week is a relatively new institution, and the idea of a four-day week is already being aired.[76]

Strictly speaking, the East European countries have never come to know the so-called calvary of capitalism. From the middle of the nineteenth century, if not before, these countries began instead to receive the *impact* of Western capitalism. Although usually regarded as a phenomenon equivalent to the "calvary of capitalism," the impact was an essentially different process. The plain fact is that the East European economies were not yet sufficiently developed to begin the calvary. The true story can be told in a few words. Increasing trade with the West revealed the existence of other economic patterns and at the same time opened up new desires for the landlords and new ambitions for the bureaucracy. Under this influence the feudal *contrat social* began to weaken. An ever-increasing number of landlords switched to the capitalist formula of maximizing profit-rent, a change which, even if it did not always increase their share, had the advantage of freeing them from their traditional obligations towards the villagers. This process was later to culminate in producing the pure absentee. From this viewpoint, the main beneficiary of the freedom of the serfs was the landlord, not the peasant. That is true also of the earliest agrarian reforms (1861 in Russia, 1864 in Rumania), which in reality sanctioned the separation of the economic interests of the landlord from those of the peasant.

75. Even in the United States the average working week was seventy hours as late as 1850. Undoubtedly, earlier it was even longer. Interesting also is the fact that the first attempt to limit the work of children under 12 to a ten-hour day was that of the Commonwealth of Massachusetts in 1842. W. S. Woytinsky and Associates, *Employment and Wages in the United States* (New York, 1956), p. 98. The ten-hour day did not become a widespread standard for the other workers until 1860. Philip S. Foner, *History of the Labor Movement in the United States* (New York, 1947), p. 218; G. Gunton, *Wealth and Progress* (New York, 1887), pp. 250–251.

76. Marx failed to see that one possible synthesis of his antithesis could be precisely this. Instead he wrote that "the relative overpopulation becomes so much more apparent in a certain country, the more the capitalist mode of production is developed in it." *Capital*, vol. III, chap. xiv, sec. 4, p. 277.

Now, to regulate production by profit maximization is probably the worst thing that can happen to an overpopulated economy, for that would increase unwanted leisure while diminishing the national product.[77] To be sure, newly imported techniques alleviated the crisis, but hardly the lot of the peasant. This is the explanation of the fact often commented upon that in Eastern Europe capitalism worsened the lot of the peasant, while in puzzling contrast increasing the prosperity of other sectors. It is this situation peculiar to the countries caught lagging behind Western capitalism that gave rise to the Agrarian ideology. And this ideology remained a regional philosophy at which the West looked as a curio, precisely because in its economic development the West had not had a similar experience.[78]

4. In a nutshell, the main tenets of the Agrarian doctrine are:

(1) Because of their geographical situation some communities will always rely on agriculture as a main economic activity. And since agriculture is an intrinsically different activity from industry, such communities cannot develop along identical lines with the industrial economies.

(2) For the countries with an agricultural overpopulation, individual peasant holdings and cottage industry constitute the best economic policy.

Evolution is subject to pure uncertainty, and the most we can do in tackling an evolutionary problem is to trust the existing evidence as a basis for meeting the future. As to this evidence, we have dealt at length with the historical refutation of the law of concentration in agriculture and with the specific differences between industrial and agricultural activities. We may add, however, that nothing as yet has happened to cast any doubt upon the validity of that analysis, and hence upon the first point of the Agrarian doctrine. Besides, the economic development of Denmark and Switzerland as well as parts of Germany and Austria prove that agriculture can provide the basis of its own economic development. Moreover, both anthropology and economic history confirm that only a substantial food production (independent of its source) has led to capital accumulation.[79] Quesnay's celebrated maxim works both ways: *riches paysans, riche royaume*. The logic is surprisingly simple: Robinson Crusoe could not be

---

77. Kautsky, *Economic Doctrines*, p. 235, recognizes the difficulties created by the adoption of profit maximization, but fails to see the real explanation of the process.

78. Only very recently have Western economists come to accept the view of the Agrarians that the East European countries had suffered the impact of foreign patterns not befitting their own cultures and conditions. Cf. *Méthodes et problèmes de l'industrialisation des pays sous-développés*, United Nations (New York, 1955), p. 141.

79. V. Gordon Childe, *Social Evolution* (New York, 1951), p. 22; Bruce F. Johnston, "Agricultural Productivity and Economic Development in Japan," *Journal of Political Economy*, LIX (1951), 498.

available for forging a sickle before Friday could gather enough fruits for both. "Industrialize at all costs" is not the word of economic wisdom, at least in overpopulated agricultural countries.

The second point of the Agrarian doctrine clearly aims at using as much labor in production as is forthcoming. It also reveals that Agrarians were the first to feel intuitively that the economic forms compatible with optimum welfare are not identical for all geo-historical conditions *even if the technological horizon is the same*. We should recall that the real novelty of Barone's work mentioned above was the proof that the controlled economy of a socialist state must imitate the capitalist mechanism, i.e., it must adopt the principles of marginal productivity theory, if it is to obtain optimum welfare. However, neither Barone nor others after him seem to have been aware of one important restriction, namely, that marginal productivity principles presuppose the existence of a well-advanced economy in order to achieve optimum welfare. And thus numerous writers have felt secure in using in their arguments the converse proposition: capitalism and controlled socialism provide the best systems for developing an underdeveloped economy. Yet this proposition is patently false, at least for an overpopulated economy.

In this light, the intuition that led the Agrarians to their double negation — not Capitalism, not Socialism — proves to have been surprisingly correct. But then, what is the theoretical schema of the Agrarian doctrine? Because Agrarians have hardly bothered with theoretical schemata, one can only attempt an *ex post* rationalization and thereby accept the risks of misinterpreting their own rationale.

5. The arguments presented in this paper unmistakably lead to the conclusion that the Agrarian schema is the feudal formula under a new form. Capitalism — as we have explained — came to Eastern Europe not as a natural phase in economic development, but as the result of cultural contamination. In the light of economic dynamics and positive welfare theory, there can be no doubt that this was a move against the grain. For feudalism was thus displaced before the respective economies had a chance to reach the calvary phase of capitalism, i.e. the *normal* gateway to the advanced stage of economic development. Only a tithe system can efficiently carry an overpopulated economy through that phase.

But, from the Marxist viewpoint, the premature disappearance of feudalism was a step in the right direction, for it represented an earlier fulfilment of what must inexorably come (a view probably shared also by most Standard economists). Into the Agrarian ideology we can read, however, a different position, even more Hegelian in spirit: no phase of economic development can be by-passed. In particular: feudalism cannot disappear before it has completely finished its job. If artificially displaced, it will return under one form or another (barring the occurrence of Malthusian holocausts). In such an alternative, the only logical attitude is then to plan rationally for the continuance of feudalism in such a way as to make it work even better. The policy of radical agrarian reforms in overpopulated countries, by which the head of each peasant family is turned into a feudal entrepreneur, responds precisely to this logic.[80]

A very interesting question now comes up: Which of the two Hegelianisms, the Agrarian or the Marxist, is supported by history? The answer must be sought in what happened in the overpopulated countries after a Communist regime took them over. Unfortunately, our knowledge of what happened is very incomplete. Because the marginal productivity formula cannot possibly work efficiently in an overpopulated economy, it is fair to assume that no Communist regime would use it in this situation. If, however, they use a formula equivalent to a tithe system, the Agrarians are fully vindicated. To be sure, in a Communist regime the distribution of income between the government and the working "groups" may follow an entirely new formula. "From each according to his abilities, and to each according to his needs" has no operational value. Only when we have learned the theoretical schema of this new formula in concrete concepts (labor productivity and labor supply) shall we be able to answer the question in a more complete way.

6. By chance, history supplies us with proof and counter-proof examples regarding the impact of capitalism. The consequences of the premature decay of the feudal formula are best illustrated by the economic situation of the Russian and Rumanian peasant which in relative terms continued to deteriorate all through the hundred years or so preceding World War I. The few timid agrarian reforms were not able to ameliorate the situation, a fact reflected by the frequent peasant jacqueries, some of exceptional intensity. The counter-proof example is provided by Hungary, the well-known bastion of feudalism. In comparison with all her neighbors also plagued by overpopulation, Hungary stood out by virtue of a better fate of the peasant (in most regions) and a conspicuous economic development in all fields. To a great extent this difference can be attributed to the fact that the Hungarian magnate did not succumb to the capitalist formula as his Polish and Rumanian colleagues had done on a large scale.[81] The praises uttered by the apologists of the paternalistic character

---

80. Engels implicitly recognized the merit of feudalism when in a letter of 1892 (Marx and Engels, *Correspondence*, p. 501) he stated that "an agrarian revolution in Russia will ruin both the landlord and the small peasant," but he clearly ignored the fact that most landlords had long since ceased to follow the feudal formula.

81. The earlier liberation from the economic yoke of a foreign power most probably constitutes another important factor of the difference. But the exploitation of national minorities for which Hungary set a paradigm cannot have been a very important

of Hungarian feudalism were more often than not a *pro domo sua* argument, but they were not entirely without basis. Its undisturbed existence, brought to an end only in 1945 by extra-economic forces, is another proof of the success of that feudalism. But precisely because of this progress, Hungarian feudalism had undoubtedly ceased to represent a necessary economic formula long before 1945.

7. The rather surprising intuition of the Agrarians, however, failed them in one important respect. They were unable to realize that in order to obtain the maximum output from given amounts of resources, the production unit must be of optimum size. Consequently, they could not foresee the danger of determining the size of the peasant holding according to extra-economic criteria. The principle "a holding for every peasant family" naturally led to a suboptimum size of the production unit. And this prevented the crystallization of the existing capital in the most efficient form compatible with the prevailing factor ratio and the available techniques. The unmistakable symptom of this situation was a relative excess of farm equipment. In Rumania, for instance, before the radical reform of 1918, there was a plow for every 26 acres; after the reform, there was one plow for every 15 acres.[82] The Agrarians discovered their error only *post partum*, and when they did, it was much too late. For in East Europe at least, the historical changes of the early 1930's prevented the Agrarian parties from forming a government again.

The facts just mentioned do not justify the prejudice of Stalinist governments in favor of large and highly mechanized farms of the North American type. This prejudice errs in the opposite direction: it leads to a size far greater than the optimum compatible with overpopulation, and hence it uses labor inefficiently.[83]

8. Poor theorists though they have been, Agrarians have never lost sight of the most elementary principle of economic development, which is that no factor should remain *unnecessarily* idle. In overpopulated economies, this may mean using labor even to the point where its marginal

---

element, for even the lot of these minorities improved in some proportion. We should also remark that, until her dismemberment in 1918, Hungary had almost as high an agricultural density as Poland, Rumania, Yugoslavia, or Bulgaria; this can be seen on the map in Moore, *Economic Demography* (cited in note 32, above), p. 73.

82. See Georgescu-Roegen, "Inventarul agricol," *Enciclopedia României*, III, 339. Instances of "inefficient" forms of capital were quite common. To mention one more: the cows represented less than 70 per cent of the entire cattle stock. This was a consequence of the fact that every peasant holding needed a pair of oxen for draught purposes, while only a few could keep more than two animals.

83. Calvin B. Hoover, *The Economic Life of Soviet Russia* (New York, 1932), p. 88, reports of farms having ten times as many workers and twice as much machinery as a farm of equal size in the United States. See also Warriner, *Economics of Peasant Farming* (cited in note 34, above), p. 169.

productivity becomes zero. From the viewpoint of positive welfare economics, we cannot do better than to cling to this principle. The question, however, is whether we can always comply with it in practice. A small farm, a small shop, can easily be run by a family or a true cooperative, and hence follow the feudal formula. On the other hand, many current products can be produced only by large plants. And large production units requiring numerous employees who have no other ties with each other than working together, lend themselves poorly, if at all, to the feudal formula. First, it is hard to see how a manager could use labor beyond the point where its marginal productivity is equal to the wage rate and still be able to prove the efficiency of his management. Secondly, once the principle that one can earn more than his own contribution to output has been accepted, the question of everyone's doing his duty becomes a thorny problem. In the continuity and closedness of the village, these problems are easily solved by the emergence of cultural patterns in which loafing is one of the worst sins.[84] There, efficiency does not need bookkeeping in order to be recognized.

The baffling problem of reconciling the requirements of modern technology with the basic principle of economic welfare is no reason for throwing the latter overboard. In every overpopulated country, at least, there are numerous sectors which either by their nature or by their tradition permit labor to be used according to the feudal formula. Agriculture is almost everywhere in this category.[85] It would be the worst economics to change the production structure of such a sector because of an ill-advised development fever. No one would dispute the truth that peasant institutions and modern industry do not fit together, but it would be a great error to sacrifice all these institutions on the altar of that truth. And if a scapegoat for failing economic policies is needed, one should find a less expensive one. Many of these institutions will still be needed if the largest output is wanted from the sustaining sectors of economic development. Besides, the iconoclast may live to regret his haste, for we should not be surprised at all if the fight of the Communist regimes against the "bourgeois spirit"[86] in reality aims at creating a "socialist man" with a peasant type of conduct.

84. Some primitive communities — we are told — hold work in such high esteem that they produce more than they need and then destroy the excess. See Richard Thurnwald, *Economics in Primitive Communities* (London, 1932), p. 209.
85. By destroying the peasant holdings altogether and replacing them by bookkeeping operated *kolkhozi*, Stalinism certainly made a losing deal with the basic welfare principle. A truly cooperative form of production on units of optimum size with *product ownership* and *tithe* paid in kind to the government would be by far the best welfare solution. Titoism seems to have realized the Stalinist error when it renounced collectivization.
86. See the resolution of the 1931 All-Russian Congress of Workers in *Report of the Ad Hoc Committee on Forced Labor*, United Nations, I.L.O. (Geneva), 1953, pp. 456–457.

9. To assume that a process that sustains the progress of advanced economies necessarily befits an overpopulated economy is an unwarranted extrapolation. The ever-growing literature on economic development, however, abounds in such extrapolations. Probably the most patent is the use of marginal productivity principles in formulating economic policies for underdeveloped economies.[87] But only a few of these economies do not suffer from overpopulation. And it is an obvious feature of overpopulated countries that enterprises operated by feudal formula exist side by side with others managed according to capitalist rules. In such circumstances, price lines are not tangent to the isoquants of every sector and, hence, the isoquants themselves are not tangent to each other. This is easily seen in Fig. 11–5(b) which represents an economy where X_1 is produced according to the feudal formula and X_2 according to capitalist rules.[88] The price line PP is tangent in M' only to the isoquant of X_2. (And only the workers of X_2 earn *wages*.) Of course, M' does not satisfy the elementary condition of positive welfare. The reason, however, is not that the national product would be definitely greater at M'' than at M', but that optimum welfare is represented by M where employment is maximum; see Fig. 11–5(a). The important fact is that neither M nor M'' can be obtained if industry X_2 works according to the capitalist formula, for in either situation the marginal productivity of labor falls below the minimum of subsistence.

That is not all: the price line may not be tangent to the X_2 isoquant either. For, in contrast with what happened in the West during the early phase of industrialization, the cities of overpopulated countries have grown to pathological sizes through a continuous immigration of rural population. The rural exodus brings into those cities not only an enormous excess of labor but also the germ of the feudal economic spirit to which practically no sector can remain immune. The social pressure of people seeking employment for their worthless leisure is so irresistible at all times that even the most convinced "marginalist" of the entrepreneurs has to yield to it and hire more people than he should according to his own rule. In these circumstances, prevailing factor prices may be proportionate to

---

87. Because of the momentous importance of the problem of underdeveloped countries in world affairs, the consequences of these extrapolations may well exceed those of mere academic licenses. All the more so when the source is as high an authority as the United Nations. In *Measures for the Economic Development of Under-Developed Countries* (New York, 1951), p. 49, the U.N. urges the use of the principle of marginal productivity which — it complains — "is frequently ignored in practice."

88. This schema is not a mere theoretical concoction. It actually corresponds to those economies with a high density of agricultural population living on family holdings and with industry operated more or less on the capitalist formula. Bulgaria, Rumania, and Yugoslavia — to mention only the cases with which I am thoroughly familiar — were precisely in this situation before the advent of Communism. It is precisely because agriculture lends itself easily to operation by the feudal formula that "too many farmers" is a rather general phenomenon, not peculiar to overpopulated economies.

anything except the corresponding marginal productivities. To compute the money equivalent of the marginal productivity of an investment on the basis of the prevailing prices is pure nonsense. The criteria of investment priority based on the results of such computation are therefore baseless.[89] Still worse: such criteria point in exactly the wrong direction. Indeed, except for the correction for external economies, these criteria are identical with those used by the private investor. And the result of private investment is a well-known paradox: although labor-intensive techniques are the only ones indicated for overpopulated countries, the industries developed there have generally been capital-intensive. The explanation is obvious: in an overpopulated country the ratio between wages and the price of other factors is greater than the ratio between the corresponding marginal productivities.

Economic development does not mean only pure growth; in the first place it means a growth-inducing process. Investment in capital-intensive industries is a wrong move in an overpopulated country not because it fails to bring about growth — for it generally does — but because they are not growth-sustaining. The power to sustain growth then is the only valid criterion of investment in undeveloped countries. The marginal productivity principles reflect this criterion very poorly, if at all. Even for a capitalist system they cannot explain more than distribution through allocation.

The path followed by the West in its economic development can help us in seeking a policy for the development of those areas that have remained behind. But it cannot show us *the* way. For, clearly, by this policy we cannot possibly aim to follow precisely the same route that the West followed. It would take us too long to reach the goal. Still more important: it would not even be feasible, for the opportunities the West has had at one time or another cannot be reproduced. The essential distinction between an historical and a dynamic process does not need to be re-examined here. But, at bottom, this is what Marxists and Agrarians quarreled about. Can we, as Standard economists, learn something from this quarrel?

89. For the investment criteria based on marginal productivity one may refer to A. E. Kahn, "Investment Criteria in Development Programs," *Quarterly Journal of Economics*, LXV (1951), 36–61; H. B. Chenery, "The Application of Investment Criteria," *ibid.*, LXVII (1953), 76–96, among others. Because these criteria are endorsed by some economists serving as consultants to various economic development agencies, one may infer that they are used as a guide for public policy (e.g., G. di Nardi, " 'Criteri' e 'Indicatori' per la scelta degli investimenti," *Rassegna Economica*, July 1957).

POSTSCRIPT (1966)*

Since the above essay appeared in 1960 the importance of the labor surplus in relation to nonlabor resources in an underdeveloped economy — especially in an economy with a preponderant overpopulated agricultural sector — has received increasing recognition among standard economists. But the best proof that principles as resilient and as extensively manifested as those expounded in that essay cannot be ignored or denied forever is that last year such recognition came from two prominent economists of Soviet Russia and Poland. The implicit admission by spokesmen for the "truest" Marxist doctrine that the very agrarians, whom Marxists have despised profoundly and fought bitterly, were after all right verges on the miraculous. For this admission came not as a mere academic matter but as a justification of some vital points of the agricultural economic policy of Soviet Russia and Poland, The purpose of this postscript is to invite the reader's attention to this highly significant development.

The first article, by G. Shmelev, treats of "The economic role of the personal auxiliary farm" ("Ekonomicheskaia rol' podsobnogo khoziaistva," *Voprosy ekonomiki; ezhemesiachnyi zhurnal,* Akademia nauk SSSR, Institut ekonomiki, n. 4, p. 27-37, Apr. 1965). It opens with the recognition that since in the Soviet economy "the social production of collective and state farms has not yet achieved a high level and the people's needs for agricultural products are still not fully satisfied, an additional source of foodstuffs is found in the personal auxiliary farms [i.e., small lots cultivated on a personal basis] of the members of the collective farms." The data quoted in support are impressive: in 1963 the vegetable production of the private lots amounted to 24 percent of the national output and that of animal production to 46 percent. And since the article appeared in the Annals of the Soviet Academy, it is beyond question that Shmelev's paper represents the present official thinking: "The groundless restrictions on the personal auxiliary farms [which in the past] not only slowed down the progress of the welfare of the laborers in villages but also affected the development of the social economy . . . are now cancelled" (p. 30 ff).

One can hardly expect yet a Soviet economist to go too deeply into the crucial question of why part of the land should be allocated to personal farms if agricultural products are scarce and if — as the official dogma proclaims — the collective farms are the embodiment of a superior technique of production. All the more significant then is what Shmelev does say. And he admits that Soviet official statistical data reveal that "on the auxiliary farms production is carried more intensively [and] their products surpass in quality those of the collective farms" (p. 32 ff). However lightly, he touches some of the roots of the problem too. The auxiliary farms provide a means for the employment of *free* labor time in the national production, so that the labor expended on them must not be regarded "as wasted labor, but as profit received through increased production" (p. 33). One of Shmelev's general statements, however, is certain to astonish many, especially the numerous advocates of economic development by industrialization and only by it: "It cannot be assumed that the change of the private farms to social production can come about in the near future. The prerequisites for the disappearance of the

---

*The Postscript was originally written for the Spanish translation "Teoría económica y economía agraria." *El Trimestre Económico*, v. 34, 1968, p. 589-638. In the present version, references have been brought up to date. The emphasis in most quotations was added.

auxiliary economies will emerge only when the returns from social production will exceed those of the private farms. *The increase in the returns of the social production will, in its turn, depend upon the nature of the development of the agricultural sector"* (p. 29 ff).

Issues are more closely grasped by the splendidly marshalled argument of Professor J. Tepicht (of the Institute of Agricultural Economy of Warsaw) in "Problems of the re-structuring of agriculture in the light of the Polish experience," presented and circulated at the Conference of the International Economic Association, Rome, 1965.[1] With an insight that contrasts with the uninspired approach of most agricultural economists of the standard school, Tepicht describes the peasant economy as a socioeconomic category "characterized by the subordination of the lot [fate] of each individual to the general interest of the family unit, and vice versa" (p. 539). In such an economy "the contribution of effort by each member of the family remains anonymous with respect to the collective income which is at the disposal of the whole family" (p. 539). This, to be sure, is a weaker statement than my description of the peasant economy as a wageless economy in which the interests of *all* members, not only of the members of the same family, are tied together by cultural bonds. But Tepicht does see and also stresses the highly important point that the peasant economy displays an "astonishing ability to survive," due primarily to its "capacity for suffering" through endless economic adversities (p. 536). That is why, he rightly concludes, "the peasant economy is a category that is valuable in analyzing more than one of the historical forms of organization of human society" (p. 537). In view of the well-known Marxian thesis concerning peasantry, this is a most surprising statement on the part of a Marxist economist. Yet it has a perfectly logical place in Tepicht's argument, which at bottom is concerned with Poland's policy towards its inherited peasant economy. For as he says, "Poland presents an instructive example of an advanced adaptation of peasant farming [economy] to the structures that surround it" (p. 537). To say that the peasant economy plays an important role in socialist Poland is perhaps a more direct way of expressing the same idea.

Tepicht argues that the factor responsible for the persistence of peasant economies in all parts of the world is "the very restricted professional mobility of which this [peasant] population can take advantage" (p. 537), which in turn is the consequence of the low economic development of the nonagricultural sectors. In this he is undoubtedly correct but also incomplete. For if this were the whole story even outside the United States, how could one explain the persistence of a peasant economy, as defined by Tepicht himself, in Germany, or in Denmark, or in France? Apparently, Tepicht does not want to become involved with the population problem and, as a consequence, overlooks the curious fact that in modern times the predominant peasant economies generally are overpopulated relative to land.[2] But he correctly interprets the high occupational ratio in countries with a large peasant economy as "a clear sign that there is regular utilization of the *marginal labor* of members of the family which adds generally in a digressive, though disguisedly digressive manner, to the total agricultural income of the family" (p. 538). Undoubtedly, by marginal labor Tepicht can only mean the surplus labor counted from the point where labor marginal productivity equals the wage rate in a marginal

---

[1] The paper has since been published in: Ugo Papi and Charles Nunn, eds., *Economic Problems of Agriculture in Industrial Societies,* London: Macmillan, 1969, pp. 534-47.

[2] For an attempt at explaining why peasant societies not only display a tendency to grow very fast but also are capable of such a growth, see my essay "The Institutional Aspects of Peasant Communities: An Analytical View," chapter 8 in this volume.

pricing economy. Thus, Tepicht too sees the merit of a peasant economy even in a socialist country, such as Poland, in that the *familial* farm offers to the surplus population an immediate opportunity of creating additional income instead of remaining idle. The reminder that in socialist Poland only "less than ten percent of the peasant families and their land was reorganized into collective farms" (p. 539) is only a simple hint. For Tepicht does not retreat in front of the conclusions of his argument: "In a large number of countries of the 'Third World', it is necessary to clear the way for expansion, *sometimes even to stimulate the creation of a peasant agriculture,* and not evade the problem by flirting with premature collectivism, or limiting attention to islands of ultra-modern agriculture that seem to deserve preservation under a new social form" (p. 536).

As to positive recommendations, Tepicht — in a far more explicit fashion than Shmelev — insists upon the necessity of trying "to obtain the most practicable increase of the social productivity of agriculture in general, rather than the maximum technical productivity of a small number of [large] enterprises" (p 544). To be sure, only enterprises modeled after the American experience can embody the most advanced techniques, but they require vast investment and, a point not touched by Tepicht, ample natural resources per capita. "But as long as the general level of economic development renders a radical population exodus from rural areas impossible, *or as long as other socio-economic elements necessitate the continuance of peasant farming,* priority must be placed on other factors for promoting efficiency" (p. 543 ff). In such a case "a combination of the past with the future [should] remain the rule with a growing domination of the latter" (p. 543).

In the course of his argument Tepicht informs us that a study of the output per acre and per family in the collective and peasant farms "with similar conditions and located in the same region," has revealed that the peasant farms are the more efficient (p. 540). Coupled with the fact that the quality of the production of the peasant farms is superior to that of the collective farms, this extraordinary piece of information complicates matters and, hence, upsets in part Tepicht's as well as Shmelev's position concerning the necessity of "tolerating only temporarily" the private family holdings. For if peasant agriculture is on the whole at least as efficient as the collective farms, then it is the collective farms that are being tolerated. Given, moreover, that the collective farms are using the most advanced techniques, the question arises why they are not superior to the peasant farms. The mystery must be solved and the argument reconsidered in the light of it.

I am tempted to repeat here the closing phrase of my essay. But to avoid monotony I can as well quote Tepicht's closing statement: "The moral of this study is an appeal for [unprejudiced] comparisons of which we have tried to show the advantages. Let us hope that it will be heard" (p. 547).

POSTSCRIPT 1975

Fifteen years have gone by since the above essay was published and, without any intention of being immodest, I find little in it that needs to be revised.

There is, first, the fact mentioned in note 73 which is no longer entirely true. In the Communist countries of Eastern Europe, industrialization has produced an impressive industrial capacity, though with efforts much greater than would have been necessary under a different regimen. Some times industrial complexes even have monster-like dimensions, designed without concern for pollution, for the availability of raw materials, and even for demand. This industrial growth for the sake of mere growth has created a remarkable shift of labor power from agriculture — now heavily mechanized — to industry. However, according to my personal observation of the situation in Romania and Yugoslavia, labor in villages may still be superfluous in some regions.

A more important point pertains to my criticism of Theodore Schultz's position that there is no such thing as agricultural overpopulation. My knowledge of the economic situation in Latin American countries, which formed the basis of his argument, being very superficial at that time, I granted that those countries might even suffer from underpopulation. Subsequently, I was able to observe on the spot the much discussed situation in the North-East of Brazil. I thus discovered that, given the fact that only a narrow strip along the coast is under cultivation, the condition of overpopulation as defined in the above essay prevails at least in that rather economically isolated region.

A conclusion of a far greater importance for the continuous struggle of economists, from Brazil as well as from other countries, to design a policy for the speedy development of that region is that the recipe of an agrarian reform — of settling the landless agricultural workers on the expropriated estates of the big landowners — would not work in this particular case (and probably in some other parts of Latin America). As I explained in a lecture for the twentieth anniversary of *Banco National de Desenvolvimento Econômico**, the peasants of Europe, Asia, and Africa, whom I had in mind in supporting the advantages of agrarian reforms, have always been skilled farmers. For, they alone had performed all agricultural tasks on the lands of the landlords or of the tribal chieftains in continuity over centuries. When agrarian reforms were implemented, from Western Europe to Asia and Africa, the settlers were thus able to continue producing without great setbacks. The only difficulties encountered were of a commercial nature, a direction in which the old serfs had had no experience. In sharp contrast, the millions of agricultural workers of the Brazilian North East have always been migrant laborers, living not on the plantations, but outside of them, moving from one plantation to another, hired only for short periods to perform tasks that varied with the nature of the crop. These disoriented *camponeses* have never had the opportunity of becoming skilled farmers for some particular crop. The upshot is that, being thus unprepared for running a farm, they would be incapable of using the lot on which they would be settled. The situation is most tragic, because millions of such *camponeses* cannot be trained, except on the spot and during a relatively long time, a task which under normal circumstances seems irrealizable. We know how to train mechanics, bricklayers, plumbers, and the like, in schools. But the school for farmers is the farm itself.

---

*Nicholas Georgescu-Roegen, "Fisiologia do Desenvolvimento econômico," *Panéis internacionais sobre desenvolvimento socioeconômico*, Rio de Janeiro, APEC/BNDE, 1974, p. 344.

Part II
Institutional Economics

CHAPTER 7

(1968)
Structural Inflation-Lock and Balanced Growth[1]

I. Introduction

"Inflation" is a notoriously plurivalent term. Its denotations display a great variability that tends to become even greater with every new writing on the subject. The main differences among the current uses regard primarily the source of the inflation, its intensity, and its duration. But the common precipitate of all these denotations is a rise in money prices consequent to an increase in the purchasing power of the community through *monetary channels alone*. The underscored expression is crucial. As unanimously conceived, I believe, inflation is essentially a monetary phenomenon. We certainly do not speak of inflation in an economy where only simple barter prevails (even if a fictive unit of account may as a rule be used): in such an economy there are no money prices. A more piercing example is the case of an economy where one regular commodity is used as *numéraire*. In this case, a rise in the price level of all other commodities with respect to

---

[1] The general ideas presented below in Sections VII-IX formed the object of a number of public lectures and seminars given by the author in Brazil and Argentina during the summer of 1966. A preliminary version of the present paper was read at the annual meeting of the Southern Economic Association (November 18, 1967) and during 1968 at several faculty seminars in the United States. A short Portuguese version, "O Estrangulamento: Inflação Estrutural e o Crescimento Econômico," appeared in *Revista Brazileira de Economia*, XXII, No. 1 (March 1968), pp. 5-14.

The author wishes to acknowledge also the help of his research assistants, Aly Alp Ercelawn and Ibrahim Eris, especially the former's tenacity in collecting and systematizing the statistical data from highly disparate sources.

the *numéraire* may represent either a cheaper method of producing the *numéraire*, the appearance of increasing costs in most of the other lines of production, or simply a shift in the tastes of the community—all nonmonetary phenomena. By the same token, most writers would not apply the term "inflation" to an increase in the price level in a predominantly agricultural country if the increase is due *solely* to a poor harvest or to a declining yield caused by the gradual exhaustion of the power of the soil.

Briefly, inflation is an event involving *money*. Although "money" is another notoriously ambiguous term, here again all definitions leave one characteristic precipitate: in the long run the normal average cost of production of a unit of any regular commodity, say, one egg or one ton of steel, is one egg or one ton of steel, respectively. (Of course, the same is true of any *numéraire*.) In sharp contrast, the cost of producing one unit of money is only a small fraction of that unit[1]. If it were not for this difference, inflation would not be the policy tool that it has been since quite old times. Also, economists would miss a highly absorbing topic. Few, perhaps, would suspect that some of the issues surrounding inflation (and which occupy a prominent place in this paper) will nevertheless subsist.

For reasons easy to understand, most economists have concentrated their attention on one particular type of inflation. It is the inflation of a moderate intensity that has been continually going on in almost all advanced economies with a well-organized money market and a monetary circuit encompassing all economic units, with only a standard guidance of the interest rate by the central banking authority, but without any direct government controls over prices, wages, and rents. That is, the standard concern has been with the inflation that propagates itself *primarily* through credit expansion—whether or not accompanied by a cheap money policy of the central monetary system. To borrow a term from

---

[1] I should hasten to add that the cost of *using* money (which normally is represented by the interest rate) is not to be confused with the cost of *producing* money. The implication of this standpoint need not disconcert us. By expanding their credit, financial institutions do not produce money; they only increase the number of simultaneous users of the *same* unit of money. For a parallel: a tenement owner may rent the same room to two, even three persons each day: he is not thereby producing rooms.

Keynes (1924, p. 90), the main concern has been with *credit-inflation*.

Cash-inflation—the other "pure" type in Max Weber's sense—is a process triggered by an increase of the volume of paper money in circulation. The type attracted some attention immediately after World War I because of its practice by the belligerent countries of continental Europe. But interest in its study quickly dwindled; as Hansen (1949, p. 39) explained, economists thought that the type is "now outmoded". This paved the way for the inflation theory built upon the particular of an advanced economy to become our point of reference for *any* type of inflation. And that is not all. More often than not, the prototype itself is attributed properties concerning the knowledge of the individual and his freedom to act according to this knowledge that, in fact, destroy some specific features of *credit-inflation* even in an advanced economy [1]. The fact that for describing any inflation we need no other analytical tools than those used by the standard theory should not mislead us. If two realities differ essentially, there must be some descriptive propositions that apply to one but not to the other; logically then, not every conclusion derived from one axiomatic foundation will necessarily apply to the other case (Georgescu-Roegen, 1966, pp. 6 f, 360 ff). That is why whenever any theory is thrust upon another reality facts appear unsettled and, hence, invite controversies.

A glaring illustration is the endless controversy over the situation in the Latin American countries which has largely fed on the uncritical extrapolation of the standard theory. In this particular case the consequences of the license have been aggravated by some special factors, one of which should be mentioned at this stage. Credit-inflation, whether caused by an increased demand from business or from the government, works through an intrinsic vehicle of the economic process, namely, the money market in the broadest sense of the term. Cash-inflation, however, implies a political act of authority external to the selfsame process. Because

---

[1] That both these tendencies are not confined to the rank and file of the economic profession is evidenced (as we shall see on many occasions) by none other than Milton Friedman in his Presidential Address to the American Economic Association (1968) as well as by the reputable economic experts who have drafted the statement on *Fiscal and Monetary Policies for Steady Economic Growth*, Committee for Economic Development, January 1969.

of this bivalent nature of cash-inflation, the analysis of any actual case must pay sufficient attention to the motives—overtly recognized or unwittingly pursued—that have presided over the political decision. The use of cash-inflation to finance a war raises no issue in this respect: since nothing can outbalance the imperative of national survival, we may say with Keynes (1930, II, p. 174) that the policy is both "inevitable and wise". But the position shared by many students of Latin American economies that a cash-inflation held within "reasonable" limits is the inevitable solution for the economic development of these countries is far from being clear. Apart from a host of rather impressionistic opinions, all one finds in the literature on this issue are some disconnected, albeit pertinent, observations and arguments.

This paper is offered as a contribution toward a systematic analysis of the problem not in terms of curves or systems of equations, but (for reasons to become apparent as I proceed) in terms of "structures" in the tradition of the French school of François Perroux. Let it be said, however, that I disclaim any virtue of this analytical piece to explain everything in every Latin American country. After all, the economic conditions vary from one such country to another and even within the same country they have often suffered the impact of sudden changes of political regimes. My only contention is that the common cultural tradition of Latin American nations is still reflected in some important common traits of their present economic, social, and political matrices, and that only an argument pitched at the level of abstract generality determined by these common traits can reveal the forces which, over the variable contingent, control the general operation of the system. And to avoid possible misunderstandings, I may add that I also contend that this argument (or any other analytical argument for that matter) cannot be a conformal mapping of the economic actuality, for the simple reason that in actuality things are not cut off from each other as sharply as our analytical concepts are.

II. Inflation and Static Analysis

1. From the highest exponent of the orthodox view of monetary matters we learn that "the quantity of money is extremely important for nominal magnitudes, for nominal income, for the level of income in dollars—important for what happens to prices, [but] it is not important at all, or, if that's perhaps an exaggeration, not very important, for what happens to real output over the long period" (Friedman, 1969, p. 46). For the real meaning of this statement we are provided with a compact argument aiming to prove that no monetary policy can have a lasting influence on a *real* coordinate of the economic system. Should one try to decrease the unemployment rate below the "natural" rate by the means of inflation, the rate will come back to its natural level by a sort of

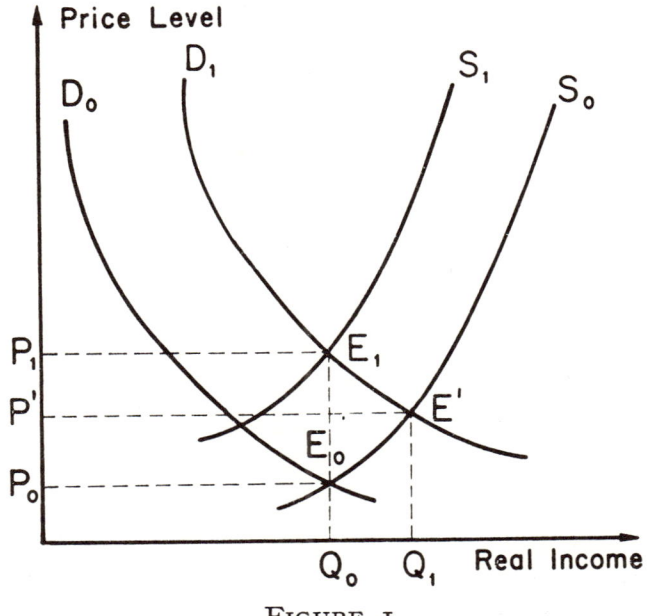

FIGURE I

pendular movement (Friedman, 1968, pp. 5-11). The quintessence of the argument is expressed by a familiar type of diagram (Fig. 1) [1]. An increase in the nominal purchasing power of the community will, at first, shift the demand from D_0 to D_1 and, hence, the

---

[1] Cf. Estela M. Bee de Dagum, « L'inflation structurelle en Amérique latine: Le cas de l'Argentine », *Tiers Monde*, VI (1965), p. 6.

"natural" equilibrium from E_0 to E', with a higher price level and a higher real income (through an increase in employment). But the consequent increase in the cost of production will inevitably shift upward the supply curve from S_0 to S_1 in such a way that the new equilibrium, E_1, will differ from the original one only by its price level, i.e., by its nominal income but not by its real income. To put it in a nutshell: like a sailing ship which causes only some temporary waves and gurgles, so inflation leaves absolutely no lasting historical trace [1].

For the accomplished proof that the system necessarily returns to its former real matrix, Friedman (1968, p. 8) invokes the Walrasian economic system garnished with a few modern "imperfections" and assumes that E_0 is the equilibrium of such a system. Considering the factual side of this procedure, we may note, first, that there are sufficient reasons against the idea that every actual system is a Walrasian analogue (Georgescu-Roegen, pp. 339, 377-379). Second, we may no longer ignore Keynes' main point against the "Classical" position, the point that even the actual systems which seem to be Walrasian analogues are prevented by a series of inexorable forces from approaching the Walrasian equilibrium to a significant extent.

The argument is even more vulnerable from the purely logical side. It apparently pays to stress the elementary fact that the Walrasian equilibrium is the solution of a system of equations (representing either physical relations or maximizing conditions) which depends on the distribution of each kind of commodity and of *money* among the members of the economy, not only on the totals of these elements. Should, therefore, the quantity of *money* and/or its distribution at E_0 be changed, the Walrasian system will yield an entirely new equilibrium in real terms. This much is clear from the most discussed chapter of Walras' *Elements*, Lesson 29. And, as if he had wanted to hammer well the point, Walras ended Lesson 33 on Fiduciary Money and Payments by Offsets (§ 305) with the announcement that "we shall see, in our study of applied economics, how far-reaching are the consequences of this law [the quantity theory of money] which places the whole equilibrium

---

[1] The usual arguments by which this position is defended imply, in fact, that in the end all prices will increase in the same proportion, a position that goes back to J. S. MILL (p. 544).

of the market at the mercy of mine operators, issuers of bank notes and drawers of cheques [1]". He referred to his 1879 study "Théorie mathématique du billet de banque" where, after showing once more that the issue of bank notes "*increases more and more the prices of all merchandises [commodities],* [2]" he argued that "*the issue of bank notes in a certain amount renders possible an increase in the quantity of capital of equal value [but does not by itself ensure it]* [3]". It should be clear, therefore, that Walras' authority lends no support to the orthodox viewpoint—on the contrary.

2. Another point concerning the use of static analysis in discussing Change (whether dynamic or structural) seems to have escaped notice completely in spite of its paramount importance. In general, let $a_1^0, a_2^0, \ldots, a_n^0$ be the initial conditions of some system (say, the distribution of various capital and income goods among the members of the economy); let also $a_1, a_2, \ldots, a_n$ be the equilibrium values of the same variables, and $c_1, c_2, \ldots, c_m$ the equilibrium values of the characteristic economic coefficients (say, prices, interest rate, tithes, etc.). Let the system of independent equations

(1) $\qquad F_i (a^1, a, c) = 0, \qquad (i = 1, 2, \ldots, m + n),$

define the static equilibrium. The question of how the system passes from the initial situation to the equilibrium position— in Walras' terms, the problem of *equilibrium ab ovo*—is full of unsuspected thorns. At bottom, the issue is whether the actual economic system by its very own functioning can solve the *mathematical* system (1). From Walras we have inherited the idea of a solution by *tâtonnement*. The idea, however, is far from providing a satisfactory answer to our query.

Indeed, let us assume that by *tâtonnement* at the end of the first "day" the system reaches the distribution $a^2 \neq a^1$. According to the very rationale of the concept of the static system, instead of (1) we now have

(2) $\qquad\qquad F_i (a^2, a, c) = 0,$

which may very well have a different solution (a, c) than (1). Let

---

[1] Interestingly, Walras' idea of inflation included also the case of a purely economic action—the increase in the production of the commodity *numéraire*.
[2] WALRAS, *Etudes*, p. 344. My translation.
[3] *Ibid.*, p. 349.

us not forget that in actuality every individual acts under imperfect knowledge and, hence, is subject to erring. This fact suffices to prove that we can no longer be certain that an actual system will *necessarily* tend toward a definite equilibrium (Georgescu-Roegen, pp. 180-183). Even the assumption that *every* member of the economy is an economic genius *and* a financial wizard (if one could be both), would not dispose of the difficulty. Only the extreme yet common assumption that every member possesses "perfect foresight", so that he would make every time the right move toward the equilibrium as defined by (1), can save the day.

Walras himself seems to have been aware of this difficulty, since he repeatedly used the artifice of provisional "tickets" to be written and rewritten until they happen to spell out the equilibrium values (*Elements*, § 207, 251). By itself, this artifice may be a substitute for the mathematical solution of (1) but it is valueless for representing the process of the equilibrium *ab ovo*. The only way to make a nonfictional sense of it, is to introduce a highly artificial restriction into the economic mechanism. The restriction, which may be read into some of Walras' remarks (§ 274), is that capital goods in the hands of individuals (as distinct from entrepreneurs) are never sold: entrepreneurs may only hire them for short periods and on the condition of maintaining them intact (not physically, but economically). In this case, the transactions of each "day" do represent a batch of *provisional* tickets. It goes without saying, though, that there is no way of ascertaining how many "days" it will take to hit upon the equilibrium transactions, nay, whether they will ever be hit upon. True, with the above restriction a change in the quantity of money will have no lasting effect on the system. However, any argument implying the restriction may win only a Pyrrhic victory for the orthodox standpoint. In an actual economic system based on private property, any private property may be bought for a money price [1].

---

[1] The point does not imply that in a socialist economy inflation poses no problems. Only, they are not as vitally important as in the other cases.

III. Inflation: A Structural Change

1. The idea that inflation operates "a violent and unjust transfer of property"—as Ricardo (p. 93) put it—is quite old. Benthan (pp. 45f, 70) adumbrated it before Ricardo [1]. While the truth of the proposition is by now incontestable (Robertson, pp. 11f; Keynes, 1924, pp. 6f), its importance for the consequences of inflation has been, in modern times, consigned to oblivion. Fundamental though it was in Marx's old input-output diagram, the social coordinate of income distribution is nowadays completely obliterated by the insipid category of "final demand." And yet, as is the main thesis of this paper, the change in income distribution is the main result and drawback of inflation.

The case of cash-inflation—which alone is under our consideration—is obvious enough. At bottom, there is no essential difference between cash-inflation and money forging. The commodities bought by the government with the money "forged" through cash-inflation obviously represent an *equal* loss for the rest of the economy *regardless of the prices at which these commodities are sold*. The whole process, however, does not reduce to the simple gain of the state or of the forger. As a rule, there are also other winners.

The point worthy of emphasis is that the ultimate consequence of an act of cash-inflation is a diffusion process which, like all such processes, requires a duration for its completion and, moreover, can be described in its broad lines but not cast into a mathematical model. To proceed systematically, let us consider the normal case in which the newly created purchasing power is distributed unevenly among certain units of the economy. For simplicity, let us assume that these units ordinarily are buyers of only one commodity, C_1. At first some amount of C_1 may sell at the old price, but soon the sudden increase in demand will make itself felt and the price will rise to a level much higher than that which will ultimately prevail after the inflationary injection has worked out its full effect. Some of those buyers of C_1 who received no share of the fresh money will therefore see their real income decreased for the benefit

---

[1] See also BENTHAM, *Sur les prix*, especially, pp. 276 f.

of some sellers of C_1. The complete point is that these buyers of C_1 are not *yet* able to increase the price of their own products or services: the newly created purchasing power has not had time to reach their clients. The same argument applies to every product as its market becomes affected by the diffusion of the fresh money. In short, the incomes of some individuals, beginning with the initial recipients of the fresh money, will increase; for others, the reverse is true. While it thus passes from hand to hand, money, as Bentham (p. 45n) hinted, changes its value continuously until the diffusion process is completed.

Clearly, the transfer of income during the diffusion process must necessarily lead to some transfer of property and, hence, to a new distribution of income. The resulting structural redistribution of demand will in turn cause a commensurate structural redistribution of resources. Consequently, save for highly improbable coincidences there will be a *new* economic structure with a *new* constellation of relative prices.

2. Two factors have been shown thus far to be responsible for the redistribution of income through inflation: one mechanical—the diffusion lags—the other political—the biased distribution of the fresh money. But the ultimate gains or losses of each economic unit are largely influenced by three other factors. Two of these are institutional: the complex of elements variously covered by the term "money illusion" and the nature of the contract regulating the income of the economic unit. The third factor is political and consists of the special government regulations usually associated with a prolonged use of cash-inflation. To these factors we must now turn our attention.

IV. Money Fetishism and Institutional Constraints

1. Because of the lattice pattern through which the effect of an inflationary injection is propagated from one market to another, every member of the economy ought to become a speculator. But in the real world not everyone sees equally well through this game and, moreover, not everyone can react with the same ease to the game.

We recall that Keynes (1936, p. 9) introduced the term "money illusion" to denote the fact that "whilst workers will usually resist a reduction of money-wages it is not their practice to withdraw their labor whenever there is a rise in the price of wage-goods". This behavior, however, is not the *privilegium odiosum* of workers alone. Any university professor will also display it if, like the worker, he has no other source of income other than his salary (and even if he has a small additional income). In effect, money illusion is on the whole stronger for the salaried personnel than for workers, the main reason being that an increase in the cost of living hits the people with lower incomes harder. At bottom, money illusion is a particular instance of *money fetishism*, a fetishism which affects every member of a monetary economy. Because of the continuous use of money in the daily transactions, we all acquire the habit of regarding it as the standard of value in our economic decisions—as Marshall (pp. 14f) rightly argued.

Now, as the value of money begins to founder under inflation, some individuals may become aware of their money fetishism before their own situation is affected by the collapse and, hence, bide their time as sellers; others, on the contrary, may not see the work of inflation even after it has involved their own transactions and, consequently, they may still go on selling. In other words, not everyone gets rid of money fetishism with equal speed. Among the most important factors of this behavior variability are some objective conditions: constraints imposed by previous contracts, the imperative of working for an income or continuing one's productive activity when no other alternatives are open.

The immediate, albeit unsuspected, conclusion is that inflation would produce a transfer of income and, ultimately, a transfer of property even if the distribution of fresh money among the members of the economy would be made so as to increase the buying power of everyone in the same proportion [1]. Inflation, to repeat, is

[1] In this respect it is instructive to recall that Henri Poincaré, (*The Foundations of Science*, Lancaster, Pa., 1940, p. 414) imagined a demon which could overnight change all physical dimensions of the universe in the same proportion. As he argued, since all physical laws are dimensionally homogeneous, nothing could reveal to us the intervention of the demon. But if a demon would change all cash and script money in the same proportion, there will be a qualitative economic residual to tell the story in spite of the homogeneity of all economic balances with respect to prices and money. Cf. note on page 562, above.

a complex diffusion process which should not be confused with a general shifting of the decimal point in all money accounts any more than the shift from francs to "new francs" in France or from cruzeiros to "new cruzeiros" in Brazil should be confused with deflation.

2. But the most peculiar form of money fetishism is reflected in the increased hoarding as inflation sets in. Peasants, as distinct from commercial farmers, have been found to be the most prone to this form (e.g., Keynes, 1924, p. 89; Bresciani-Turroni, p. 166). W.A. Lewis (p. 222) attributes this fact to the well-known thriftiness of the peasant. The explanation may work in part only if—as is most likely—he had in mind the peasant of Western Europe who has some large contact with the monetary circuit and has a sufficiently high economic standard to enable him to save in money [1]. For other types of peasants, the explanation does not work. While all peasants must be thrifty to survive, not all save in money. The poor peasants everywhere and the numberless peasants who have only a weak contact with the monetary circuit are not money savers. The reason why money exercises such a strong attraction for these peasants is that to obtain cash for paying taxes, buying a few industrial products, or repaying debts, constitutes for them the torment of daily life. The same continuous and hard pressure for cash is felt by all low income earners. And the poorer the country, the harder is this pressure. For poor people everywhere cash represents a *summon bonum*.

This, however, is not the only explanation of why the poorer are the people the more resilient is their money fetishism. One correlated reason is that poor people have a low intellectual level combined with a complete lack of economic acumen. And if, at times, they may amaze us by their bearing with long stretches of constant money wages while the value of money dwindles away, it is because in addition to everything else the economic and political condition of the lower classes in that situation is extremely weak (if it counts at all). Postponing a more ample discussion of this last point for a more appropriate place (Sec. VI. 2-3), we may note another, more substantial, deterrent for the wage-earners (in

---

[1] Lewis' passage contains some textual difficulties, for he attributes the quality of thriftiness to the entrepreneurs as well. But then the entrepreneurs should be the victims of money fetishism as much as the peasants.

general) to adapt themselves to the inflation game even if they begin to see clearly how it is run. This factor, common also to other categories of individuals, pertains to the type of contract by which wage-earners obtain their share of the national income [1].

3. Analytically, some exchanges may be regarded as *instantaneous*—such as when one buys something in a shop or goes to his dentist and pays the current price (or fee) forthwith. Other exchanges involve a *lag* and, normally, a *duration* as well. There is a lag because the money price to be paid *later* is established *now*. There is a duration, because either what is sold is some service or the money price consists of an annuity. It is natural that we should refer to the money income arising from an instantaneous transaction as *noncontractual*. For these incomes—profits and fees—there is no contract in the strict sense. Indeed, as we have learned it from the controversies raised by Euler's theorem, profit is the residual after all other services have been paid according to their contracts. The category of *contractual* incomes includes wages, salaries, rent, and interest. A special case, typified by pensions, is formed by the annuities paid for some past sales.

Even among the contractual incomes there are differences that bear upon the effect of cash-inflation. To begin with, in the case of an annuity contract the income receiver cannot possibly do anything about the loss due to the devaluation of the currency: he can neither revoke the past nor change the future. Other contracts have as a rule a limited duration, so that the service in point may be recontracted on a new basis. (In extremis, the contract may be denounced by sheer force.) But there are still some important differences. For the individual who has leased his money-capital inflation means a loss of substance, not only a loss of income over the period of the contract. The crucial point for the analysis of inflation is the fact that for wages and salaries the duration of the contract is determined (and only vaguely so) by the prevailing institutional relations between employers and employees, not by the contract itself. For this reason, and also because hired people

---

[1] Institutionally, the salaried personnel is expected to abide by the old « contract » longer than the workers, i.e., money illusion is, as we have said, stronger for the former than for the latter although the reverse relation applies to their money fetishism. But in advanced economies, even some salary earners may strike, occasionally.

generally have no other source of income, they cannot defend their own position from inflation as much as the landlord, the money lender, or the businessman. Even during very trying conditions —such as the German inflation after World War I—workers are usually unable to make a bid for wage increases as often as they might want. For one thing, striking every day or even every week is not technically feasible; for another, the greater burden of the cost involved will ultimately fall on their shoulders [1].

Nothing need be added to see that only the recipients of noncontractual incomes have the freedom of adapting themselves almost instantaneously to the inflationary current—a privilege which, though extremely important, is not completely decisive (Sec. VI. 7). But it is important to note that this category of people has other gambling advantages deriving from the fact that, as a whole, it represents the upper income class. In sharp contrast with those at the other end of the socio-economic scale, the well-to-do people have a broader economic experience and are, in general, better educated. As a result, they possess a keen sense of business and follow and understand better the march of politics. This is why, even though they ordinarily are the main saving element of the society, they shake off their money fetishism without any delay [2]. In addition, their relatively high income, ordinarily derived from multiple sources, gives them enough freedom in the choice of a good strategy. As the history of the post-World War I inflations in Europe (e.g., Bresciani-Turroni, pp. 292-298) proves

---

[1] What organized labor such as that of the United States would do in case of a high cash-inflation is anybody's guess. But let us not be mistaken: the collective contract with the escalator clause, now current in advanced economies, is not the natural consequence of the wearing out of workers' money fetishism, but the product of the ideas of labor organizers who succeeded only after the workers' standard of living enabled them to strike.

[2] The fact that some members of this category have occasionally increased their hoarding substantially even during an inflation of the proportions of that of Germany after World War I does not change these institutional considerations. More often than not, such behavior represents a speculation on the reevaluation of currency of the same nature as any other speculation. One can cite numberless other speculations which failed because of a political flair that can be censured only after the fact. However, the case of those who now buy the new government bonds of Brazil (which are only in part protected against inflationary devaluation) is not as clear. It may be only a symptom of the well-known point that people would save for old age even at a negative interest; but it may equally well correspond to inability to play the inflation game better.

beyond doubt, business savoir faire enhances tremendously the chances of inflation gains.

4. To summarize the findings of this section. Inflation introduces into the normal economic process a speculative game such that: 1º some participate with *free* initial stakes, 2º its rules vary with the economic status of the individual, and 3º the ability to play it also varies with this status. From this viewpoint, there can be no doubt that, as President Johnson wittingly put it, inflation is "the pickpocket of the poor". The whole truth is that inflation is the pickpocket of anyone who is earning a contractual income and, especially, of those who have already purchased a future contractual income. To be sure, in an advanced society most labor contracts include the escalator clause; but even with such a clause some loss cannot be avoided since wage adjustments are discontinuous and only with regard to the day's level of the cost of living. It is because of this institution (in particular) that in advanced economies inflation spirals up very quickly—a reason why it is rightly frowned upon as a policy tool.

The situation changes completely wherever it is possible to introduce those regulations that on the basis of historical induction we may expect to be decreed by any government resorting to cash-inflation as a continuous economic policy in an underdeveloped country. As I shall argue presently, then the pickpocketing of the poor not only reaches deeper but may also be prolonged almost *sine die*.

V. Inflation: A Passable System of Taxation?

1. That any taxation means a shift of income from one part of the community to another is a tautological thought. From what we have seen thus far, we can regard inflation as a system of taxation (Ricardo, p. 334). A great deal of gunpowder is added to this view by those who claim that inflation is a quite acceptable method of taxation (Keynes, 1922; Lewis, p. 404). Keynes even argued that the "efficiency [of the method] cannot be disputed". First—he said—"the burden of the tax is well spread, cannot be evaded, costs nothing to collect, and falls, in a rough sort of way, in proportion to the wealth of the victim". Second, "even the

weakest government can enforce [it] when it can enforce nothing else".

The first thought echoes Mill's idea (p. 552) that only the issuer of money (i.e., the forger) reaps a gain at the expense of all possessors of money. But as we have seen in the preceding sections, when all accounts are settled the burden of the inflation tax is borne mainly by those who have hardly any wealth and the benefit generally goes to the wealthy.

As to the second thought, we find it implicit in the frequent opinion that because of "the inexperienced, corrupt, and generally cumbersome bureaucracies" in Latin American countries cash-inflation is a more efficient method than taxation proper for achieving the same goals [1].

The danger of these rationalizations (as well as of others to be considered later on, Sec. VII) is that they blot out the real reason why governments have ordinarily resorted to cash-inflation: to protect the real income of the classes that were their debtors at the time—as Keynes (1924, p. 13) was to observe later. And the brute fact is that political acumen has seldom been at a loss to make the public believe that the move was in the "general" interest. To wit, cash-inflation has frequently been used to buy political stability through the creation of an excessive bureaucracy or of a superfluous army, however inadequately paid in comparison to the private sector. At times, cash-inflation has been used in Latin America to thwart the socio-economic reforms adopted by a semi-feudal society for display or demagogic purposes [2]. More pithy is the case of cash-inflations aimed at supporting the price of a "national" product when its international price has fallen —wheat during the early 1930's in Eastern Europe, coffee in

---

[1] BAER, 1962, p. 90; also LEWIS, p. 404. Arguments of this nature often advanced in connection with a precarious historical heritage are downright frustrating; any policy based on them perpetuates the ills. To wit, because of the ease with which deficits of public budgets can be written off by cash inflation, public administrators acquire the illusion that the public revenue is limitless. Naturally, this increases « the spirit of dissipation and neglect » (Bresciani-Turroni, p. 52). So that the longer cash-inflation is used, the stronger its « necessity » becomes—a political lock of the same nature as that we shall find in the economic mechanism (Sec. IX).

[2] An eloquent example is offered by Argentina where a policy aimed at increasing wages was initiated in 1945. Labor's gain was soon wiped out by the greater increase in the cost of living. *Argentina Económica*, 1966, p. 102.

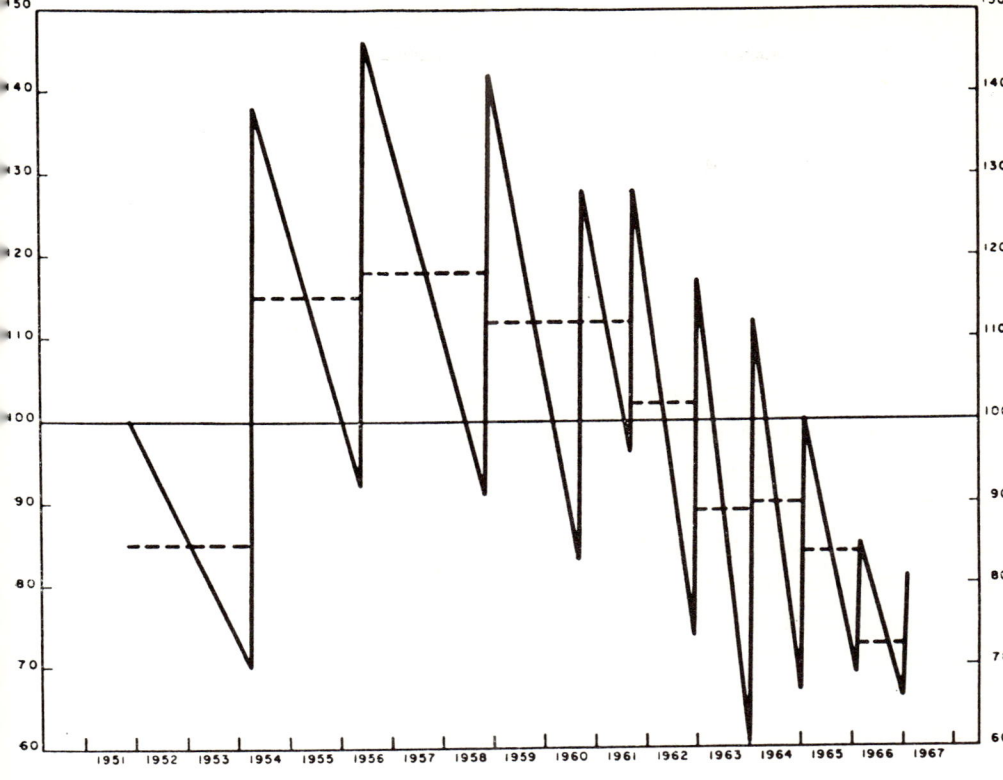

FIGURE 2. — The movement of real minimum wage rate in Brazil.

Brazil, and beef in Argentina. It is elementary that since the real incomes of these particular producers were protected during a period of declining national income, other classes had to bear a loss.

It cannot be denied that ordinarily each cash-inflation move achieves its *immediate* goal: the superfluous public servants get their salaries and the producers of the "national" product receive a higher price. But the drawback of any inflationary move is that while it fills one hole in the system, it creates others; often, it also increases the size of the first. That is why one cash-inflation *move* breeds another one [1].

[1] This algorithm, it should be noted, does not constitute a complete theory. As with a theory of life or of business cycles, we need a cause for the *first* event in time, i.e., the first cash-inflation move. In opposition to most Latin American economists who relate the prime cause to geo-historical conditions (CAMPOS, pp. 82-85), the contention of this paper is that the motive is in essence political.

VI. The Inflation State: Its Losers and Its Winners

1. Governments that find it politically expedient to resort to cash-inflation for solving some "disequilibrium" of the real economic coordinates are normally prey to the illusion—shared also by some economists—that this can be done successfully and at no opportunity cost provided the right steps are taken to correct the side-effects of inflation: 1º the pressure on wages by the continuous rise of the cost of living, and 2º the effect on business activity of the disorganization of the normal process by which savings are transferred to investors. The "corrective" measures create what I propose to call a *state of inflation*. It is this state, not the condition created by a single inflationary injection or a succession of such injections into an advanced economy unrestricted by extraordinary regulations and working close to full employment, that is relevant for the analysis of the Latin American economies.

An inflation state is generally characterized by the following four controls: 1º price ceilings for the basic food items, usually, for some other necessities also; 2º ceiling on house rents; 3º regulated wages and salaries; and 4º ceiling on interest.

2. As is always admitted by the advocates of inflation as a policy tool, the control of wages and salaries is the *sine qua non* condition for the success of the policy. Politically, however, the measure is presented under a different rationalization. First, as the usual official proclamations maintain, the ceilings on prices and rents are introduced for the protection of the workers and the salaried personnel, who by their numerical weight and their location in vital urban centers represent one of the most critical elements of the political problem. Wage control is presented as the price of this protection. Unfortunately, the workers pay the price without receiving the goods. Given the practical impossibility of controlling even part of the wage-goods prices continuously, real wages drop systematically between one adjustment and the next [1]. Consequently, a worker loses some real income, namely, the amount

---

[1] The prices of wage-goods would be under pressure even if they were bought only by the workers (with fixed wages): the producers would no longer be satisfied with the old profit now devalued by inflation.

represented by the difference between the pre-inflation real income and the average real income between two wage adjustments (Fig. 3a, Sec. VII).

As noted earlier, even an escalator clause would not avoid this loss. But in countries where the politico-economic matrix still

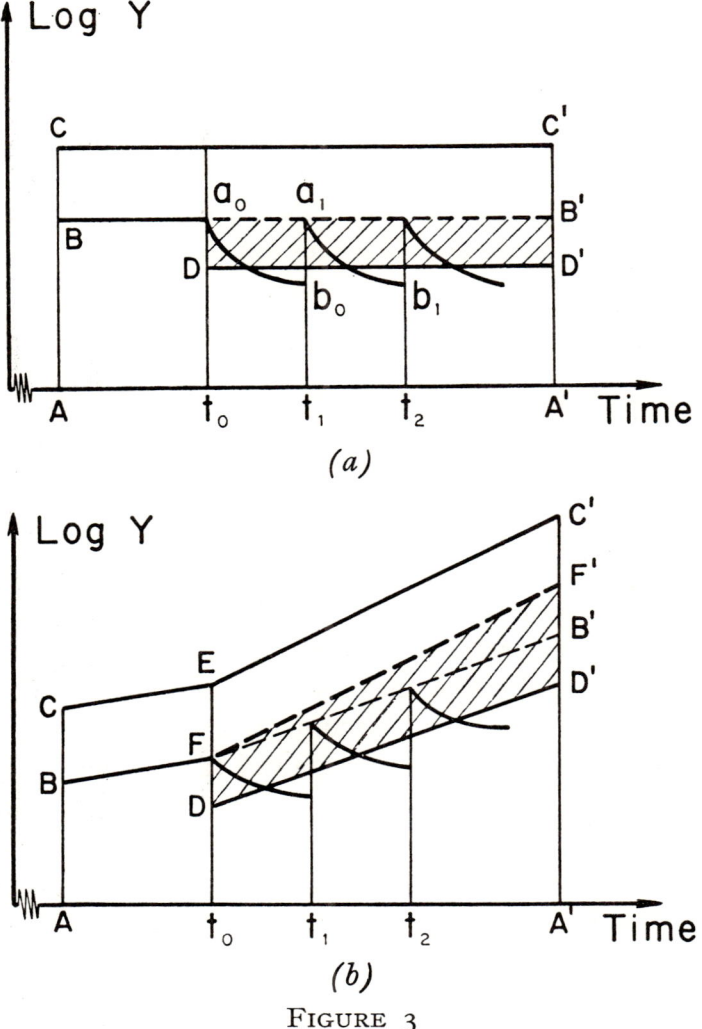

FIGURE 3

harbors vestiges of feudal relations—as is almost everywhere the case in Latin America—labor's bargaining power, whether vis-à-vis the government or the private management, is so weak that it cannot even obtain a collective contract. The little power the workers have must be spent on making the government consent to a readjustment. The question of the real level at which the readjustment is made—not to mention that of an increase in proportion to labor productivity—is thus thrown into the remote background. It is not excluded, therefore, that the real wages should be readjusted at a lower level than that of the previous readjustment—as is evidenced by Table I and, especially, by Table II and Fig. 2.

Some generally observed facts confirm these points. Exceptionally, a political regime may find it politically expedient to increase real wages by a deliberate action (as Peron tried to do in the mid-1950's). But apart from such unusual interventions, the usual trend of real wages is to remain constant or, especially during periods of high inflation rates, to fall[1]. Of course, the tendency to fall weakens as real wages approach the bare minimum of subsistence and is idle if the initial standard of the workers' mass was near that bottom. Apparently, the concern with the worker's receiving a subsistence income turns during an inflation state into the idea that he should not have more than that. The "packing" of all wages toward the lowest level, the wage rate of the unskilled worker, is a phenomenon found to be associated with inflation even in countries with a far better organized labor than in Latin America[2].

3. Unsophisticated though the poor are, even they realize, as inflation goes on year after year, that after all money is not a fixed measure of value. But as we have seen, there is little they can do given the prevailing political circumstances. Nor can they upset the economic apple cart by running away from money. Their only income being not much above the subsistence level[3] and being received at small intervals, they can neither concentrate all their purchases immediately after the pay day nor follow the advice "buy anything to protect your money against depreciation." This

---

[1] This behavior of real wages has been very conspicuous in Europe after World War I (BRESCIANI-TURRONI, pp. 304 ff.). For Latin-America, see Tables I and II.

[2] See *Fluctuations des salaires dans différents pays de 1914 à 1921*, International Labor Organization, League of Nations, Geneva, 1922, and BRESCIANI-TURRONI, p. 313. The data for Argentina confirm this fact (Table I). For the other Latin American countries, the necessary data are either unavailable, incomplete, or patently unreliable.

[3] See, for instance, Maria J. VILLAÇA, *A Força de Trabalho no Brasil*, São Paulo, 1961, p. 181.

strategy in the inflation game may be adopted only by those who, having a high income, can save. Nor would it make any sense for the poor to borrow money from the money lenders in order to play the game in a speculative spirit: the "black market" interest they would have to pay is prohibitive in real terms. Instead, because of the periodic wage readjustments they end by acquiring a roller coaster complex: to find solace in the pleasurable anticipation of the next lift of their money income. In fact, this complex is enhanced by the fact that as the inflation rate increases, adjustments have to be made more frequently [1]. In addition, with a somber life horizon such as their's, the low income people are inherently prone to take at its face value every government pronouncement that inflation is on the way out and that "this time" their real income will no longer dwindle away.

The foregoing argument and evidence notwithstanding, one may insist that, since inflation generally increases the demand for labor, real wages must in the end increase if entrepreneurs compete with each other for labor. The crucial point is that this argument cannot apply to an underdeveloped economy, perhaps not even to all developed economies. For even in developed economies there is a rural-urban differential. As to the underdeveloped economies, their very hallmark is the existence of a "reserve army". Specifically, historical conditions are such that a mass of people lives on the land or in the urban slums on a lower standard than the minimum wage and, generally, is willing to work for that minimum, *but not for less*. True, as long as the reserve army exists, there is no purely economic reason for real wages to increase in any case. The inflation state, however, causes a decrease in real wages below the pre-inflation level (itself determined by some noneconomic forces as well). Above all, the regimen of wage control normally implies great, perhaps total, restrictions to labor organization with the result that labor is prevented from fighting for a share of the increased productivity (Sec. VII. 4).

4. The farmer's condition may vary greatly. While some commercial farmers are penalized by the ceilings on food, others may be favored by the price support of export goods. The discriminatory treatment accounts for the frequently observed increase in the

[1] See the dates of the successive wage readjustments for Brazil in Table II.

activity of the subsidized and the quasi stagnation of the penalized sector. With the increase in population, this stagnation ultimately renders the price ceilings impossible to enforce; at that point, the tables may turn in favor of the formerly penalized activities. Needless to add, the lot of the poor or self-sufficient peasants—who, as a rule, form an important part of the population in underdeveloped countries—is hardly affected by this price see-saw. For a peasant who eats his own corn it does not matter much whether the price of corn is subsidized or controlled.

5. Because a pure rentier is only an income claimer for past services, there is hardly any political interest to protect him from inflation. For the owners of pensions, life insurance policies, bonds, *"Krone ist Krone"*—as Schumpeter allegedly declared while he was Austria's Finance Minister after World War I. This category is the surest loser; as a result, it fades away as quickly as biology permits.

Those deriving an income from rent must also lose some real income during the life of the lease. But since they can recontract, they should, under normal circumstances, be able to readjust the rent to the price level. Rent control—however poorly applied or cleverly avoided—works definitely against this category which also tends to fade away with disastrous consequences for the housing situation [1]. Money lenders, however, being in a position to operate under the counter, are greatly favored by the ceiling on interest which (as we shall see) excludes almost all personal loans and loans to small firms from bank credit.

6. It is a highly symptomatic fact that, in contrast to the other three controls, the reason for the ceiling on interest does not seem to arrest the attention of the supporters of cash-inflation. An excellent example is such a learned and multifarious collection of essays as that edited by H. S. Ellis on *Economic Development for Latin America*. In the whole volume only one discussant, Dorival Teixeira Vieira, has some brief but interesting remarks on the topic (pp. 192-197). The rationalization that the measure protects the low income earner is advanced only rarely, so plain is the opposite

---

[1] For which see *Socio-Economic Progress in Latin America : Social Progress Trust Fund Sixth Annual Report, 1966*, Inter-American Development Bank, pp. 9 ff.

truth. The ceiling on interest destroys the normal money and capital markets, a fact which provides the government with a justification for taking over the task of supplying the capital needed by the business sector. The inevitable solution is to funnel a substantial part of the fresh money through the commercial banks and other financial institutions into the business sector. Naturally, the interest rate for the credit made available in this manner is tied to the legal interest rate, with the result that the *true* interest after the depreciation of money is negligible, often negative. And since the demand for such "cheap loans" is practically infinite, the banks must ration their credit in some way. The "obvious" solution is to restrict personal credit [1]. Needless to add, it is to the low income people, for whose protection the ceiling on interest was supposedly introduced, that credit is completely denied. Actually, discrimination does not stop here: it is the large enterprises or those with "good connections," which enjoy the highest priority (Baer and Simonsen, p. 286).

The most important category of winners is already contoured in this picture. But the main mechanism by which the business class as a whole benefits from inflation needs to be clearly understood. The run of the mill—that firms buy now and sell later at inflated prices [2]—is totally inept. Any gain must come from a difference in *real* terms. True, since inflation reduces the real value of contractual incomes and since real income cannot disappear into the air, the difference must accrue to the other members of the economy. But this argument alone cannot explain the continuous increase in real terms of noncontractual incomes. Such an increase can come only from economic development—inflation or no inflation. Only, in an inflation state the entire increase in productivity is pumped into profits through continuous bank loans at a very cheap or even negative interest rate. "Borrow now and repay later in devalued currency" is the whole secret.

7. Two additional observations will complete our scoreboard of losers and winners. The first concerns a previous thought (Sec. IV.

---

[1] In Brazil, for instance, mortgages have represented less than two, often less than one percent of all credit. *Boletim*, Banco Central do Brazil, August 1967, p. 113.

[2] E. g., A. L. MEYERS, *Elements of Modern Economics*, New York, 1948, p. 310.

3), that the noncontractual incomes are not necessarily in a privileged position: with a decrease in real demand nothing could prevent a businessman or a self-employed person (whether a physician or an electrician) from being a loser. For a telling example, during the German inflation professionals as a group suffered a loss of real income because the demand for their services followed the decrease of the worker's real income (Bresciani-Turroni, p. 170). On the other hand, because in underdeveloped economies the masses seldom use such services, the same decrease has no influence on the income of professionals. Instead, professionals benefit indirectly from the increase in real income of the direct beneficiaries of inflation to whom they cater almost exclusively.

The second observation pertains to a new group of beneficiaries, namely, the upper echelon of the bureaucracy. Under the pretext that a managed economy—especially one with a "managed" cash-inflation—calls for an increasing number of experts in planning, public finance, management, etc., one special advantage after another has been created for them: additional fringe benefits (some in kind) and extraordinary fees or honorariums in addition to the regular salaries. Abuses of this sort, however, grow easily in an inflation state, so that in some public or mixed economic enterprises the specialists' salaries have actually reached "multiples of those paid for similar work in the private sector" (Kafka, 1967, p. 609). Being largely self-generating, this phenomenon should not be discarded as economically negligible and politically irrelevant.

VII. Inflation and Development

1. The position that an increase in the flow of money is "the oil which renders the motion of the economic wheels more smooth and easy," was first elaborated with many interesting details by Hume (pp. 37 ff). Against it, we remember, Ricardo raised his authoritative voice in "The High Price of Bullion" and especially, in his comments on Bentham's essay *Sur les prix*. It was the voice of monetary orthodoxy: "I cannot agree that any addition to the money of the country produces riches ... No sum of money carried into Switzerland would enable the possessor to drain a marsh and render it productive ... It was not by money, but by capital that

Scotland has been improved". All inflation can do is to enable "A to carry on part of the business formerly engrossed by B and C, [but] nothing will be added to the real revenue and wealth of the country [1]". Yet in the end Ricardo felt obliged to bow to Hume's and Bentham's arguments:

> an increase of money no matter how it be introduced into the society, can augment riches, viz. at the expense of the wages of labor; till the wages of labor have found their level with the increased prices [of] commodities [it] can employ an *additional* number of hands, so that the real riches of the country will be somewhat augmented. A productive laborer will produce something more than before relatively to his consumption, *but this can be only of momentary duration* [2].

Moreover, this admission of Ricardo does not deny the possibility of repeating the feat by a new inflationary injection. And as we have seen (Sec. VI. 3), both facts and analytical considerations tell us that it can be repeated for as long as there is a "reserve army," provided the inflation rate does not cause a general run away from money [3].

The thesis that cash-inflation is the only expedient policy for accelerating the growth of an underdeveloped economy [4], therefore, is not entirely without foundation. Its first step is in the clear: given that the *spontaneous saving* of the masses is for all practical purposes inexistent, *forced saving* must be exacted through taxation to speed up the accumulation of capital. The second step is

---

[1] The quoted sentences are from RICARDO, pp. 93, 286 n, 301 n. MILL (pp. 550 f), too, takes a similar position.

[2] RICARDO, p. 318 n (italics added). Cf. HUME, p. 38.

[3] A run away from money is less likely to occur if the overwhelming majority of incomes cover only current expenditures or if obstacles of some sort hinder the use of foreign currency or gold coins as a substitute for paper money. It should also be noted that a run away from money may be triggered by a psychological snowball even if fresh money is not printed at an excessive rate. The frequent declarations of Latin American governments that the inflation state is on its way out are not aimed only at exploiting a psychological effect on the masses (Sec. VI. 3) but also at stifling any potential source of panic elsewhere. The justification of the opinion that cash-inflation can go on provided that its rate is decreased now and then (Campos, p. 102), may be that of preventing these declarations from becoming totally ineffectual if never backed up.

[4] E. g., Lewis, 404; Campos, 1962, p. 82; Baer, 1962, p. 87: 1963, p. 405: 1965, pp. 113 ff; Bruton, pp. 160 ff.

that given also the inherent and hard-to-cure faults of the administrative apparatus in almost any underdeveloped country, the only efficient method of taxation is cash-inflation, which by one stroke taxes those who only consume and puts the proceeds at the disposal of "the dynamic elements of the society" for investment. Since the order of the day is "Grow!" and since some of the most spectacular growths have been associated with inflation, we need not look further. So, the argument rests here. At most, it admits that there is also the thorny problem of how to keep the various monetary circuits adjusted so as to save such a wonderful instrument from self-destruction.

Let us nevertheless look further, not only behind the monetary veil but also behind the screen raised by aggregation, to see what happens in real terms to the structure of investment and consumption during an inflation state.

2. Two facts are essential for the understanding of the inflation state as a tax mechanism. First, the tax imposed by inflation is a very peculiar tax. It is levied neither in money nor in kind, for there is no tax collector. The cunning of it is that an equivalent amount of money, T, is nevertheless put at the disposal of some privileged units in the form of higher salaries or cheap bank loans. For the purpose of this section, however, we may safely ignore the first category.

The second point is that—contrary to what the advocates of inflation affirm—the taxed away income is transformed neither directly nor entirely into investment. After the cheap loans are entered into the books of the firms, T is metamorphosized into "additional" investment, $I = \alpha T$, and *extra* profit, $P = (1-\alpha)T$, with $0 \leqslant \alpha < 1$. The manner in which this is achieved constitutes the crux of the whole matter. No doubt, any firm desiring to continue the same level of business must increase its working capital to the level required by the inflated prices. In an inflation state, however, firms use additional cheap loans for this purpose. And indeed, the data cited by Baer and Simonsen (p. 285 f) for Brazil prove that bank loans increased even faster than working capital. But the same authors argue that firms should increase their working capital by using part of their "book" profit. If this part, which they call "illusory profit," is distributed, firms—they argue—

"eat into their capital" (p. 282). This argument is outright suspect: Brazilian business would achieve the super-miracle of eating the cake and having *more* of it. The authors err in reasoning as if the cheap loans were to cease suddenly. It is precisely because inflation permits these loans to go on uninterruptedly that the "illusory profit" is as real as anything can be. In fact, it is the substantial form in which the main beneficiaries receive the lion's share of the taxed away income [1].

True, the individuals to whom P accrues may not consume it all. And since in an inflation state we may count on practically any saving to be forthright invested, the total investment should be $\delta T > \alpha T, \delta < 1$. The size of δ depends on a complex of factors, the most important being the consumption habits of the upper classes (in the long run) and the business outlook (in the short run) [2]. But given the traditional proclivity of these classes in Latin America for grandeur in living and also the deep impact which the material progress of other countries normally has on such a society, their marginal propensity to save must have been extremely low even before inflation became the order of the day [3]. The effect of the long therapy of cash-inflation on this propensity has been, if anything, to lower it. The result is that the efficiency of cash-inflation as a strategy of capital accumulation is objectively very poor. In terms of *subjective* cost, it is much poorer. The occasional specta-

---

[1] The counterproof, that a decrease in the ratio between bank loans and the volume of productive activity causes a decrease of P and, hence, a decrease in demand, is provided by the slowing down of Argentina's economy in the last years. See the partial data in *Argentina Económica*, p. 338, even though they do not seem to tell the whole story.

[2] For a model in which capitalists behave « normally, » i.e., they save an increasing proportion of the surplus value and invest in constant capital an increasing proportion of their savings, see « Mathematical Proofs of the Breakdown of Capitalism, » in Georgescu-Roegen, Ch. 12. The model could be easily extended to the situation under discussion by dividing the production sector into three departments (capital goods, luxury goods, and wage goods), but only at a tremendous loss of flexibility.

[3] Clearly, only an extremely low average propensity to save of the upper classes can explain why, with unusually favorable ecological conditions and a relatively disturbance-free history, Latin America has not been able to achieve substantial economic development since colonization. On another plane, one may note the tale-telling fact that in Brazil the demand for automobiles was from the outset so high that it became profitable to establish an assembly plant as early as 1918, when countries with a far greater income per capita were still importing automobiles completely assembled.

cular growth of some Latin American countries—Brazil is the often-quoted example—has been achieved by even more spectacular sacrifices imposed upon the masses (cf. Simonsen, p. 73).

3. Now, for anyone to invest part of his income the demand in that sector must be increasing *in real terms*. But where does the continuous increase in demand come from? The simple answer is that at first it comes primarily from the profit incomes which, because of the first inflationary injection, increase in real terms at the expense of contractual incomes. And should real wages be periodically adjusted at a decreasing level, profits will naturally keep increasing. But, if real wages are always readjusted to the same level, there is no *additional* transfer of income to account for the continous increase in demand. The answer to this new puzzle is that demand keeps increasing because there is development and that the corresponding increase in real income is funneled into the incomes of the upper classes through the device of cheap bank loans.

To be sure, the device is idle if there is no room for economic development or growth. This peculiar situation, no doubt, is that which Ricardo had at first in mind and which nowadays is implicit in the orthodox position. For if there is no room for economic development, the real income rate, Y, cannot be increased in any case. Inflation only shifts the income flow rate B'D' from contractual incomes to the other sector of the economy whose income rate now becomes C'D' (Fig. 3a) [1]. As a result, there will also be some business reshuffling—from producer X to producer Y. But that is all.

By definition, an underdeveloped economy cannot fit the Ricardian model. In such an economy there *is* room for development. And from what we have seen, the inflation state, with one hand, increases the demand for some commodities and, with the other, supplies easy money for the expansion of the corresponding industries. Abstracting from the usual frictions, we have an ideal push-pull mechanism that reminds us of Say's law. The abstract

[1] For simplicity, the graph is drawn on the assumption that wages are readjusted to the initial level, AB, and also abstracts from the disturbances immediately following the introduction of the inflation state (at t_0). The *actual* labor share follows the wave curve $a_0\,b_0\,a_1\,b_1\ldots$, but DD' represents the *average* share over time.

basis of the position that an inflation state promotes economic development is, therefore, in the clear. In actuality, however, because of the atmosphere of uncertainty associated with such a state, business expectations may easily miss the target by a very long shot. Exaggerated optimism at one time, for instance, may lead to excess capacity and, hence, be followed by a stagnant period. The push-pull mechanism between increased demand and increased investment, therefore, does not work as smoothly as it looks on paper.

The irregular association of growth with inflation over time noticed by Kafka (1962, p. 162) for the Latin American economies should not then intrigue us [1]. Nor should Kafka's other observation, that monetary stability has been almost invariably associated with stagnation, do so. For contrary to some opinions (e. g., Lewis, p. 236), profits (not to be confused with windfalls) cannot grow above their normal level without some *kind* of monetary expansion. Given the prevailing historical heritage of Latin America, cash-inflation seems the only available kind [2].

4. To go now beyond the veil of aggregate demand we need to observe that human wants display a dialectical hierarchy (Georgescu-Roegen, 1966, pp. 190-201) which is, in turn, reflected in a hierarchy of commodities. That is, each commodity is included in the budget of an individual only after his real income has exceeded a certain level. In each particular situation, we may divide all commodities into "wage goods"—those consumed ordinarily by the workers and the low income earners—and "luxury goods". It thus becomes obvious that the result of any increase in the income of the upper classes is practically confined to an increase in the demand for luxury goods.

Moreover, because of the same hierarchy, this demand not only increases quantitatively but also expands qualitatively, i.e., it continuously extends over new products. Old lines of industrial activity must therefore expand and new ones be introduced.

---

[1] SIMONSEN (p. 73), too, noted that "economic development may be compatible with a state of inflation, but no positive correlation exists between the two phenomena."

[2] The counterproof is provided by Mexico where an improved system of financial institutions can now support growth through normal credit expansion.

In both cases, methods with higher capital-output, viz. lower labor-output ratios are introduced. This is the very essence that distinguishes *economic development* from mere *growth*. In plainer words, wherever there is economic development employment must increase in a smaller proportion than the flow rate of total output. The statistical data of practically every Latin American country confirm this elementary point at least as regards the manufacturing sector (Table III). However, since in an inflation state the trend of the real wage rate is, at most, constant, the total share of labor must increase in a much smaller proportion than the value of the total product. As shown in Fig. 3 b, the introduction of the inflation state boosts the growth rate, from CE to EC'. The labor share, after dropping by the amount FD, follows the rate of growth DD', smaller than EC'. Were labor able, however, to fight for a proportionate share of the increased productivity, its share would follow FF' (FF' being parallel to EC'). The gain over time of the upper classes, therefore, is the shaded area FDD'F'.

The question arises whether the data which we have been able to obtain confirm the last point. For Argentina and Colombia, although the percentage share of labor fluctuated appreciably, its general trend is unquestionably decreasing over the periods considered (Table IV) [1]. The Brazilian data show, on the contrary, a steadily increasing trend. Some (e.g., Kafka, 1967, p. 603) accepted the last evidence on its face value; others doubted it (e.g., Baer, 1965, pp. 119-125). My own position is that a far greater degree of statistical accuracy and sophistication than that prevailing in Latin America (and even in the more advanced economies) is indispensable for a conclusive test of the point in question. A simple glance at Table V suffices to see how flimsy such data may be. It is abundantly plain that all incomes other than wages and salaries paid to *regular personnel*—especially those of professionals and executives—have been greatly underestimated. As we are told these incomes have been estimated either from tax returns or on the basis of the inflation rate [2]. However, fiscal

---

[1] An index which I constructed for Argentina from data that are generally more reliable shows even a slightly declining trend of the industrial real wage bill (Table III, col. 4).

[2] *A economia Brasileira et suas perspectivas*, APEC, Rio de Janeiro, May 1963, pp. 22-24.

evasion (which in Latin America is notoriously high) operates only for incomes other than regular wages and salaries and, as is fair to guess, is progressively aggravated by inflation. For this as well as for other reasons, on which I need not dwell, we simply cannot get to the real facts. This is one of the special cases in which we must fall back on the elementary "theory" as presented above and leave the "facts" alone.

5. With real wages being kept down, any increase in the demand for wage goods can come only from an increase in employment. Moreover, it may involve no qualitative expansion. If an individual real income cannot, by the necessity of the hierarchy of needs, afford an icebox, even millions of additional incomes of the same size cannot increase the demand for iceboxes by an iota. In addition, the individual real wages being as low as they generally are in Latin America, the individual demand for *industrial* wage goods does not amount to much. The increase in the total demand itself is, therefore, insufficient to attract large entrepreneurial ventures, especially with the contrasting picture across the fence. Across the fence there is the continuous expansion of demand for luxury goods as well as the government's various price supports and, above all, there are no price ceilings. The result is elementary: on the one side, a developing industry, on the other, a quasi-stagnant one. All advantages of large scale production are *ipso facto* denied to the wage goods industries [1].

6. There is then economic development, but this economic development is confined to particular areas of economic activity and benefits almost exclusively one particular section of the society. Production data speak loudly on this point (Table VI). It all amounts to an economic square-dance, as it were, in which only a privileged group participates and which moves faster and faster to the increasing delight of the participants: the masses

---

[1] The point finds a strong confirmation in the case of shoes and clothing. Except in the case of Argentina, whose income per capita and relatively high standard of the working class places her in a category by herself, only a very small part of these products comes from industrial organizations; the rest is still produced by small craft shops.

remain the gloomy onlookers they were at the outset [1]. As Marx would have perhaps put it, the inflation state is the craftiest device by which all gain from increased productivity is turned into surplus value.

Because nowadays positive economics dominates our discipline, such considerations are easily pushed aside as "value judgments" (e.g., Bruton, p. 161). Yet we do turn around to decry the "positive" evils of cash-inflation: the excessive investment in the luxury goods industries (including the personal investments in luxury apartments) as well as the neglect of those lines which require a longer "waiting"—the capital goods industries and the public utilities. I am, however, at a loss to see how one can prove any of these features to be an evil from a purely positive viewpoint, that is, without invoking the progress of the masses as an ultimate criterion. Be this as it may, if they are evils, then it is highly relevant to observe that they are the direct consequences of a development whose flywheel is the demand for luxury goods which is continuously fed by cheap, at times gratuitous, loans [2].

Campos (p. 82) is right therefore in admitting that the development of Latin American countries has not been the product of a spontaneous drive of Schumpeterian entrepreneurs. As we have seen in this section, Latin American entrepreneurs had to be induced by heavy subsidies to take over the job. But Campos distorts the other side of the story by saying that the inflation state is the inevitable answer to "the aspirations of the masses to improve their standard of consumption." The truth is rather that

---

[1] To recall a caveat of the introductory section, I do not mean to deny all vertical movement from the lower to the upper strata. My only contention is that the intensity of this movement is so small that it cannot spoil, in the short run, the analytical picture outlined in the foregoing paragraphs. Income distribution data for Latin American countries are scanty, incomplete, and, of course, most unreliable of all. The most complete ones, for Argentina, are patently suspect because of their exceptionally large fluctuations. Between 1953 and 1959 they show an appreciable displacement of income from the lowest two deciles to the higher deciles, especially to the highest one; between 1959 and 1961, the movement is in reverse (*Argentina Económica*, p. 98).

[2] The fact that import substitution has generally produced results opposed to those expected from its rationale is, in all probability, due to the same complex. The substitution of imported luxury goods has been pushed so far that the imports necessary to support their industry are now greater then the initial import of such goods.

this state is the answer to the aspirations of the upper classes for a more luxurious life [1].

VIII. Inflation and Balanced Growth

1. There is no positive criterion for defining balanced growth. All we can do is to take as a point of departure the definition of a balanced structure of consumers' expenditures as that which prevails, on the average, in the countries that are most developed at the time of discourse. Admittedly, this concept is historically parochial; but it is not the only such concept that economics may use advantageously.

An underdeveloped country, therefore, can have only an unbalanced structure. And if by economic development we understand —as we should— the process by which the structure of an underdeveloped country is changed so as to move gradually toward the developed structure, then clearly economic development implies a continuous structural change. While this change need not be rigidly determined, it cannot be arbitrary either. There is a penumbra, however wide, that separates the cases of balanced growth (i.e., *a growth tending to diminish the differences with respect to a balanced structure*) from those of unbalanced growth. For a homely analogy: a newly born infant has not the same biometric proportions as the average mature person; but for his growth to be normal these proportions must change in a more or less definite direction, not in an arbitrary manner. The head of the infant, for example, should grow in a smaller proportion than his legs.

The upshot should be obvious: the growth induced by an inflation state of the kind that has prevailed in Latin America is highly unbalanced. In fact, the structures of the large majority of these countries may even be as unbalanced as they were before the process of industrialization began.

---

[1] One often hears the argument that, with the increase of public investment this situation would be corrected. The truth is that for the countries for which data were available (Argentina and Mexico) and for the period 1950-1965 the share of public investment shows no growing trend. Moreover the proclivities and class interests outlined above preside also over public investment which, more often than not, creates new « utilities » mainly for the upper classes.

2. The argument—likely to be interjected at this juncture—that without some heavy industrialization there can be no economic development does not, of course, justify a growth concentrated mainly in the luxury goods industry. Another connected argument, however, should arrest our attention. It is the old tenet that the only spur of progress is the demand of the rich for luxuries. To cite its originator,

> . . . . . . whilst Luxury
> Employ'd a Million of the Poor,
> And odious Pride a Million more;
> . . . .
> That strange ridic'lous Vice was made
> The very Wheel that turn'd the Trade
> . . . .
> To such a hight, the very Poor
> Live'd better than the Rich before,
> And nothing could be added more [1].

Though in verse, this is an analytical argument which, like most analytical arguments in economics, is valid only for the reality that its author had in mind. Given the institutional privileges of the upper classes during Mandeville's time in Europe, it is elementary that there could not have been any development of the arts without the demand for luxuries of the privileged classes. It is equally elementary that the additional employment was then a powerful developmental factor for labor because of the tremendously high labor-output ratio [2]. These two essential conditions implied by Mandeville's thesis no longer apply to Latin America, nor to any other country for that matter. Feudal institutions having been legally abolished, there is no longer any circumstantial necessity for the national income to be under the control of a minority and for development to depend on the well-being of this minority. On the other hand, there are two reasons why the present economic expansion in Latin America works with the lowest known

---

[1] Bernard MANDEVILLE, *The Fable of the Bees: or Private Vices, Public Benefits*, F. B. Kaye ed., 1924, Oxford, vol. II, p. 25.
[2] Surprising though it may seem, in the socialist systems, too, one group of people—the present generations—works to build a luxury goods industry for another group—the future generations. The only issue that matters, therefore, is whether each living generation may benefit at least in part from its own toil.

labor-output ratio. The new techniques are imported ready-made from the most advanced economies where they represent the last word of technical progress. Second, the expansion occurs primarily in those lines for which the labor-output ratio is relatively the lowest—automobiles, refrigerators, television sets, etc. The increase in employment cannot therefore have the same developmental acceleration as three centuries ago. In contrast with what happened at that time in Western Europe, in Latin America the "reserve army" is showing little sign of shrinking. Actually, at the slow rate at which employment has been increasing in Latin America it would take an irrelevantly long time to absorb the reserve army even if population were to remain stationary (Table III, col. 7).

IX. THE STRUCTURAL LOCK

1. Although opinions concur in that it would be dangerous to try to end an inflation state by simply stopping the supply of fresh money (cf. Hansen, p. 163), when it comes to practical advice they usually place the accent on some fiscal, financial, or monetary rehabilitation. As to actual policies, they seem dominated by the general idea that all one has to do to bring inflation to an end is to operate the monetary levers properly. The idea ignores the warning by Hansen (p. 157) that "exclusive reliance on monetary policy ... is a dangerously one-sided weapon."

The validity of this warning is easily proved. If the government begins to apply the brakes on inflation, two things will happen with necessity. First, there will be a "credit crisis"—as we have so often heard it put whenever one of the Latin American governments tried to slow down the printing of fresh money. As a result, it will no longer be possible to inflate the distributed profits by the whole amount of "illusory profits." The drop in the income of the business class will cause a fall in the demand for luxury goods by almost the same amount. And then we may see the laid off workers demonstrating in the streets—as they did peacefully in São Paulo during the summer of 1964—against the decision of the management, indirectly, against that of the government [1]. The only *imme-*

[1] It may be well to note that the different explanation offered by FRIEDMAN (1968, p. 9) for the same fact reflects, time and again, his clinging to a model moulded on an advanced economy without government controls.

diate solution for the government is obviously to return to the practice of oiling the wheels of trade with fresh money.

It is, therefore, the very unbalance of the economy that defeats any attempt at decreasing the dose of inflation. On the other hand, to persist with the inflationary therapy accentuates the structural unbalance (Fig. 3 b). This seems to vindicate the notion that the Latin American countries are confronted with a dilemma, a notion that constitutes a subtle leitmotiv of many an analysis of the situation (e.g., Prebisch). The truth is that there is no dilemma, only a lock that can be removed by correcting the unbalanced structure that goes with it. This operation, however, should not be conceived as a reallocation of the employed resources. The simple reason is that resources once crystallized into physical capacity cannot be shifted from one sector to another with the same instantaneous ease with which we are accustomed to shift the tip of the pencil from one "equilibrium" position to another in our analytical discussions.

2. Some years ago, I presented a model to represent the idea—rather novel at the time—that the economic system does not follow the same laws during the upswing and the downswing of the trade cycle (Georgescu-Roegen, 1966, Ch. 8). I wish now to add that the basic reason for this asymmetry is the fact that *decumulation* is not the reverse process of *accumulation*, with the equally important consequence that *deflation* is not the reverse process of *inflation*. Contrary to the usual implication of the monetary orthodoxy, there is no pendulum.

In spite of its being the terror of a planner's mind, accumulation of productive capacity is a simpler process than the decumulation of extant capacity. For one thing, accumulation is a much speedier process once the means are available. To be decumulated in the proper economic sense, any industrial installation must be *effectively* used until all its parts become scrap *at the same instant*. This may take an extremely long time (as well as some additional investment), so long that for all practical decisions the structure is rightly assumed perdurable. To be sure, an industrial installation may be wrecked or abandoned, but this is not the sense in which we speak about capital moving from one industry to another. Nor does the sale of an industrial installation at a great loss represent decumulation in real terms. Yet both these means have currently accompanied the painful process by which excess investment in one direction has been "decumulated." An apropos illustration: in the early 1920's after the German industrialists came to realize that the inflation had produced over-investment in some sectors (mainly those of luxury goods), they promptly rallied behind the surprising slogan "Demolition! We must consider as finally lost the capital unwisely invested in [such] factories" (Bresciani-Turroni, p. 390). No Latin American country, however, is sufficiently rich to afford the correction of its industrial unbalance by this extraordinary means. Besides, in Latin American countries the production capacity for luxury goods is excessive only relative to their present real income per capita.

3. Short of a radical political revolution, the only solution for correcting the unbalanced structure of such an economy and setting it on the track of balanced development is to stop the growth of the luxury goods industries and to develop the other sectors, in particular the wage goods industry in the hierarchical order of these goods. After all, it is the wage goods industry that has provided the largest and the most solid basis for the past and the present development of the more advanced economies. In fact, there is lasting economic development only if more and more luxury goods become wage goods, or to rephrase a pointed remark by Schumpeter[1], only if more and more factory girls wear silk stockings, not if the queen alone wears more of them.

Having a relatively higher labor-output ratio, the wage goods industry creates its own demand and by offering a larger number of opportunities to the would-be entrepreneurs it also supports the demand for the luxury goods. The gradual weaning of the luxury goods industries from cheap credit can then proceed without producing a lock. The corrective plan need not observe any criterion of profitability: there is yet no effective demand, or practically none, for the industries to be developed by the plan. This does not mean that the problem of "where the demand comes from" should be ignored. Real wages must necessarily be increased,

---

[1] J. A. SCHUMPETER, *Capitalism, Socialism, and Democracy*, 2nd edn., New York, 1947, p. 67.

but only in step with the growth of the wage goods industry, so as to avoid a new source of inflationary pressure.

In closing, let me stress the point that the argument of this paper does not deny the importance of a monetary policy well-adjusted to the process of readjustment. But it does contend that monetary wizardry alone cannot cure an inflation state. For if we fail to act on the real coordinates so as to direct the structure toward a balanced one, the structure bred by inflation will not only remain locked in the manner described above, but the key will turn in the wrong direction. The lock will become double-locked.

REFERENCES

1. Werner BAER, " Inflation and Economic Growth: An Interpretation of the Brazilian Case ", *Economic Development and Cultural Change*, XI (1962), 85-97.

2. Werner BAER, " Inflation and Economic Efficiency, " *Ibid.*, XI (1963), 395-406.

3. Werner BAER, " The Inflation Controversy in Latin America: A Survey, " *Latin American Research Review*, II (1967), 3-25.

4. Werner BAER and Mario Henrique SIMONSEN, " Profit Illusion and Policy-Making in an Inflationary Economy, " *Oxford Economic Papers*, XVII (1965), 279-290.

5. Jeremy BENTHAM, *Works*, John Bowring ed., Edinburgh, 1838-1843, vol. III.

6. Jeremy BENTHAM, *Sur les prix*, in David RICARDO, *Works*, P. Sraffa ed., Cambridge, 1951, vol. III pp. 267-341.

7. E. M. BERNSTEIN and I. G. PATEL, " Inflation in Relation to Economic Development, " *International Monetary Fund, Staff Papers*, II (1951/52), 363-398.

8. C. BRESCIANI-TURRONI, *The Economics of Inflation*, London, 1937.

9. Henry J. BRUTON, *Principles of Development Economics*, Englewood Cliffs, N. J., 1965.

10. Roberto de Oliveria CAMPOS, " Inflation and Balanced Growth, " in *Economic Development for Latin America*, Howard S. Ellis, ed., London, 1962, pp. 82-103.

11. Roberto de Oliveria CAMPOS, " Economic and Financial Policies of the Brazilian Government, " in *A Economia Brasileira e suas Perspectivas*, APEC, Rio de Janeiro, 1966, pp. 317-321.

12. Milton FRIEDMAN, " The Role of Monetary Policy, " *American Economic Review*, LVIII (1968), pp. 1-17.

13. Milton FRIEDMAN and Walter W. HELLER, *Monetary vs. Fiscal Policy: A Dialogue*, New York, 1969.

14. Nicholas GEORGESCU-ROEGEN, *Analytical Economics: Issues and Problems*, Harvard University Press, 1966.

15. Alvin HANSEN, *Monetary Theory and Fiscal Policy*, New York, 1949.

16. David HUME, *Writings on Economics*, E. Rotwein ed., London, 1955.

17. Alexandre KAFKA, " The Theoretical Interpretation of Latin American Economic Development " in *Economic Development for Latin America*, Howard S. Ellis ed., London, 1962, pp. 82-103.

18. Alexandre KAFKA, " The Brazilian Stabilization Program, 1964-1966 ", *Journal of Political Economy* LXXV (1967), 596-631.

19. J. M. KEYNES, " Inflation as a Method of Taxation, " *The Manchester Guardian Commercial*, July 27, 1922, pp. 268-269.

20. J. M. KEYNES, *Monetary Reform*, New York, 1924.

21. J. M. KEYNES, *A Treatise on Money*, New York 1930.

22. J. M. KEYNES, *The General Theory of Employment, Interest, and Money*, New York, 1936.

23. W. Arthur LEWIS, *The Theory of Economic Growth*, London, 1955.

24. Alfred MARSHALL, *Principles of Economics*, 8th edn., New York, 1949.

25. J. S. MILL, *Principles of Political Economy*, London, 1920.

26. Raul PREBISCH, " Economic Development and Monetary Stability, " *Economic Bulletin for Latin America*, VI (1961), 1-25.

27. David RICARDO, *Works*, P Sraffa ed., Cambridge, 1951, vol. III.

28. D. H. ROBERTSON, *Money*, New York, 1922 (rev. edn., 1948).

29. Mario Henrique SIMONSEN, *A Experiencia Inflationaria no Brasil*, Rio de Janeiro, 1964.

30. Léon WALRAS, *Elements of Pure Economics*, tr. W. Jaffé, Homewood, Ill., 1954.

31. Léon WALRAS, *Etudes d'économie politique appliquée*, 2nd edn., Lausanne, 1936.

STATISTICAL SOURCES

Anuario Estadístico, Buenos Aires.
Anuario Estatistico do Brasil.
Anuario General de Estadística, Bogotá.
Argentina Económica y Financiera, OECEI, Buenos Aires.
Boletín do Banco Central do Brasil.
Boletín Estadístico, Banco Central de la República Argentina.
Boletín Mensual, Dirección Nacional de Estadística y Censos, Buenos Aires.
Boletín Mensual, Banco Central de Chile.
Conjuntura Económica, Rio de Janeiro.
Demographic Yearbook, United Nations.
A Economia Brasileira e Suas Perspectivas, APEC, Rio de Janeiro.
Economic Bulletin for Latin America, United Nations.
Informe E. S., Dirección Nacional de Estadística y Censos, Buenos Aires.
Importaciones, Industrialización y Desarrollo Económico en la Argentina, Tome I, FIAT, Buenos Aires, 1963.
Monthly Bulletin of Statistics, United Nations.
Revista del Banco de la República, Bogotá.
Statistical Bulletin for Latin America, United Nations.
Statistical Yearbook, United Nations.
Statistics on the Mexican Economy, Mexico, 1966.
Yearbook of Labour Statistics, United Nations.
Yearbook of National Accounts Statistics, United Nations.

Note: Some data have been made available by the courtesy of *Instituto Torcuato di Tella* and Mr. H. Biggs.

STATISTICAL APPENDIX

A star has been used to show that a series is not comparable, or was reconstructed from partial information. Data between parenthesis are incomplete.

TABLE I. — *Cost of Living, Wages, and Nonwage Incomes*

	Argentina						
	(1)	(2)	(3)	(4)	(5)	(6)	(7)
1950	100	—	100	100	100	100	100
1951	137	37	119	113	155	87	82
1952	190	39	152	147	159	80	77
1953	197	4	160	154	190	81	78
1954	205	4	187	173	206	91	84
1955	230	12	192	177	256	84	77
1956	261	13	263	249	321	101	95
1957	325	25	270	255	415	83	78
1958	428	32	396	374	578	92	87
1959	914	114	673	624	1204	74	68
1960	1164	27	794	730	1541	68	63
1961	1320	13	985	922	1761	75	70
1962	1691	28	1231	1155	2137	73	68
1963	2098	24	1540	1443		73	69
1964	2563	22	2033	1869		79	73
1965	3295	29	2764	2508		84	76
1966							
1967							

	Brazil				Chile			
	(1)	(2)	(3)*	(6)*	(1)	(2)	(3)*	(6)*
1950					100	—	100	100
1951	100	—	100	100	122	22	110	90
1952	116	16			150	23	142	95
1953	133	15	132	102	187	25	161	86
1954	163	23	217	138	323	17	205	63
1955	201	23	260	139	565	75	355	63
1956	243	21	290	132	881	56	563	64
1957	283	16	410	154	1173	33	752	64
1958	324	14	451	133	1429	22	1292	90
1959	451	39	637	143	1986	39	1569	79
1960	583	29			2214	12	1703	77
1961	777	33	1019	131	2386	8	1957	82
1962	1179	52	1488	131	2714	14	2314	85
1963	2009	70	2362	125	3914	44	3078	79
1964	3852	92			5700	46	4683	82
1965	6383	66			7357	29	6526	89
1966	9019	41			9029	23	9452	105
1967								

TABLE I *(Continued)*

	Colombia				Mexico			
	(1)	(2)	(3)*	(6)*	(1)	(2)	(3)*	(6)*
1950	100	—	100	100	100	—	100	100
1951	107	7	97	91	113	13	111	98
1952	105	—2	89	85	129	14	116	90
1953	113	8	92	81	127	—2	123	97
1954	123	9	—	—	132	4	136	103
1955	122	—1	94	77	154	17	156	101
1956	130	7	104	80	161	5	169	105
1957	150	15	138	92	170	6	176	104
1958	172	15	157	91	189	11	194	103
1959	183	6	171	93	192	2	215	112
1960	191	4	224	117	204	6	236	116
1961	207	8	258	125	206	0,9	247	120
1962	213	3	296	139	209	1.5	269	129
1963	280	32	411	147	209	0	316	151
1964	330	18	468	142	215	3	347	161
1965	341	3	520	152	223	4	370	166
1966	408	20	591	145	232	4	384	166
1967	442	8						

(1) Index of Cost of Living (Annual Averages).
(2) Annual percentage change in (1).
(3) Index of Nominal Wages of unskilled workers.
(4) Index of Nominal Wages of skilled workers.
(5) Index of Nominal Incomes other than wages and salaries.
(6) Index of Real Wages of unskilled workers — (3) (1).
(7) Index of Real Wages of skilled workers — (4) (1).

Notes : For Chile, Colombia, and Mexico, (3) refers to all categories, skilled and unskilled. For Brazil and (3), *average* wage in July, 1951-1954, *median* wage in April, 1955-1963.

TABLE II. — *Minimum Wages*

Year	Brazil				Mexico	
	(1)	(2)	(3)	(4)	(1)	(2)
1950 . . .					100	100
1951 . . .	100 (XII)	100	100		100	88
1952 . . .					160	124
1953 . . .					160	126
1954 . . .	200 (V)	138	70	85	190	144
1955 . . .					190	123
1956 . . .	317 (VII)	146	92	115	217	135
1957 . . .					217	128
1958 . . .	500 (XII)	142	91	118	243	129
1959 . . .					243	127
1960 . . .	800 (X)	128	83	112	296	145
1961 . . .	1120 (X)	128	96	112	296	144
1962 . . .			76	102	372	178
1963 . . .	1750 (I)	117			372	178
1964 . . .	3500 (II)	112	61	89	479	223
1965 . . .	5500 (II)	100	67	90	479	215
1966 . . .	7000 (III)	85	69	84		
1967 . . .	8750 (II)	81	66	75		

(1) Index of Minimum Nominal Wage (at the Adjustment Decree).
(2) Index of Minimum Real Wage.
(3) Index of Minimum Real Wage in the month preceeding the decree.
(4) Index of Average Minimum Real Wage between decrees.

Note: Roman Numerals show the month of decree.

TABLE III

Employment, Wage Bill, and production in manufacturing.

Year				Argentina			
	(1)	(2)	(3)	(4)	(5)	(6)	(7)
1950	100	100	100	100	100	100	100
1951	102	101	101	88	87	100	99
1952	105	102	104	82	79	102	97
1953	107	96	104	78	75	108	90
1954	109	101	118	92	78	117	93
1955	111	108	137	90	66	127	97
1956	113	115	140	116	83	122	102
1957	115	120	176	100	57	147	104
1958	117	125	201	116	58	161	107
1959	119	130	200	96	48	154	109
1960	121	122	217	83	38	178	101
1961	123	118	234	88	38	198	96
1962	125	114	221	83	38	194	91
1963	127	106	210	78	37	198	84
1964	129	102	245	81	33	240	79
1965	131	106	272	89	33	257	81
1966							

Year				Brasil			
	(1)	(2)	(3)	(4)	(5)	(6)	(7)
1950							
1951							
1952	100	100	100	100	100	100	100
1953	102	109	102	103	101	93	107
1954	105	113	109	119	109	96	108
1955	107	116	122	125	102	106	108
1956	110	116	133	126	95	114	105
1957	112	110	137	126	92	125	98
1958	115	119	167	137	82	141	103
1959	118	135	198	149	75	147	114
1960	128	—	—	—	—	—	—
1961	131	—	—	—	—	—	—
1962	135	150	229	184	80	153	111
1963	140	142	245	193	79	172	101
1964	145	151	253	204	81	167	104
1965	149	146	259	188	73	178	98
1966							

TABLE III. — *(Continued)*

Year	(1)	(2)	(3)	Chile * (5)	(5)	(6)	(7)
1950	100	100	100	100	100	100	100
1951	102	100	120	91	76	120	98
1952	104	—	133	—	—	—	—
1953	106	101	143	87	61	142	95
1954	109	105	148	67	45	141	96
1955	112	—	143	—	—	—	—
1956	115	—	153	—	—	—	—
1957	118	112	150	72	48	134	95
1958	121	(113)	154	(102)	(66)	(136)	(93)
1959	123	(115)	182	(91)	(50)	(158)	(94)
1960	127	(113)	172	(87)	(51)	(152)	(89)
1961	129	114	183	93	51	161	88
1962	132	116	202	99	49	174	88
1963	135	121	214	95	44	177	90
1964	138	128	225	105	47	176	93
1965	141	(129)	236	(114)	(48)	(183)	(92)
1966	145	(128)	252	(134)	(53)	(197)	(88)

Year	(1)	(2)	(3)	Colombia (4)	(5)	(6)	(7)
1950							
1951							
1952							
1953	100	100	100	100	100	100	100
1954	103	—	91	—	—	—	—
1955	106	110	110	116	105	100	104
1956	110	118	126	129	102	107	107
1957	113	120	132	145	110	110	106
1958	117	120	133	154	116	111	103
1959	121	124	141	169	120	114	102
1960	125	129	157	193	123	122	103
1961	129	131	165	212	128	126	102
1962	133	134	187	258	138	140	101
1963	137	135	188	254	135	139	98
1964	141	136	194	255	131	143	96
1965	146	135	—	253	—	—	92
1966	151	138	—	246	—	—	91
1967							

TABLE III. — *(Continued)*

Year				Mexico			
	(1)	(2)	(3)	(4)	(5)	(6)	(7)
1950 . .	100	100	100	100	100	100	100
1951 . .	103	102	106	100	94	104	99
1952 . .	106	102	105	92	88	103	96
1953 . .	108	98	106	95	90	108	91
1954 . .	112	95	115	98	85	121	85
1955 . .	115	100	129	101	78	129	87
1956 . .	118	108	142	113	80	132	92
1957 . .	122	109	161	113	70	148	89
1958 . .	125	110	176	113	64	160	88
1959 . .	129	110	192	123	64	174	85
1960 . .	135	122	208	142	68	171	90
1961 . .	140	127	216	152	70	170	91
1962 . .	144	126	231	162	70	183	88
1963 . .	149	110	252	166	66	229	74
1964 . .	154	129	285	208	73	221	84
1965 . .	160	136	306	226	74	225	85
1966 . .	165	—	340	—	—	—	—
1967 . .	171	—	368	—	—	—	—

(1) Index of Population.
(2) Index of Employment in Manufacturing.
(3) Index of Gross Value of Manufacturing Output at Constant Prices.
(4) Index of Real Wage Bill.
(5) Index of Labor Cost in relation to Gross Output—(4)/(3).
(6) Index of Gross Output per worker—(3)/(2).
(7) Index of Labor Absorption—(2)/(1).

Notes: Argentina: (3) = (6) × (2); (4) = (2) × Index of real wages of unskilled workers.

Colombia: includes all employees; 1965-1966, (4) = (2) × Index of real wage.

Mexico: (2) wage earners only: (4) = (2) × Index of real wage.

TABLE IV. — *Distribution of National Income*

	Argentina				Brazil				Colombia		
Year	(1)	(2)	(3)	Year	(1)	(2)	(3)	Year	(1)	(2)	(3)
1950	100	100	49.6	1947	100	100	29.5	1950	100	100	53.5
1951	141	131	46.2	1948	113	117	30.6	1951	117	108	49.3
1952	164	167	50.5	1949	130	142	32.2	1952	129	115	47.9
1953	186	183	48.9	1950	153	171	32.9	1953	146	131	48.1
1954	210	211	49.9	1951	182	198	32.0	1954	163	142	46.6
1955	253	239	47.0	1952	210	237	33.2	1955	206	156	40.5
1956	312	292	46.4	1953	257	288	33.1	1956	232	170	39.2
1957	400	362	44.8	1954	324	366	33.3	1957	271	199	39.3
1958	581	542	46.3	1955	412	495	35.4	1958	302	233	41.4
1959	1096	892	40.4	1956	522	681	38.5	1959	350	269	41.2
1960	1332	1074	40.0	1957	622	810	38.4	1960	400	313	41.9
1961	1586	1348	42.0	1958	741	966	38.4	1961	460	369	42.9
1962	1994	1652	41.1	1959	1010	1310	38.2	1962	518	432	44.7
1963	2416	1960	40.2	1960	1352	1754	38.3	1963	651	566	46.5
1964	3298	2890	43.5					1964			
1965	4533	4229	46.3					1965			

(1) Index of National Income.
(2) Index of Wage Bill.
(3) Percentage share of Labour in National Income.

Notes: Brazil: (2) Excludes salaries of Executives or professionals.
Colombia: (2) Includes all personnel on public payroll and all employee compensations.

TABLE V. — *Distribution of National Income in Brazil*
(In Million N Cr.)

Source	1947 (1)	1960 (2)	(2):(1)
1. Wages and Salaries	41	719	1754
2. Independent workers	16	167	1044
Subtotal: (1) + (2)	(57)	(886)	(1554)
3. Liberal Professions	4	37	925
4. Executive Salaries	20	154	770
5. Non-incorporated firms	2	14	700
6. Subtotal: (3) + (4) + (5)	(26)	(205)	(788)
7. Profits, Interest, and Rent	19	275	1447
Total (Urban Sector)	102	1366	1345
National Income	139	1879	1352

TABLE VI. — *Annual Production of Some Selected Commodities*

A. *Per Capita*	1950-1954	1955-1959	1660-1964
Argentina			
Sugar (tons)	0.038	0.041	0.040
Cotton Fabrics (metric tons) .	(0.0041)	0.0041	0.0035
Shoes (Pairs)	1.31	1.50	1.35
Cigarettes (000)	1.077	1.122	1.138
Brazil			
Sugar (tons)	0.033	0.043	0.044
Cotton Fabrics (metres) . . .	(20.2)	(19.9)	—
Cigarettes (000)	(0.746)	(0.793)	(0.812)
Chile			
Cotton Fabrics (metres) . . .	12.2	(11.1)	11.0
Cigarettes (000)	(0.823)	0.756	0,796
Colombia			
Sugar (tons)	0,016	0,018	0,023
Cotton Fabrics (metres) . . .	14.2	15.0	18.0
Cigarettes (000)	0.974	0,997	1.010
Mexico			
Sugar (tons)	0.028	0.034	0.043
Cotton Fabrics (metric tons) .	0.00129	0.00135	0.00122
Cigarettes (000)	0.954	0.962	0.889

TABLE VI. — *(Continued)*

B. *Total*	1950-1954	1955-1959	1960-1964
Argentina			
Cement (ooo tons)	1603	2222	2780
TV Sets (ooo)	—	(74)	129
Refrigerators (ooo)	69	185	188
Passenger Autos [a] (ooo)	(1.5)	11.1	85.3
Telephones [b]	5.23	6.05	6.46
Brazil			
Cement (ooo tons)	1545	3414	4992
Radio sets (ooo)	—	408	1047
TV Sets (ooo)	—	81	284
Passenger Autos [a] (ooo)	—	(18.7)	87.6
Telephones [b]	(1.22)	1.44	1.54
Chile			
Cement (ooo tons)	706	771	1065
Telephones [b]	(2.28)	2.27	2.74
Colombia			
Cement (ooo tons)	753	1214	1702
Telephones [b]	(1.02)	1.57	2.12
Mexico			
Cement (ooo tons)	1678	2432	3531
Radio Sets (ooo)	(169)	308	633
TV Sets (ooo)	—	70	114
Autos [a] (ooo)	(17.4)	18.2	43.1
Telephones [b]	1.13	1.28	1.59

Notes: [a] Includes Assembly: Mexico: only assembly;
[b] in use per 1000 persons.

CHAPTER 8

(1965)

The Institutional Aspects
of Peasant Communities:
An Analytical View[1]

OVER THE LAST 20 years especially, evidence of the failure of policies founded upon orthodox or Marxian economics to solve the difficulties of the underdeveloped agricultural economies—whether within the Communist world or outside it—has been continuously mounting. As a result, a reaction against the received doctrine has begun to spread slowly but persistently among traditional economists. This very volume is one relevant symptom of the new orientation which has already produced a fast-growing literature on peasant economies and communities.

Peasant communities, we now begin to realize, constitute a social category distinct from the urban, bourgeois societies upon which both orthodox and Marxist theories were molded. Consequently, in order to understand the problems that beset the peasant class in many parts of the world and to increase our chances of improving the lot of the vast masses of the unfortunate, it is absolutely necessary to arrive at some understanding of the peculiar institutions by which most peasant communities still live. Like all understandings, this one too requires more than a mere recital of facts; it calls for discovering a rationale behind the facts, which in the case of institutions means to discover their internal logic. This paper is offered as a modest contribution toward this particular end.

The historical side of peasant institutions with particular emphasis on Europe has been covered elsewhere [Georgescu-Roegen, 1969]. Therefore the primary focus in the present paper will be upon peasant institutions at the community or village level, examined with a substantial amount of pure economic analysis so that these "strange" institutions may become more intelligible to students formed at the school of traditional economic theory. For in the ultimate analysis, as the oldest philosophers taught, everything must rise from a cause or a reason. In the light of this

1. The version published here differs from that presented at the A/D/C Seminar: some sections of the original version have been left out, others have been shortened. The version presented at the seminar, amplified and completed with some closely related topics, will appear as a separate monograph by this author.

reason, nothing, whether the "irrationality" of traditional economists or of the peasant institutions, is irrational.

The Peasant Economy and Traditional Economics

Since interest in the study of peasant economies and village communities is most likely to emerge in a country with an overwhelming peasant population, it is normal that the pioneering in this direction should have been done by European economists. But why should the economists of a nation like the United States, which is the most urbanized and industrialized in the world and which moreover has never possessed a peasantry, suddenly become interested in the economy of peasant societies? What are the factors responsible for the tardy discovery of peasant economics and sociology?

Bridgman, a Nobel laureate for physics famous also for his writings on the philosophy of science, argued that the major handicap of economics is the characteristic intellectual opportunism of its servants [1950, pp. 303-305]. Now, no one can deny that, economic historians being excepted, every economist who has won a place in the history of thought —from Cantillon and Quesnay to Schumpeter and Keynes—has been intellectually opportunistic in the sense that each has been exclusively preoccupied with the contemporary economic problems of the society in which he happened to live. And the same is true of all those numerous Eastern economists who are little, if at all, known in the West precisely because they were interested only in the peasant economy.

But, I submit, the economic profession should take pride in being opportunistic in the above sense. Surely it would have been everybody's loss and nobody's gain if, for instance, Keynes had applied his talent to studying the problems of a country with an agriculture ravaged by prolonged wars—as was the case of France in Quesnay's time—instead of those posed for a modern government by periodic spells of high industrial unemployment. And, as we can now appreciate, it would have been uneconomical and even absurd if the social scientists in nineteenth-century Russia had been preoccupied with the industrial working class instead of Russia's peasantry. By the same token, we cannot but applaud the fact that the economists of the countries with an advanced economy have shifted their main interest to the problem of how to speed up the development of the underdeveloped economies as soon as this problem became vital for their own countries.

On the other hand, there is no denying that the intellectual opportunism of the economic profession has had some undesirable and unnecessary consequences. In contrast to natural scientists, students of human society have been prone to theorize about any social form without a direct, material knowledge of it, by merely extrapolating the laws they have established for the society known to them immediately. Some even claim a merit for doing so. But no other profession has committed the sin of theorizing in a vacuum as often and with as complete an ingenuousness as the economist of the Classical and, especially, the Neo-Classical tradition. The circumstances speak for themselves: the more he learned about the pasticular economy in which he lived, the more absorbed he became in its study. Needless to add, the sin became even more stubborn as economic science developed along Ricardian lines in becoming more quantitative and less institutional, hence less historical. For although numbers possess some unique powers, they also have a hidden vice: they tend to lure us into ignoring form and qualitative factors.

Surprisingly enough, denunciations of the sin came primarily from the ranks of economists themselves. The celebrated *Methodenstreit*—the strife of methods during the last decades of the nineteenth century among German economists—comes naturally to mind. Unfortunately, it turned out to be mainly a denunciation of theory in general, which was tantamount to blaming the soda for the intoxicating power of a mixed drink. That is why the *Methodenstreit* seems to us so frustrating. Only later on do we find the real root of the difficulty touched upon occasionally. For an illustrious example, the lament that "Ricardo and his followers ... work[ed] out their theories on the tacit supposition that the world was made up of city men" ran through all editions of the modern bible of traditional economics, Marshall's

Principles [1949, p. 62]. Yet the lament, or rather the denunciation, has caused no stir.

The situation is most intriguing: although the fact that life can take on an infinite variety of biological forms is unanimously recognized, there is appreciable disagreement concerning the qualitative variability of social forms. In fact, traditional thought in economics is characterized by the position that the economy of every society is only a particular instance of a unique pattern, that of the civil society. The phenomenon can best be explained by the attraction economists have felt, many still feel, for urban society and town life, perhaps because the towns display greater institutional uniformities than any other communal organization.

From the earliest times towns have tended to become increasingly alike to each other; contact between distant settlements, whether within the same or different cultural groups, has always been achieved through the exchange of commodities and ideas, i.e. through media attached to urban life. The result is that by now towns all over the world present an almost identical spectrum of institutions. This was already the case for the urban centers of Western Europe at the time of the founders of Classical economics. And since the picture an observer can draw of the world or of the universe cannot possibly include elements other than those perceived through the particular window from which he happens to contemplate the outside, no Classical economist can be indicted for not having discovered the variability of social and, hence, economic forms. As Marshall [1949, p. 62] correctly explained, "the people whom they [the Classical economists] knew most intimately were city men." One may feel pretty sure that Adam Smith's confidence in his own economics must have become stronger after his visit to France. This applies with even greater force to Marx's orientation, since he had occasion to verify directly the uniformity of the bourgeois society in more urban centers than any other nineteenth-century economist. Along the same line of thought, one should also find it natural in retrospect that the first impulse to economic anthropology should have come from Lewis H. Morgan, who "spent a great part of his life among the Iroquois Indians ... and was adopted into one of their tribes."[2] Had the English peasantry survived as a *significant* social stratum for England's productive capacity or had there been a peasantry in the United States, the science of economics would have, most probably, developed on broader tracks than those of the Classical school.[3]

In connection with the preceding remarks, the case of Marx deserves a special note at this juncture. For Marx, unlike all other Classical economists of his time, was fully aware (through Hegel's influence) of the importance of institutional differences for economic science. He thought of economists as strange creatures precisely because for them "the institutions of feudalism are artificial institutions, those of the bourgeoisie are natural institutions" [Marx, 1900, p. 120f]. Another spelled-out example is found in one of his earliest writings, where Marx takes Ricardo to task for his "bourgeois horizon" in making the primitive hunter and fisher "consult the annuity tables [of] the London Exchange" [Marx, 1904, p. 69f].[4] Yet no Classical economist has had so decisive a role as Marx in spreading among both the orthodox and the unorthodox chapters of traditional economics the tenet that peasants do not even constitute a social class and, hence, it is senseless to speak of a peasant economy as a distinct analytical category.[5]

We should not fail to note another, equally significant, case. The important place of agriculture in the economy of the United States has created an intense interest in agricultural economics among American students and policy-makers alike. The symptoms are seen in such government initiatives as those of Theodore Roosevelt, who in 1908

2. For which see Engels [1884, p. 25]. Professor Raymond Firth drew my attention to the fact that Morgan's living among the Indians is a mere legend. However, Morgan's lengthy contacts with the Indian tribes as well as his adoption by the Senecas are historical facts.

3. Tradition, however, is a powerful master. Quite recently the majority of British economists received with immense satisfaction Walter Eucken's attempt—perhaps the ablest of all—to justify the Classical position in his *Foundations of Economics* [1940].

4. See also Marx's emphatic remarks on the institutional differences bearing upon the writings of British and French economists [1904, p. 56n].

5. David Mitrany [1951] offers a most complete and able analysis of the impact of this article of Marxist faith upon the ideology, strategy, and policy of the Communist Party in Russia and other East European countries. Some additional thoughts on the struggle between Communists and Agrarians are found in the author's essay [1960, pp. 1–40].

created the Country Life Commission, and of Woodrow Wilson, who in 1919 established the Division of Farm Population and Rural Life. However, the most interesting outcome is the American school of agricultural economics, whose admirable scientific achievements stand above all others in the same domain. That this domain has been confined to *farm economics* is averred by the very title of the foremost periodical of the school.[6] To point out that there is an immense difference between a *farmer* economy—regulated by cash profits and resting on a granular texture of individual interests—and a *peasant* economy is not to belittle in the least the merits of that school, but only to note a relevant fact for our topic: because of its particular objective, the American school of agricultural economics has been the staunchest preserver of the traditional viewpoint concerning the universal validity of Classical and Neo-Classical theory.[7]

As we observed earlier, a sustained interest in a study of the village community is most likely to emerge in a country with an overwhelming peasant population. And since any sociological study—be it of the industrial worker or of the peasant—can be undertaken only by a prepared mind, a second necessary condition for the formation of such an interest is the existence of a sufficiently sophisticated intelligentsia. The economic conflict, which in such a setting is apt to germinate around the condition of the peasant, would then easily provide the necessary intellectual motive. These conditions suffice to identify Russia during the early decades of the last century and to account for the dominant concern with the condition of the Russian peasants among the leading intellectuals from the mid-nineteenth century on —Herzen, Chuprov, and Chaianov, to mention a few among the best known.[8] (For further details see Georgescu-Roegen [1969].)

First the Slavophiles and then the Narodniki set their hope for the economic salvation of Russia on the resilient qualities of the Russian peasant villages. In fact, it was the Narodniki who, in reaction to the extreme romanticism of the Slavophiles, proclaimed—a strange idea for that time—that no adequate agrarian policy can be devised without a *direct* study and understanding of the peasants' social conditions. Their outspoken scorn for "theory" was the ultimate consequence of this excessive positivism. But whatever one may think about this characteristic bent of the Narodniki or about their platform, the fact remains that it was their ideology and determination that opened the field of peasant sociology.

The Narodniki as well as their ideological heirs, the agrarians, used to accuse their political adversaries of viewing the peasant problem not as a problem concerning people but as one of land. The same accusation applies equally well to the long line of students who, for decades after Haxthausen's work [1847–1852], reduced the sociology of peasant communities to the problem of the origin of landed property. The basic problem, that of the institutions peculiar to the peasant communities, was thus set aside immediately after Haxthausen revealed its importance. A long time elapsed before the tide turned. British functionaries in India, like Haxthausen in Russia, were in a position to observe closely the life in the native villages. Some were greatly surprised by the contrast between life in the Indian villages and what they had thought to be the normal, the rational. The result was a series of most interesting studies, which not only aroused the interest of other European scholars but also set village sociology again on its natural track [Maine, 1861, pp. 252–261; Baden-Powell, 1892, 1896, 1899]. They inspired at least one British scholar, F. Seebohm [1896], to turn to the living villages of Great Britain and their documented history with the idea of searching for some relevant vestiges of the unknown past.

Thus, although the peasant community as a distinct social category was discovered by the intellectuals of a peasant nation, the Narodniki, the inspiration source of the method now considered the most appropriate

---

[6]. A parallel orientation is observed in sociology. Witness the standpoint adopted by numerous writers such as D. Sanderson [1917], C. R. Hoffer [1926], and W. Gee [1929].

[7]. Perhaps the only interlude worth mentioning was marked by the interest J. D. Black and M. L. Wilson, in particular, manifested in the late 1930's for problems closely related to the economy of peasant societies [Black, 1939a, 1939b; Wilson, 1939a, 1939b].

[8]. It is instructive to observe that the situation of the peasant class developed into a conflict for the first time in the West, to wit the *jacqueries* and the peasant wars of the earlier ages. But the intelligentsia of that time lacked adequate sophistication.

for village sociology has been provided by British scholars. Given Great Britain's interest in *and* also her peculiarly sophisticated attitude toward her colonies, there is nothing paradoxical about this mutation.

Village Typology and Taxonomy

The greatest predicament of the student of the peasant community was incisively formulated by the Rumanian sociologist Stahl [1946, 40]: "There is no 'Rumanian' village, but only 'Rumanian villages.'" What that scholar, who had spent long years of observation among the Rumanian peasants, wanted to impress upon us is that village communities do not form a homogeneous universe even if we confine our attention to a region with a fairly uniform history. On the other hand, we must admit that there are differential aspects—perhaps more important for the policy-maker than for the analytical student—justifying a division of peasant villages along their ethnical origins: the Russian *mir*, the Swiss *allmend*, the Scandinavian *allmenning*, the Saxon *tun*, the English *vill*, the Irish *tuath*, the German *Dorf*, the Rumanian *sat*, the French *communal*, the Indonesian *ndesa*, the Indian *pueblo*, and the like. Some sort of classification, be it for the sake of exposition alone, is indispensable for any further morphological study of the villages within each of these broad classes. One of the criteria frequently adopted in the literature is the topographical location: plain village, hill village, mountain village. Another criterion distinguishes between the compact and spread-out villages, still another between villages with irregular fields and villages with strip fields. At times villages are divided into wheat-growing, rice-growing, potato-growing, and so on. Nothing is basically wrong with these and other similar classifications. But they come very close to classifying mammals, for instance, into black-haired, brown-haired, and so forth. In other words, they are mere catalogs. Whatever its practical use, a mere catalog does not satisfy the urge of the understanding any more than that of a mail-order house does.

To be revealing, a classification must be taxonomic—that is, it must reflect some force function, as does the Mendeleev table or biological taxonomy, for example. It is natural, then, for one to think that because village institutions are subject to evolution, it should be possible to arrive at a taxonomy of village communities which, like biological taxonomy, would map out, however imperfectly their evolution. Unfortunately, criteria for such a taxonomy seem so far unavailable. In all probability, even a new Linnaeus could not change the situation.

The reason for the immense variability of forms is the same in the social and biological domains. It only works more powerfully in the former than in the latter. With the first living cells that emerged from *inorganic* (viz. inert) matter, a new phenomenal domain, the *organic*, came into being. Thereafter life could propagate itself through an entirely different process than the original one—that is, life begot life. That is why life has assumed forms without number which could not arise directly from a calm sea of warm mud, as the first living cells presumably did. Moreover, in their evolutionary process life-bearing forms have not followed a unilinear direction, nor even a treelike pattern. Often lines which branched out at one point came to meet again in another form.

The story repeated itself as the first social forms emerged among the herds of men living until that time as any other animals do. The *superorganic* domain—to use the well-chosen term of Kroeber [1917]—came into being at that moment. Social forms, like biological forms, have ever since evolved from other social forms with a flexibility far greater than that of organic evolution. Whatever might be the reason for it, the fact is that the necessity of the laws indisputably weakens as we pass from the inorganic to the organic and from the latter to the superorganic level.

Diffusion of mutations in the organic world cannot go beyond the limits set by sexual reproduction: no mutation of drosophila, for example, can pass on to the mammal class. On the other hand, once some form of individual property in land emerged in one community, it propagated itself easily to other communities. Clearly, individual landed property among the contemporary Iroquois or Incas has not come about in direct line from the institutions of the old Indian tribes. As Maitland [1897, p. 345] nicely put it, the Anglo-Saxons did not arrive at the alphabet or the Nicene Creed by following all the initial stages; they got them from others. Everything points to the fact that mutations, in both directions, not only occur more easily at the superorganic than at the organic level, but also represent greater discontinuities. There is no question that revolutions and counterrevolutions are phenomena specific to the superorganic domain.

Some authors—like Stahl, for instance [1946, pp. 48, 156]—have classified village communities according to the degree that the land of the village is held in individual property. Yet such a classification can hardly represent a relevant taxonomy for all village forms. At most, it may serve as a convenient device for analyzing the social transformations over a relatively short historical span and within some definite area. To use an analogy, the classification in point is comparable to that of a biological species according to its varieties.

Property rights in land have been advocated so frequently as the best classification for a taxonomy of peasant villages that a brief digression to buttress my objections is in order. There is more than one reason why the degree of individual property cannot provide a force function for a general taxonomy of villages. To begin with, the change from individual to communal property has not been a rare phenomenon—Communist revolutions apart. Engels himself had to admit [1884, pp. 133–143] that this "back mutation" occurred on a large scale after the collapse of the Roman empire under the barbarian invasions. In France, the same change has been noticed as late as the eighteenth century [Lafargue, 1901, p. 70]. According to some authorities, the Russian *mir* itself, far from being a direct continuation of the primitive communism of the Slavs—as the Narodniki and the Marxists believe—is a relatively recent back mutation imposed from above for the purpose of insuring better collection of taxes and providing stricter control of the agricultural manpower.[9]

As Vinogradoff observed [1920, p. 42], during the expansion of the Roman Empire "the Romanisation of outlying provinces [was] at the same time the barbarisation of Rome." How many times land thus passed from one regime to another and then back to the first during the long periods about which we know very little is a moot question. What is certain is that the extensive anthropological studies undertaken recently show that some definite form of personally inheritable property in land is rather common among extant societies that are far less evolved than were the village communities of Tsarist Russia or sixteenth-century England.[10] Moreover, some of these "primitive" societies are still practicing polygyny, a fact that does not fit into Engels' thesis.

The earliest human settlements may have differed little from an animal herd, but authorities on this difficult subject assure us that each one deserving the name of *social* community was established by a "successful brigand" of some sort or other. Indeed, it seems impossible to think of a tribe or clan, however primitive, without a chief. But the connection between a leader and a settlement is found long after individual property emerged. Documents attest that the "successful brigand" who founded Rome endowed every settler with individual landed property —of the Roman type, to be sure. Throughout the Middle Ages, and even much later, numerous villages or domains were founded on the same basis by "enterprising barons" and even by "rich, bourgeois-like entrepreneurs," not to mention kings and princes.[11]

---

9. The original author of this thesis is B. N. Chicherin, who advanced it in two articles (1852) not available in translation. Chicherin summarized his argument in his article "Leibeigenschaft in Russland" [1861]. But see also the works cited in Robinson [1949, p. 274n24]. Robinson is right in observing that "in the whole range of Russian history, there is perhaps no subject so obscure and so highly controversial as that of the *mir*'s origin."

10. See the summarizing conclusions of a survey covering some two hundred societies hardly touched by outside civilization in Murdock [1949, p. 82]. Also the monographs on the aborigines of Vietnam, Luzon, Borneo, Java, Ceylon, Formosa in Murdock [1960]. One should also note the observation made by Baden-Powell [1896, pp. 7ff, 131, 402; 1899] concerning the absence of communal lands in the village communities of Dravidian tradition in South and Central India and among the Tibeto-Burman (Kolarian) settlers in Eastern India. The point that all these instances pertain to non-Aryan populations may have some significance.

11. Bloch [1952, pp. 3–20]; Orwin and Orwin [1954, p. 19]. The phenomenon was not confined to the nations with a nobility organized after the Germanic tradition. Villages known to have had a "lord" as far back as one can go or founded by a princely grant as a reward for some military feat were not uncommon in the history of Rumania [Stahl, 1946, p. 49]. After the forced exodus of the Tartars under Catherine II, the vast area of Southern Russia thus emptied was populated among others by German and Serbian immigrants to whom land was given—as is true for all colonists—in complete individual property.

Numerous also are the villages in the southern half of Europe that descend from a Roman *castrum* or *vicus*, where only the Roman form of property once existed. Consequently it would be absurd to contend, as Marx did, that in every extant rural or urban agglomeration the Iroquois "shows through unmistakably" [Engels, 1884, p. 90].

No doubt it would be equally absurd to maintain that every extant village in Southern and Western Europe began as a *villa nova* founded on the principle of individual property, just as Rome or Boston was. Of course, if one trusts only the written evidence, one can never discover the existence of villages having no birth certificates. "I have read *all* these documents, not once, but several times, not in extracts, but all through from beginning to end", declares Fustel de Coulanges [1885, pp. 172f, 171 n1], but "one cannot find there one single word, before the tenth century, meaning community." To proceed otherwise, he insists, is not good history.[12] Yet history—especially, the history of peasant communities—is not always recorded or faithfully recorded by documents.

For this very reason most students of village communities have abandoned the old methods for reconstructing the past either by extrapolating the Iroquois or by trusting the written testimony. Nowadays they first search for the concrete traces of the past in the institutions of our own era, even in the landscape of the countryside, and interpret the written sources or accept anthropological parallelisms only in the light of such contemporary remains—in some of which, luckily, the past is often very much alive. They thus aim at a reconstruction of the past by a method akin to paleontology.

The Anatomy of the Peasant Village

A general principle of scientific procedure is that each special science should build its analytical framework on those elements which represent atomic units within its particular domain. They are the elements that, if divided further, cease to reflect the very phenomena in which the corresponding discipline is interested. In chemistry, for example, the atomic unit is the molecule, not the atom or the intraatomic particle, because chemical properties are borne by the former, not by the latter. There are incontrovertible reasons, I submit, why the village community constitutes the analytical atom in the phenomenal domain of peasant sociology.

To begin with, the village, next to the human individual, is the most clearly delimited social entity. Like the human individual, it is a perfectly natural, atomic, social unit. As history shows, kingdoms, provinces, counties, and even cities, can be split into several other kingdoms, provinces, and the like. The same is true of tribes and clans. But a peasant village, as long as it remains peasant, is indivisible. Whole clans or parts of them have often left their home village and founded a new village elsewhere, but this does not mean that the latter is a part cut from the former. Nor can this swarming be taken as proof that the peasant village is not an organic unit in the strictest sense of the term: all biological organisms multiply by the same process as peasant communities—that is, through internal growth.

Secondly, a study of peasant institutions must focus its attention first of all on that social entity which is certain to display their entire spectrum. Villages within a geographical region may have identical institutions and, moreover, be connected with each other in different ways so as to form a social entity, some *terra*, as its name goes in medieval tradition (for which, see Stahl [1939]). Its study is of no little value. But if one would decide to exclude the village altogether from the picture, the analysis of this higher social organization will bring the village back in full force. A simple formalization of the structure will show this without difficulty. Let $A_1, A_2 \ldots$ be the individuals belonging to such a *terra* and $R_1, R_2 \ldots$ be all institutional relations that may exist between an A_i and an A_j. The analytical map of the true relations $A_i R_k A_j$ will immediately separate the whole structure into several distinct nuclei, each corresponding to one of the villages. The analytical separation results from the fact that the number of relations true for any pair A_i, A_j of the same nucleus exceeds by a significant magnitude the number of

---

12. It should be noted, however, that Fustel de Coulanges had the better of those of his opponents who also invoked written testimonies. See, for instance Fustel de Coulanges [1885, chapter V] and E. D. Glasson [1890].

relations applicable to internuclear pairs. Of course a whole group of villages may be related so as to form a tribe; or the families of the same village may be associated in clans which in turn may cut across a number of villages. Yet the relations applicable to families belonging to the same village outnumber by far those between the families of the same clan but of different villages.

Thirdly—a reason that should carry great weight with the student of peasant economies —the economic activity of the village forms a unit of production as close-knit as a simple workshop. A peasant household can perform practically no economic activity independently of those of others. On the contrary, as has been repeatedly emphasized, all must move in step, whether it is for cultivating the fields, mowing the meadows, cutting wood from the forest, or depasturizing the animals.

In all economic respects, not only in respect to production, the village is not a granular mass of households, much less of individuals, loosely connected through anonymous markets, factories, banks, or other similar urban institutions. Above all, it is not a civil society. On the contrary, it is an indivisible social and economic whole, "an organized and self-acting unit"—as Sumner Maine, among many, characterized it [Maine, 1871, p. 125; Vinogradoff, 1920, p. 325].

Some of the preceding thoughts are easily recognizable in almost any definition proposed for the traditional (nondegenerate) village community. Such a community, says Baden-Powell [1896, p. 9], rests upon

the connection which a group of cultivators must have when located in one place, bound by certain customs, with certain interests in common, and possessing within the circle of the village the means of local government, and of satisfying the wants of life without much reference to neighboring villages.

Similarly, Chuprov [1902, p. 4], whose attention was fixed on the Russian *mir*, describes it as

a totality of households disposing of a territory and connected by traditional relationships characterized by the principle that the totality has the right of interfering in the economic activity of every household.

More recently, Ruopp [1953, p. 4], while recognizing the analytical difficulties raised by the concept of community, admits—a very significant fact—that these difficulties do not exist for the village, which alone constitutes a definite whole cemented by multiple integrative forces.

To gain, however, a greater insight into these definitions, two questions should be considered in some detail: What is the primary basis of the oneness of the traditional village community? By what means has it preserved this specific quality over a period reaching far back into prehistory?

THE UNITY OF PEASANT VILLAGES

No observer, however casual, of the life in a peasant village has failed to be strongly impressed by the exceptional feeling of unity binding all members together. Most systematic students of the problem, however, have been content with describing the outward manifestations of this spirit, i.e. the rules of conduct which the village community itself developed over time. Some have suggested either that there is nothing behind these manifestations or that the spirit of unity has grown out of the conduct rules. A few, however, have tried to find the cause of this spirit.

As previously mentioned, the most popular thesis is that of Maine [1861, p. 64ff] and Engels [1884]: the universal bond uniting the members of a village community derives from their blood relationships, i.e. from an evolutionary residual of the primitive family group or human herd. An idea which Ruopp [1953, p. 3] notes goes back to Aristotle. The thesis is plausible enough for villages descending directly from a group family or a clan. But how many actual villages, past or present, fall into this category? Besides, kinship on a large scale does not seem to be a *necessary* condition for the typical bond among village members. If it were, we could not explain this bond in numerous villages founded in Europe during the early Middle Ages by settlers we positively know to have been Christian, hence monogamous. Some authorities maintain that even the early village community of the Germans was not a *gens* [Maitland, 1897, p. 349]. Still more perplexing would be the case of those villages which initially were Roman settlements and later were invaded by barbarian people, or vice versa. Maine, no doubt, realized that the kinship thesis fails to explain the universal bond of the village communities which no longer form a group family or a clan. In his own time, that was the case for every village in Europe, if not in India as well. For these, Maine [1875, p. 87] turned to another principle: they are "bodies of men held together by the land they cultivated [together]."

Maine's thesis raises several questions. In the case of a village consisting of a single group family, we are especially certain that its land was cultivated in common. Yet why should the sociologist then choose kinship rather than communal cultivation as the explanatory principle of the village bond for this particular situation? Communal cultivation would certainly suffice to cover both the tribal and the familial (but communalistic) village. However, it still could not account for another salient phenomenon, which seems to have escaped Maine's attention. Even after all land in a village came under individual property and, as a consequence, came to be cultivated in severalty—as had happened in England by Maine's time—the village community more often than not continued to function as a social and economic whole. Equally significant, villages newly created by various colonization plans or land reforms, though amorphous and anarchic at first, in time acquired all the qualities of an organic structure little different from that of older communities. There is an obvious danger in elevating any temporal form of society to a universal working principle. If either kinship or land communalism were such a principle, then in the absence of exogeneous cataclasmic forces we should expect every tribal village to remain tribal and every *mir*-like organization to preserve its communalism forever.

Since manifestations of a community bond are found as far back in time as any meaningful evidence is available, the primary cause of this bond must be sought in some very primitive phase of human evolution. The answer then seems inescapable: the tap root of the village spirit of unity must be that instinct which man shares with many other living creatures, the cooperative or, as Veblen preferred to call it, the gregarious instinct. Needless to add, this instinct in turn is the product of natural selection through the Darwinian advantage of group action, first in defense and second in livelihood.[13]

VILLAGE TERRITORY AND OPTIMUM SIZE

The fact that in the traditional village (meaning the peasant village) habitations are clustered densely in one place—the village hearth, as it is known in many a peasant vernacular—may very well be explained by the gregarious instinct. But the same instinct would equally well explain the nomad horde, which had neither the net individuality of the village community nor its peculiar cohesion. The gregarious instinct, therefore, does not suffice by itself to produce these qualities. They are qualities that cannot be divorced from the problem of size (of unit size, not of quantum size).

A bond as inclusive and as staunch as that of the traditional village community cannot be a phenomenon indifferent to size. Not even the physical bonds of inert matter, we should note, are indifferent to this variable. As I have argued in a paper already cited, towns, cities, and metropolises did not turn into civil societies because the institution of the universal bond had proved intrinsically deleterious or because the *Weltanschauung* of the townee had come about by a biological mutation. For purely physical reasons alone, the old bond could not remain operative at the large size of urban agglomerations [Georgescu-Rogen, 1960]. But there is also another element of the problem to which biologists who have studied the influence of population density upon the behavior of many species would invite our attention: cooperation is a Darwinian fitness when the group living on a given territory is small but intraspecies competition appears with the growth of the population [Allee, 1940]. What then kept the village community from outgrowing that size at which the bond would be operative or constitute a Darwinian fitness? For the answer to this question we must take a second look at the role played by land.

Land has indeed played from primeval times the most decisive role in the development and preservation of the unity of the village community, but not as *mere land*. It is only recently, after repeated studies of the

---

13. Cf. J. B. S. Haldane [1935, p. 131]: "altruistic behavior is a kind of Darwinian fitness, and may be expected to spread as the result of natural selection."

village *in vivo*, that we have come to notice that no peasant village is settled on just a slice of land. The village territory is in relation to the village hearth as a complete garden in relation to its homestead. Perhaps that is why villages in Western Europe came to be known by such names as *villa* or *tun, tuin,* which originally all meant "garden." In addition to the house lots (each with its own courtyard) and to a well-planned net of roads, the village territory comprises not only the cultivable acreage, but also some woodland, some grazing land, and, most vital of all, a body of water—a creek or a river, a pond or a lake. Not infrequently, it includes some orchards or vineyards.[14]

This structure clearly reflects the gamut of the basic, perennial needs of human life. The fact that a village territory has such a splendidly balanced composition of all vital land resources may seem simple and ultraobvious. But it has caused many a thoughtful student to marvel at the wisdom of the original settlers who could not possibly be guided by any scientific knowledge proper [Orwin and Orwin, 1954, p. 27; Denman, 1958, p. 61]. Their intuition may appear all the more incredible if one takes into account the fact that what we see now on the map of almost any village is the work of numberless generations on a site which initially was nothing but thick forest or thicket. It thus seems that Oswald Spengler's likening the peasant to a plant with its roots deeply spread into the right kind of soil is not, after all, mystical nonsense [1929, p. 89f]. For nowhere does the biological intertwine with the economic so intimately as in that activity by which man confronts directly the living sector of his environment.

A site suitable for a permanent settlement of cultivators must fulfill some very restrictive conditions. It must contain all the land resources enumerated above; in addition, these resources must be roughly balanced in the same proportions as the basic and, hence, nonsubstitutable needs they severally satisfy. That is not all. Since at no time before the turn of the last century was transportation a relatively easy task, the various resources had to be conveniently located around some suitable spot for the village hearth to be settled. These conditions together greatly limit the choice of a suitable site. The third condition limits also the optimum size of the village territory and, as a consequence, the optimum size of the village community for any past state of the arts.[15]

We need no sophisticated theory of location to understand why—given that the gregarious instinct moved the first settlers to build their habitations in a cluster—the optimum size of the village territory could not exceed some relatively small area (especially in the earliest times). The textbook illustration of the gardener who must walk an increasingly greater distance after each refilling of the spray can suffices. Also, its nature is so elemental that we need not wonder why its principle was immediately felt even by the earliest cultivators.

These, then, are the reasons why village communities are neither so small as to include only a couple of households nor larger than a few thousand people.[16] The problem now is to see what helped them to remain at the optimum size in spite of the continuous growth of population. For as population increased, a point was inevitably reached where an additional individual could no longer be supported by the village resources at the prevailing state of husbandry arts. From what has been said earlier, it should be clear why expanding the village territory could not be a definite remedy, *even if there still was "free" land beyond*. The solution is

---

14. For maps of villages from different regions of Europe, see Seebohm [1896, passim]; Bloch [1952, p. 267ff]; Orwin and Orwin [1954, passim]; Robinson [1949, p. 217]; Stahl [1946, p. 289ff].

15. The relation between the state of the arts and the optimum size—well known to the economist—finds an indirect confirmation in the variability of village size according to the main activity practiced by the village. Murdock [1949, p. 81] reports that for the communities surveyed the fishing and hunting villages had on the average 50 people, those practicing agriculture and husbandry, about 450 people.

16. In connection with the above point, it is instructive to mention an idea applied by the Czarist government shortly before World War I and which later found many champions among Danish agricultural economists. It consists of consolidating the open land of the village and resettling every family on one single lot, house and all, as in the pattern of American farms. See Robinson [1949, p. 217]. Since an alternative and, moreover, much simpler operation is to consolidate only the field strips, the idea is totally inept, unless one can prove that the optimum size of a "village" is always that of a single household. For our own thesis, it suffices to note that for this last proposition to be true, transport facilities have to be so well developed that very few countries besides the U.S.A. would meet the condition.

spelled out by the numerous instances of several neighboring villages having the same basic "last name" and a different "first name" —Altdorf, Neudorf, Hochdorf, Niederdorf, or something of the sort. As population pressure approached the critical level, one group of the village community migrated and founded a new village on a nearby site [Vinogradoff, 1920, p. 146f].

One point, however, needs to be sufficiently stressed: migration has at all times constituted the last resort of the village community in avoiding overpopulation. The bond uniting the various members of such a community—which must have been the stronger the farther back we go in time—should have made the thought of migration repelling even to a barbarian—nay, especially to a barbarian. So, the peasant communities first bent their efforts to discover means by which the village resources could be made more productive so that people may not have to migrate. This technological development, about which I shall say more later on, gradually increased the optimum size of the village communities, but the increase was very slow. In any case, it did not keep pace with the increase in population. Thus, even though it was the last resort, migration had to go on, and it did so as long as suitable sites were available over the hill or even farther. Thereafter, the peasants had to assault the cities—as they are still doing in many parts of the world. For the village must preserve its optimum size one way or the other, lest it perish: the agrarian problem, as Chayanov first observed [1923, p. 131], is primarily a population problem. (See also Georgescu-Roegen [1960].)

To sum up: it is the economic optimum size of the material basis—the village territory —that accounts for the individuality of the village community. And it is the peculiar, almost invariable structure which this material basis must possess in order to sustain the basic needs of its occupants that has held the people of the same village together in one economic and social unit. Migration, as long as there were new sites available, helped the village communities in preserving their optimum size with respect to the state of the arts and the conditions of each locality.

As I have hinted earlier, the relative smallness of its optimum size made it possible for the village community to acquire an organic structure—that is, a structure in which every individual part is subservient to the activity of the whole. And, like any organism, the village community could not survive without a pronounced tendency to stability. Oral tradition has served as the most important element in this respect. But we should not fail to note the important fact that even this tradition—the only means available until quite recently for preserving the conduct rules from one village generation to another—could not have fulfilled this role as well as it has done if the size of the village community could have grown beyond a certain limit. For as I shall argue later, the stability of village institutional matrices derives primarily from the fact that, the village population being small, an oral tradition can be imparted to all.

Property Rights in Land and the Peasantry

A careful review of the literature would show that property rights have been and still are regarded as the main pivot of the institutions of peasant communities. Students of peasant communities repeatedly emphasize the relationship between the peasant and the land. The institutional patterns of control over land—collective, communal, individual—are frequently cited as the most basic factor affecting the economic process in a peasant economy. Yet this accepted doctrine regarding the influence of property rights upon economic behavior and economic development is spurious. In my view, the doctrine ignores a fundamental fact, namely that the institutions of peasant communities have never sought to control the "fund" factors of the economy (land) but the "flows" factors (the incomes from land). Since the alternative view has prevailed in the literature so long, the root of this error in the sociological analysis of peasant communities must first be exposed before we can proceed to describe the economic physiology of peasant communities.

In retrospect, one fact in the development of village sociology appears clear. It was the intellectual forcefulness of Maine's work that impressed a long line of writers to reduce village sociology to one proposition: "The historical passage from collective to individual property," as Kovalevskii formulated it most clearly in the title of one of his works [1896].

I have already cited some factual evidence which casts great doubt upon Maine's theory of unilinear and uniform evolution of village institutions and, all the more, upon the sweeping generalization of the same idea by Lavelye, Engels, Kovalevskii, and many others. To speak against all these, there are the back mutations mentioned earlier.

There is the Chicherin-Seebohm argument that in most parts of Europe, at least, land communalism is the original shell of the serfdom imposed upon an earlier community of free landholders.[17] The very evidence adduced by the advocates of the thesis that every instance of communal administration of land in any civilization is a direct survival of primitive agrarian communism, often lends support to this argument. For example, Laveleye's analysis of land tenure in Egypt traces the Moslem tradition of a single land proprietor uninterruptedly to the Pharaohs' era [Laveleye, 1878, pp. 44, 327]. Clearly, then, if some sort of land communalism was still observed in Egypt at the end of the nineteenth century, it could not have been the direct descendant of "natural communism."[18] The problem of what was before the first pharaoh, emir, or satrap therefore cries for an answer. And the answer cannot be given by simply visualizing what we would do if we, modern *Homines sapientes*, had to live in a primitive era.

A different way of answering the problem, though not wholly free from a somewhat cognate sin, is the interpretation of archeological remains of vanished settlements. Some of these tend to show that long before man turned to settled agriculture, the cemetery was the first fixed site to which man felt attached. The landmarks around these early cemeteries support the hypothesis that land was then divided in lots, each one cultivated separately by a "family" [Denman, 1958, p. 7f]. And if one also accepts Toynbee's argument that some loose attachment of man to land (as a swidden cultivator) was a necessary prerequisite for nomadism proper to develop [Toynbee, 1956, pp. 167-169], then separate "possession of land in use" must be a very old practice. This does not necessarily mean that either then or later on such a practice conformed to some innate human instinct. Laveleye, in arguing that land communalism is the natural state because it embodies "the juristic instinct of people" [Laveleye, 1878, p. 23] simply forgets that the juristic instinct of the primeval man was, if anything, the law of the jungle.

It must be admitted that the search for the historical passage from "agrarian communism" to individual property has brought to the surface a wealth of sociological facts which otherwise might be still unknown. Yet the searchers have failed to bring back what they set out to discover because their vision was marred by the purely juristic perspective that, with unparalleled talent, Maine blended into it from the outset. In one place Maine did argue that such terms as "*command, sovereign, obligation, sanction, right,*" become empty of empirical content if applied to the traditional village communities of India [Maine, 1871, pp. 67-70, 164]; but when it came to "property," he did not hesitate to use the term in describing the institutional setup of the same communities or in comparing it with the British system of property rights. Under Maine's influence, the study of the peasant communities became for many a study of land laws with the accent only on "property."

Today, anyone who says "property," whether collective or not, ordinarily means an almost irrevocable and easily transferable *title* of juridic value to a *fund* coordinate of the economic process—to a slave, to a field acreage, to a piece of technical equipment, to a house and so on. But the very *raison d'etre* of such a title, which is the existence of at least two parties potentially opposed to each other, seems to be glossed over. No doubt, to speak of the "collective property" of mankind over our planet would constitute what Alfred North Whitehead denounced as "the fallacy of misplaced concreteness." It is

---

17. "The Russian land community is the outcome of serfdom," [Chicherin, 1861]. See also Seebohm [1896, pp. 78, 368ff]. Baden-Powell [1896, pp. 184, 433; 1899, p. 51f], too, attributed the origin of the joint ownership of land in the North Indian village to the submission of the older agricultural communities by Aryans.

18. The case of the Javanese *ndesa* constitutes an excellent example of the difficulty in determining the evolution of village institutions. Little is known about the life of the Javanese communities before the Dutch administration. All subsequent accounts come from Dutch government officials, occasionally from Dutch scholars. They were the only source for Laveleye's analysis, which differs essentially from that of R. M. Koentjaraningrat in Murdock [1960].

some time now since Maitland pointed out the muddle in saying that the public land is the "collective property" of the nation [Maitland, 1897, 342f]. The same applies to using the term "property" in describing that village institution typified by the *mir*. To rationalize such an institution by saying that "the soil still remains the collective property of the clan, to whom it returns from time to time, that a new partition may be effected" [Laveleye, 1878, p. 4; Kovalevskii, 1891, p. 92], is not only a perplexing use of terms but also results in a loss of essence. On the other hand, we do not have to go farther back in history than the last century to see that "individual property" could not apply even to the fief of a landlord in many parts of Europe; there, landlords and former serfs got their titles to land at the same time, through one agrarian reform or another. In Hungary some landlords never became *landowners*; they ceased to be landlords with the advent of the Communist regime (1945).

The Economic Physiology of Peasant Communities

The considerations of the preceding sections suggest two guide lines for any attempt at delineating the physiology of the peasant economies.

First, we must reexamine the wealth of available information on past and present village institutions without any preconceived thesis and, especially, without any prepared etching traced by the property meshes of civil society. Secondly, we should not insist upon reaching a vast evolutionary synthesis embracing all village communities. This does not mean to renounce all analysis. For even though it does not seem possible to fit all village institutional matrices into a finite number of taxonomically relevant boxes, we can discern in the multitude of their forms some features which appear with striking frequency both in time and space. An analytical sifting of these features reveals, as I hope to prove next, that the economic physiology of the traditional village community, in general, is governed by a few principles of extreme simplicity.

Before proceeding, one word of caution is necessary in relation to these principles: it would be a great mistake to attribute their genesis to constitutional preoccupations in all cases. The more relevant ones, indeed, came into being either as a result of the struggle of the village community with its natural environment or because some indifferent feature happened to last a long time during the history of the village. However, it would be equally wrong to deny that once such a principle becomes part of a tradition, it generally turns into an independent sociological agent. As such, it may serve not only as a guiding light for constitutional preoccupations but also as an influential factor of economic evolution (see below).

We may begin with those principles, two in number, that are more transparent. The first principle is that only labor creates value, and hence, labor must constitute the primordial criterion in the sharing of the community's income. Its tradition goes back to the margin of the economic (exosomatic is the right term) evolution of mankind. For only on that margin did labor alone matter: land was not limitatively scarce, and tools were so simple and of such short life that they were not yet capital.[19]

The second guiding principle is that of equal *opportunity* for all, and—we should insist—not equal *income* for all. The Narodniki and the Marxists notwithstanding, nothing would distort more the general picture of the village than a background of thoroughgoing communism with equal shares for all. That would mean that personal merits and toil come last if at all, or as Rousseau indirectly put it, "the fruits go to all, the land belongs to no one" [Baden-Powell, 1896, p. 401]. A close examination of the factual evidence will reveal that the opposite is true of village philosophy. *Equal opportunity to toil should be open to all, but the fruits go to him who has applied his labor and industry*;[20]

---

19. It is instructive to note that every argument by which Marx [1932, p. 46ff] justifies his fundamental tenet that "as values, all commodities are only definite masses of congealed labor-time" implies either conditions prevailing on that historical margin—where "coats" were produced without any constant capital—or a cumulative regress to the same margin. Marx, therefore, was as much of a "marginalist" as his famous predecessor, Ricardo.

20. A most eloquent illustration is provided by a peasant custom in eighteenth century Switzerland. All men would line up, as in a race, on the village meadow; at a signal, they all would start to mow, each over an equal track. The race ended, each takes home the hay he mowed; the meadow is then opened to the cattle of all [Lafargue, 1901, p. 54]. Similar customs have been noticed elsewhere, even as far as the Urals [Laveleye, 1878, p. 24].

"a fair start in life" for all, as Vinogradoff suggested [1920. pp. 30, 326], with freedom for everyone to shoot ahead in proportion to his own efforts. Needless to say, with the passage of time the last part of this principle must ultimately breed loopholes tending to weaken the power of the first clause. The interesting fact is that the leading thought of the principle survived nevertheless in social attitudes even after the village community had lost part of its faculty for self-government. We can recognize it plainly in the custom of periodic reallocation of the land use—the battle horse of those who read "primitive communism" into every village institution—as well as in several other institutions. For even though the principle, just like the "unseen hand" of Adam Smith, seems to have guided the common life of the village, circumstances varying with time and place molded it into a wide range of efficacious institutions.

There is hardly any doubt that the primeval man, with no fixed abode, lived on the forest. Even after he became a cultivator, the forest continued to support man in his economic struggle. For a very long time man raised pigs and cattle only on pannage and leaf fodder, a practice which has survived in some areas down to our own time [Stahl 1946, p. 100ff]. But the most important thing man has wrung from the forest is land itself. Almost the entire area now under cultivation in Europe was initially woodland, which man gradually cleared. We know of vast operations of *défrichement* which were undertaken systematically in France from the beginning of the eleventh to the end of the fourteenth century and in England during the reigns of the two Charles. Rumania's present fertile plains were covered with thick, forbidding forests until the second half of the last century [Bloch, 1952, p. 3ff; Orwin and Orwin, 1954, p. 15; Stahl, 1946, pp. 91, 111]. But man's war against the forest began with the first cultivators and became more intensive with every season. It had to go on even in the absence of any increase in population; the achievable yield per acre is limited at any time by the state of the arts and, moreover, *inevitably decreases if this state does not advance.*

At the time when the only agricultural implement was a hard stick, only virgin land could yield some surplus over the seed.[21] New land had to be cleared continually, cultivated for one or perhaps two seasons, and then abandoned to the wild weeds and brush. How many people labored together in one complete enterprise from land clearing to harvesting is hard to say. The answer depends on how large was then the optimum size of such an enterprise, which in turn depends on many factors impossible to assess in retrospect. One thing, however, seems certain: as long as a settlement consisted of relatively few people, they all worked together, whether they lived as a group family or in several large families bound by multiple kinship ties. But in the case of large settlements—and almost any settlement ultimately became large as population increased—people sooner or later came unwittingly to associate in squads of optimum size. After all, it may be this economic cleavage that led to the social cleavage into clans and further into households, not vice versa.

Be that as it may, as long as land clearing was to be repeated every year, or almost so, and woodland was still plentiful, there was no need for submitting the use of land, cleared or not, to any restrictive rules. The crop alone mattered. As to this matter, there was a most natural solution in those circumstances: the crop belongs to those who have labored for it. Its distribution among coworkers raised no difficulty, any more than the distribution of the family income does nowadays in the overwhelming majority of families. No doubt, the people most likely to be associated together in one primitive economic enterprise were already related to each other by family ties.

With the gradual improvement of the "plowing" stick and the discovery of the advantage of burning the brush between crops, it became possible for a piece of newly cleared land to bear crops for an increasing number of years before it had to be abandoned.[22] A question now comes

21. Bloch [1952, p. 26] reports that even in seventeenth-century France the average harvest was only three to six times the seed.
22. Because of the extraordinary fertility of virgin land, the practice described above has survived to this century in some parts of Europe where forests are still plentiful, e.g. in the French provinces of Ardennes and Vosges [Bloch, 1952, p. 27] and in Rumania [Stahl, 1946, pp. 104, 129].

immediately to the mind of the theoretical economist: in these circumstances, who had the right to use a piece of land after the first crop? For a modern economist is apt to think that any economic clan, after exhausting the powers of the land it had itself cleared, would have sought to cultivate next the land cleared by others. It is more plausible, however, that no clan thought of this possibility. The tradition that anybody who has cleared a piece of land is entitled to use it undisturbed by others must have acquired through its long life such a grip on the minds of those early peasants that none could even think of a different arrangement in life. Besides, there was no extraneous force to compel those peasant communities to cast away the old tradition: forests were then still plentiful and remained so for a very long time to come.

As numerous written documents of relatively recent times attest, the principle just discussed continued as a pillar of tradition in village communities for an amazingly long time, long after forests became scarce. We also find it very much alive in the living fossils of village sociology, the village communities that have preserved many of their archaic institutions [Laveleye, 1878, p. 21; Baden-Powell, 1896, pp. 205–207; Baden-Powell, 1889, pp. 129; Kovalevskii, 1898, p. 143; Kovalevskii, 1896, p. 183f; Stahl, 1946, p. 163]. The significance of this survival can hardly be overemphasized. In the survival of this principle—in all probability the oldest of all economic rules—we have a striking illustration of how a rule of conduct may originate from economic practices, not by necessity, and how, once it becomes a part of tradition, it may acquire immunity to economic change and thus become a sociological factor of a new nature.

Nothing could be more erroneous than to see any trace of "property" in the principle that he who clears a piece of land has the exclusive right to use it for crops. The point may be related to the fact that the oldest written law of India, the Manu Law, says that each *works* his land, not that each *owns* his land [Kovaleskii, 1896, p. 186].

As mentioned above, in early times cultivators abandoned of their own will a piece of land as soon as its powers were exhausted. Even to a mind accustomed to the idea of individual property it would appear nonsensical to think of ownership of a valueless thing. Thus, once a piece of land was abandoned, anyone else could later "vivify it"—to borrow the Koran's expression—and use it again for crops [Kovalevskii, 1898, pp. 143, 320; Kovalevskii, 1896, p. 186f]. The principle was so sternly followed from generation to generation that it gave rise to many local proverbs: "You occupy land by the plow, not by the hatchet," one runs.

The idea that sprang up from the early art of tillage is now clear; since he who cleared a piece of land in those times knew that the result of his labor did not go beyond a couple of harvests, he could not be entitled to any other benefit than that toward which he had labored. The principle is embodied in many written laws, not too old, which stipulate that he who clears or vivifies a piece of land can have the exclusive use of it for only a small number of years, usually three [Kovalevskii, 1898, p. 164].[23] We find it also at the basis of many peasant institutions.

One such institution is the free use of the natural pastures, woodlands, and fishing waters by all members of the community. Since such natural gifts were not the creation of anybody's labor, no one was entitled to the exclusive use of them for any period of time. This institutional setup is not only one that has survived longest, but is found in lands as far apart as England, Russia, and India [Nasse, 1871, O'Curry, 1873; Laveleye, 1878, p. 116ff; Kovalevskii, 1891, pp. 76, 106ff; Baden-Powell, 1899, pp. 18, 129; Stahl, 1946, p. 161]. Even the successive agrarian reforms in Russia and Rumania, for example, though intended to establish a new order on the principle of individual landed property, refrained from completely abolishing the communal use of pastures, woodlands, and fishing waters.

Another institution sets in even stronger light the idea that land is to be used, not to be owned by exclusion. It is the custom called "open field pasture" in England, *vaine pâture* in France, and *Gemenglade der Felder* in Germany. According to it, after harvest all fields are opened for pasture to everyone's

---

23. It is most interesting to note that in Colombia, a country where land is still plentiful in proportion to the population, a 1946 law stated that any person can become the owner of up to 2,500 hectares of *baldios* (virgin land) provided he vivifies it and actually uses it.

cattle and remain so until the next tilling.[24] Once the practice was common to all European countries; it survived in many parts of Western Europe until the nineteenth century and all over Eastern Europe well into our own time [Nasse, 1871, p. 15; Seebohm, 1896, p. 12; Bloch, 1952, pp. 40–49; Kovalevskii, 1896, p. 193; Stahl, 1946, p. 131]. There is no need to insist in detail on its economic rationale. Old meadows were gradually brought under the plow and new ones torn from the forest. As the limit was reached where no further expansion of meadows could be made at the expense of the forest, the stubble became the only additional means by which the animals of a growing community could be supported. Open field pasture may have also been induced by the discovery of the advantage of manuring. In any case, it came about as a natural extension of the depasturing of animals in common on the undivided grazing land or in woods. But its survival comes from the fact that it fitted so well with the old principle concerning the use of land.

Even after land came to be held as individual property in severalty, the right to keep a field closed remained restricted to the period during which the field was bearing a crop. Once the crop was harvested, the field became "land" again. In the judgement of Maine [1871, p. 86], the institution of open field pasture had such deep roots that in his time the courts would not have dared to rule against that right. Also, the right of gleaning after harvest remained open to all; as the French saying went, "the corn ear belongs to him who sowed, the stubble to all" [Bloch, 1952, p. 48]. This admirably expresses the traditional principle that one has an inviolable right only to what he sought at the outset to obtain by his labor; the *windfalls of any sort belong to all*. But splendid though this principle appears from an ethical viewpoint, its application met with increasing difficulties as the progress of technology lengthened the period between the investment of labor and its ultimate reward.

As population increased, additional plow land was needed. It was obtained by occasional campaigns in which the entire village took part in clearing a large track at a time, as the lasting prints left by each such track on the village territory now reveal. As a result of these successive clearing operations, the plow land cultivated by each household came to be distributed in several strips—as a rule one strip in each field cleared in one campaign. This is, no doubt, the real origin of the intermixed strips that have characterized the distribution of arable land in almost all peasant villages. True, since land quality generally varied from one area cleared in one campaign to another, strip mixing represented an equalitarian distribution as well. But, as we have just seen, its origin had no connection with constitutional preoccupations; these came only later.

With population continuing to grow, the limit was ultimately reached where further clearing of woodland would have upset the balance of necessary resources within the village territory. Each household had to go on cultivating the same land that its ancestors had cleared. On the surface, the continuous use of the same land strips by the same family may look to the modern mind as if they were held as individual property. The village community, however, did not think so. The proof lies in the fact that at a later date the community was able without much ado to introduce the periodic redistribution of plow land.

The picture is clear. As some families branched out more than others, the traditional distribution of the plow land ceased to represent a balance between earned income and the size of the household. In a small community of people living close to each other, so that everyone knows exactly what everyone else does and has, inequality cannot be ignored for long, especially if property in land is an unknown idea and the tradition is that all should have a fair start in life. One may remain indifferent to a starving family but hardly so if the family is his next-door neighbor.

To repeat, as long as it was still possible to clear land from the forest, there was no

---

24. "Open field" is occasionally used to denote open field pasture. In its discriminate use, however, it denotes a setup of which open field pasture is only a part. It includes, in addition, the separation of the arable land into two or three rotation fields, the distribution of each field in intermixed strips for each household, the compulsory cultivation of the fields according to some general rules, and the communal use of the woods, grazing land, and waters. Arable land may or may not be subject to periodic reallocation [Vinogradoff, 1920, p. 165f].

reason for any constitutional preoccupations with equality—whether we have in mind a primeval village or one founded anew in later times. The point is that "agrarian communism," as some writers label the institution of periodic and equalitarian distribution of land, is a feature of an evolved type of village community, not of the most primitive one. Moreover, it emerged from constitutional preoccupations alone, the aim of which was to preserve a principle of long tradition—equal opportunity for all. Once can hardly think of a better example to show that the social philosophy embodied in tradition may be an effective factor of social evolution.

That a periodic redistribution of plow land may have also come about because of the increasing economic demands of a "lord" who had become the master of the village, as Chicherin and Seebohm argued, cannot be denied. Doubtless, heavy taxation by the lord, or by the state too, could kindle the peasants' preoccupations with equality. But it seems far more probable that even a conqueror of a village, for political expediency, only copied the form other villages had already developed. For it is difficult to believe that a master of land tracts large enough to form a village had any economic reason to think of dividing them in intermixed strips. But the best proof that the system of intermixed strips is not necessarily connected with a period of servility are those villages where a land distribution *per stirpes* (i.e. in proportion to some ancestral quotas) survived to our own time—as happened in thousands of cases in Rumania.[25]

We should also note that at the time when periodic redistribution was introduced, the village community must have already reached a strong control over not only the plow land but also over almost all farming activities. What had been learned through a very long practice, namely that the powers of the soil are rejuvenated by letting it lie fallow for a while, had already led to the division of the entire plow land into rotation tracts. This institution in turn gradually led to that of the compulsory cultivation, i.e. to the *Flurzwang*, by which to eliminate the possibility that a prepared strip should be damaged during the tilling of the adjacent ones. Therefore, the village community as a whole had already achieved a substantial degree of authority in farming matters before the advent of land communalism; we do not need to assume that only the personal power of a master enforced the complicated practice of rotation and *Flurzwang*.

The institution of periodic redistribution of land, once believed to be a specific feature of the Russian *mir*, existed all over Europe, though its form varied greatly from one place to another. In the highlands of Scotland, in some parts of France, not to mention Eastern Europe, it survived into the last century [Haxthausen, Vol. I, 1847, p. 52, p. 34; O'Curry, 1873; Maine, 1875, p. 101; Seebohm, 1896, p. 15; Bloch, 1952, p. 47].[26] Among the Anglo-Saxons it was known as *run-rig* or *rundale*. In some cases, which seem to be correlated with a hilly or mountaineous location, it was made *per stirpes*. In others, the redistribution was made on a more equalitarian basis, in equal shares per family or in proportion to family size [Maine, 1875, p. 195; Laveleye, 1878, p. 21; Stahl, 1946, p. 36]. A highly significant fact is that, whatever the distributive criterion, in most regions each family was assigned, not a particular lot, but an abstract coefficient, as it were, which only determined the area of all strips for each family. The location of each strip was determined by lottery, held now and then or even every tillage season. This procedure, together with the fact that in all cases each family was attributed one strip in every land tract of different quality, obviously reflects some eager equalitarian preoccupations. Intermixed strips equalize not only the average soil quality, but also the distance one has to travel to his fields.[27]

Since crop rotation with fallow in between

---

25. In 1852 in some mountainous districts of Rumania two-thirds of the village communities were still *free* villages, with a *per stirpes* organization. A 1912 census reveals that in all regions their number had hardly diminished in the meantime. These communities were reputed for their stubborn resistance to any attempt to encroach on their traditional rights and freedom of self-government by the old or the modern state authority. An eminent chronicler of the early eighteenth century describes the *terrae* formed by these villages as "republics" [Stahl, 1938].

26. Periodic repartition was occasionally practiced in India as well [Baden-Powell, 1899, pp. 67f, 104; Kovalevskii, 1896, p. 188].

27. Both the lottery drawing and the allocation of strips in various locations were practiced in East as well as in West Europe [O'Curry, 1873; Maine, 1875, p. 101; Stahl, 1946, p. 118].

reduced the productive capacity of a village to one-half or at best to two-thirds of the arable land, increased population pressure and decreasing availability of new sites for swarming compelled the peasants to seek new ways of increasing the fertility of land.[28] Frequent manuring removed the necessity of periodic fallow. Irrigation in dry-climate regions assured a higher average and more stable yield. Earthworks sheltered the fields from floods, soil erosion, or landslides. Operations such as these made it possible for him who undertook them to justify and obtain the right to a prolonged use of the piece of land improved by his own effort and industry. Thus, the very principle that everyone is entitled to the fruits of his labour, and only to these, prepared the stage for landed property in severalty.[29] Yet the principle itself did not disappear altogether from the social matrix, even after the institution of property crept into the life of the village. There still remained the tradition formed during millenia around that principle as a backbone.

Before stopping to have a look at this tradition, let us point out one extraordinary object lesson of the long history of the village community. Brief though the analysis of this section had to be, it shows the village always intent on winning the arduous battle with an exacting nature and concomitantly on keeping a balance of economic shares among its members. But the point deserving special emphasis is that the traditional village community has never been concerned with the distribution of titles to economic *funds*. Instead, as its economic institutions show, it has constantly been concerned with the distribution of the *flow of comprehensive income*—that is, with the distribution of both the fruits and the burden of labor. And since the village constitutes thus far the social organization with the longest life in human evolution, the conclusion, in my opinion, is inescapable: the economic conflict at bottom turns upon the distribution of the *income flow*, not upon the distribution of *funds*. Equally significant is the amazing diversity of the institutions through which the village has sought to direct the income flow during its long history. All this proves that the institution of property is neither the first nor the last artifact man has devised and will devise in order to rationalize this or that pattern of distributing the income flow. Current trends in many parts of the world suggest that it is worth pondering on this object lesson of village economic physiology.

Peasant Traditions and Attitudes

Nowadays not only town people but also many social scientists think of the peasant as a rock of irrational traditions. That an optical illusion, so to speak, is responsible to a great extent for this opinion is beyond doubt. Whenever the observer belongs to another tradition than the observed, an optical illusion is inevitable. The fact that even a Londoner, say, finds the tradition of a Parisian unintelligible—and vice versa!—is part of a very general phenomenon. And we can be sure that the peasant, too, thinks of the townee as a slave of a tradition he cannot understand. We usually consider it utterly inept, for instance, that in many village communities living on the verge of starvation a great deal of food is wasted at the numerous festivals held with punctual regularity throughout the year. On the other hand, the peasant—were he sufficiently sophisticated—would certainly decry, for instance, the even more numerous feasts in town at which much food is wasted while there are still many hungry mouths around. Actually, he could cite many other instances of the same sort; for the

---

28. Incidentally, Bloch [1952, p. 64ff] finds that there is some correlation between the type of plow used and that of crop rotation. A two-crop rotation prevailed mostly in the southern regions of Europe, the French *midi* in particular, where also the Roman plow (i.e. the swing plow) was predominant. On the other hand, a three-crop rotation was practiced almost exclusively where the wheel plow—probably an invention of the people living in Northern Europe—was the current implement. Very likely this regional distribution had something to do with the soil being lighter in the south and heavier in the north. Ordinarily the swing plow had to be pulled by eight oxen on the fields of England. Orwin and Orwin [1954, pp. 30–33] give a very useful description of all plowing implements, but their argument [1954, p. 12] that the swing and the wheel are equally efficient is faulty: they say that the swing plow cuts as deeply and as uniformly as the other provided the plowman constantly steadies it and controls the depth "by throwing his weight upon the stilts."

29. Naturally it was much easier for man to discover the advantage of irrigating than of systematic manuring, so that the earliest instances of individual property rights some from dry-climate areas. See Baden-Powell [1896, pp. 180, 400, 408; 1899, p. 105]. They are found at an early date also in the areas of North Germany, where the technique of marling eliminated the need for crop rotation [Seebohm, 1896, p. 372].

town, too, has its own tradition—a tradition which is not as free from strange peculiarities as one may think. No human society could survive without a tradition, any more than it can live without a common and, for all practical purposes, stable vocabulary.

THE RESILIENCY AND POWER OF PEASANT TRADITIONS

The tradition of a peasant community stands out in two important respects: first, it has an extraordinary resiliency, and second. it encompasses almost every action and reaction of every member of the community.

There are two factors one may immediately think of in explaining the first characteristic of village tradition. As explained above, because of the small size of the village community, a tradition preserved by word of mouth was known in detail by every adult person. Everyone learned it gradually as he grew up and took part in various village activities. In addition, the communalist and equalitarian essence of the principles discussed in the preceding section could hardly allow a system of authority in which the voices of some members would not be heard. True, in a later phase of development we find some village communities where authority is exercised only by a council of elders within which, exceptionally, a headman may have had a preponderant role. But for a long time before this aristocratic form of government came about, the same villages knew only the authority of the village assembly. In most parts, however, the village assembly never lost its constitutional power completely.

In a political setup, where every member of the village community knows in detail the received tradition and also has a full voice in the debates, tradition has almost as many defenders as there are people in the assembly. Proposals for changing an old rule have little chance of being accepted.How strong the resistance to change may be in this situation is illustrated by the fact that on certain matters a single "nay" used to be sufficient to defeat such a proposal [Gomme, 1890, p. 262]. In later periods, the village elders merely proclaimed what had been from old [Maine, 1871, p. 68].

Tradition, whether in a peasant village or in any modern society, has a distinctive property to which not enough attention has been paid in the literature. Tradition not only embodies the rules of conduct for one individual in relation to others but also dictates the attitude of the individual towards tradition itself. It is this reflexive property of tradition that accounts for the individual of modern society being less attached to tradition than the peasant. Also, the fact that in modern societies the written part of tradition, i.e. that embodied in the system of laws, has a greater force than that carried orally, is a dictate of tradition. The same is true of the fact that in a peasant society only oral tradition has value.

No species, social or biological, could survive for long if it could mutate each second: natural selection would not have the time required for separating advantageous from deleterious mutations. The peculiar chemical stability of the biological gene—a property at which natural scientists still marvel—prevents chaos at the organic level; the reflexive property of tradition prevents it at the superorganic level. That is why, as I stated earlier, no society can live without tradition, or more exactly, without some traditional attachment to its institutions.

Nor can an organism survive if it is unable to adapt itself to the inevitable changes in its material environment. Now, tradition seems to be an impediment in this respect. From all we know, the longer an institutional matrix has been in use, the stronger becomes the traditional attachment to it. However, this law of cumulative inertia must not be interpreted rigidly. It only explains why principles of a very long tradition never give in completely under the pressure of material changes and why even some institutions that have no connection with the material basis of human activity seem to have an extraordinary tendency to survival. Principles of long tradition always restrict the extent of a necessary adaptation because, as we have seen in the preceding section, every new institution must fit into the traditional *Weltanschauung*.

Why peasant institutions have always had "a tendency to a more lasting duration than other human institutions" [Nasse, 1871, 13]—as many students have noted—should now be clear. During its long history up to recent times, the village community was seldom under pressure to change its mode of life. Migration over the hills, a short average life of the individual, as well as the fact that basic human needs have a low saturation limit, took the sting out of population increase. Technical innovations, highly significant though they were, came at distant intervals of time. Attachment to tradition could thus grow to the point that one can hardly discern any changes in the history of a given village community or in those forming a *terra*. It is this fact, one may guess, that led Oswald Spengler [1929, p. 26] to argue that the peasants form the most durable—eternal, he said—class.

The resiliency of the institutional matrix of a peasant community, it should be stressed, does not prove that all institutions spring up with a rigid necessity from material conditions. As the example of handshaking shows, institutions may have diverse origins. The point is that an institutional matrix has the peculiar power of transforming into an institution some rather indifferent event that has by accident occurred with some regularity over a period of time. The evidence of this peculiar power lies in the extreme variability of numerous institutions from one area to another. As we know, some village communities are exogamous, others endogamous;[30] in some the bride is expected to bring a dowry into the new family, in others it is the groom who must pay a nuptial price to the bride's family [Kovalevskii, 1891, p. 28]; in some, immediate relatives keep and work their lands jointly, in others every new family gets possession of its share(s); land redistribution is made at times per head, at others per household, at still others *per stirpes*.

Most students have also remarked that the institutional matrix of a village community is far from being as sharply defined as the rules embodied in a written system of law. It would be, however, a mistake to think that, because of this imprecision, village institutional matrices are as soft with regard to change as urban institutions that do not have a legal status. We must not confuse the softness with which institutional principles are applied in each concrete instance with lack of resiliency. It is incorrect to argue that, in countries where nonstatutory law serves as a basis for court decisions, the common law is an easily mutable body of principles because of the flexibility with which the courts interpret them. And if the principles of village institutions are applied with even greater flexibility, it is because the bond between the people of a village makes them more fully aware of the fact that genuine justice is incompatible with an interpretation of the law as rigid as that of mathematical theorems. Authority on law though he was, Maitland [1897, p. 349] was greatly mistaken in arguing that the village community lacked social cohesion on the ground that "there was no form of speech or thought in which [the communal feeling] could find an apt expression [because] *it evaded the grasp of the law*," or that the village assembly had no jurisdiction because it "would be comparable rather to the meetings of shareholders than to sessions of a tribunal." To be sure, *some* institutions of a village community are so differently interpreted by various villagers that an observer may easily be confused about their reality. Such institutions are on the verge of becoming obsolete or, having become so, are merely part of the village folklore. But in regard to vital economic or constitutional matters, village tradition has at all times contained a core as hard in its force and sanctions as British common law.

Little remains to be said about other factors that have something to do with the power of tradition in peasant communities. One may mention, first, the human proclivity to conformity—an evolved manifestation of the primitive instinct of imitation—which accounts for the survival of spontaneous attachment to unwritten tradition everywhere. However, in the case of village communities, conformity was further enhanced by the fact that from the outset all people had to conform to an economic activity the rhythm of which is dictated by the solar system, and later on to an even stricter schedule, that of the *Flurzwang*. Because every household had to produce the entire gamut of life's necessities for itself, all were engaged in almost the same activities. Nor was there any room for a differentation of techniques within the small and closed community of one village. Whatever the prevailing state of the arts, in the

---

30. As a rule, the traditional Rumanian village was endogamous [Stahl, 1946, p. 35]; while the neighboring southern Slav communities were strongly exogamous [Maine, 1886, p. 254]. The same applies to India's tradition [Baden-Powell, 1899, p. 26].

same village everyone used the same technique for plowing, harvesting, spinning, weaving, building, and the like. It is understandable that this uniformity bred conformism in every other respect, beginning with how one gets a bride and ending with how high the fences around the house should be. Lack of contact between villages lying in different *terrae* helped conformism to grow such deep roots that even when, in time, contact between adjoining *terrae* became more intense, this could not affect the proclivity to conformity. The only outcome was that institutions, especially those of an indifferent nature, became uniform over increasingly large areas.

As I have said earlier, the variability and diversity of peasant institutional matrices are so wide and irregular that the thought of covering it by a relevant taxonomy seems utterly hopeless. Agrarians have argued, in the Narodniki tradition, that the only way to disentangle some general picture from such a complex of individual forms is to focus one's attention upon the *Weltanschauung* of the peasant. For peasant institutions, though greatly different, may have a common substratum of rationalizations and attitudes. The Agrarians have been right, I think, in their claim that the discovery of this special substratum calls for another method than that of the so-called objective sciences. What one needs is delicate touch in interpreting opaque facts. A sympathetic attitude is not enough: what is needed is a trained faculty of empathic understanding. Contrary to current thought in many quarters, this position is far from being silly. Indeed, if we deny man's faculty of empathy, there really is no game we can play at all, whether in philosophy, literature, science, or the family [Georgescu-Roegen, 1966, p. 129]. On the contrary, it is an absurd asymmetry to maintain, on the one hand, that matter can be studied only by instruments of the same essence—material, that is—and, on the other hand, to deny that mind is a legitimate, in fact the most essential, instrument for studying mind.

No one denies that only analysis can bring into full light the various connections between the elements of a problem. But some may go a step further and contend that, since we can discover what lies deep below the polychromic surface of institutional details by an analysis such as above (pp. 211ff), we should dismiss empathic interpretation from our thoughts completely. Let me then hasten to disclaim that I could have built that analysis without knowing the empathic interpretations of peasant institutions at which numerous earlier authors had arrived after observing life in villages directly. The reader should have no difficulty in seeing that there is an intimate connection between the principles set forth by my analysis and the attitudes which, according to the interpretative school of thought, constitute the distinctive characteristic of the peasant's *Weltanschauung*.

EQUALITY OF OPPORTUNITY WITH
DIFFERENTIAL REWARDS

Most students of peasant communities agree that the traditional peasant village is a little world complete in itself. For the peasants living in such a world generation after generation, it is natural that the world beyond, i.e. that of the town, should be a strange unknown [Baden-Powell, 1899, p. 14]. The converse is not less true. As Kautsky [1900, p. 3] once admitted, for the earliest Socialists —and we may note, for Marx as well—who were all town people, "the peasant was a strange and mysterious, nigh disquieting, creature." One would expect then some important differences between the peasant's and the townee's ideas about the purpose of society. But the reason why the two visions are directly opposed, as they indeed are, is not so immediately apparent.

The most important difference between the two visions hinges on the opportunity one has to earn a livelihood through his own efforts and industry. According to his notion of social justice, the peasant generally does not mind earning a bare minimum of subsistence, but a world in which the opportunity to labor for it does not exist for everyone is completely unintelligible to him. In the preceding section I have insisted on this point, in order that we may now see where the root of the conflict between the economic philosophies of the village and of the civil society lies.

The allocation and the employment of human resources in the urban society is governed by the principle of marginal productivity, which we consider to be the normal criterion because "it is the foundation of the businessman's policy in buying productive power" [Knight, 1933, p. 104; Schumpeter, 1934, p. 77]. Now, this "icy water of egotistical calculation ... has resolved personal worth into mere exchange value," as Schumpeter noted [1951, p. 293]. And the individualistic peasant—as we shall find him to be—may resent this degradation. But this resentment, like all resentments, is only relative and, hence, of secondary importance for our problem. The real evil—the greatest of all, in fact—is that calculation on the margin is incompatible with equal opportunity to work for all, because it results in a smaller employment of labor than any formula found in the history of the village. As an analytical proposition, this is elementary if one pauses to consider the two alternatives. But there is also the convincing evidence that as landlords in the peasant countries of Eastern Europe turned into capitalistic calculators, *genuine* unemployment made its first appearance in the countryside [Georgescu-Roegen, 1960].

The town strives to satisfy the *effective* demand for consumer goods by employing as few men as possible. As Carl Menger [1950, p. 170] bluntly admitted, "labor services do not have value as a matter of necessity." Moreover, individual merit comes first in determining who is to be employed. The traditional village, on the contrary, wants to enable as many of its people as possible, preferably all, to labor for a livelihood within its ecological niche without primary consideration of individual merit. Merit determines not who can labor but only how much one's earnings shall be. The peasant knows only that he has a *real* demand for the necessities of life and that he is eager to toil for them; he cannot make any sense of a principle according to which someone else's effective demand should decide whether or not he can earn a living.

Some 40 years after Engels and Marx in the Communist Manifesto denounced the "idiocy" of the peasant, Engels [1884, p. 121] had to admit in so many words that when the Irish peasants "find themselves in one of the big English or American towns among a population with completely different ideas of morality and justice, they easily become confused about both morality and justice."

Surprisingly enough, that is precisely what the Agrarians have preached at all times: the soul of the village is solidarity and social justice, that of the town is treachery, shrewdness, and "every man for himself" [Mitrany, 1951, p. 40]. In this connection it is highly significant that one of the earlier Marxist deviationists, completely unaware of the coincidence, took exactly the same stand as the Narodniki: "To the village, not to the town, we must turn for the elucidation of the notion of association in the sense of the Socialist program" [Sorel, 1901, p. 35]. The line between the Agrarian ideology and Marx's ideas about peasantry could not be more sharply drawn. And one should not treat lightly the thought that the great similarity between the socialist principle "from each according to his ability, to each according to his needs" and the economic philosophy of the village is responsible for the rather puzzling fact that Marxism sold its first ticket to a peasant, not to an industrial, society. Craft and wile alone could not have achieved this tour de force.

However, we must not commit the error of believing that the village community is the paradigm of economic equality or the very expression of the much lauded "natural communism." That the social organization of the traditional village is communalistic—a term coined precisely for avoiding the verbal source of confusion—does not mean that the majority of such villages about which we have some reliable evidence did not possess their "aristocracies" [Baden-Powell, 1896, p. 335; Baden-Powell, 1899, p. 16; Bloch, 1952, p. 49]. In the villages practicing distribution of land *per stirpes*, a "landed aristocracy" emerged as a natural process. In many others we find a "cattle aristocracy," which established itself with even greater ease because chattels seem to have been always personally owned.[31] Nor should we overlook the economic differentiation resulting from owning slaves when slavery existed. Orchards and vineyards, because they did not come under the rule of the plow, also constituted a basis for wealth differentiation. But, to repeat, what the village has striven to offer to its people, including the landless immigrant, is not undifferentiated equality,

---

31. For the Bo-Aires, the cattle nobility of the ancient Irish, see O'Curry [1873, p. ci.].

nor even the guarantee of a continuous minimum of fair subsistence, but security in the long run for all who are willing to toil.

The peasant has also been described by some as strongly individualistic. There is great truth in this opinion, provided the term "individualistic" is properly qualified for the occasion. Living in a society where no one needs a name badge when people meet, the peasant naturally has a total respect for the individual person. In a village *anyone*, even the poorest fellow, is *someone*, not a mere name or a number. Nothing could be more resented by a peasant than one's failure to recognize the individuality of each member of the village. The peasants would immediately volunteer to help the field worker learn to know each villager by his own individual traits, good or bad. There is, then, no need to insist on the reaction of the immigrant peasant to the general anonymity of large agglomerations. But we should note that this feeling of being *someone* does not speak against the peasant's identifying his interest with those of the community as a whole, i.e. against the bond of which I have spoken in earlier sections. As Maine had occasion to note in connection with India's peasant communities, a peasant would not even voice a grievance unless he could express it as a wrong for the entire village [Maine, 1871, p. 68]. Nor has the "individualistic" peasant been unwilling to associate and even march in perfect step with others. The practices related to the *Flurzwang* reflect a willingness to cooperate equal, at least, to that required by any industrial plant. Besides, in western Europe, where the traditional village first began to disintegrate, the practice of compulsory cultivation survived for a long time thereafter.

On the other hand, the spontaneous association of several independent households for cultivating their fields in common has not been so general a phenomenon in settled agriculture as the theorists of "natural communism" claim [Baden-Powell, 1899, p. 16; Kovalevskii, 1891, p. 182; Bloch, 1952, p. 156]. Coaration alone seems to be a fairly common institution at some stages. By the Brehon Laws, for example, it was mandatory for the whole village. In many places it would not have been possible to plow all fields within the short period imposed by climate without bringing out all draft animals. These were the cases where the soil was unusually heavy for the plow then used, so that a large team of oxen was needed to pull it. The real reason for coaration, therefore, was the relative shortage of animals and, perhaps, plows, not—as Kovalevskii [1898, p. 131] liked to argue—some special love for work in common.[32] The same economic necessity led the poorer peasants to pool their plowing chattels together even if the whole village did not [O'Curry, 1873; Seebohm, 1896, p. 121; Denman, 1958, pp. 126, 131; Arensberg and Kimball, 1940, p. 73]. The individualistic peasant has always shown a marked preference for being alone responsible for what he does and earns. But as concerns other forms of aiding the less fortunate, he seldom failed to respond to the call [Kovalevskii, 1898, p. 132; Arensberg and Kimball, 1940, p. 75; Murdock, 1960, p. 94].

To sum up the broad lines of the general picture: village life is dominated by a strong feeling of unity, reflected in an oral tradition which varies from one case to another and is in some respects very specific, in others highly diffused. But in spite of this unity, the village community is not an undifferentiated association of people within which the individual loses his own personal worth and entertains no ambition for personal affirmation.

There are several specific attitudes which should be of great interest to the economist, especially to the economist concerned with the economic development of underdeveloped economies with a numerous and suffering peasantry. The first group of these attitudes pertain to the economic behavior of the peasant. Two other attitudes should be discussed because of their particular relevance for a sound, feasible policy for solving the difficulties of the economies of the sort just mentioned. They are, first, the love peasants have for raising a large family and, second, their deep distrust of every idea that the town tries to sell them.

PEASANT ECONOMIC BEHAVIOR

Private enterprisers and public policy advisers alike have repeatedly expressed their exas-

---

32. Bloch [1952, p. 61] mentions the case of the peasants in Brittany who rejected the idea of constituting a single sheep flock for the whole village.

peration with the peasantry because, as they have generally explained, the peasants are "proverbially indolent" and also have no desire whatsoever beyond securing a bare subsistence. Briefly, they are economically inert [McCulloch, 1825, p. 353; Starcs, 1939].[33] But if the peasant had generally been inherently indolent and economically inert, instead of hard-working, frugal, and thrifty, there would have been no basis from which urban civilization could develop. Before anyone could even think of devoting his time and talent to observing the stars, or building temples and palaces, or producing other works of art and gadgets for comfort and amusement, or even teaching, he had to be fed by others. The point, which goes back at least to Xenophon, is that there can be no nonagricultural activity before agriculture has reached the stage where the work of one can feed two [United Nations, 1951, p. 58]. And there can be no doubt that the initial relation between towns and villages was not that of symbiosis but of parasitism. Moreover, a great deal of this parasitism has subsisted almost everywhere; the town has maintained its hold on the countryside through the fiscal power and military authority of the state, a hold which even the progress of democratic institutions has not eliminated entirely [Kautsky, 1900, p. 314ff].

The apparent indolence of the peasant—as I have argued elsewhere—may after all be only unwanted leisure imposed by the limitativeness first of land and later, as a consequence, of capital equipment [Georgescu-Roegen, 1960]. There are, however, obvious differences between the eagerness to work of peasantries which are equally overcrowded but have a different political history. The explanation of these differences must be sought in this history. Where the villages happened to be exploited for long periods by the state to the limit of mercilessness, the peasant first tried to appear poor to the tax collectors. Continuous exploitation made him really poor. Ultimately he discovered that working just enough to stay poor was the best strategy for making the most of his life in the struggle with his exploiters. The cumulative inertia of tradition did the rest.

But where his fate had not been so harsh,

we find the peasant holding toil in the highest esteem. That this represents an old, normal tradition with cultivators of the soil is seen in the fact that some primitive communities used to produce more than they usually needed and destroy the surplus during a festival just before the next harvest [Bancroft, 1886, p. 192; Thurnwald, 1932, p. 209]. This custom may seem utterly antieconomical. Let us note that the freakish nature of the weather in most parts of the globe must have impressed upon many cultivators the fact that one is never safe in toiling just what would suffice for an average year.

Concerning the peasant's economic decisions we find two opposing opinions. Many of those who have watched closely the peasant maintain that most of these decisions are entirely determined by traditional, hence inflexible, rules, and consequently the village economy more often than not violates the economic principle of product or utility maximization.[34] On the other hand, traditional economists as a rule maintain that, whatever friction may exist in a village economy, this economy must conform to the general principles embodied in the analytical apparatus of the standard theory. For example, Schumpeter [1934, p. 80], who was not speaking out of complete ignorance of the peasant, insisted that "the peasant sells his calf just as cunningly and egotistically as the stock exchange member his portfolio of shares."

We must admit that this was true of all peasants of the Hapsburg Empire in Schumpeter's time and also of any peasant who nowadays trades produce and wares in an urban market. Yet what matters for village sociology is how the peasant behaves, not outside, but within his own community. We could not commit a greater enormity than to assume that in his own community the peasant behaves just as the stock exchange dealer. To the peasant it does matter whether it is a poor widow who must sell a calf under the pressure of necessity. The stock exchange dealer, on the other hand, does not care whom he corners when he buys cheap; to

---

33. But see, by contrast, the insight of Marshall [1949, p. 226].

34. One of the earliest allusions to this economic "backwardness" is found in Gomme [1890, p. 18], who mentions the repeated complaints of the British economic reformers about "the unreasoning folly of the peasant farmers, who love to do only what their fathers had done."

repeat an earlier thought, he has no means of knowing from whom he buys and, hence, whether the latter needs the money badly or is actually a smarter dealer.

As a touchstone example of the difference between the two patterns of economic behavior, let us mention a feature of American Indian life reported by a famous scholar as clear evidence of the spirit of solidarity in a peasant community. Any villager, he observed, "may help himself to his neighbor's store *when needy*" [Bancroft, 1886, p. 191]. Now, the last qualification is significant. For obviously no society could last long if *anyone* had free access to his neighbor's store. The condition that one must be *needy* to deserve help raises no difficulty in principle. The town, too, seems to recognize that the needy should be helped by the more fortunate. But the difference between town and village is that the latter is in a position to know who really is needy in every particular instance. Circumstances vary greatly, and only the intimate knowledge everyone in a village has about everyone else makes it possible to apply a moral principle with the necessary flexibility. In proportion to the means at its disposal, the village seems indeed far more efficient in taking care of the poor, in preventing crimes, in assessing penalties, and at the same time making everyone respect the oldest of all commandments—to earn one's bread by the sweat of one's brow.[35] To recall an earlier remark, it is the small size of the village community that both imposes and maintains an ethical temper which Hegel would have regarded as the only genuine one: the individual is actual only in the identity of all its interests with the total.

The preceding observations bring to mind an important issue of methodology (rather, of epistemology) over which the Agrarians and their critics fought in vain. Agrarian ideologists insisted that theory is the surest way for an economist to commit scientific suicide. But, as I have pointed out earlier, this was a wrong conclusion from a correct datum. The datum, for which they could vouch better than the economists of the orthodox school, is that the theoretical apparatus constructed by that school is useless, even disastrous, for the study of village economy. This economy having the characters of an organic whole, the use of the standard analytical tools, which have been designed for handling the isolated parts of the civil society, can only result in the destruction of the very phenomenon one wants to study. As Whitehead [1919] once pointedly remarked, murder is the prerequisite for using the test tube of physicochemistry in biology. In fact, at present nothing justifies the hope of constructing a strictly quantitative model of the traditional type for the village economy. The ultrarationalist dogma according to which every phenomenal domain can be exhaustively described by such a model has long since suffered blow after blow from none other than the science of physics. Much less can it then be upheld or even nursed in the domains where social or biological life takes more forms than there are numbers or, in contrast to numbers, interpenetrate each other. These caveats do not imply that one should neglect the possibility of representing some of the general aspects of the village economy by a quantitative *simile*. For a simile is helpful in many ways; above all, it may detect errors in our reasoning, just as the rule of casting out nines may signal arithmetical mistakes.[36]

One can, as the author once did [Georgescu-Roegen, 1960], represent the choice-function of a peasant by $\psi(Y; Y_s)$, where Y_s merely marks the fact that village institutions work their way through the behavior of every village member. But it would be foolhardy to think that such a formula quantifies this behavior. Since we have no means to denote a concept other than using a symbol, one must guard against thinking that a symbol necessarily represents an arithmomorphic concept.

The problem of quantifying behavior is far more complex than is generally thought. The received doctrine notwithstanding, the Paretoan ophelimity, $\psi(Y)$, does not tell the complete story, even for an individual of a civil society, precisely because its formula pays attention only to the quantitative elements of the problem.

---

35. Baden-Powell [1899, p. 141] also thinks that the Indian village is more efficient even as sanitation is concerned. Most likely, this was the case in his own time. But in some parts of the world even nowadays there are urban slums far more pestiferous than the poorest villages.

36. For a more elaborate discussion of the points of methodology touched above, see Georgescu-Roegen [1966, pp. 114–129].

It is indisputable that the *outcome* of any economic choice is expressible as a vector Y, the coordinates of which represent quantities of commodities. But in actuality the choice itself is not between two such vectors, Y_1 and Y_2, but between a set of complex pairs (Y_1, α_i) and (Y_2, β_j), where α_i and β_j denote the various actions by which Y_1 and Y_2 are obtainable. One may beg for a dollar, or pinch the cash register, or ask his brother to give him a dollar for keeps. What one would most likely do depends upon the institutional matrix of the community to which he belongs. The point is that whether the outcome of choice is Y_1 or Y_2 is not independent of the cultural *value* which the actions α_i and β_j have according to the institutional matrix of the particular economic agent. To leave an employer with whom one has been for some long years only because another would pay better or, conversely, to let out an old employee because business is slack is not compatible with every cultural tradition.

Even in urban societies there are cases where an individual's choice is determined not only by the purely economic coordinates Y_1 and Y_2, but also by the cultural values of α_i and β_j. Such instances are, however, rare. The contrary is true of a peasant community; more often than not it is the cultural value that weighs more in the peasant's decision if this decision concerns other village members. Consequently, when one asserts that the peasant's economic behavior is irrational, the assertion implies that a choice is rational if and only if it is made on the basis of commodity quanta alone.

SIZE OF FAMILY

Of all the attitudes prevailing among peasant societies, one alone poses a really difficult problem of which we are becoming increasingly aware: it is the desire of the peasant to raise a family as large as it might come.[37] But as any geneticist would tell us, the mere desire for a large family does not suffice to produce it. Large families can exist only where there is also a high fertility. The point I wish to stress is that the desire for a large family by itself could not explain why the peasant population has grown and is still growing faster than the urban one, in spite of its enormously greater tribute to wars and pestilence.

One may be tempted to argue that high fertility must always prevail over low fertility and in the long run eliminate it completely, and hence, the present high fertility of peasant populations is self-explanatory. But a biologist again would instruct us that high fertility alone does not represent a Darwinian fitness. It may be associated with multiple disadvantages which would make the individuals possessing that gene less fit to survive than those who do not possess it. The most instructive illustration pertinent to our topic is the fact that in urban societies, especially in an urban society with the institution of private property, high fertility represents an appreciable economic disadvantage [Fisher, 1930, pp. 228-255].

The question of why the gene of high fertility has eliminated that of low fertility within peasant populations—instead of merely continuing to survive alongside the other, as is the case for urban populations—has hardly been entertained. No doubt it is a difficult question because of the lack of sufficient data, but one can at least offer some plausible speculations. Of all the writers on the village community, Laveleye [1878, p. 31] alone saw some connection between a peasant economic institution and population growth. He argued that the Russian *mir*, by distributing the plow land equally per capita, "removes every obstacle to the increase of population, and even offers a premium for the multiplying of offspring." But after he thus almost touched the core of the problem, he went off on a wrong track.[38]

Simple arithmetic suffices to show that if *at all times each individual shares an equal opportunity with all others and in all respects*, the relative frequency of the high fertility gene must tend toward 100 per cent. This is the intrinsic advantage of that gene. Even though the village community has never been a

---

37. This particular bent is by now a commonplace. Yet the reader may find it interesting to peruse the detailed evidence in Arensberg-Kimball [1940, p. 136f]. This evidence is all the more revealing since the Irish peasant is the only one to have adopted a custom aimed at controlling the increase in population: the custom of refraining from marriage when the land owned is small.

38. Laveleye immediately turned to proving that, on the contrary, peasant nations (Russia and France) cannot grow as fast as the others (England and Prussia). It should be noted that his data, first, are incomplete and, second, do not bear on fertility alone.

model of the absolute equality required by the preceding theorem, it has approximated it closely enough for us to explain the present high fertility of peasant populations by the equalitarian structure of the traditional village. However, other factors may have increased the relative advantage which high fertility has under conditions of equal opportunity for all.

One such factor is the cunning role of the laws of returns in a community which continuously redistributes its land so as to maintain some equality between all households with respect to the man-land ratio. From the viewpoint of the village, land had become scarce long before the institution of equalitarian redistribution of land came about. Consequently, in the villages practicing this sort of redistribution the size of the economic unit of every household must have been well below the optimum scale, i.e. every economic unit was operated at a size where the returns to an additional dose of land and labor were increasing. In addition, all units were operated with practically the same man-land ratio. In these circumstances, the result of an increase in some family was twofold. First, the income per head in that family *increased* because of the increasing returns to the additional dose of land and labor. Second, the same income *decreased* in every family that did not grow, because land was taken away from it to be given to the other. What was on the surface an equalitarian system actually represented a systematic discrimination against the gene of low fertility.

A similar discrimination, perhaps even more potent, has its roots in the special nature of the agricultural process which, except in some very rare spots of the earth, follows a very unequal rhythm. For long periods during the year labor power and implements find no use; during some phases of the vegetation cycle one needs all the draft animals one can get hold of and during others all the hands. In particular, harvesting by hand is an operation that requires the mobilization of the whole family, including children, for it must be performed by every household within a very short time at the same critical moment for the entire village [Arensberg and Kimball, 1940, pp. 46, 74]. To bring in all the grain safely and as quickly as possible necessitates a multiple division of labor. This requirement is all the more pressing for a bumper crop. Let us also note that, at least on the Eurasian continent, climate conditions seem to lead to a skew distribution of crop yields such that the most frequent yield is far smaller than the average. In these circumstances a small family cannot take full advantage of a bumper crop because it cannot provide the required division of labor. The economic loss thus incurred places such a family in a disadvantageous position in comparison with the large family. No wonder, then, that the peasant's thoughts have been continuously focused on the best years and on how to take full advantage of them. For a very long period, which ended only recently as villages no longer had any safety valve against the pressure of population, the peasant had a ground for thinking that "if you don't have [many] children, you are no good," as the Irish say [Arensberg and Kimball, 1940, p. 136f]. The rationale behind the peasant's desire for a large family—as I have suggested in a paper read before the Agricultural Economics Society of Thailand (February 1963)—does not differ in essence from that of the modern industrialist who also wants an excess capacity so as to be able to take advantage of any increase in demand when it comes.

Undoubtedly the problem is more complicated than the preceding analysis might suggest. But this analysis at least points out the sort of factors that are responsible for the present situation in which the attribute of high fertility has come to be associated with the peasant's traditional desire to raise as large a family as may come.

DISTRUST OF THE TOWN

The peasant views any idea or anyone coming from town with suspicion, and rightly so. Political history tells of numberless stratagems, one more ingenious than another, by which the town has repeatedly inveigled the village into accepting a losing deal. Over the years the peasant has thus learned by his own misfortunes to distrust the voice that comes from town. He has even preferred to align himself with his direct masters, the landlords, against the forces of the bourgeoisie, not because he is reactionary (as some have charged), but because the landlords opposed the encroaching by the town. A townee, too, ultimately comes to distrust anyone who has tried constantly to deceive him. It is natural that at present in most parts of the world the peasant is "unwilling to accept advice" from the urban authority [Starcs, 1939]. And it is highly significant that the poorer the peasantry in a country, the stronger is this unwillingness.

It does not matter to the peasant whether the urban authority comes to him with some taxation scheme or with some technical advice. As far as that authority is concerned, the peasant wants "to be left alone"—a conclusion at which even as casual an observer as Carlo Levi, physician and painter, arrived in his sociological novel *Christ Stopped at Eboli*. For even "honest" advice has often proved disastrous for the peasant. Doreen Warriner, for instance, reports [1939, p. 160] that in Hungary during the 1930's she saw much mechanized equipment abandoned though hardly used. During a field study I once came across the same situation in Rumania. Many middle-income peasants had followed the counsel of enthusiastic experts concerning the superiority of tractors over draft animals. But the buyers soon discovered, first, that to pay for fuel, parts, and repairs they had to sell more produce than the animals would eat and, second, that they still had to keep the animals for the indispensable task of transportation.

The Present Impasse

The contrast between the economic situation of the peasants in the overpopulated areas of the globe and that of the farmers in the thinly populated countries—the United States, Canada, Australia, and even Argentina—has led some to believe that overpopulation is the disease of peasant farming [Wilson, 1939b, p. 55]. The analysis presented in this essay sets this belief on a deeper basis: the world is now confronted with a vast peasant population which possesses both the biological potential and the strong wish to raise families of unlimited size. The obvious impasse is that this situation may ultimately endanger the food supply of the cities before resolving itself into a Malthusian holocaust at the source.

From the plans now aired or carried out, we seem to think that the cure of the disease is extremely simple: just tell the peasants to practice birth control. Unfortunately, for the reasons explained above, the peasant will hardly listen—at least not in sufficient number for the solution to work in time. In any case, the solution exceeds the power of economic science. So, the economist can do nothing more than attend to the most urgent problems of a short-run nature. But even for this narrower task the standard methods with which he is ordinarily acquainted are not of much help.

First, the economist must take into consideration also the constraints deriving from the traditional economic institutions of the particular peasant society with which he happens to deal. The point that "man is not a passive instrument, with a movement determined by a simple law: one must therefore know how man [in each society] adapts himself to his task"—as Sorel [1901, p. 8] put it—is not new.[39] Maine repeatedly denounced the tremendous loss resulting from the fact that the British wanted to superimpose on India's tradition an administrative and fiscal system imported from Britain [Maine, 1861, p. 252; Maine, 1871, p. 115; Marshall, 1949, p. 762]. An interesting counterproof is supplied by the Austrian legislation of the last century which, after some groping, was adapted to fit the southern Slavs' institution of *zadruga* (a patriarchal household economic unit) [Stahl, 1946, p. 151].

Second, in the impasse reached by the peasant economies the biological has burst through its economic shell and now demands recognition by whoever approaches the problem. And this demand cannot possibly be satisfied with the aid of standard economic analysis alone. For some reason or other this analysis views the economic process as a mechanical analog, i.e. as a circular motion between production and consumption. A biological process, on the contrary, is irreversible, not circular. Moreover, if instead of artificially reducing the economic process to a closed mechanical system, as we have done ever since Jevons and Walras, we carefully consider all its material aspects, we must arrive at the con-

---

39. A. G. Richey, too, insisted that "the good and evil effects of any law depend upon its being applicable or inapplicable to the social condition of the society into which it is introduced."

clusion that this process is only an extension of the biological evolution of the human species. Like any biological process, the economic process, too, cannot create or destroy energy matter. Both are irreversible processes because both are only consumption processes as far as their material nature is concerned.

Any material process associated with life consumes low entropy—the term by which thermodynamics covers free (usable) energy and material structures arranged in some regular patterns. It transforms the input of low entropy into an output of high entropy, i.e. into dissipated (unusable) energy and valueless waste. The important point is that the real product of such a process is not material, but a pure *flux*—the enjoyment of life by the corresponding life-bearing entity. Moreover, the material transformation of low entropy into high entropy is irrevocable. Actually, the Second Law of Thermodynamics states that low entropy, even if left to itself, continuously turns into high entropy.

For the economist the moral is twofold. First, no scrap campaign can be completely successful; in other words, it is infinitely more profitable (in terms of entropy) to obtain gold from a mine than to gather it from the sands of the seas. Second, the economic process is sustained only if it includes a continuous tapping of the environment for low entropy.[40]

There are two distinct types of activities by which this tapping is mainly done. In mining, man simply helps himself, as it were, to low entropy from the *stocks* existing in the earth's crust. In husbandry, he mainly catches the low entropy which reaches us as a *flow* of solar energy. In this activity, land plays a role completely analogous to that of the fisherman's net. If the globe were bigger, we could catch a greater amount of solar energy. But given the size of our planet, the maximum amount of this energy that can be caught annually is rigidly determined. This is the real reason why we feel that land is scarce in a different way than other factors. To be sure, the total amount of coal-in-the-ground, for example, is also limited. The important difference is that it lies in our power to decide how much of it we may consume in any given year.

There is an equally vital difference between husbandry and mining: husbandry must slavishly follow the unequal seasonal rhythm in which the energy radiated by the sun determines the climatic conditions in each spot of the earth. The difference is even more conspicuous if we compare husbandry with manufacturing. In manufacturing, a process can, in principle, go on uninterruptedly at our will as long as the other two sectors supply the necessary inputs of low entropy. It is precisely because of this latitude that man has been able to shorten radically the time necessary to weave an ell of cloth but hardly at all the time needed to grow corn or raise a domestic animal. The same freedom of choice is responsible for the factory system, an *economic* invention of an importance which has not been sufficiently appreciated. Indeed, only in the factory system is it possible to eliminate completely the periods of idleness imposed on practically every agent by the elementary process which transforms the input(s) into products [Georgescu-Roegen, 1965; 1967]. The point is that, contrary to Marx's claim that we can transform the entire agricultural sector into "open-air factories," husbandry will, in all probability, remain a discontinuous sequence of annual activities.[41]

We may now understand why the work of the Entropy Law, although the very root of man's struggle for life, has been ignored by an economic science interested almost exclusively in the economy of the town. The peasant, on the other hand, has never lost sight of the problem. "Whoever could make two blades of grass to grow ... where only one grew before, would deserve better of

---

40. I am aware of the fact that these ideas as well as some of the subsequent paragraphs may seem esoteric to many an economist. Yet it is impossible for me to elaborate them within the space or the scope of this essay. I have attempted to do this in Georgescu-Roegen [1965; 1966, Part I, Chapters 2-5; 1967], to which I refer the reader desirous of greater detail.

41. I can think of two counterexamples; however, in the ultimate analysis they strengthen the above argument. First, there are spots on the globe—Bali Island is one—where, because of the small seasonal variations of the climate, crops could be raised in an assembly-line fashion; but such spots are highly exceptional. Second, chickens are now produced in the United States by chicken "factories." The spectacular decrease in the real cost of production brought about by this innovation needs no complicated argumentation: the famous "chicken war" suffices as a proof.

mankind," observed Jonathan Swift.[42] This is precisely what the peasant communities have succeeded in doing in their long history to our own times. The basic principles of modern agriculture have been laid out by none other than the peasant [Orwin and Orwin, 1954, p. 32]. Not even the principles by which the plow is nowadays constructed have another origin. It would therefore be foolish to believe that the peasants have not accumulated any economic wisdom during their long struggle with the hardest of all economic problems. The reality is that the peasant communities in an overwhelming number have ultimately reached a point when the problem has no longer a solution within their own reach, even if they would stop growing at the same rate as in the past.

From across the fence there comes the widely supported idea of absorbing the agricultural labor surplus through industrialization and of concomitantly preventing those remaining on the land from increasing their average consumption of food [Nurkse, 1953, pp. 37, 43]. It should be obvious to every economist that this scheme solves neither the present food scarcity nor the marasmus of the village: it merely proposes to implement an old idea that the peasant is only a special kind of draft animal entitled to a subsistence ration of food and nothing else. This is not the proper place to dwell on the contradictions inherent in all policies of industrialization and only industrialization, but I cannot refrain from inviting the reader to ponder over the case of India and also over the picture of a world in which all underdeveloped countries will have completely achieved their present dreams of economic development through industrialization alone.

Of course, the "industrialization" of agriculture is an idea entirely different from and largely independent of that of the industrialization of the entire economy. And, as I shall explain presently, it is the only rational economic solution for the impasse. But in connection with its application we hear the objection (or the complaint) that the peasant stubbornly clings to his old techniques. This is undoubtedly true. We would err, however, in believing that the peasant needs to be instructed on the advantage of mechanization. However vaguely, he is aware of the fact that the mechanization of agriculture originated within his own society with the substitution of the foot plow for the swidden stick. Nor can there be any doubt in his mind about the fact that land yields more if one uses the plow instead of the caschrom (the foot plow). His opposition to the new changes derives from an internal logic that looks at the net rather than at the gross advantage. According to this logic, he knows perfectly well that if pastures are not freely available from a still unconquered forest, then one has to share the gross product with his draft animals, and his net product may not be much greater. Wherever land reached the limit of absolute scarcity, feeding the animals became one of the most agonizing problems of the village community. "The horse eats people" is an adage in which the Rumanian peasant has crystalized a substantial dose of economic "analysis." We must rest assured that if the antiquated caschrom has survived until this century in parts of the British Isles—as reported by Orwin and Orwin [1954, pp. 30, 153]—it is a sign of economic wisdom for the circumstances, not a symptom of "rural idiocy." And if the wooden plow, similarly, has survived in many East European villages, the reason is only that, in exchange for the iron plow, the town demanded a share of the crop larger than the difference made by such a plow.

Other implications of the mechanization of agriculture beyond the present level are apt to make the peasant apprehensive. He finds it very strange that, grain being so scarce, one should wish to use machines which often result in a greater loss of grain than if the same task is performed by the old methods (harvesting and threshing are good examples). Nor can he make good sense of using machines while the village has so many free but idle hands. Thoughts such as these and a long history of poor deals with the town explain why the peasant fears that the proposal for increased mechanization is only another crafty device by which the town seeks to increase its share of agricultural produce at the expense of the toiler of land.

---

42. The Entropy Law of Thermodynamics in fact says that even one single blade of grass cannot grow on the same spot year after year on end. Yet the heresy, inherited from William Petty and James Anderson through Marx and Engels, that agricultural production "can keep pace with human population whatever that might be," lingers in many minds at this late hour.

The situation calls for some sober, down-to-earth rethinking that would, for once, recognize the distorting screen raised by any criterion of profitability between us and the most important economic issues. One point is beyond the shadow of a doubt: mechanization must go on, for the simple reason that man shall no longer share the crop with the draft animals. There also are elementarily obvious reasons why we must definitely stop equating mechanization with the introduction of giant combines and huge tractors and, implicitly, advocating kolkhozation of one kind or another. The contrast between the achievements of Japan with the garden tractor and the family farm and those of the Socialist regimes speaks loud enough. The increased use of artificial fertilizers must also go on, not only because of the deficit of manure created by the elimination of the draft animals, but also because the soil has long since been impoverished by the work of the Entropy Law within the closed entropic process of peasant farming over millennia. In a nutshell, the "buffaloes," their "fodder," and the "manure" must now be supplied by the town, but not at the prices based upon the present living comfort of the townees. Otherwise we dodge the issue and, implicitly, vindicate the traditional mistrust of the peasant for whatever the town proposes. The town has the choice between two alternatives, both involving a sacrifice in personal welfare for some time to come. One alternative is to allocate part of the industrial capacity, now used to produce goods of urban comfort, to the production of tractors, fuel, and fertilizers. The other is to increase the working day and redistribute resources so as to produce these goods while preserving the present level of *real* income.[43]

No doubt the peasants, too, will have to be induced to cooperate in the proper manner with this new relationship between the town and the countryside. To win the confidence of the peasant, a task which is the hardest of all, may require the mobilization not only of our knowledge, but also of an army of educated people capable of educating the peasant—a Peace Army instead of a Peace Corps. The author, however, is unwilling to air his view concerning another problem, namely whether the town is capable of educating itself so as to raise such an army and accept the sacrifices that go with the whole scheme.

References

ALLEE, 1940. W. C. Allee, "Concerning the Origin of Sociality in Animals," *Scientia*, Vol. LXVII (1940), 154–160.

ARENSBERG and KIMBALL, 1940. C. M. Arensberg and S. T. Kimball, *Family and Community in Ireland* (Cambridge, Mass.: Harvard University Press, 1940).

BADEN-POWELL, 1892. B. H. Baden-Powell, *The Land Systems of British India* (3 vols.; Oxford: Clarendon Press, 1892).

BADEN-POWELL, 1896. B. H. Baden-Powell, *The Indian Village Community* (London: Longmans, Green, 1896).

BADEN-POWELL, 1899. B. H. Baden-Powell, *The Origin and Growth of Village Communities in India* (New York: Scribner's, 1899).

BANCROFT, 1886. A. L. Bancroft, *The Works of Hubert Howe Bancroft*, Vol. I (San Francisco: A. L. Bancroft, 1886).

BLACK, 1939a. J. D. Black, "The Problem of Surplus Agricultural Population," *International Journal of Agrarian Affairs*, Vol. I (1939), 7–24.

BLACK, 1939b. J. D. Black, "Discussion," *Proceedings of the Fifth International Conference of Agricultural Economists*, (London: Oxford University Press, 1939), 86–87.

BLOCH, 1952. Marc Bloch, *Les Caractères originaux de l'histoire rurale francaise* ("French Rural History; An Essay on Its Basic Characteristics"), (Paris: Librarie Armand Colin, 1952), English translation by J. Sondheimer (Berkeley: University of California Press, 1966). First published in Oslo: H. Aschehoug, 1931.

BRIDGMAN, 1950. P. W. Bridgman, *Reflections of a Physicist* (New York: Philosophical Library, 1950).

CHAYANOV, 1923. A. V. Tschajanow (Chayanov), *Die Lehre von der bauerlichen Wirtschaft* (Berlin: P. Parey, 1923).

CHICHERIN, 1861. B. N. Chicherin, "Leibeigenschaft in Russland" in Bluntschli and Brater, eds., *Deutsches Staats-Wörterbuch*, VI, Stuttgart: Expedition des Staats-Wörterbuchs, 1861), 393–411.

Chuprov, 1902. A. A. Tschuprow, *Die Feldgemeinschaft: Eine morphologische Untersuchung* (Strasbourg: K. J. Trübner, 1902; quotations translated by Georgescu-Roegen). For a brief information on Chuprov see J. M. Keynes,

---

43. The point recalls the concept of the working day upon which Marx alone insisted. But even Marx did not realize the full importance of it as a coordinate of the economic process.

"Professor A. A. Tschuprow," *Economic Journal*, Vol. XXXVI (1926).

DENMAN, 1958. D. R. Denman, *Origins of Ownership: A Brief History of Land Ownership and Tenure in England from Earliest Times to the Modern Era* (London: Allen and Unwin, 1958).

ENGELS, 1884. Frederick Engels, *The Origin of the Family, Private Property and State in the Light of the Researches of Lewis H. Morgan* (New York: International Publishers, 1942).

EUCKEN, 1950. Walter Eucken, *The Foundations of Economics* (London: William Hodge, 1950). The original, in German, published in 1940.

FISHER, 1930. R. A. Fisher, *The Genetical Theory of Natural Selection* (Oxford: Clarendon Press, 1930).

FUSTEL DE COULANGES, 1885. N. D. Fustel de Coulanges, *Recherches sur quelques problèmes d'histoire* (Paris: Hachette, 1885).

GEE, 1929. W. Gee, "Rural Sociology as a Field of Research in the Agricultural Experimental Stations," *American Journal of Sociology*, Vol. XXXIV (1929), 832–46.

GEORGESCU-ROEGEN, 1960.
  Chapter 6 in this volume.

GEORGESCU-ROEGEN, 1965
  Chapter 5 in this volume.

GEORGESCU-ROEGEN, 1966. Nicholas Georgescu-Roegen, *Analytical Economics: Problems and Issues* (Cambridge, Mass.: Harvard University Press, 1966).

GEORGESCU-ROEGEN, 1967. Nicholas Georgescu-Roegen, "Chamberlin's New Economics and the Unit of Production," in R. E. Kuenne, ed., *Monopolistic Competition Theory* (New York: Wiley, 1967), 31–62.

GLASSON, 1890. E. D. Glasson, *Le Communaux et le domaine rural à l'époque franque: Réponse à Mr. Fustel de Coulanges* (Paris: F. Pichon, 1890).

GOMME, 1890. G. L. Gomme, *The Village Community* (London: W. Scott, 1890).

HALDANE, 1935. J. B. S. Haldane, *The Causes of Evolution* (Ithaca, N.Y.: Cornell University Press, 1935).

HAXTHAUSEN-ABBENBURG, 1847–1852. August F. L. M. von Haxthausen-Abbenburg, *Studien über die innern Zustände, das Volksleben und insbesondere die ländlichen Einrichtungen Russlands* (3 vols., Hanover: Hahn, 1847–1852).

HOFFER, 1926. C. R. Hoffer, "The Development of Rural Sociology," *American Journal of Sociology*, Vol. XXXII (1926), 95–104.

KAUTSKY, 1900. Karl Kautsky, *La Question agraire* (Paris: V. Giard and E. Brière, 1900). The original, in German, published in 1899.

KNIGHT, 1933. F. H. Knight, *The Economic Organization* (Chicago: University of Chicago Press, 1933).

KOVALEVSKII, 1891. M. Kovalevsky, *Modern Customs and Ancient Laws of Russia* (London: D. Nutt, 1891).

KOVALEVSKII, 1896. M. Kovalevski, "Le Passage historique de la propriété collective à la propriété individuelle," *Annales de l'Institut International de Sociologie*, Vol. II (1896), 175–230.

KOVALEVSKII, 1898. M. Kovalewsky, *Le régime économique de la Russie* (Paris: V. Giard et E. Brière, 1898).

KROEBER, 1917. A. L. Kroeber, "The Superorganic," *American Anthropologist*, Vol. XIX (1917), 163–213.

LAFARGUE, 1901. P. Lafargue, *The Evolution of Property from Savagery to Civilization* (London: S. Sonnenschein, 1901). Originally published as a series of articles in *Nouvelle Revue*. The first English translation published in 1891.

LAVELEYE, 1878. Emile L. V. De Laveleye, *Primitive Property* (London: Macmillan, 1878). The original, in French, published in 1874.

McCULLOCH, 1825. J. R. McCulloch, *Principles of Political Economy* (Edinburgh: W. and C. Tait, 1825).

MAINE, 1861. H. J. Sumner Maine, *Ancient Law, Its Connection with the Early History of Society, and Its Relation to Modern Ideas* (London: Dent, 1861).

MAINE, 1871. H. J. Sumner Maine, *Village Communities in the East and West* (London: J. Murray, 1871).

MAINE, 1875. H. J. Sumner Maine, *Lectures on the Early History of Institutions* (London: Holt, 1875).

MAINE, 1886. H. J. Sumner Maine, *Dissertations on Early Law and Custom* (New York: Holt, 1886).

MAITLAND, 1897. F. W. Maitland, *The Domesday Book and Beyond* (Cambridge: Cambridge University Press, 1897).

MARSHALL, 1949. Alfred Marshall, *Principles of Economics* (8th ed., New York: Macmillan, 1949). First edition published in 1890.

MARX, 1900. Karl Marx, *The Poverty of Philosophy* (London: Twentieth Century Press, 1900). The original, in French, published in 1847.

MARX, 1904. Karl Marx, *A Contribution to the Critique of Political Economy* (Chicago: C. H. Kerr, 1904). The original, in German, published in 1859.

MARX, 1932. Karl Marx, *Capital*, Vol. I (Chicago: C. H. Kerr, 1932). The original, in German, published in 1867.

MENGER, 1950. Carl Menger, *Principles of Economics* (Glencoe, Ill.: Free Press, 1950). The original, in German, published in 1871.

MITRANY, 1951. David Mitrany, *Marx Against the Peasant* (Chapel Hill, N.C.: University of North Carolina Press, 1951).

MURDOCK, 1949. G. P. Murdock, *Social Structure* (New York: Macmillan, 1949).

MURDOCK, 1960. G. P. Murdock, ed., *Social Structure in Southeast Asia* (Chicago: Quadrangle Books, 1960).

NASSE, 1871. E. Nasse, *On the Agricultural Community of the Middle Ages and the Enclosures of the Sixteenth Century in England* (London: Macmillan, 1871). The original, in German, published in 1869.

NURKSE, 1953. Ragnar Nurkse, *Problems of Capital Formulation in Underdeveloped Countries* (New York: Oxford University Press, 1953).

O'CURRY, 1873. E. O'Curry, *On the Manners and Customs of the Ancient Irish* (London: Williams and Norgate, 1873).

ORWIN and ORWIN, 1954. C. S. Orwin and C. S. Orwin, *Open Fields* (2nd ed., Oxford: Clarendon Press, 1954). First edition published in 1938.

ROBINSON, 1949. G. T. Robinson, *Rural Russia Under the Old Regime* (New York: Macmillan, 1949).

RUOPP, 1953. Phillips Ruopp, ed., *Approaches to Community Development* (The Hague: W. Van Hoeve, 1953).

SANDERSON, 1917. D. Sanderson "The Teaching of Rural Sociology," *American Journal of Sociology*, Vol. XXII (1917), 433–60.

SCHUMPETER, 1934. J. A. Schumpeter, *The Theory of Economic Development* (Cambridge, Mass.: Harvard University Press, 1934). The original, in German, published in 1911.

SCHUMPETER, 1951. J. A. Schumpeter, *Essays*, R. V. Clemence, ed., (Reading, Mass.: Addison-Wesley, 1951).

SEEBOHM, 1896. F. Seebohm, *The English Village Community* (4th edition, London: Longmans, 1896). First edition published in 1883.

SOREL, 1901. G. Sorel, Introduction to G. Gatti, *Le Socialisme et l'agriculture* (Paris: V. Giard et E. Brière, 1901; quotations translated by Georgescu-Roegen).

SPENGLER, 1929. Oswald Spengler, *The Decline of the West*, Vol. II (New York: Knopf, 1929).

STAHL, 1938. H. H. Stahl, "Organizarea socială a țărănimii" (Social organization of the peasantry), *Enciclopedia României*, Vol. I (Bucharest: Enciclopedia României, 1938), 559–576.

STAHL, 1939. H. H. Stahl, *Nerej: Un Village d'une region archaïque*, Monographie sociologique dirigée par H. H. Stahl (3 vols., Bucharest: Institut des Sciences Sociales de Roumânie, 1939).

STAHL, 1946. H. H. Stahl, *Sociologia satului devălmaș romănesc*. Vol. I: *Organizarea economică și juridică a trupurilor de mosie* (Bucharest: Institutul de Stiințe Sociale al României, 1946).

STARCS, 1939. P. Starcs, "The Problem of Surplus Agricultural Population," *International Journal of Agrarian Affairs*, Vol. I (1939), 79–90.

THURNWALD, 1932. R. Thurnwald, *Economics in Primitive Communities* (London: Oxford University Press, 1932).

TOYNBEE, 1956. Arnold Toynbee, *A Study of History*, abr. D. C. Somervell (New York: Oxford University Press, 1956).

UNITED NATIONS, 1951. United Nations, *Measures for the Economic Development of Underdeveloped Countries* (New York: United Nations Publications, 1951).

VINOGRADOFF, 1920. Paul Vinogradoff, *The Growth of the Manor*, (3rd ed., London: Sunnenschein, 1920). The first edition published in 1904).

WARRINER, 1939. Doreen Warriner, *Economics of Peasant Farming*, (London: Oxford University Press, 1939).

WHITEHEAD, 1919. Alfred North Whitehead, "Time, Space, and Material," *Problems of Science and Philosophy*, Aristoteleian Society, Supplement, Vol. II (1919), 44–58.

WILSON, 1939a. M. L. Wilson, "The Problem of Surplus Agricultural Population," *International Journal of Agrarian Affairs*, Vol. I (1939), 37–48.

WILSON, 1939b. M. L. Wilson, "The Social Implications of Economic Progress in Present Day Agriculture," *Proceedings of the Fifth International Conference of Agricultural Economists* (London: Oxford University Press, 1939), 41–56.

Part III
Epistemology and Methodology

CHAPTER 9

(1974)
Dynamic Models and Economic Growth*

Summary. — Standard theory of economic growth has followed the usual approaches—the mathematico-imaginative and the mechanico-descriptive. Money fetishism made it so that it is dominated by the formula 'save-invest-grow'. The fact that economic growth involves not only quantitative changes but also qualitative transformations is generally ignored. The understanding of growth calls for a new approach, the analytico-physiological. The inadequacy of standard theory is admirably illustrated by the proof that the Leontief dynamic model, even if considered only as a purely quantitative expansion model, rests on utterly unrealistic assumptions.

If I am to know an object, though I need not know its external properties, I must know all its internal properties. Ludwig Wittgenstein, *Tractatus Logico-Philosophicus* (2.01231)

I. INTRODUCTION

In this paper I propose to examine critically the validity of standard dynamic models as adequate representations of actual processes and hence as safe instruments of economic planning. Specifically, I will deal with the question of whether these models cover all the important factors involved in the process of reiterative equilibria, as each situation emerges from past decisions and activities. My answer to this new question—as I believe the question to be—is in the negative. But the paper presents a second, yet not subsidiary, interest in that the method by which the question was conceived and treated also is off the beaten path.

Let me begin by observing that the manners in which economic problems have been treated may be divided into three distinct, but not discretely distinct, categories. There are, first, the studies erected on one or several assumptions having no operational value whatsoever outside the paper-and-pencil concatenation. This category, to which I propose to refer as *mathematico-imaginative,* is illustrated by the mathematical exercises which assume that future demand is known to the end of Time, or presuppose that the discount factor of all future degrees of utility (disutility) is also known, or assert that technological progress measured by some aggregate economic coordinate always proceeds at a known exponential rate. How far such flights of fancy may depart from actuality is revealed by some studies which assume that there are as many traders as the real numbers (Aumann [1966]); others assume even that there are more traders than that (Brown and Robinson [1972]). Being skilled mathematicians, the authors of these studies must have known however that even an infinite universe cannot accommodate a continuum of three-dimensional objects. It is because of the fact that most of the mathematico-imaginative studies deal with such unreal structures that these studies have been repeatedly denounced as revealing a mathematical interest—often, not even of a high level of difficulty—rather than an interest in the economic aspect of the problem (cf. Georgescu-Roegen [1966, pp. 114–24; 1970a, pp. 117–27; 1971, pp. 300–40]). On the other hand, we would be mistaken to overlook their purely didactic service or to deny that mathematical economics of a sober drive is an indispensable tool of the economic discipline.

---

* This article is copyright by D. Reidel Publishing Co. in *Equilibrium and Disequilibrium in Economic Theory* (Dordrecht, Holland, 1975), a volume of a Conference of the Institute of Advanced Studies, Vienna, and appears here with the kind permission of D. Reidel Publishing Company. The present version, however, involves a few additions.

One phase of the work required for this paper was completed under the auspices of a C. N. R. research project directed by Professor Giacomo Becattini, University of Florence.

The second category consists of the *mechanico-descriptive* studies. They are mechanical because, in common with mechanics, these studies reduce the essence of all phenomena to some motions—economic motions, to be sure, but in essence still motions because of their reversibility. And they are descriptive because they show no interest in going beyond the description of these motions. As a rule, they even seem to percolate a denial of the usefulness, perhaps of the possibility as well, of any inquiry beyond that level. An exemplar of the contentment with a mechanical description is supplied by the Phillips curve. But precisely because these studies are built around an empirical (or seemingly so) scaffold, some of them may do even more harm than the most eccentric mathematico-imaginative exercises. Indeed, nothing could mislead more than a mechanical model erected on such a scaffold, for a study of this sort is the most likely to create an impression of finality free from any omissions.

The economic literature of the last hundred years abounds in examples of this category. The situation is the inevitable consequence of the mechanistic epistemology of our Neoclassical forefathers, who succeeded in convincing almost every subsequent ecconomist that, if economics is to be a science at all, it must be set up as 'the mechanics of utility and self-interest'.[1] We may mention, first of all, the picture of the economic process as a self-sustained circular movement between production and consumption (indifferently, between consumption and production) which adorns the most respected manuals (e.g., Bach [1957, p. 60], Samuelson [1970, p. 42], Heilbroner [1972, p. 177]). Perfect reversibility is present everywhere. It constitutes the main pillar of the theory of market equilibrium. According to the ultra-familiar picture, if demand shifts from D to D', the market moves from E to E' (figure 1); and should, later, the factor responsible for the shift disappear, the market would return to E, in a manner perfectly similar to that of a mechanical pendulum which can swing back and forth with equal ease. True, no economist has even suggested that a process of production may be reversed so as to convert pieces of furniture back into trees. However, the classical theory of business *cycles*—as this traditional name indicates—rests on the idea that the entire economic process may come back to any previous position by following the same path in reverse.[2] We should also note that the entire theory of production is still based on the simple formula known as the production function, which (as we shall see presently) is not a satis-

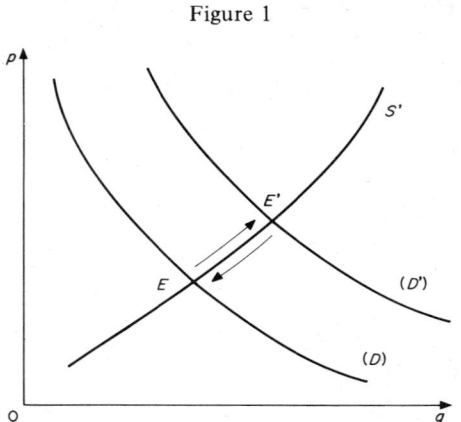

Figure 1

factory description even of the reproducible process of production, i.e., of the simplest possible arrangement. But the most pertinent examples of the shortcomings of the mechanico-descriptive approach are the standard dynamic models beginning with that of Harrod and Domar and ending with those of Solow and Leontief.

Finally, there is the *analytico-physiological* approach. The difference between this category and the previous one may seem only one of degree, for, as long as we insist on proceeding *analytically* (as opposed to *dialectically*), we must reduce the representation of any partial domain of inquiry to a number of analytical relations. But if we set aside the borderline cases, which always provide a ground for endless quibbling, we must come to the conclusion that the difference is one of essence. It stems from the detailed manner in which the interrelations between the various components of the process under consideration are described and from the particular attention paid to the qualitative change inherent in any actual process. In a nutshell, this third category is characterized by a manifest endeavour to submit the economic phenomena to a physiological analysis akin to that of biology. It was Alfred Marshall [1898, pp. 42–4]—we may remember—who first argued with unique insight

---

1. This splendidly terse formulation of the programme is Jevons's [1879, p. 21]. But the idea moved even Vilfredo Pareto *qua* economist [1906, chap. III, §4. 36 *bis*).

2. For a dissent from this position, see Georgescu-Roegen [1951].

that 'the economic problems are not mechanical, but concerned with organic life and growth', and hence 'the Mecca of the economist is economic biology rather than economic dynamics'. Yet even Marshall did not subordinate the analysis of his *Principles* to this epistemological position continuously.

Not all analytico-physiological studies, however, are descended directly from Marshall; some preceded by long his explicit rationalization. Instructive examples, however, are not many; in modern economics they are exceptions. A clearly visible concern with qualitative change does not seem to have existed before Malthus's theory of population, which envisaged such changes both in the mode of living and the methods of production. Francois Quesnay's famous *Tableau économique*, Adam Smith's classical description of the operations in a pin factory, and Karl Marx's discussion of some important aspects of commodity production may be cited as the earliest attempts at a physiological analysis. But Karl Marx's analysis of the evolution of the capitalist system and, perhaps, in a still more stringent way. Joseph A. Schumpeter's theory of economic development, constitute the most illustrious examples of analytico-physiological economics (Georgescu-Roegen [1974]).

The crucial point to note about these last works is that they are not presented in the form of a mathematical dynamic model. According to the arithmomorphic temper now dominating the economic science, this constitutes an unpardonable drawback. Actually, the works have often been denounced as 'vague and impressionistic' (Baumol [1970, p. 35]). But those judges only proved thereby that they were unable to see the deep-seated reason for the omission, namely, that no analysis which, instead of assuming away the qualitative change associated with an actual process, focuses on that very change can attain its aim through an arithmomorphic model *alone*. The reason is that there is an irreducible incompatibility between qualitative change, i.e., between essential novelty, and arithmomorphic structures (Georgescu-Roegen [1966, ch. ii; 1970a, pp. 18–47; 1971, ch. iii]).

It is this last point that shatters the generally accepted validity of the standard dynamic models as adequate representations of actual processes. And the lever of the proof will be provided by physiological analysis. A few extremely simple structures, beginning with that represented by a stationary (reproducible) process, have been found to constitute a sufficient working base for a series of results.

As it will be shown, *mere* growth—i.e., change confined to *quantity*—cannot exist in actuality continuously. The same is true even for the so-called stationary state.[3] Briefly, continuous existence in a finite environment necessarily requires qualitative change. And it is this qualitative change that accounts for the irreversibility of the economic process, of any actual process for that matter.

Irreversibility and reversibility are the very properties that distinguish actual processes (which all are evolutionary in some sense or another) from those governed only by the laws of mechanics (Georgescu-Roegen [1966, pp. 83–7; 1970a, pp. 85–9; 1971, pp. 196–200]). We may therefore define a purely dynamic system as a system capable of returning to any of its previous positions. Certainly, all dynamic economic systems fulfil this condition: the fundamental notion behind dynamic economics is that investing and disinvesting, growth and contraction, are absolutely symmetrical operations.[4]

Another point to be brought home is that even with the extreme assumption of unlimited resources a dynamic state presupposes an 'origin'. At any one time, such a state is inherited from the past, from a previous planner or from a Prime Planner, who, like Aristotle's Prime Mover, set the system in motion at the beginning of Time. The characteristic task of a present planner (or of the concerted actions of the enterprising agencies) is to change the dynamics of the inherited system. And although here again one may quibble over the difference, this change is not of the same nature as that represented by a dynamic system of equations. In a dynamic system—*conceived as a true system which determines all phases over the entire time from $-\infty$ to $+\infty$*—the change is internal, from one such phase to another. The

---

3. Incidentally, this conclusion exposes the fallacy of those topical programmes which see the ecological salvation of mankind in a stationary state—as the Club of Rome, for instance, does. See Donella Meadows *et al.* [1972, pp. 156–184], and *contra* Georgescu-Roegen [1975, sec. 8].

4. In actuality, however, contraction does not take place according to the same law as expansion, only in reverse. The two laws are entirely different, their difference stemming from some discontinuities analogous to those involved in relaxation phenomena (Georgescu-Roegen [1951; 1970a, pp. 205–20]; also sec. V, below). Some issues raised by these particular phenomena are recognizable in Schumpeter's theory of economic development.

other change is external, that is, from one system to another. In a certain sense, it constitutes a discontinuity, but one which, as we shall see, necessarily implies a qualitative change.[5] Admittedly, calculations of the common sort are involved in the planning operation as well; but each such operation must be taken for what it actually is—in the words of Schumpeter, 'a historic individual'. It is an individual because, by discontinuous nature, it cannot occur continuously at every moment in time.

II. THE PHYSIOLOGY OF PRODUCTION PROCESSES

Physiological analysis must begin at the level of a stationary process. The reason is that such a process constitutes the basic element of reference for any dynamic change, just as uniform motion, the simplest form of movement, is for the analysis of any other movement.

In sharp contrast with the physiological orientation adopted by Adam Smith and Karl Marx (mentioned earlier), standard economists have generally felt no need to go beyond the mechanico-descriptive representation of a production process by what is known as the production function. The vapid, uninformative way in which Wicksteed [1894, p. 4] introduced this formula eighty years ago still passes as a wholly satisfactory definition: 'The product being a function of the factors of production we have $P = f(a, b, c, \ldots)$'. Apparently, we do not even pause to ask what corresponds to these symbols in actuality.

Almost at every turn of an argument we speak of 'process', without realizing however that as long as the term is not endowed with an analytical definition, we are abusing it. It goes without saying that a process must be, first of all, identified, i.e., separated from the rest of actuality. This is done with the aid of a boundary and a duration which can always be represented by $[0, T]$. For the analytical description of the physiology of a material process—i.e., of what the process does and of what it needs to do it—we must consider the elements (supposed measurable) which cross the boundary in one direction or another.[6] Against this background the terms 'output' and 'input' are no longer imprecise notions at most referred to their etymological construction. *Output* is now defined as any element that crosses the boundary of the process from the inside to the outside; *input* is any element that crosses the boundary from the outside to the inside. The complete analytical picture includes for each element a function which for every instant t, $0 \leqslant t \leqslant T$, shows how much of the corresponding input or output has crossed the boundary over the closed interval $[0, t]$.

The physiology of any material process is thus represented by a point in a *functional space*

$$\left[I_i{}_0^T(t); O_k{}_0^T(t) \right], \qquad (1)$$

where I is an input and O an output co-ordinate. This is a far cry from the standard representation of a process by a point in an ordinary space, i.e., by a vector whose co-ordinates are numbers. And it would serve no good purpose to deny that the difference matters. First of all, the time factor—without which we cannot think of any happening—is completely absent from the standard representation. Second, we can no longer now fail (or refuse) to see that the physiology of any production process involves an inflow of natural resources and an outflow of waste (Georgescu-Roegen [1969; 1970; 1971]).[7] The immediate upshot is that the production function, understood as an analytical catalogue of all possible processes by which a particular process may be obtained, is a *functional*,[8] i.e., a relation between *functions*

$$Q_0^T(t) = \mathcal{F}\left[I_i{}_0^T(t); W_0^T(t) \right]. \qquad (2)$$

---

5. See note 22, *infra*.

6. Analysis cannot do more than that. In analysis, as opposed to dialectics, a process is treated as a *black box*. Should one be interested in what happens *inside* the box, one must introduce new boundaries, which will decompose that process into others, and others and others, . . . but not beyond any limit. For in the ultimate limit, everything would be reduced to simple motion.

7. Neoclassical economists, just like most of their classical predecessors, have generally ignored the economic role of mineral resources. The only element of nature that appears occasionally in standard production functions is Ricardian land. We should note, however, that Ricardian land belongs to the same category as a fisherman's net, which catches fish; Ricardian land catches solar energy, rain, etc.

8. The suggestion that functional equations may be useful in dynamic economics (in an unspecified way, however) seems to have been made for the first time by Paul A. Samuelson [1947, pp. 260f, 286].

In this formula $Q_0^T(t)$ represents the product co-ordinate(s), $I_i^T{}_0(t)$ stands for the input co-ordinates (including that of natural resources), and $W_0^T(t)$ is the co-ordinate for the *output* of waste.

The foregoing physiological analysis also prompts us to realize that the participating elements are divided into two distinct categories according to their role in the process. On the one hand, there are the *agents*, and on the other, the *objects* of the agents' action and the end products. The corresponding analytical co-ordinates in (1) can be accordingly divided into *fund* and *flow* co-ordinates (Georgescu-Roegen [1969, pp. 512–20; 1970, p. 4; 1971, pp. 224–31; 1972, p. 283]). This distinction is particularly important for the analytical fiction of a process ordinarily called stationary, although Marx's term 'reproducible' describes it more adequately. For the essential property of a reproducible process is that the agents are 'maintained intact' by that very process, so that at the end of the process (i.e., at T), the same process may be started again *providing the necessary input flows are still available*. We are dealing here with a fiction, to be sure, but a fiction which is not as remote from actual processes as it may appear to some.[9] In any case, according to one of the basic principles of analysis—on which the entire mathematical analysis is built— any process whatsoever, even one involving mutations, may be viewed as a sequence of stationary processes, each of an infinitesimal duration.

Another new concept will allow us to penetrate further into the physiology of productive processes of any dimensions. It is the *elementary process*, defined as the process by which, in each particular situation, a unit or an appropriate batch of the product (one chair or one pound of nylon, for example) is produced from some specific materials by some specific agents. This process constitutes the atomic element of commodity production, in the sense that any production process can be decomposed into elementary processes. The only difference between one process and another lies in the manner in which the corresponding elementary processes are arranged (Georgescu-Roegen [1969, pp. 516–18; 1970, pp. 3f; 1971, pp. 236–8; 1972, pp. 284f]).

If the process in point is stationary, naturally, the corresponding elementary process must also be stationary (rather, reproducible). In figure 2, the three most important structures are represented schematically. There is, first,

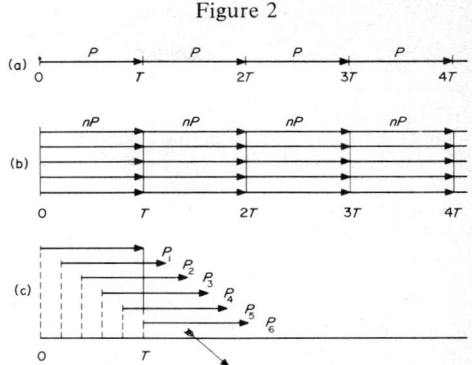

Figure 2

the case in which the elementary processes are arranged *in series*, one following another, as in a small artisan shop or in the production of bridges, transatlantic Queens, and other such immense structures (figure 2a). The elementary processes may also be arranged *in parallel*, as in most bakeries or in ordinary farming (figure 2b). Finally, they may be arranged *in line*, as they are arranged in a factory and only in a factory (figure 2c).

It is only in the case of a factory that the functional (2) degenerates into a point function. But even this degenerated form differs in several important respects from the standard formula. First, the new form reveals that the physiology of a factory is described not by one, but by several such functions. The most important ones show what the agents can do and what they require for doing this. The first relation is

$$Q = tF(L, K, H), \qquad (3)$$

where Q is the daily output, L and K are the *amounts* of land and capital equipment, H is the *number* of workers, and t is the length of the *working* day.[10] The second is

$$q = f(x, y, \ldots ; w), \qquad (4)$$

---

9. This is not to deny that the notion raises some subtle logical questions when applied to individual processes (Georgescu-Roegen [1970, p. 4; 1971, pp. 229f; 1972, p. 283]). But the restriction concerning the flow elements is crucial even for the all-inclusive process. Section III, *infra*.

10. As can be seen now, the absence of any time element from the standard formula is tantamount to writing $s=v$, instead of $s=tv$, for the relation between space and velocity in the uniform motion!

where q is the rate of product flow, w the rate of waste outflow, and $x, y, \ldots$ are the rates of flow of material inputs (Georgescu-Roegen [1969, p. 521; 1970, p. 7; 1971, p. 238–44; 1972, pp. 285–8]).

Contrary to what one may be inclined to think at first blush, the relevance of the foregoing analysis is not limited to pure theory.[11] If it were so confined, it would be poor theory. Actually several important object-lessons emerge immediately. The first is that in the factory system (and only in such a system) no agent remains idle while the production process goes on. Every agent passes without any pause from one elementary process to the next one in line.[12] We can now see that, because idleness of economic agents is the worst form of economic waste, the factory system represents an *economic* innovation of momentous importance (similar to, and just as anonymous as, the innovation of money).[13] This innovation, which was born in some artisan shops as they were confronted with an ever increasing demand, provided the primary impulse to the great wave of technological progress known as the Industrial Revolution. The elimination of idleness of agents through the substitution of arrangements in series by arrangements in line is still fostering economic development not only in underdeveloped but also in developed economies. A splendid example is the 'chicken war' of yesteryear. Chickens imported from the USA could be sold on the European market at highly competitive prices only because the chicken farms in the USA had been transformed into 'chicken factories'. Since in an ordinary farm elementary processes are set in parallel, one must wait several months for the product to become available and, in addition, almost every agent is idle at one time or another during this period. In the factory, things are entirely different. *Once the factory is built and primed*, there is no waiting; the product starts to flow out the instant the factory opens. This point is true in a more general sense which will prove useful later on in our analysis (section VII).

III. THE ANALYTICAL REPRESENTATION OF MULTI-PROCESSES

In passing now to the analytical representation of the economic process considered in its entirety, we must first emphasize the point that only a flow–fund structure can represent actuality as adequately as is possible in analysis. This admonition is necessary because some of the most famous models–that of J. von Neumann, in particular–involve only stocks. Growth, in this case, must come from spontaneous generation, and contraction, from complete annihilation. Others involve only flows– Marx's, because Hegelian Dialectics equates Being with Becoming, and Leontief's, because of the general propensity to ignore some issues in order to simplify problems. This does not do either. For in the everyday world one cannot possibly cross a river only on the flow of main maintenance materials of a non-existent bridge.

Now, for the scope of this essay we need only dissect the economic process into some broad sectors, as shown by the multi-process matrix of table 1. The process going on in the environment is indirectly and only in part represented by the flow elements that it provides to and receives from the economic process. The first process, P_1, is a process which Marx expressly left out of his reproduction diagram and has also been ignored by every standard economist. It corresponds to the so-called primary activity, which produces, say, coal-on-the-surface (M) from coal *in situ* (R). P_2 produces capital equipment (K); P_3 produces consumer goods (C); and P_4 is the consumption process by people (H_4).[14] Ricardian land is represented by L, and waste by W. The co-

---

11. The traditional attachment to the standard formula, which raises no nettling question, is so strong that when some of the above results were first presented, nine years ago, they were met with prompt protests. True, one critic, D. Patinkin, recanted at the first opportunity (Georgescu-Roegen [1969, p. 528]). But a more recent critic (W. D. Fisher in *The Annals of the American Academy of Political and Social Science* (July 1973) p. 189) thought of settling the issue by simply proclaiming that in my arguments I am 'highly critical and contentious'. Apparently, dogmatic slumber has its own delights.

For the blow inflicted by the new representation to the edifice of marginal productivity pricing, see Georgescu-Roegen [1972, pp. 288ff].

12. See the physiological diagram of this characteristic property of the factory system in Georgescu-Roegen [1969, p. 517].

13. That this innovation is economic, not technical, is proved by the point that we may set up a factory system for the cloth-weaving technique practised during the time of the Pharaohs. On the other hand, transatlantic Queens are not produced in a factory system, even though the newest techniques are used in producing them.

14. Because the issue does not matter for our immediate purpose, all social classes are aggregated in this process.

ordinates of each process are listed in the corresponding column. The flow co-ordinates are *rates of flow* at the particular moment to which the matrix refers and are entered with a positive or negative sign according to whether they are outputs or inputs for the corresponding process. The funds M_i and C_i are technically necessary inventories. For the production sectors, K_i includes 'goods in process', or what I have called 'the process-fund' (Georgescu-Roegen [1970, p. 7; 1971, p. 239]). H_4 represents the entire population, parts of which, H_1, H_2, H_3, are active; the rest represents the non-active population. The flows r, m, and c, represent material consumption in the strict sense of the term; k_i represents the necessary flow for maintaining K_i constant.[15] Among the flow co-ordinates there exist the following physically necessary equalities:

$$m_1 - m_2 - m_3 - m_4 = 0$$

$$-k_1 + k_2 - k_3 - k_4 = 0 \quad (5)$$

$$c_3 - c_4 = 0$$

It must also be observed that the true product of P_4 being the enjoyment of life, it cannot be represented in a scheme involving only material co-ordinates.

If the economy represented by our multi-process matrix is to be stationary (reproducible), it is necessary and sufficient that the matrix *as a whole* should be a periodic function of time. The particular case in which all co-ordinates are constant over time corresponds to the ordinary definition. This case, however, presupposes that all production processes consist of factories. But since this aspect is immaterial for the argument of this paper, for the sake of simplicity we may assume that there are only factories operating continuously.

We have thus reached a physiological picture which lays bare all the elements involved in any economic process in which exosomatic organs, i.e., produced means of production, are used (Georgescu-Roegen [1966, pp. 98f; 1970a, pp. 100f; 1971, pp. 307f]).[16] Needless to add, this representation does not prejudice the issue of which elements should have any value in exchange. This issue is related to parochial–geographical, technological, or historical–factors. For under certain conditions, things that have a value for the continuous existence of mankind may have no exchange value at all–as is dutifully explained in some old-fashioned manuals. But table 1 makes us aware of some crucial truths of development which

Table 1. *A multi-process representation of a stationary economy*

	P_1	P_2	P_3	P_4
	Flow coordinates			
(R)	$-r_1$	*	*	*
(M)	m_1	$-m_2$	$-m_3$	$-m_4$
(K)	$-k_1$	k_2	$-k_3$	$-k_4$
(C)	*	*	c_3	$-c_3$
(W)	w_1	w_2	w_3	w_4
	Fund coordinates			
(K)	K_1	K_2	K_3	K_4
(L)	L_1	L_2	L_3	L_4
(H)	H_1	H_2	H_3	H_4
(M)	M_1	M_2	M_3	M_4
(C)	*	*	C_3	C_4

the parochial constellation of values in exchange may conceal.

We may recall that Schumpeter [1934, ch. i] has rendered popular the use of the term 'the circular flow of economic life' to describe the economic process as a whole. Our physiological picture shows that, although there are some circular movements of a material nature inside that process, the process is not a circular affair. Only the circulation of money may be considered circular, and only to a certain extent, for even money of any sort ultimately wears out. The economic process is, instead, grafted on nature in a unidirectional way. Nature's con-

---

15. To be sure, maintaining capital constant also requires other kinds of input flows, which would appear explicitly in a narrower classification. In such a classification, most m's and k's would necessarily be zero in the household process. (Incidentally, one may certainly assume that there is an input of R in every process; the argument of this paper would not be affected thereby.)

16. This picture differs from the ultra-popular input–output table in several important respects. Among other things, it leaves no room for the fallacy of internal flows (Georgescu-Roegen [1971, pp. 258–61]). If one eliminates (R) and (W) from the flow matrix of table 1 and scrambles the remaining flow co-ordinates, one arrives at the basic input–output table of Leontief (Georgescu-Roegen [1971, pp. 254f]). But a difference still remains in that Leontief does not distinguish between capital and consumer goods. *Infra*, section V.

tribution is r_1, and the ultimate material product of the entire economic process (which is the total flow of waste, $w = w_1 + w_2 + w_3 + w_4$) returns to nature. And, according to the Entropy Law, which is one of the most solidly supported by facts, this last flow is irrevocably irreversible (Georgescu-Roegen [1966, p. 97; 1970a, p. 99; 1971, ch. x]). It is, therefore, beyond any doubt that the economic process may remain stationary only as long as the constant flow rate r_1 can be obtained with the constant effort represented by the production coefficients of the table. We may encounter, by chance, an immense lode at a uniform depth and with a uniform mineral concentration. But such a lode must sooner or later—rather sooner than later—come to an end. At that moment, we can no longer produce the same m_1 (hence, the same consumer income) with the same cost in energy terms, unless production efficiency happens to be increased by some innovation, i.e., by some qualitative mutation. Unfortunately, even this solution cannot go on forever. Not only are all accessible resources finite, but also the Entropy Law sets a definite limit to the efficiency which technological progress can achieve. The most advanced technology cannot derive more useful work from a piece of coal than the free energy contained in it—in truth, not even as much as that.[17] These points deserve unparsimonious emphasis in order that we should not persevere in the error, inherited from the Classical School, of arguing that the size of the population is kept in check only by the finiteness of land, with the result that the world economy will ultimately reach a stationary but everlasting state. The greatest (yet unnoticed) error of Malthus was that he believed that our environmental dowry can support even a growing population, provided that the growth is not faster than an arithmetic progression.

to pass them in silence (Baumol [1970, p. 53n]). Yet even works which stand aside the trodden path have clung to this definition of economic growth (Kaldor [1960, p. 240]; Kaldor and Mirrlees [1962, p. 174]), although, curiously, some have in the same breath denounced as inept any thought of constructing a measure of the equally multifarious stock of capital (Kaldor and Mirrlees [1962, p. 174]). A few attempts to go beyond this crude mechanico-descriptive conception have usually succeeded in introducing baffling complications. W. W. Rostow [1962, p. 81n3], for example, defines growth as 'a relation between the rates of increase in capital and the working force, on the one hand, and in population, on the other, such that *per capita* output (not necessarily consumption) is rising'.[18] Another example of analytical incongruity is the definition of growth by J. E. Meade [1961, p. 1], which covers a mosaic of alternative situations, not all independent and some redundant: an increase in the stock of capital equipment as a result of savings from current income, or an increase in the working population, or some technical progress which allows 'more output to be produced by a given amount of resources'.

To be sure, the economic developments about which we have sufficiently accurate historical records normally display three distinct features usually operating together—technological progress (equated with an increase of productivity per man), increased accumulation of capital, and population growth (Kaldor [1960, p. 233]). But to take this as a definition of economic growth is to abide by the mechanico-descriptive approach. For a clarifying analogy, consider the commonplace that the development of a child is normally accompanied by some growth in height, weight, physical power, etc. A simple recognition such as this could not possibly contribute anything substantial toward our understanding of the biological develop-

IV. GROWTH VERSUS DEVELOPMENT

We may now turn our attention to the physiological analysis of economic change. This endeavour requires, before anything else, some suitable analytical definitions. The literature offers the ultra-familiar definition which equates economic growth with increasing real income (alternatively, real gross national product), at times as an aggregate amount, at others as a *per capita* ratio. The unlimited aggregation on which all these co-ordinates rest raises thorny issues, so thorny that even works devoted entirely to macro-dynamics find it hard

---

17. Cf. notes 3 and 7, *supra*.

18. This confusing amendment may reflect another position of Rostow's [1962, p. 15]—namely, that 'the rate of growth $[dQ/Q]$ is a function of the rate of change in these stocks [the working labour force, H, and the capital equipment, K]'. He does not seem aware of the fact that this can be true only if $Q'_H H/Q = b$, $Q'_K K/Q = c$, where b and c are positive constants, from which it follows that the aggregate production function is necessarily of the form $Q = aH^b K^c$, a being again a positive constant. And the rub is that such a form cries out for some justification.

ment of a human individual. This understanding has increased only with the discoveries of the physiological interrelations between phenomena that are not all as conspicuous as the outward manifestations of growth. Similarly, the understanding of the ways followed by economic change can come only from a physiological picture of that process; the arithmetical relations alone that may exist between the traditional macro-co-ordinates would not do.

The numberless dynamic models that have grown and are still growing out of the seemingly perdurable fashion set up by R. F. Harrod's paper, 'An essay on dynamic theory', can at most indicate the direction in which these co-ordinates may move, but not what causes the movement itself. The observation that 'economic theory in the traditional mechanistic sense contributes next to nothing' to our understanding of the evolutionary economic process is just as valid today as in 1911, when Schumpeter [1934, p. 59] made it. In writing down the typical relations between some elementary, essentially outward characteristics of the economic process—say, between the savings from current income and the increase in real income (Harrod [1939; 1948])—the Harrod-type models assert nothing false. On the contrary, what they assert is only too obvious. Their deadly sin is that by stopping at the mechanico-descriptive level they overlook all the things that are truly important (Schumpeter [1934; p. 68]).

But in the literature we also find the right starting point for a physiological analysis of economic change. It is a distinction upon which Schumpeter, with his characteristic flair for what is analytically relevant, repeatedly insisted. The idea is that any economic change consists of two entirely distinct types of phenomena—*growth* and *development*. As Schumpeter defined it, development consists of a 'spontaneous and discontinuous' change that comes from *within* the economic process because of the very nature of that process. This change consists of some entirely new ways of combining the productive forces and materials, briefly, of new methods of production. Such a novelty changes the face of the economic world forever, that is, in an irreversible and irrevocable manner (Schumpeter [1934, p. 58, 62–5; 1935, p. 4]).

And, if we may abandon for the occasion the mental habit of believing that all factors of production can be substituted one for another without necessarily undergoing a qualitative change, we must inevitably come to face the truth that any change in the combination of factors implies a qualitative change (Georgescu-Roegen [1964]). Undoubtedly, a more capital-intensive method for manufacturing shoes uses different machines, different personnel, and, probably, even different materials than a less capital-intensive technique. The two types of shoes, too, are qualitatively different. It thus follows that, if by commodity we also understand the intermediate goods and the various kinds of labour services—as Schumpeter certainly must have done—the introduction of a new method of production necessarily changes the spectrum of commodities and, conversely, such a change implies some innovation in the production technique.[19] In the ultimate analysis, therefore, Schumpeter's discrimination between development and growth has its roots in whether or not the spectrum of commodities is altered qualitatively. A metaphor which he was fond of repeating illustrates the difference incisively. 'Add successively as many mail coaches as you please, you will never get a railway thereby' (Schumpeter [1934, p. 64n; 1935, p. 4]).

In relating economic development to a qualitative change of the spectrum of commodities, Schumpeter was certainly well inspired. There is no denying that one objective sign of the economic evolution since the times of the Romans, for example, is the fact that they travelled in chariots and we travel in jets. Actually, one single novel commodity suffices to establish the fact of economic evolution (as any archaeologist, whose only data consist of material artifacts, would instruct us). The point becomes still clearer as Schumpeter [1934, p. 65] explains that the seat of economic development is in the 'industrial and commercial life'. The spark that generates development, therefore, is anything that changes the technology prevailing at the time. Fundamentally, this is what Schumpeter called innovation—a term which has since become an important article of the economic trade. And a simple look at our physiological table 1 suffices to make us see why Schumpeter included the conquest of a *new* source of natural resources among the various cases of innovation. Such a conquest, if it is a true conquest, must supply a flow rate of resources greater than r_1 but with the same effort as previously, which means that it must

---

19. From Schumpeter's list of the possible origins of developmental changes [1934, p. 66], it appears, however, that he overlooked this factual equivalence.

necessarily change the old methods of doing things.[20]

The logical classification is rounded off by including under *growth* all cases in which the change is confined to a quantitative redistribution within a qualitatively invariant spectrum of commodities. These changes are all caused by factors external to the economic process, which may include such important social disturbances as wars, natural catastrophes, famines, etc. (Schumpeter [1934, p. 62; 1935, p. 2f]). This position does not mean that events of this sort cannot *induce* qualitative changes in technology. The logic is that these events by *themselves* can give rise only to a quantitative adaptation to a given qualitative structure—as Schumpeter [1934, p. 62f] neatly illustrated the point by the case of changes in tastes.[21] Or to put it differently, economic development would still go on in the world even if we were to abolish wars completely and also control all other calamities. In this manner, Schumpeter placed in a distinct analytical category those changes which, as he repeatedly explained, may be handled by the tools of the traditional analysis, by the economics of the circular flow, in essence by what in my own works I have termed standard economics. The nature of these changes—as explained in section I—is purely mechanical; hence, potentially, they are reversible. The point that all standard dynamic models pertain only to aspects of mere growth no longer needs an argument. For, as we can see it now, the point rests on simple logic.[22]

In retrospect, it appears that, in setting up the proper tools for the analysis of economic change, Schumpeter must have been moved by the same considerations that led Aristotle to insist on the essential difference between change of place (i.e., pure motion) and change of quality. Moreover, one must not fail to note the additional fact that Schumpeter developed his argument on lines followed by the most stringent arguments about biological evolution. The fact is not surprising, yet highly noteworthy. Among the biological analogies which one can detect after the fact (Schumpeter himself did not argue on their basis, nor did he seem to be aware of them at that time), one presents some interest for our topic. While he insisted on the fundamental difference between growth and development, he nevertheless included some changes under growth. These are changes that occur in small steps, ordinarily as adaptations to a change of data within the same commodity space (Schumpeter [1934, pp. 61f. 65, 81]). The illustration used by Schumpeter [1934, p. 62] is the gradual change of a small shop into a department store. The reasons for this apparent concession are relegated to a brief footnote (Schumpeter [1934, p. 81n1]), and are not sufficiently explicit. With this concession, it would appear that the ordinary kind of contemporary shoes does not constitute any novelty in respect to a Roman sandal. Probably, Schumpeter wanted to allow some changes after all, provided that they did not represent a discontinuous qualitative change. But his point, that 'one can assert of "small quantities" under certain circumstances what one cannot assert of "large quantities" ', merely shifts the issue on to what is small and what is large.

On the other hand, one should be reminded that one cannot discourse about evolutionary changes without resorting to dialectics (Georgescu-Roegen [1966, pp. 31–5; 1970a, pp. 32–7; 1971, pp. 62–5]). In the case under discussion, the small changes considered by Schumpeter are implicitly characterized by the property of being reversible, just as is assumed to be the case for the biological micromutations. In this way, the irreversibility of economic development—on which Schumpeter

---

20. However, not all the items include in Schumpeter's list [1934, p. 66]—which in his intention was exhaustive—seem as well justified as that just examined. See also the preceding note.

21. However improbable the misunderstanding may be, to avoid its happening we should observe that, although the definitions of growth and development are mutually exclusive, the events themselves may occur simultaneously. In fact, this is the normal case—as we shall argue in the following section.

22. The contrasting mechanistic view of economic change is most clearly reflected by Harrod's position that dynamics is specifically concerned with 'the effects of continuing changes and with rates of change', and, consequently, as 'the correspondent concept of velocity in Physics [we have] a steady rate of change (*of increase or of decrease*)'. (Harrod [1948, pp. 4, 8]. Italics added.) Purely quantitative (dynamic in our sense) and qualitative (evolutionary) changes are explicitly placed in the same bag. 'A continuing stream of new inventions, a continuing change of taste ... or a continuing increase of capital available at a given interest rate', for example, constitute the proper domain of economic dynamics. What matters for Harrod's position is the continuity of change: 'problems arising from a once-over change can ... be satisfactorily handled by the apparatus of static theory' (ibid., pp. 7, 21). For Schumpeter, on the contrary, the qualitative change, which may occur only intermittently, escapes the meshes of dynamic analysis.

[1934, pp. 58, 68] insisted with even greater force than Marshall [1898, p. 51]—is assured by the discontinuous novelty in the realm of commodities. Highly interesting is that Schumpeter's position on the small changes anticipates that of the biologist R. Goldschmidt [1940, pp. 390–9], who argued that the unidirectional character of biological evolution cannot rest on the small, reversible mutations, but on the occasional emergence of a 'hopeful monster'. In economics, the case is illustrated by the railway engine, a successful innovation, but a 'monstrous' one in relation to stage-coaches.

This point brings us to another entrenched view of economic development. Biologists have variously tried to relate the direction of biological evolution to increased efficiency, to increasingly more successful forms of life. But this attractive idea collapsed because no objective criterion of any kind could be found for evaluating the superiority of a biological species; species are successful in quite diverse ways. Economists have also been attracted by the idea that economic development necessarily means greater efficiency, measured not in money, but in some absolute terms. Accordingly, we often read that technological progress means that a larger output can be obtained from the same given factors (e.g., Meade [1961, pp. 1, 10]). However, unless some plain waste (say, keeping a furnace burning when not necessary) exists and all the new method does is to eliminate it, the definition is vacuous. The *same* factors (in quantity and quality) may only produce either the same amount of output as before or, perhaps, a different product, physically noncomparable with the previous one.[23] Schumpeter carefully avoided the pitfall. He consistently identified development only with new ways of doing things and associated a successful innovation only with an increase in the receipts of an enterprise. Apart from this pecuniary superiority, the merit of an innovation derives from a greater satisfaction of wants (Schumpeter [1934, pp. 68, 91]). Unfortunately, wants are notoriously parochial and also hard to compare historically.

But if one stops to look at this problem more closely, one would discover that there are innovations efficient in some objective sense. For example, there is the innovation which improves combustion in an engine in such a way that from the same amount of energy (including the difference which might be required for producing the new engine) we could derive more work. Some innovations may, however, cause even a waste of energy. Surprising though it may seem, the continuing substitution of terrestrial energy for solar energy—of the tractor for the beasts of burden—represents development, but not progress in a global economic sense (Georgescu-Roegen [1975]).

The peculiar nature of economic development being such as that expounded in the foregoing critical discussion, it stands to incontrovertible reason that the phenomenon of development constitutes the pulse of economic change even when it is barely felt at the scale of human lifespan. Yet its inherent dependence on quality puts it well beyond the reach not only of the dynamic models, but of any other conceivable arithmomorphic apparatus. The irreducible obstacle, as I see it, is this. We can crystallize the relations between the various elements of an arithmetical constellation in simple laws only because numbers being endowed with an intrinsic order lead to a definite graph for any relation. By contrast, pure qualities, which by their very nature are infinitely many, lack this analytically superb property.

V. MATHEMATICAL DYNAMIC SYSTEMS AND GROWTH

As a basis for the next argument—namely, that the standard dynamic models cannot represent adequately even pure growth—I propose to use the Leontief *open* model. The reasons for this choice are twofold. First, there is the popularity of that model, which because of its extreme mathematical simplicity and close relation with actual data has understandably swept the economic profession off its feet and has become the only *ars operandi* of almost every planning agency. Second (and the far more important reason), the Leontief model has the unique merit of adopting a physiological approach to the national economy, by taking the circular flow between the various individual sectors of such an economy for its scaffold.

As has been pointed out repeatedly, but apparently without impressing its promoters or its users, the Leontief model involves several critical assumptions.[24] Not all of them are rele-

---

23. No innovation would be possible if some factors were not flexible, if all factors were specific.

24. One analytical curiosity of the Leontief system seems to have completely escaped attention. It is illustrated by a famous French cook-book which has certainly baffled many culinary enthusiasts. Its recipe for a stock A requires stock B, and the recipe for stock B requires stock A.

vant for our argument, but those that are lead to some highly enlightening insights. The only modifications we need to introduce here concern some basic elements of the complete representation of a multi-process system (table 1) that have been ignored by Leontief, some in general, others only in his static model. Furthermore, to prevent purely mathematical complications from obscuring the issue at stake—which is at bottom epistemological—we shall consider a model of only two processes, producing commodities C_1 and C_2 (table 2).

In this table x_i is the *daily* output flow of C_i; x_{ki} is the *daily* input flow of C_k into P_i; X_{ki} is the *daily* service of the fund of C_k used in P_i; L_i is the *daily* labour service in P_i; t_i is the *working day* of P_i; and s_i is a *pure number* expressing the scale of P_i relative to a unit-scale process, which for $i = 1$ is

$$P_1^0 (a_{11} = 1, -a_{21}; B_{11}, B_{21}, H_1). \qquad (6)$$

This formulation reflects Leontief's characteristic assumptions which within our own analytical framework boil down to saying that a_{ki}, B_{ki}, and H_i are physical constants specific to each state of the arts.[25]

Two differences are immediately noticed between Leontief's input–output table for the static model (Leontief [1953, pp. 18f]) and table 2. Incomplete though this last table is in comparison with table 1, it expresses at least the fact that any production process includes some funds other than labour power. Leontief forcibly repaired the omission later in his dynamic system (in which he omits instead the labour fund).

But like every standard economist, Leontief has completely ignored the working day, which is an important co-ordinate of any production process, and even more so of a linear process as assumed by his system. Unfortunately, this second omission, which is a general characteristic of standard economics, is far from being a matter of abstract theory only. It has caused us to ignore the elementary fact that over-all growth can be achieved without any ado other than a simple, *proportional* lengthening of the working day. To be sure, this proportionality no longer operates for more realistic systems, which include mining and agricultural sectors. In mining, output cannot remain for long proportional to the working day; in agriculture, there is no working day in the usual sense of the term. Yet the truth is that the most directly available way of bringing about growth without waiting is in all cases the lengthening of the working day in some appropriate way in each

Table 2. *The multi-process table for an open Leontief system*

	P_1	P_2
	Flow coordinates	
C_1	$x_1 = t_1 s_1 a_{11}$	$-x_{12} = -t_2 s_2 a_{12}$
C_2	$-x_{21} = -t_1 s_1 a_{21}$	$x_2 = t_2 s_2 a_{22}$
	Fund coordinates	
C_1	$X_{11} = t_1 s_1 B_{11}$	$X_{12} = t_2 s_2 B_{12}$
C_2	$X_{21} = t_1 s_1 B_{21}$	$X_{22} = t_2 s_2 B_{22}$
Labor Services	$L_1 = t_1 s_1 H_1$	$L_2 = t_2 s_2 H_2$

sector.[26] This point constituted one of the most important 'secrets' by which the Western world achieved its spectacular economic advance of the eighteenth and nineteenth centuries—as was correctly assessed by Marx and confirmed by other observers.[27] In the ultimate analysis, the effect of lengthening the working day is to telescope time, as it were, so that *the future is brought nearer* in terms of the average human lifespan. The point is a corollary of the elementary principle that the reduction of idleness of agents is the main recipe for growth. One may debate the issue of the proper

---

25. It is well to note that a_{ki} is a flow *rate*, B_{ki} is an *amount* of C_k, and H_i is a *number* of workers. However, in P_1^0 the same numbers represent a flow and some services respectively, because in that process t_1 is assumed to be equal to a whole day. It is with a view to keeping in focus the dimensional aspect that I have used above the explicit notation a_{ii}, although the numerical value of this dimensional co-ordinate is unity and, hence, numerically, *but not dimensionally*, $x_i = t_i s_i$.

26. To be sure, the lengthening of the working day by itself could not cause any increase in production: the appropriate increase in the flow rate of natural resources should also be feasible. This point uncovers an important, yet overlooked, flow in the Leontief system resulting from the fact that the source of all material flows—the natural environment—is not included in the system (at least, not explicitly). In its static form—the most frequently used by planners—the system tells us only *how much are the inputs needed for a given shift in the net product, but not how we may actually achieve that shift.*

27. Even in the liberal Commonwealth of Massachusetts only in 1842 did a law protect children under twelve from working more than 10 hr. per day. For others, the 10 hr. day did not become a general practice in the USA before 1860, and the 8 hr. day not before the 1920s.

length of the working day endlessly; but one cannot deny that, if a speedy growth constitutes the *top* priority (and even if it is not quite so), to build an additional plant while the other plants of the same line are not used around the clock, by shifts, is a patent economic waste (Georgescu-Roegen [1969, pp. 527f]).[28]

The standard dynamic models not only ignore this elementary lever of growth—valid both historically and analytically—but they also reduce the process of growth to the familiar, simple formula 'Save—Invest—Grow'—the Open Sesame of current dynamic theories. Even the Leontief model is not an exception in this respect. The result is that neither the intellectually curious student nor the policy adviser can learn from the usual dynamic systems what growth, let alone development, really implies.

To isolate issues, in the argument concerning the second of the difficulties just mentioned, we may assume that the working day is fixed and put $t_1 = t_2 = 1$. It will also suffice to present the argument for the case of growth from an initial stationary state. For this state, we have the familiar system, equivalent to (5),

$$a_{11}s_1 - a_{12}s_2 = y_1,$$
$$-a_{21}s_1 + a_{22}s_2 = y_2, \quad (7)$$

where (y_1, y_2) represents the net income (which now is equal to consumption) and well-known condition

$$a_{11}a_{22} - a_{12}a_{21} > 0. \quad (8)$$

prevails. A new state in which the flow of the net income is increased by[29]

$$(\Delta y_1, \Delta y_2) \geq 0, \quad (9)$$

can now be reached only if the scales of production are increased by $\Delta s_1, \Delta s_2$. These increases must satisfy the system derived straightforwardly from (7):

$$a_{11}\Delta s_1 - a_{12}\Delta s_2 = \Delta y_1,$$
$$-a_{21}\Delta s_1 + a_{22}\Delta s_2 = \Delta y_2. \quad (10)$$

But for the scales of production to be increased, the fund elements of each sector must be increased proportionally (Georgescu-Roegen [1951, pp. 124–38; 1970a, pp. 212–14]; Leontief [1953, pp. 56f]). Specifically, the total funds

$$B_1 = s_1 B_{11} + s_2 B_{12},$$
$$B_2 = s_1 B_{21} + s_2 B_{22}, \quad (11)$$

must be increased by the following quantities:

$$\Delta B_1 = B_{11}\Delta s_1 + B_{12}\Delta s_2,$$
$$\Delta B_2 = B_{21}\Delta s_1 + B_{22}\Delta s_2. \quad (12)$$

According to Leontief's argument [1953, pp. 55–8], this can be achieved by the same old formula, 'Save–Invest–Grow'. But the particular way in which this formula is interpreted must be well understood. *The mere accumulation of the cuts in the flows of consumption suffices to increase instaneously (i.e., without any additional waiting) the capacity to produce (i.e., to bring about growth).* The only waiting is that necessitated by the process of accumulation itself.

Let Δt be the duration of the accumulation and let z_1, z_2 be the reduced flow rates of consumption during that interval. The accumulated savings from the current income (old consumption) are

$$\Delta B_1 = (y_1 - z_1)\Delta t,$$
$$\Delta B_2 = (y_2 - z_2)\Delta t. \quad (13)$$

By eliminating $\Delta s_1, \Delta s_2, \Delta B_1, \Delta B_2$ from (10), (12), (13), we obtain the system

$$z_1 = y_1 - M_{11}\left(\frac{\Delta y_1}{\Delta t}\right) - M_{12}\left(\frac{\Delta y_2}{\Delta t}\right),$$
$$z_2 = y_2 - M_{21}\left(\frac{\Delta y_1}{\Delta t}\right) - M_{22}\left(\frac{\Delta y_2}{\Delta t}\right), \quad (14)$$

where

$$\|M_{ik}\| = \|B_{ik}\| \times \begin{Vmatrix} a_{11} & -a_{12} \\ -a_{21} & a_{22} \end{Vmatrix}^{-1} \quad (15)$$

This system determines the level of consumption (z_1, z_2) during the 'gestation' period Δt for

---

28. In this connection, I may cite the highly pertinent explanation of the President of the Twin Buttes Copper Mine, who told me that he operates the mine around the clock because he cannot afford to let the immense capital equipment required for the operation lie idle.

29. The notation $(a, b) \geq 0$ excludes the case of both $a=0$ and $b=0$.

any properly chosen $(\Delta y_1, \Delta y_2)$, and conversely.[30] As the result of the accumulation during that period, at $t + \Delta t$ the level of the national net income jumps from (y_1, y_2) to $(y_1 + \Delta y_1, y_2 + \Delta y_2)$—which may now be all consumed if no further growth is envisaged.

If we consider a sequence of 'gestation' periods, for each period we have a system (14),

$$z_1^i = y_1^i - M_{11}\left(\frac{\Delta y_1^i}{\Delta t}\right) - M_{12}\left(\frac{\Delta y_2^i}{\Delta t}\right),$$

$$z_2^i = y_2^i - M_{21}\left(\frac{\Delta y_1^i}{\Delta t}\right) - M_{22}\left(\frac{\Delta y_2^i}{\Delta t}\right),$$

(16)

where $y_j^{i+1} = y_j^i + \Delta y_j^i$, $y_j^0 = y_j$. With the aid of these systems, we can determine step by step the sequence $(z^i{}_1, z^i{}_2)$ from any given sequence (y_1^i, y_2^i), and vice versa. But if C_1 and C_2 are continuous substances, it would be antieconomical to let the accumulated stocks remain idle between t and $t + \Delta t$; instead, we should add them to the productive funds as they are being accumulated. This is tantamount to making Δt tend toward zero in (16), which yields

$$z_1(t) = y_1(t) - M_{11}\dot{y}_1(t) - M_{12}\dot{y}_2(t),$$

$$z_2(t) = y_2(t) - M_{21}\dot{y}_1(t) - M_{22}\dot{y}_2(t),$$

(17)

where the dot indicates the derivate with respect to t. Again, $z_j(t)$ is determined by simple algebra if $y_1(t), y_2(t)$ are given, and $y_j(t)$ is determined by integrating the differential system if $z_1(t), z_2(t)$ are given.[31]

The foregoing argument looks fine on paper; it runs into some serious snags if pushed beyond that stage.

The first such snag comes from the fact that, according to the characteristic rationale of the Leontief systems, an increase in scale $(\Delta s_1, \Delta s_2)$ requires a proportionate increase in *all funds*, hence in the labour power as well,

$$\Delta L_1 = H_1 \Delta s_1, \quad \Delta L_2 = H_2 \Delta s_2. \quad (18)$$

One question now cries for an answer, for the benefit of the theorist as well as for the planner's. 'Where does this additional labour come from?' For, if this labour is not available, the system cannot be in equilibrium; the market of producer goods would not clear. Several answers are possible, but none saves the day for the Leontief model, or for any other open model for that matter.

One possible answer is that the necessary additional labour existed (unemployed) before the planned change. That would restrict the operationality of the model to a special situation. Another answer is that production suddenly becomes more capital intensive at $t + \Delta t$. But if we remember a point made earlier (section IV), this means that growth is necessarily accompanied by development, for which there is no proper place in a strictly dynamic model. Indeed, if there is development, the systems that represent the processes during the successive growth periods refer to a continually changing spectrum of commodities. These systems, therefore, can no longer be subsumed into a single dynamic system such as (17). Grant all the information one would ask for, no dynamic system can cover the 'growth' process of England from Cromwell to the present, for example. The same conclusion follows from a third possibility—namely, that the additional labour (which must possess a modicum of skill) is produced by a *new* process P^* which begins at t and ends at $t + \Delta t$. Finally, there is the answer that P^* was already in existence at t, so that no development occurs. But this means that at t there was a process not represented in the analytical matrix of the system. Moreover, since this process produces more labour power than was necessary at t, the system could not have been then in a truly stationary state.

A second snag is related to the already mentioned fact that, according to the logic of the Leontief system, the accumulation of the differences resulting from abstinence is both a necessary and sufficient condition for growth. The accumulation of saved bread, for example, is an indispensable element of the physiology of growth. But that is not all. According to the same logic, *household consumption must necessarily include, say, bull-dozers;* for otherwise it would not be possible to achieve growth by mere abstinence. The upshot is that in a realistic approach we must separate producer from consumption goods, as Marx did so neatly

---

30. Obviously, M_{ik} is non-negative. This confirms the elementary fact that, if $\Delta y_1, \Delta y_2$ are chosen, the inherent lower limits of consumption flows impose a lower limit on Δt. And if Δt is chosen, there is an upper limit for $(\Delta y_1, \Delta y_2)$.

31. The standard system used by Leontief [1953, p. 56f] involves the gross product (x_1, x_2) instead of the net product (y_1, y_2). That form may have more appeal for a planner, but (17) serves better a discussion of the growth issue.

in his model of production (Georgescu-Roegen [1971, pp. 262–5]). Accordingly, if C_2 is only a consumer good and C_1 only a producer good, we must put

$$y_1 = 0, a_{21} = 0, B_{21} = B_{22} = 0. \qquad (19)$$

In this case, however, Leontief's formula for growth no longer works. To put it plastically, it is no longer possible to increase the number of bull-dozers by merely accumulating the abstained consumption of, say, yogurt. Actually, this is *the* question any growth theory must answer in real (as opposed to monetary) terms.

An entirely different recipe from Leontief's comes easily to mind. It consists of reallocating some capital, say δB_1, from P_2 to P_1. Accordingly, during the gestation period, the level of consumption becomes

$$z_2 = y_2 - a_{22} \frac{\delta B_1}{B_{12}} \qquad (20)$$

and the additonal stock of C_1 accumulated at the end of the gestation period is

$$\Delta B = \left(\frac{a_{11}}{B_{11}} + \frac{a_{12}}{B_{12}} \right) \delta B_1 \Delta t$$

$$= \frac{A}{B_{11} B_{12}} \delta B_1 \Delta t. \qquad (21)$$

If this is distributed in the proper proportions between P_1 and P_2, it increases the net income by

$$\Delta y_2 = \frac{a_{11} a_{12}}{B_{11} B_{12}} \delta B_1 \Delta t. \qquad (22)$$

As before, the continuous assumption yields

$$\dot{y}_2(t) = \frac{a_{11} a_{12}}{B_{11} B_{12}} \delta B_1(t), \qquad (23)$$

and (20) becomes[32]

$$z_2(t) = y_2(t) - \frac{B_{11}}{a_{11}} \dot{y}_2(t). \qquad (24)$$

Undoubtedly, the problem of how to bring about growth is not simple. Even this second recipe, which seems to avoid one snag of the Leontief system, raises a troublesome question of practical operationality. Often, capital equipment is specific and hence cannot be used in a different process from that for which it has been designed. This obstacle compels us to think of still another recipe, which consists of allowing B_{12} to decumulate so that the entire net output of P_1 may accumulate in an increased fund of producer goods.[33]

This third recipe, however, confronts us with still another stumbling block (the most important of all) in the dynamic representation of growth. The new stumbling block is the fact that decumulation is not accumulation in reverse (Georgescu-Roegen [1951; 1971, pp. 226f]). If the object in point is a stock of readily consumable materials, decumulation may proceed at any rate we may choose. Conceivably, any stock may be decumulated (i.e., consumed) almost instantaneously or at a zero rate, for example. The pace of decumulation of equipment, however, is rigidly determined by the nature of the equipment. To give a homely example, the only way to decumulate a pair of boots is to use them until they are completely worn out—a process which normally requires more time than the production of the boots from leather and other materials. Accumulation may constitute the greatest worry for a planner in an underdeveloped country, but decumulation of durable goods is in fact a far more refractory issue in practice and, perforce, in analysis.

Through decumulation we aim to turn back past history, to reverse past actions, which is an impossible feat. But, again, because of the mechanistic epistemology of standard economics, the models which constitute the pride of economic dynamics ignore the difference

---

32. There is no surprise that this formula cannot be obtained from (17) by introducing the conditions (19); the operational recipe is different.
Another way of treating the above problem is to consider that, during Δt, $B(t) = B_{11} s_1(t) + B_{12} s_2(t)$ is distributed between the two processes so that $\delta B_1(t) = [s_1(t) - a_{12} B(t)/A] B_{11}$.

33. Not to miss a subsidiary but new snag, we should point out that this recipe may result in labour unemployment if, as seems to be the case in actuality, the capital goods industry is more capital intensive than the consumer goods industry (Georgescu-Roegen [1971, pp. 266f]).

between 'forward' and 'backward'—the most visible symptom of the irreversibility of Change. In these models not only what we abstain from consuming immediately increases the capital funds—the $s_k B_{ik}$'s in the Leontief system—but also, in reverse, what we consume above our income immediately decreases the same funds. The same simple algebra that led us from (9) to (14) stands correct whatever sign Δy_1 and Δy_2 may have. And if Δy_1, Δy_2 are negative, system (14) shows that $z_i - y_i > 0$ and the entire 'machine is set in reverse: we eat up today exactly what we have saved and invested yesterday.[34]

VI. DIFFERENTIAL EQUATIONS AND ECONOMIC DYNAMICS

Difficulties of a concrete nature, similar to those mentioned above, may come from other directions, but are perhaps not of equal importance. Be this as it may, the use made in economics of differential systems (or of finite difference systems) often involves a theoretical misapplication. The detailed manner in which systems (17) and (24) have been developed here will readily expose the slip. From basic mathematical calculus we know that a differential system of the form (17), for example, has a solution which involves two indeterminate constants. These constants can be determined from the condition that for some arbitrary t_0 the solutions must coincide with the actual position of the system at that moment. However, builders of dynamic models seem to ignore the rationale underlying this mathematical rule. In mathematics, a system of differential equations represents a *particular family* of functions, such as the family of curves (C) shown by solid lines in figure 3. The so-called initial conditions, y_0 for t_0, help us to identify the individual curve, C_0, of that *particular* family. The important point is that, if the system of differential equations represents an *actual* structure, the identification rule is valid *only if we know beforehand that the actual structure follows one of the curves of that particular family* (and we wish only to identify that curve). If the actual structure is such that its dynamic movement is described by a differential system whose solution consists of the family of curves (C')—represented in figure 3 by interrupted lines—the identification procedure must be applied to this last family. The behaviour of the individual system in point is then represented by C'_0. The point deserving emphasis is that, if an economic system

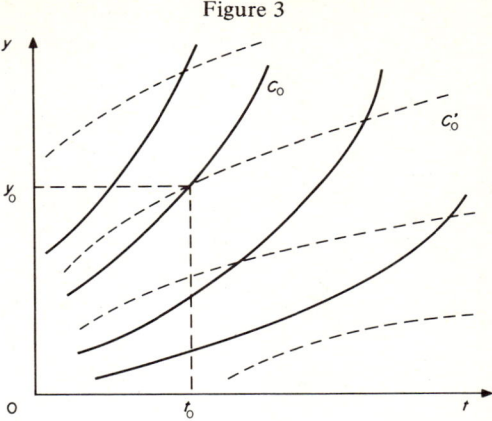

Figure 3

happens to be in a stationary state, for example, we would be mistaken to use its initial conditions for determining, say, the path on which an exponentially growing system with the same initial conditions would move. Before determing the constants of integration we must make sure that the mathematical differential system used is that which describes the behaviour of our system. Interestingly, even Marx completely dodged the question of how growth may come about. For he considered expanded reproduction only as a completely separate case from simple reproduction. His expanded reproduction diagram reflects only the notion that growth comes from growth. That diagram corresponds to what we would now call a *particular* solution of a system of differential equations, i.e., a particular trajectory of a growth system set in motion by a Prime Planner at some remote time in the past.

---

34. Among all authors of dynamic models, Leontief alone has the great merit of raising the issue of irreversibility in relation with his dynamic system [1953, pp. 68–72]. To include the case in which some $\dot{s}_i < 0$, i.e., some product flow must be decreased and hence the corresponding funds $s_i B_{ki}$ must be decumulated, Leontief proposed to replace the system (17) by one derived from it by putting $B_{ki} = 0$, in one situation, and $B_{ki} = 0$ as well as $a_{ki} = 0$, in another. In this last case, $s_i B_{ki}$ is decumulated at the flow rate of $s_i a_{ki}$, i.e., as a simple stock. As a sign of the difficulty of the problem, even this apparently reasonable amendment leads to contradictions, as shown by McManus [1957]. For a very simple illustration of the contradiction see Georgescu-Roegen [1951, pp. 126–8; 1966, pp. 310–12; 1970a, pp. 215–17], where the diagrams show that the relaxation change to $b_{11} = 0$ (i.e., to $B_{11} = 0$) when $\dot{x}_1 = 0$ does not lead to $\dot{x}_1 < 0$.

As the considerations of the earlier paragraphs show, the last point does not mean that a stationary system cannot be changed into a growing one. It means only that this change cannot be achieved by the indiscriminate use of the rule of identification. A dynamic system, *once established,* is kept at all times in dynamic equilibrium by its own physiological structure. This is by definition true for any dynamic system *(including the stationary one).* The point is that all dynamic systems must have a beginning, at $t = -\infty$ (the simplest analytical form) or at some known moment. But the problem of how a given dynamic system came into being is not only essentially different from that of how that system functions, but also far more intricate. It is the old problem of the cause of motion, only in a different context. To adapt a metaphor of Schumpeter [1934, p. 85] to this occasion, a dynamic system is like a movement along a road built by someone at a moment of no interest for the physiology of the system. To transform a given dynamic system into another involves, however, an act of creation, namely, the building of a *new* road. Some dynamic systems may grow as a matter of their particular physiology. But the most discriminate meaning of 'growth' in economics is the passage from one physiology to another of higher intensity. And as I have argued in the preceding pages, this operation implies a qualitative change—an innovation in Schumpeter's sense.

The passage also implies a dynamic manipulation consisting of a reduction of consumption from its initial level. As the new system keeps growing, consumption eventually reaches the old level and even surpasses it. The system grows because at one point in the past some event set it growing. After that event, and only after it, the movement of the system is described by a new differential system. To see all this clearly, let us return to the argument beginning with the stationary system (19), which is described by the simple differential $\dot{y}_2 = 0$. But if at t_0 we shift an amount of capital δB_1^0 from P_2 to P_1 and then redistribute continuously the increased capital fund in the adequate proportion between the two sectors, the behaviour of the *new* system from t_0 on will be given by (23), i.e., by

$$y_2(t) = \frac{a_{11} a_{22}}{B_{11} B_{12}} (\delta B_1^0) t + c, \quad (25)$$

the constant c being determined by the condition that $y_2(t_0) = y_2^0$. And from (20), it is immediately clear that $z_2(t) < z_2{}^0$ for some interval of time after t_0.[35] This is the *cost* of bringing the economy from a stationary state to one of arithmetical growth. But the argument is applicable to any change consisting of an increased speed with which a system grows. It is only after $z_2(t) > z_2(t_0)$, that the population may benefit from such a change. One may speak in this case both of abstinence and waiting.

VII. COMMODITIES VERSUS PROCESSES IN GROWTH

But there is another category of waiting in any passage from one state to a faster growing one, a category which is generally ignored. To be sure, lags have been introduced in some dynamic models. But their existence does not seem to be justified by concrete reasons. Thus, Leontief [1953, p. 83] is of the opinion that their causation is 'a somewhat mysterious relationship'. The source of difficulty is another implicit, yet analytically crucial assumption of practically all standard dynamic models—the assumption that the capacity of production increases simultaneously with the accumulation of producer goods and in the same amount. This assumption, in turn, is the consequence of the failure to realize that in any *non-stationary* economic system the productive activity is aimed at two distinct objectives—to produce *goods* and to produce *processes.* Without this distinction, there can be no difference of essence between the materials in a lumber yard and the home that may be built with them. The point is that commodities are *not* produced by commodities, but by processes. Only in a stationary state is it possible for production to be confined to commodities. In such a state, the same processes produced some time in the past may simply be maintained intact, i.e., in the same good order of technical efficiency.

But if there is growth, not only must the old processes be maintained intact but also new processes in running order must be added to the old ones. Now, to bring into being a running process means not only to build a physical plant but also to prime it (Georgescu-Roegen [1971, pp. 239f, 273]). How this complete project is realized may be described by an elementary process, which from the analytical

---

35. It is precisely for setting this point in bold relief that I have chosen to describe the movements of the illustrating systems in terms of net income and actual consumption, instead of consumption and gross product, as Leontief does. Compare (17) with (3, 3) of Leontief [1953, p. 57].

viewpoint is in no way different from the process by which one piece of furniture or one automobile is produced. Both involve a definite duration as well as some periods of (relative) idleness of the participating agents. This means, in the first place, that a definite waiting period must follow the period necessary to accumulate the goods necessary to build and prime the new process. And this additional period of waiting can no longer be made to tend to zero, as was possible in the case of the gestation period Δt (section V).[36]

Without the special act by which the necessary new individual processes are created and without the necessary waiting involved in it, we obviously cannot pass from the system (7) to that in which the net income is $(y_1 + \Delta y_1, y_2 + \Delta y_2)$. or from the system (19) to that described by (22). However, we may recall that any elementary process may be arranged *in line*, an arrangement in which there is no idleness of agents and (still more important for the point I wish to make now) no waiting. The product of a (primed) factory starts to flow out the instant the factory begins to work. We may, therefore, imagine a new kind of factory. Π_1, which produces processes in line. Once such a factory is built and primed, the corresponding economy is capable of growing at a constant speed without waiting.

Needless to add, this new process Π_1 must be included explicitly in the multi-process matrix alongside with the processes P_i that produce only goods. And if we add still another Π_1^* which produces in line additional people with the right proportion of prepared workers, we arrive at an economic system that actually grows *without waiting*. This system is in *true dynamic equilibrium*, which sustains its own arithmetical scale of growth (subject to the perennial limitation set by natural resources). It is a system in equilibrium because it possesses within itself all that is needed for its functioning.

The foregoing analytical argument may be pushed still further, if we note that an increase in the prevailing speed of growth of a dynamic system can be obtained by adding more processes Π_1. Then, we may again visualize a new type of process, Π_2, which produces processes Π_1 in line. With Π_2 built and primed, the economy can grow at an *accelerated* speed without any waiting. In the limit, we arrive at an abstract model of an economy that grows at a constant rate of growth (i.e., exponentially) without waiting and also is in complete dynamic equilibrium.[37]

To be sure, the world of facts does not fit into this analytical model point by point. Even in the most advanced economies which grow without any apparent waiting we do not find factories that produce factories that produce factories that produce factories... Yet the model is instructive in that it draws our attention to a crucial scaffold of any developed economy. This scaffold is a complex and extensive network consisting of general contracting firms, of architectural firms, of various engineering firms, of firms that make machines to make machines to make machines... Even though not related into a single organization, these enterprises operate separately or in association like a factory—a flexible factory but still a factory in its broad lines.

It is this Π-sector, as we may call it to distinguish it from the *P*-sector (commodity production), that constitutes the fountainhead of the growth and further growth which seems to come about as by magic in the developed economies and which, precisely for this reason, has intrigued the economic theorist and puzzled the planner.

Admittedly, the distinction between commodities and processes by which commodities are produced is dialectical, i.e., not discretely distinct (Georgescu-Roegen [1966, p. 21–4; 1970a, pp. 24–6; 1971, pp. 43–7]). Dialectical also is the distinction between growth and development—as we have learned from Schumpeter's endeavours to clarify the notion of innovation. But without distinguishing between commodities and processes and without a physiological analysis of the still mysterious transmutation of yogurt or soya beans into bull-dozers or blast-furnaces (for example), we have only a surrogate of growth theory, such as standard dynamic economics offers us.[38]

---

36. From what I have said in these pages about dynamic economics we should not be surprised to see Harrod [1948, p. 13] explicitly stating that his type of dynamics has nothing to do with lags.

37. Since the point was touched on above, we need not ask again, the question of where the additional natural resources come from. At least for some relatively short period we may accept the possibility that all the necessary resources are accessible at constant cost, as they seem to be during the current bonanza.

38. A last-minute word for the reader. From a different viewpoint, Louis J. Junker's 'Capital accumulation, savings-centered theory and economic development', *Journal of Economic Issues*, vol. 1 (June 1967) pp. 25–43, contains a very interesting critical survey of the standard position that savings is the only prerequisite of capital accumulation.

REFERENCES

Aumann, R. J. (1964) 'Markets with a continuum of traders', *Econometrica*, vol. 32 (January) pp. 39–50.

Bach, G. L. (1957) *Economics*, 2nd ed. (Prentice-Hall, Englewood Cliffs, N.J.).

Baumol, W. J. (1970) *Economic Dynamics*, 3rd ed. (Macmillan, New York).

Brown, D. J. & A. Robinson (1972) 'A limit theorem on the cores of large standard exchange economies', *Proc. Natn. Acad. Sci.*, vol. 69 (May) pp. 1258–60.

Georgescu-Roegen, N. (1951) 'Relaxation phenomena in linear dynamic models', in T. C. Koopmans *et al.* (eds.), *Activity Analysis of Production and Allocation* (New York: Wiley) pp. 116–31. Reprinted in Georgescu-Roegen [1966, pp. 300–15]. French translation in Georgescu-Roegen [1970a, pp. 205–20].

—— (1964) 'Measure, quality, and optimum scale', pp. 232–246 in C. R. Rao (ed.) *Essays on Econometrics and Planning Presented to Professor P. C. Mahalanobis* (Pergamon Press, Oxford) pp. 232–46; also *Sankhya*, Series A, vol. 27 (March 1965) pp. 29–64.

—— (1966) *Analytical Economics: Issues and Problems* (Harvard University Press, Cambridge, Mass).

—— (1969) 'Process in farming versus process in manufacturing: a problem of balanced development', in Ch. Nunn and U. Papi (eds.), *Economic Problems of Agriculture in Industrial Societies*, Proceedings of a Conference of the International Economic Association, Rome, 1965 (Macmillan, London), pp. 497–528.

—— (1970) 'The economics of production' (Richard T. Ely Lecture), *Amer. Econ. Rev.*, vol. 60 (May) pp. 1–9.

——(1970a) *La science économique: Ses problèmes et ses difficultés* (Paris: Dunod).

—— (1971) *The Entropy Law and the Economic Process* (Cambridge, Mass.: Harvard University Press).

—— (1972) 'Process analysis and the neoclassical theory of production', *Amer. J. Agri. Econ.*, vol. 54 (May) pp. 279–94.

—— (1974) 'L'economia politica come estensione della biologica'. *Note Economiche* (Monte dei Pasch di Siena), No. 2, pp. 5–20.

—— (1975) 'Energy and economic myths', *Southern Economic Journal*, vol. 41 (January) pp. 347–81.

Goldschmidt, R. (1940) *The Material Basis of Evolution* (Yale University Press, New Haven).

Harrod, R. F. (1939) 'An essay on dynamic theory', *Economic Journal*, vol. 49 (March) pp. 14–33.

—— (1948) *Towards a Dynamic Economics* (Macmillan, London).

Heilbroner, R. L. (1972) *The Economic Problem*, 3rd ed. (Prentice-Hall, Englewood Cliffs, N. J.).

Jevons, W. S. (1879) *The Theory of Political Economy*, 2nd ed. (Macmillan, London).

Kaldor, N. (1960) *Essays in Economic Stability and Growth* (Duckworth, London).

Kaldor, N. & J. A. Mirrlees (1962) 'A new model of economic growth', *Rev. Econ. Studies*, vol. 29 (June) pp. 174–92.

Leontief, W., *et al.* (1953) *Studies in the Structure of the American Economy* (Oxford University Press, New York).

Marshall, A. (1898) 'Distribution and exchange', *Economic Journal*. vol. 8 (March) pp. 37–59.

McManus, M. (1957) 'Self-contradiction in Leontief's dynamic model', *Yorkshire Bulletin of Economic and Social Research*, vol. 6 (May) pp. 1–21.

Meade, J. E. (1961) *A Neo-Classical Theory of Growth* (Allen & Unwin, London).

Meadows, D. H., *et al.* (1972) *The Limits to Growth* (Universe Books, New York).

Pareto, V. (1906) *Manuale di economia politica* (Milano: Società Editrice Libraria).

Rostow, W. W. (1962) *The Process of Economic Growth*, 2nd ed. (Norton, New York).

Samuelson, A. (1947) *Foundations of Economic Analysis* (Harvard University Press, Cambridge, Mass.).

—— (1970) *Economics*, 8th ed. (McGraw-Hill, New York).

Schumpeter, J. A. (1934) *The Theory of Economic Development* (Harvard University Press, Cambridge, Mass.).

—— (1935) 'The analysis of economic change', *Review of Economics and Statistics*, vol. 17 (May) pp. 2–10. Reprinted in Richard V. Clemence (ed.), *Essays of J. A. Schumpeter* (Cambridge, Mass., Addison-Wesley, 1950) pp. 134–42.

Wicksteed, H. (1894) *The Co-ordination of the Laws of Distribution* (London, Macmillan). Reprinted in Scarce Tracts in Economic and Political Science, London School of Economics and Political Science, 1932.

CHAPTER 10

(1966)

Further Thoughts on Corrado Gini's *Delusioni dell'econometria*

One of the many academic enterprises which Corrado Gini, the scholar whom we are honoring today, encouraged or supported with his scientific prestige is the Econometric Society. For Gini, it is proper to remember on this occasion, was a founding member of that society. But Gini did not have to wait for the Econometric Society to be founded. He had been an «econometrician» in the true sense of the word for a long time before this term was coined. Therefore all the greater is the significance of the message which Gini, during the later part of his life, wanted to send to his fellow econometricians and which he crystalized in the title of a short article: «Delusioni dell'econometria» [1]. Coming from a scholar who had already devoted many good years to devising new quantitative tools and to applying them successfully to the analysis of many social phenomena, the message cannot be interpreted as a denial of the value of quantitative analysis in economics. Gini wanted only to impress upon us the danger created for the economic science by the ostentatious yet decidedly false claims econometricians ordinarily make for the scientific superiority of many of their procedures. In one place [2], he even alluded to an additional danger, of far greater consequences, which may confront the student of economics if the ostentation of the econometric *Akademia* turns into scientific intolerance — a thought which though not absurd

---

[1] *Giornale degli economisti*, Anno XV, 1956, pp. 174-177. See also his "Au sujet de l'utilité et de la limitation de l'emploi du calcul des probabilités en économie politique", *Economie appliquée*, t. X, n. 1, 1957, pp. 49-55.
[2] "Delusioni", p. 176.

seems highly unrealistic. Be this as it may, the consummate scholar wanted to warn us for the sake of the very science which he served with such high honor.

Gini's message about the delusions of econometrics has already been taken up and amplified by a few of my distinguished colleagues from Europe. In support of the same message I wish to add a few observations concerning some specific points which by their nature belong to four distinct aspects of the problem. These observations may suggest that my message is stronger than Gini's — and perhaps it is. But I wish to make it perfectly clear that they are offered only with the hope that they will orient further discussions toward the constructive end which certainly Gini had in view.

I. — EXCESS OF MATHEMATICAL FORMALISM.

It is certainly not because of some fancy of modern mathematicians that formalism has fared in mathematics as splendidly as it has. Its success derives from the immense economy of thought it created for mathematicians themselves, who are continously invited by special sciences to solve this or that particular problem. Mathematicians have thus begun to study, for instance, the properties of relations in unspecified terms, $x R y$, because these properties could serve equally well to a sociologist for whom x and y may stand for individuals and $R = $ « related by blood », or to an economist for whom x and y may represent productive processes and $R = $ « more efficient than ». This very idea implies that the sociologist and the economist should fill the empty boxes of mathematics with some specific empirical content proper to the one's own field of endeavor. But because of the well-known difficulties of getting down to empirical brass tacks in economics, many a student has found it more comfortable to continue the formalism of mathematics or, as this is often put, to substitute mathematical exercises for economics. This type of pseudo-economics is particularly prevalent in the most recent contributions on utility theory where often the word utility appears only in the title. The text itself speaks only of an undefined relation, of its being upper or lower continuous, or of some other abstract properties which have hardly any relevance for the study of consumer's behavior in the real world. Curiously, none of the respective authors seem to be aware that a proposition based upon the continuity of a mathematical system cannot be tested empirically any more than the irrationality of $\sqrt{2}$ can be established on the workbench. Whether the consumer can be indifferent between two different commodity combinations is an issue that can be settled only by general introspection, not by testing theorems on lower and upper continuity through laboratory experiments on behavior. Such absorbing preoccupations with the mathematics called for by a particular problem are responsible in a great degree for the fact that econometricians in general tend to forget that, apart from engineering economics, economic models are not blueprints of the reality. They are only schemata or similes [3], which are useful only if handled with the delicate touch demanded by the very nature of economic phenomena. There is certainly no harm but gain in using a simile in which the entrepreneur is assumed to know the probability of every future market coordinate, for the purpose of illustrating the main thread of one's argument. But to identify such a simile with actual behavior in a world of pure uncertainty and, moreover, to use the mathematical expectation formulae as a guide for « rational » behavior is an irrational position against which John Maynard Keynes, Frank H. Knight, and Corrado Gini have, apparently in vain, raised their protests. To quote Gini:

> En économie politique [il y a] des problèmes (et des nouveaux en surgissent tous les jours) dont les données ne peuvent pas être chiffrées et mesurées comme l'économétrie le présuppose, mais dont l'importance est incomparablement supérieure, au point de vue théorique et pratique, aux raffinements, sans doute élégants et parfois aussi notables, que l'économétrie est en condition d'apporter [4].

---

[3] Cf. GINI, " Delusioni, " p. 174. See also NICHOLAS GEORGESCU-ROEGEN, *Analytical Economics: Issues and Problems*, Harvard University Press, 1966, pp. 117f.

[4] GINI, " Au sujet de l'utilité et de l'emploi du calcul de probabilités," p. 53. For a proof of the nonmeasurability of uncertainty see my *Analytical Economics*, pp. 208-211, 263-275.

Admittedly, formalism has been part and parcel of mathematical economics from its inception, but its excess has become alarming only in modern times. Surprising though it may seem, the most glaring illustration is the ultra-familiar concept of production function, which even in the most recent works is described indifferently as the relationship between input and output *rates* or between input and output *quantities*. This patent indifference towards the empirical content of the symbols used has prevented economists from seeing that if the two definitions are equivalent — as they have been always regarded — then all production functions must be homogeneous and of the first degree.

Let

$$q = f(x, y, \ldots, z) \tag{1}$$

be the « production function » expressed in terms of *rates of flows per unit of time*, q for output, and $x, y, \ldots, z$ for inputs. Let

$$Q = F(X, Y, \ldots, Z) \tag{2}$$

be the « production function » expressed in terms of *quantities*. Since the process these are supposed to describe is a steady-going (static) process, we have $Q = tq$, $X = tx, \ldots, Z = tz$, where t represents an arbitrary time interval. Clearly, if the modes of representing a production process, (1) and (2), are equivalent, we must have

$$Q = qt = tf(x, y, \ldots, z) = F(tx, ty, \ldots, tz) \tag{3}$$

for any nonnegative values of $t, x, y, \ldots, z$. Making $t = 1$, we obtain

$$f(x, y, \ldots, z) \equiv F(x, y, \ldots, z). \tag{4}$$

From (3) it then follows that all production functions are homogeneous of the first order, as said. However, traditional satisfaction with the formal representation is so deep-rooted that at two recent international meetings I have encountered unusual difficulties in making some of my fellow econometricians see this most elementary point.

Still more important is that because of the same penchant to formalism, in standard — as opposed to Marxian — economics the time element, i.e., *the working day*, is completely absent from the description of a productive process. From what has been said above it can be shown that its correct formula is

$$Q = t f(x, y, \ldots, z),$$

where Q is the « daily » output, $x, y, \ldots, z$, the input flow rates and service rates of which the productive system is capable, and t is the length of the working day. This is not the proper place to dwell on the far reaching consequences which the omission of the time factor has for pure theory and especially for policy recommendations ([5]). Nor is it within the scope of this communication to cite other minor instances where the gale of formalism has driven sound economics aground. But one particular consequence of the curious fact that even for a non-mathematical economist mathematics comes easier than economics, should invite our attention next.

II. — The Loss of the Social Dimension.

It was Irving Fisher who first argued that the modern utility theory is not based upon a hedonistic philosophy of human behavior ([6]). The argument, challenged by a few, is long since an article of faith of the neoclassical economist. But if one tries patiently to penetrate the veil of words — at times, mangled, at times, empty — in which the argument is enveloped, one would discover that the whole edifice is rather specious. Indeed, to say with Fisher that utility is not synonymous with pleasure but only with the desire of pleasure does not absolve utility theory of hedonism. And to say, also with Fisher as well as with

---

([5]) For further details the interested reader is referred to two of my papers, "Process in Farming vs. Process in Manufacturing: A Problem of Balanced Development," chapter 5 in this volume, and "Chamberlin's New Economics and the Unit of Production," ch. 2 in *Monopolistic Competition: A Study in Impact*, Essays in Honor of Edward H. Chamberlin, ed. R.E. Kuenne, New York, John Wiley & Sons, 1966 (in press).

([6]) Irving Fisher, *Mathematical Investigations in the Theory of Value and Prices*, Yale University Press, 1925, pp. vii, II.

Pareto, that « each individual acts as he desires » ([7]), is a type of empty talk which leaves one in complete empirical darkness. What one would certainly like to know is *what* the individual desires. One reads the answer of the utility theory to this question only as he gets to the second page: the individual desires *only* commodities. Hence, an ordinal measure of his desire is a function $U(x, y, \ldots, z)$ only of the amounts of the various commodities he may possess, and with this the way is clear for all kinds of mathematical elaborations and conclusions by which we now generally swear.

No doubt, we do so because, like the great founders of utility theory, we have in mind a Civil Society where the actions of the individual are determined only by utility, in the ultimate analysis, by commodities. But the exceptions to this rule, few though they are, in the urban stratum of the Western world should have made us see that in its general form economic choice is not between two commodity vectors, X and Y, but between two complexes $(X; A)$ and $(Y; B)$, where A and B denote the actions by which each vector may be obtained. After all, contrary to what utility theory assumes, the economic choice is not a culturally *free* choice. One chooses between available complexes on the basis of the values which the actions have in the corresponding cultural matrix *and* the values which the commodities have on the personal utility scale. The fact that in some societies — which are generally referred to as « traditional » societies — the first element weighs heavier in the balance than the second, does not mean at all that their members are irrational. Only their behavior is not susceptible of being cast into a purely arithmomorphic model ([8]). But this is no reason for us to proceed like Procrustes or to throw up our arms in despair at their « irrationality », should we be called to make policy recommendations for a traditional society.

III. — THE ABUSE OF STATISTICAL THEORY AND TOOLS.

Every econometrician knows that all statistical inferences and tests are based on some special assumptions in addition to

---

([7]) FISHER, *op. cit.*, p. 11; VILFREDO PARETO, *Manuel d'économie politique*, Paris, Giard, 1927, p. 62.

([8]) See further *Analytical Economics*, pp. 124-129.

one general assumption, that of randomness. Yet in many cases there seems to be a great gulf between what one knows and what one does. I shall discuss these cases one by one.

1. — As we all know, the entire edifice of statistical theory rests on the general assumption that the relation between any sample and the parent population is homeomorphic to that produced by a random mechanism. Most econometricians have assumed all along — implicitly as well as explicitly — that *all* economic data fulfill this condition and yet no justification other than mere verbalism has been offered in support of this position. Perhaps the predicament of the econometrician is that, since in the domain of social sciences only in a few cases can we point to the parent population, a proof of the randomness of economic data is impossible. In agronomy, for istance, we are justified in regarding any group of observations as a random sample because we can experiment with the *same* type of fertilizer on as many plots selected at random as we wish. But what is the reason for treating, say, the occupational ratios in all the counties of a state as a random sample or, still worse, as a random sample from a single universe ([9]) ? *Irregularity*, it should be repeated again and again, is not necessarily the same as *randomness*. The last decimal digits in a five decimal logarithmic table certainly form an irregular pattern. But it is equally certain that they are not determined at random. Besides, the idea that the distribution of the economic features of the counties in a state is the result of God's tossing some special dice over the state is certainly bizarre. I need not mention the more obvious fallacy of treating time series data as a random sample because this fallacy after being duly exploded seems now extinct, at least as far as respectable works are concerned.

2. — No theorem on which a statistical test is based is valid in a vacuum. All such theorems — even those pertaining to the so-called nonparametic tests — make some assumptions con-

---

([9]) To be sure, hardly any economic data is not affected by errors of observation resulting from the imperfection of statistical registration. But this is not what an econometrician means by the economic data being a random sample.

cerning the parent population. For instance, the most popular tests invoked in support of the realiability of an econometric model, the t-test, the F-test, and the z-test, all require that the sample be chosen at random from a *normal* population. Consequently, even if one would deal with data that can be safely regarded as constituting a random sample, before applying any of these tests one also needs to make sure that the parent population is normal. Unfortunately, the extremely few cases analyzed in the literature do not encourage us at all to expect economic data to be normally distributed [10]. A number of doctoral candidates, who at my insistence have tested the normality of some of the data used in their dissertations, have all obtained decisively negative results. In this situation, to claim the validity of an econometric model on the basis of, say, the F-test is tantamount to claiming that a patient does not have cancer because his blood test for sugar has come out negative.

3. — Because this point concerns a muddle for which statistical theory shares part of the blame, it deserves to be discussed at some length. For a while statisticians and econometricians were troubled by the question of which of the two regression lines of a bivariate distribution represents the « true » relation between two variables. The question has as much sense as if one would ask which of the two polar circles is the equator. It is not surprising therefore that it led to the idea that the difficulty of the answer lies in the existence of more than *one* regression line. Apparently, everyone ignored the fact that the existence of regression lines does not necessarily imply the existence of a « law » between the theoretical values of the variables. Or to put it differently, the point that was ignored is that a scatter of observations may reflect different stochastic structures.

A two-dimensional scatter may be the random image of a *point*, as is the case of the gun shots spread around the target point. It may also represent the results of shooting at a target which shifts on a given *curve*. Such a scatter is the random image

---

[10] E.g., IRVING FISHER, *The Making of Index Numbers*, Boston, Houghton Mifflin, 1927, pp. 408-10.

of a curve, in general of a k-dimensional variety ([11]). We may refer to the element of which the scatter is the random image as the kernel of the scatter.

Any bivariate scatter, whatever its kernel, has two regression lines. In case the kernel is a curve, neither of these coincides with the kernel curve — except in quite special cases. The point that seems to need stressing is that it is the kernel curve that represents the relation between the two variables as this relation is conceived in any theoretical construction. For example, for a scatter of prices and quantities observed in different actual market situations where demand is the same, it is the kernel that represents this demand, $p = f(q)$. The idea is that each observation p', q' corresponds to some p, q representing the theoretical equilibrium which the market would have reached in the absence of any imperfections. That is, $p' = f(q) + \xi$, $q' = q + \eta$, where ξ and η are random variables with zero mean.

Let us take the simple case where the « true » (or the equilibrium) position $P(X, Y)$ shifts at random on the straight line $Y = aX$ and that X is a variable with zero mean and variance σ^2. Let (ξ, η) be the deviations from P:

$$\overline{\xi} = \overline{\eta} = 0, \overline{\xi^2} = \sigma_1^2, \overline{\eta^2} = \sigma_2^2, \overline{\xi\eta} = \rho\,\sigma_1\sigma_2. \qquad (5)$$

For the observations

$$x = \xi + t, \; y = \eta + at \qquad (6)$$

we have

$$\overline{x} = \overline{y} = 0, \overline{x^2} = \sigma_1^2 + \sigma^2, \overline{y^2} = \sigma_2^2 + a^2\sigma^2, \overline{xy} = \rho\,\sigma_1\sigma_2 + a\sigma^2 \qquad (7)$$

If the regression of y on x is linear, we have

$$BG(y/x) = \frac{\rho\,\sigma_1\sigma_2 + a\,\sigma^2}{\sigma_1^2 + \sigma^2} \, x \qquad (8)$$

([11]) NICHOLAS GEORGESCU-ROEGEN, "Sur un problème de calcul des probabilités avec application à la recherche des périodes inconnues d'un phenomène cyclique," *Comptes Rendus de l'Académie des Sciences*, July 7, 1930, and "Further Contributions to the Scatter Analysis," *Proceedings of the International Statistical Conferences*, 1947, vol. V, pp. 39-41.

which represents the « best guess » of y for any *given* x. It is immediate that this line represents the true law if and only if

$$\rho \, \sigma_1 \, \sigma_2 + a \, \sigma^2 = a \, (\sigma_1^2 + \sigma^2), \tag{9}$$

i.e., if $a = \rho \, \sigma_2/\sigma_1$ or if $\sigma_1 = 0$. The first condition means that $BG \, (\eta/\xi) = a \, \xi$, the second that *the observed x is the true value X of the corresponding equilibrium position*. Needless to add, both these conditions represent very special cases. But one widespread fallacy connected with the second case must be denounced. The fallacy confuses the fact that, as happens often, x is measured with a high degree of approximation with the fact that x is the « true » value. This confusion is at the bottom of the prevalent position that (8) represents the theoretical law.

The famous « regression law » of Francis Galton provides an excellent illustration of the points made above. As biologists later learned, for a population consisting of a pure biological line, the height, for instance, of the parents, x, and that of the off-springs, y, are uncorrelated identical normal distributions:

$$\overline{x} = \overline{y} = M, \; \overline{(x-M)^2} = \overline{(y-M)^2} = \sigma^2, \; \overline{(x-M)(y-M)} = 0 \tag{10}$$

The kernel of the distribution is a *point*, $X = M$, $Y = M$. Now, if for the sake of simplicity, we take a population formed by a normal distribution of pure lines, M becomes a normal variable with

$$\overline{M} = M_0, \; \overline{(M-M_0)^2} = \Sigma^2, \tag{11}$$

The kernel of the new distribution of x and y is $X = Y$ which, clearly, represents *the true heredity law*. The regression line is

$$BG \, (y/x) = \frac{\Sigma^2}{\Sigma^2 + \sigma^2} (x - M_0) + M_0, \tag{12}$$

the slope of which is $\Sigma^2/(\Sigma^2 + \sigma^2) < 1$. It is thus seen that Galton's error was precisely that of the modern econometrician: to take the regression line for the true law. A far more instructive way of describing this error is this: *ex post* we can say that Galton reasoned as if the *observed* height of the parent represented the characteristic height of the pure line to which the parent belongs. But in a mixed population the individual's height, be it measured with perfect accuracy, does not reveal to which pure line he belongs, the reason being that such a height is the sum of two unknown random components. Similarly, an observed price, for instance, even if known with perfect accuracy should not be confused with the equilibrium price at the time of the observation.

Let us consider the scatter (ξ_1, η_1) such that its kernel is a point, the origin, and such that

$$\overline{\xi_1^2} = \sigma_1^2 + \sigma^2, \ \overline{\eta_1^2} = \sigma_2^2 + a^2 \sigma^2, \ \overline{\xi_1 \eta_1} = \rho \sigma_1 \sigma_2 + a \sigma^2. \quad (13)$$

Obviously, without outside information, it is absolutely impossible to distinguish between the structure of this scatter and that of (7) the kernel of which is a straight line. Also, without some outside information — seldom available in the domain of economics — we cannot discover the true law $Y = a X$, for the simple reason that the last three equations of (7) cannot *identify* five unknowns: $\sigma, \sigma_1, \sigma_2, \rho$, and a. We cannot even affirm that $Y = a X$ lies within the acute angle of the two regression lines. All the more useful therefore is to mention a particular situation which may be encountered in some special problems where $\sigma_1^2, \sigma_2^2, \rho \sigma_1 \sigma_2$ are known except for a constant factor of proportionality λ ([12]).

In this case the last three equations of (7) become

$$\overline{x^2} = \lambda K_1 + \sigma^2, \ \overline{y^2} = \lambda K_2 + a^2 \sigma^2, \ \overline{xy} = \lambda K_{12} + a \sigma^2, \quad (14)$$

and a is determined by

$$\begin{vmatrix} 1 & K_1 & \overline{x^2} \\ a & K_{12} & \overline{xy} \\ a^2 & K_2 & \overline{y^2} \end{vmatrix} = 0 \ ([13]). \quad (15)$$

([12]) See the references given in footnote 11.

([13]) To take care of the fact that we generally do not know *ex ante* whether the kernel passes through the origin, relation (7) must be replaced by

$$\overline{x} = x_0, \ \overline{y} = y_0, \ \overline{x^2} = x_0^2 + \sigma_1^2 + \sigma^2,$$

$$\overline{y^2} = y_0^2 + \sigma_2^2 + a^2 \sigma^2, \ \overline{xy} = x_0 y_0 + \rho \sigma_1 \sigma_2 + a \sigma^2. \quad (7a)$$

Moreover, in economics it is more realistic to assume that the distribution

The difficulty in determining the true law with the aid of a scatter is analogous with that of the identifiability of the coefficients in the so-called shock-models. Now, if according to logic a certain system of data cannot provide the kind of information we would like to have, that is the end of the matter. There is no good purpose in overrunning logic by a makeshift and presenting the makeshift as the product of the highest form of scientific procedure. Curiously, the point has been immediately accepted in the case of nonidentifiable coefficients but completely ignored in the case of nonidentifiable parameters of scatter distributions.

IV. — Curve Fitting Fetish.

This last point pertains to the confusion between discovering a quantitative law from a series of data and merely fitting a mathematical formula to the same data. The confusion thrives on the characteristic fluidity of the phenomenal domain of economics: almost any economic phenomenon is a *potential* element of change for almost any other such phenomenon. That is why we profess the highest esteem for general equilibrium theories. In this we are, no doubt, right. But the case of econometric models — which generally aim at formulating precise quantitative macroeconomic laws — is quite different. One should endeavor to explain a macroeconomic phenomenon by only a few variables chosen for their theoretical affinity and their predictive qualities surmized from simple analysis. The contrary happens in many an econometrician's shop. There, as I once put it, we see him selecting his tools with a single purpose in mind: to cut the log on his workbench in such a way that he

---

of the equilibrium positions on the theoretical law, instead of following a random law, is *irregular*. For this second alternative,

$$\bar{x} = T_1, \bar{y} = y_0 + a\,T_1, \overline{x^2} = \sigma_1^2 + T_2,$$

$$\overline{y^2} = y_0^2 + \sigma_2^2 + a^2\,T_2, \overline{xy} = \rho\,\sigma_1\,\sigma_2 + a\,T_2,$$

(7b)

where $T_1 = (\Sigma\,t_i)/N$, $T_2 = (\Sigma\,t_i^2)/N$, and N is the number of observations. The necessary modifications of (15) present no difficulty.

may be able at the end to exclaim triumphantly « I told you that inside that log there was a beautifully carved Madonna ! ». Since I used this metaphor I came across a research project which, sad to say, is equally piquant : to find which of the numberless ways in which « money » items can be combined in one item fits Keynes' system of equations. The project differs little from the more conspicuous form of pseudo-scientific endeavor, namely, that of seeking among a large set of variables those that would make the regression equation pass with flying colors the « statistical test ».

Admittedly, not all econometric shops follow exactly this recipe. But with the increasing facilities of the computer centers the practice is likely to become predominant, at least by numbers.

There is no denial that progress in many special sciences has often consisted in adding a new variable in the functional relationship expressing a law. A most convincing illustration is the addition of lunar nutation in the formula for the position of the earth. But econometricians seem to ignore the fact that a better fit obtained by adding a new variable does not mean at all that the formula is also a better law. For a formula to represent a law it is not sufficient that it should fit well the available observations : the acid test is the fit for all other observations. Perhaps, another predicament of the econometrician is that in the case of a stochastic law with a high variance this acid test requires a large number of additional observations, which in turn requires a new census or a great number of annual observations. But then one should display greater skepticism and far less assurance as regards the significance of the quantitative laws derived by mere fitting. On the other hand, if, as has appened in a not small number of cases, the acid test came out negative, the excuse that meanwhile evolutionary changes took place boomerangs with disastrous effect : it is tantamount to admitting that evolutionary factors play a substantial role and yet cannot be caught in an arithmomorphic scheme.

V. — CONCLUDING REMARKS.

Some of the observations made in this paper seem to point to the illusions of those who profess the science of econometrics, others to the delusions of the discipline itself as it appears in the actual endeavors. Now, illusion is the very thing upon which a scholar's devotion to science feeds. He must have the illusion that what he thinks is right. Should he begin considering all the doubts that may exist about what one can say « correctly » in his own domain of study, progress would come to a standstill. After all, mathematicians themselves at one time entertained such illusions as the duplication of the cube or the differentiability of any continuous function. And no one can say what other illusions are still hidden within the body of modern mathematics. The point which seems to characterize the econometric profession and which has created a source of irritation between that profession and the rest of the economists, is that, contrary to what happens in other analytical sciences, the econometricians tend to cling to their illusions even after they have been duly exploded. It is because of this attitude that illusions — a necessary companion of any scholar — turn into delusions. Recalling an earlier metaphor, we may compare the resulting consequences with those which would prevail if medical science would not fight — as it does — its delusions as they become certified. Perhaps, we, the econometricians, can get by with our delusions either because most of our models pass into oblivion as soon as we have finished dressing them up elegantly or because — and this is the saddest part of the story — the consequences of their adoption by a policy maker, even if they can be traced back to the real culprit, become manifest only after such a long time that there is no longer any purpose in indicting him. By then the culprit may be busy extolling the qualities of another model. An epistemology which disregards entirely the acid test of a formula is certainly strange for a discipline which at the same time claims to believe only in objective science. The consequences have been so admirably described by Corrado Gini that I find it appropriate to close my communication by quoting him:

> La condition essentielle pour faire des progrès par le procédé des modèles, et en général par la méthode hypothético-déductive, est que les conséquences qu'on en deduit soient dûment vérifiées. [Les modèles] méritent con-

fiance seulement si on peut s'assurer que les hypothèses schématiques qu'ils impliquent ne portent pas à des conclusions trop divérgentes des faits. A mesure que des faits nouveaux sont mis en évidence, le contrôle doit être renouvelé. ... C'est ce qu'on a fait et ce qu'on fait systématiquement dans les sciences physiques ; dès qu'on a construit un modèle, des milliers d'expérimentateurs, dans les laboratoires du monde entier, se hâtent de le contrôler. Ce n'est pas ce que l'on fait et que l'on a fait en économie politique, et en particulier en économétrie, et c'est bien à cause de cela que les sciences physiques font des progrès par la construction des modèles, tandis que l'économie politique et en particulier l'économétrie n'en font pas, ou, du moins, pas autant qu'elles pourraient. Les économistes, et surtout les économètres, n'hésitent pas à lancer sur le marché scientifique de nouveaux modèles sans se donner la peine d'écarter ceux qui sont déjà en circulation, ni de contrôler sur les faits si les nouveaux modèles sont préférables aux anciens. De cette façon, les modèles se multiplient, s'amoncellent, s'enchevêtrent, se contredisent et suscitent la méfiance et le scepticisme [14].

---

[14] GINI, « Au sujet de l'utilité et de la limitation de l'emploi du calcul des probabilités », p. 54. Also " Delusioni ", p. 176.

CHAPTER 11

(1964)

MEASURE, QUALITY, AND OPTIMUM SCALE

It is difficult to contemplate the evolution of the economic science over the last hundred years without reaching the conclusion that its mathematization was a rather hurried job. Undoubtedly, even social sciences could not remain completely indifferent to the loud cry "science is measurement." But the haste in adopting the method supposed to elevate political economy to the rank of "true science" had some unfortunate consequences: many epistemological issues, which ought to have been clarified before any attempt at using the new tools for the old tasks were ignored or avowedly by-passed. The most important of these issues is that of the relation between quality and quantity.

However we may look at the matter, quality is our most basic concept. It is definitely prior to that of quantity, for before we can speak of a measure of A relative to B we must distinguish between A and B in some way or other. And as we do not yet have a measure of A this way can be but qualitative.

A very eloquent example of the importance of the problem of quality versus quantity is offered by Professor Chamberlin's penetrating analysis of the relationship between divisibility and efficiency of factors of production.[1] For, indeed, I believe that the whole argument of that article turns upon the manner in which the passage from quality to quantity has been handled by prominent writers on the problem of scale. There is little doubt that we much too often take for granted that the quantification of quality leaves no *qualitative residual*. The arguments advanced against Chamberlin's paper by both A. N. McLeod and F. H. Hahn[2] constitute excellent illustrations in point.

As we all know, the question of whether or not there is an optimum scale for the unit of production is far from being purely academic. In a free enterprise economy, it constitutes the basic issue in any rational policy concerning monopolistic competition; in a planned economy, it has a decisive bearing upon allocation policies. Last, but not least, it constitutes the main issue in the controversies over management by central authority.

---

[1] Edward H. Chamberlin, "Proportionality, Divisibility and Economies of Scale," *Quarterly Journal of Economics*, **lxii** (1948), pp. 229 ff. All subsequent references, however, will be made to the sixth edition of Chamberlin's *Theory of Monopolistic Competition*, which includes a more recent version of that paper as Appendix B.

[2] "Proportionality, Divisibility, and Economies of Scale: Two Comments," *Quarterly Journal of Economics*, **lxiii** (1949), pp. 128-143.

Chamberlin's thesis, therefore, is much too important to be left without an analysis by formal logic, the only one that can prevent the argument from mazing inconclusively between the multiple connotations of the same word. Such an analysis is attempted in the present paper. Its first part deals with the general problem of measurability, which has to be clarified before anything else. The second part takes up Chamberlin's problem of divisibility and proportionality versus efficiency.

To avoid at least some *ab initio* misunderstandings, some preliminary remarks appear necessary. First, I should explain that by "the passage from quality to quantity" I do not mean "the dialectical relation between quality and quantity" as taught by Hegel and understood by Hegelians of all strains. While I do not deny the existence of this philosophical problem, I proceed from the assumption—unquestionably borne out by the brute facts—that we can actually perform the specific operations required by the theoretical concept of measure. The paper being intended as a piece of *analysis* cannot enter into the deep waters of *dialectics*. The principle, to which we overtly subscribe, that quality cannot be quantified without a qualitative residual may sound as an echo of the Hegelian dogma that quantitative variations bring about qualitative changes; and actually, there are points of contact between the two. But nothing prevents us from accepting the former as a factual truth based upon the structure of the laws of the most quantitative of all sciences, physics.

Second, I should also explain that, following the teachings of Ernst Mach, by "theory" I understand a logically— as opposed to taxonomically or lexicographically —organised *description* of a definite *reality*. To the point: theorems regarding magic squares do not constitute a theory in this sense as long as they do not help us describe a definite reality. In addition to *logical* propositions, such as "If A, then B," a theory must contain also at least one *factual* proposition of the type "There exists some element in reality that corresponds to A," even if such propositions are not always explicitly listed. It is rather puzzling that economists need be continually reminded of this, while other scientists need not. Indeed, no physicist would ever regard as physical theory a piece of analysis proceeding, among others, from the proposition that light travels along a conical helix—for instance.[3]

I. Measurability and Measure

1. *General considerations.* The problem of measurability of utility has by now been distilled and redistilled. In contrast with this situation, the measurability of the elements involved in production has received—to my knowledge—no systematic treatment. The anachronism is probably due to the fact that the protests against the mathematization of political economy have been exclusively directed against the idea of submitting human actions to mathematical rigor. Naturally, mathematical economists have been wont to strengthen precisely that position against which their opponents concentrated their attack. Speaking in retrospect, one should have expected the problem of measurability to be more intricate in production than in

---

[3] The reader will have no difficulty in seeing that in their criticism both McLeod and Hahn (*op.cit.*) adopted a different meaning of "theory" than the Machian one.

consumption theory. The comparability of utility has a solid basis in the concept of "the *same* individual," who chooses without ever changing his mind. Moreover, in the case of utility measurability concerns only one element : all others, the commodities involved, are assumed to possess a definite measure — an assumption generally though not absolutely true to facts. Both these simplifying conditions are not present in the productive sector where we need a measure for both products and factors and where comparability must rest on an objective instead of a subjective basis. It is, therefore, clear that even the simple concept of production function, as this is used throughout the literature, is far more delicate than that of ophelimity function. The contrast between the two is accentuated by the fact that, while the most important propositions of the theory of consumer's behaviour are unaffected by whether utility is ordinally measurable or only ordered, there are no orientation marks for the behaviour of the producer unless factors and products are measurable in a particular sense.

By now, we are familiar with the most general concept of measure, i.e., the ordinal measure. This rests on the following definitions :

Definition 1 : A set , S, is said to be ordered by a relation, R, if for $a, b, c \in S$,

(i) either aRb or bRa;

(ii) aRb and bRa implies $a = b$;

(iii) aRb and bRc implies aRc.

Relation R is referred to as an ordering relation of S.

Definition 2 : A set, S, ordered by a relation, R, is ordinally measurable if there exists a numerical function, $f(x)$, $x \in S$, such that for $a, b \in S$

$$f(a) \geqslant f(b) \iff aRb.$$

Any function $f(x)$ satisfying this condition is an ordinal measure of S. It is elementary that if $f(x)$ is an ordinal measure of S and $F(t)$ an increasing function of t, then $F(f)$, too, is an ordinal measure of S. Or, as this is frequently stated, ordinal measurability is indifferent to monotonic transformations of the scale.

The usual conclusion reached by most studies of measurability by both economists and psychologists seems to be that all measures are ordinal and there is no reason whatever for attaching any special significance to the concept of cardinal measures. According to this view the old controversy between cardinalists and ordinalists turns around a bogus. However, these studies fail to take into account the manifold aspects of reality. While it is perfectly true that any usual scale can be converted into the most bizarre one by a monotonic transformation, it is far from certain that we could always live by the latter without first reconverting it into the former. And if in some cases we must reconvert, the proposition that any scale is as good as any other becomes a purely analytical statement without any significance as far as scientific theory is concerned.

For a psychologist as well as for an educator it does not matter at all whether certain ability is graded on a scale from 1 to 4 or from -2 to $+2$, or even on an alphabetical scale. One may speculate about a world in which a carpenter says "two" after laying down his yardstick four times, and "three" after laying it down nine times.

But it is inconceivable that in such a world money should be counted in an arbitrarily different manner. That is not all. If we maintain that any scale is as good as any other, then such fundamental notions as *decreasing marginal rate of substitution, constant returns, efficiency*, etc., loose any meaning whatsoever. A most simple example will show this clearly.

Example 1 : The production function $q = \sqrt{xy}$ exhibits constant average returns to scale as well as decreasing marginal rate of substitution. Let us adopt a new scale for x obtained by the "smooth" monotonic transformation

$$(T) \quad x = e^{-2/u} \quad \text{for} \quad 0 \leqslant x \leqslant e^{-2}, \quad x = e^{u^2 - 3} \quad \text{for} \quad e^{-2} \leqslant x.$$

If we use the same transformation for changing y into v, the production function in the domain $u, v \geqslant 1$ is given by $q = e^{\frac{1}{2}(u^2 + v^2) - 3}$ In that domain (at least) the new isoquants, $u^2 + v^2 = \text{const.}$, are concave towards the origin. Constant average returns, too, have disappeared from the picture. Which ordinal scale shall we then consider as proper for describing the essential properties of the production function? Most certainly, ordinal measurability does not fit even the simplest picture of a production process.

Nor can we follow the advice offered by ordinalists and choose that scale for which the formula of the given production function would be as simple as possible. For the same factors enter into a host of other production functions, and the ordinal scale that would simplify one particular production function is apt to render others forbiddingly complex. We are not in the same position as the physicists; they were able to choose, for instance, a particular ordinal scale for temperature in order to obtain the simplest possible expression for the basic equation of thermodynamics. On the contrary, we are bound to some specific procedures of measurement that are applied to commodities without number, and our task is to find out a notion of measure that would bring to the surface the common essence of all these procedures. For our behaviour as producers (or planners) has been shaped on that basis over a very long time and, consequently, cannot be separated from it. The point is that these procedures could not have resisted the test of time if they, in turn, were not rooted in some basic property of matter. And this property is that of being *cardinally* measurable.

2. *Cardinal measurability*. Instead of writing down at once all the axioms that a cardinally measurable set, C, fulfils, it is preferable to introduce them one by one. In this way we shall be able to discuss every axiom from two viewpoints : epistemological, in order to establish its status as an *operational* description of reality, and logical, in order to test the theoretical construction of the final concept.

One of the most conspicuous facts of everyday life is that we can perform a special kind of operation with some variables but not with others. We can, for instance, pour one cup of water and one glass of water into a pitcher : the result is another concrete instance of the same logical category, "water". Clearly, in other cases this operation is not feasible. We cannot always subsume two words into another word. There is no sense in which utility, calendar dates, or temperature-levels, can be subsumed. However, we can subsume two colours by mixing two pints of liquids having those colours. Another case of subsumption will interest us later on: if a bicycle frame and a set of bicycle tires are both products, then a bicycle, too, is a product. This subsumption corresponds to the fact that production processes can be consolidated (cf. Sec. 3, B4). What counts is that *some operation exists by which subsumption is carried out*. Moreover, in all cases where such an operation exists, its result is not affected by the order in which the elements involved are manipulated. We are thus justified in introducing the following

Definition 3: An internal operation, $\perp$, of a set, S, such that

(i) if $a, b \in S$, then $a \perp b = b \perp a = c, c \in S$,

(ii) if $a, b, c \in S$, then $(a \perp b) \perp c = a \perp (b \perp c)$,

is called subsumption.

It is also characteristic of many variables in the real world that "nothing" has an unambiguous meaning in connection with them, i.e., a particular concrete instance corresponds to the basic notion of "nothing". (The latter should not be confused with the number *zero* which — as illustrated by the ordinary temperature scale — does not necessarily mean "nothing".) One property of "nothing" is reflected by

Definition 4: If there is an element $z \in S$ such that if $a \in S$, then $a \perp z = a$, z is a neutral element.

It is instructive to observe that though colours can be subsumed by mixing, the set of all colours has no neutral element. There is no colour that can be mixed with *any* colour without changing the latter.

Definition 5: A set that possesses a subsumption and a neutral element with respect to it, is a subsumptive set.

As the first property of a cardinally measurable set, we shall list

A1: The set, C, is subsumptive.

This property does not permit the analysis to be carried very far. But we can easily prove the following

Theorem 1: *The set C has only one neutral element.*[4]

We may also introduce a relation, $G =$ "greater", through

Definition 6: If $a = b \perp c$, and $c \neq z$, then $a\,G\,b$.

At this point one may be tempted to believe that C is ordered by G. Unfortunately, proposition A1 does not exclude the possibility that both $q\,G\,p$ and $p\,G\,q$ should be true for some $p \neq q$. (See Example 2 below.)

The next three propositions pertain to the following abstract problem:

If $a, b \in C$ does the equation $a \perp x = b$ have a solution $x \in C$?

---

[4] This is elementary: if z and y are neutral elements, then by Definition 4, $y = y \perp z$ and $z = z \perp y$, and by Definition 3, $y \perp z = z \perp y$; hence $y = z$.

Or simpler, can the elements of C be *subtracted* as well? We know that we can take goods out of a storehouse or a pail of water from a lake, so that the problem raises unavoidably. Let us begin with the simplest properties:

A2: If $a, b \in C$ and $a \perp b = a$, then $b = z$.

A3: If $a, b \in C$ and $a \perp b = z$, then $a = b = z$.

Intuitively these are so obvious properties of "nothing" that one may be inclined to believe them to be direct consequences of A1. The following example proves that they are not.

Example 2: Let S be the set of all ordered pairs (m, n) where m, n are real numbers and $m \geqslant 0$. Let the operation $\perp$ be defined by

$$(m, n) \perp (m', n') = [\max(m, m'), n+n'], \qquad \ldots (1)$$

where $\max(m, m')$ stands for the greater of m and m'. It is immediate that S satisfies A1 with $z = (0, 0)$. Yet, according to (1),

$$(m, n) \perp (m, 0) = (m, n),$$

which for $m > 0$ shows that S does not satisfy A2. Also

$$(0, n) + (0, -n) = (0, 0), \qquad \ldots (2)$$

which for $n \neq 0$ shows that S does not satisfy A3.

We can use this example to prove a point made earlier, namely, that A1 does not warrant the irreflexiveness of G. Indeed, if $n \neq n'$ we have both $(m, n) G(m, n')$ and $(m, n') G(m, n)$, for $(m, n) = (m, n') \perp (m, n-n')$ and $(m, n') = (m, n) \perp (m, n'-n)$.

The following two examples are designed to prove that neither A3 can be derived from A1-A2, nor A2 from A1 and A3, so that A1, A2 and A3 are not redundant.

Example 3: Let S' be a subsumptive set satisfying A1, and let y be its neutral element. Let $z \notin S'$, and let S be formed from S' by the inclusion of z. Let us also put $a \perp z = z \perp a = a$, for any $a \in S$. It is easy to prove that S is subsumptive and satisfies A1 and A3, but not A2. Indeed, if $b \in S'$ we have $b \perp y = b$, although (by Theorem 1) $y \neq z$.

The example also proves the following

Theorem 2: *The set obtained by eliminating the neutral element from a set satisfying* A1 *and* A2, *has no neutral element.*

The theorem shows that A2 expresses the absoluteness of the notion of "nothing"

Example 4: Let $S = \{a, b, c\}$ and let $\perp$ be defined by the table

	a	b	c
a	a	b	c
b	b	c	a
c	c	a	b

The structure thus defined satisfies A1 and A2 with $z = a$, but not A3. Indeed, $b \perp c = z$, although $b, c \neq z$.

This example also shows that A3 expresses the fact "nothing" cannot be greater than any other "something" : from $b \perp c = z$, it follows that zGb.

We need also to prove that A1-A3 are not inconsistent. That can be done by referring to the set of non-negative numbers with $\perp$ defined by the usual addition. However, an example from the familiar field of economics will prove more instructive.

Example 5: Let S be the set of all ordered pairs (m, n), where m and n are non-negative numbers. Let $\perp$ be defined by

$$(m, n) \perp (m', n') = (m+m', n+n'). \qquad \ldots \quad (3)$$

It is obvious that this set satisfies A1-A3. Moreover, in this case relation G is irreflexive, for $(m, n) G(m', n')$ implies $mn > m'n'$, $m \geqslant m', n \geqslant n'$. Therefore we cannot have also $(m', n') G(m, n)$. In addition, G is transitive as well. Yet, S is not ordered by G; for if $m > m'$ and $n < n'$, neither of the equations $(m, n) \perp (x, y) = (m', n')$, $(m', n') \perp (x, y) = (m, n)$, has a solution in S. The solution of the first equation is $(m'-m, n'-n)$, of the second, $(m-m', n-n')$; they do not belong to S since $m'-m < 0$ and $n-n' < 0$. As we usually put it, the set S cannot be ordered because not all of its elements are *comparable* through the only available relation, G. The purpose of the "preference" concept in consumer's theory is precisely to introduce a more inclusive relation than G, so as to render any two elements comparable. In the case of a cardinally measurable variable, comparability derives from the *homogeneity* of the elements covered by it. The latter property is defined as follows:

Definition 7: A subsumptive set, S, is homogeneous if for any $a, b \in S$, at least one of the equations

$$a \perp x = b, \qquad b \perp y = a, \qquad \ldots \quad (4)$$

has a solution.

Equations (4) may have a solution for some elements even though the corresponding set is not homogeneous. However we can prove

Theorem 3: *If S is a set having the properties A1-A3, and if both equations (4) are satisfied for given $a, b \in S$, then $a = b$.*

Indeed, if (4) are both satisfied we have $a \perp x \perp y = a$. By A2, $x \perp y = z$, and by A3, $x = y = z$. By Definition 4, $a = b$.

A4. The set, C, is homogeneous.

We may note that Example 5 shows that a set fulfilling A1-A3 does not necessarily fulfil A4; indeed, if $m > m'$ and $n < n'$, neither of the equations (4) has a solution for $a = (m, n)$, $b = (m', n')$. On the other hand, the set of Example 2 satisfies A1 and A4, but not A2 and A3; this proves that neither A2 nor A3 follows from A1 and A4.

Theorem 4: *If S is a set having the properties A1-A4, and if, given $a, b \in S$, the equation $a \perp x = b$ has a solution, then this solution is unique.*

Let us assume that we have $a \perp x = b$ and $a \perp y = b$; by A4 we have either $y = x \perp t$ or $x = y \perp v$. Because of the symmetry present, we need consider only the first alternative. We have $a \perp x \perp t = b \perp t = b$. By A2, $t = z$; hence $y = x$.

Definition 8: If $a, b \in S$, and $a \perp x = b$ has a solution $x \in S$, then x is the difference between b and a, and we write $x = b \top a$.

From Theorems 3 and 4 we derive immediately

Theorem 5: *If a set S has the properties A1-A4, and $a, b \in S$, then there is only one difference connected with the pair (a, b) that is, only one of the following alternatives are possible*:

(i) $a \neq b$, $(b \top a) \in S$, $(a \top b) \notin S$,
(ii) $a \neq b$, $(a \top b) \in S$, $(b \top a) \notin S$,
(iii) $a = b$, $(a \top b) = (b \top a) = z$.

From Definition 6, it follows that if we have (i), then bGa, and if (ii), then aGb. If we now introduce

Definition 9: Relation $a\Gamma b$ means either $a = b$ or aGb, we can easily prove the following

Theorem 6: *A set having the properties A1-A4 is ordered by Γ.*

This result shows that A4 is a truly critical step in our argument. We can subsume "apples" and "pears" into "baskets of apples and pears." But subsumption alone does not guarantee that we can always perform the inverse operation, that of subtracting one element from another. No one can subtract a "basket of two apples and five pears" from a "basket of four apples and three pears," or vice versa. The reason is that while "apples" and "pears" are, each by itself, homogeneous substances, "baskets" are not so. *It is subtraction, not subsumption, which provides the acid test for homogeneity.* A4, therefore, can be properly called the *Homogeneity Axiom*.

Now that we know that a set fulfilling A1-A4 is ordered, the question arises of whether such a set lends itself to some sort of measure. As the following example shows, not every set which satisfies A1-A4 is ordinally measurable. (On the other hand, it is hardly necessary to point out that not every ordered set satisfies these same conditions.)

Example 6: Let S be the set of all ordered pairs (m, n), where m and n are real numbers such that either $m > 0$ or $m = 0, n \geqslant 0$. Let operation $\perp$ be defined by (3). It is immediate that S has the properties A1-A4 and, hence, is an ordered set. The order is lexicographic:

$$(m, n)\Gamma(m', n') \quad \text{if} \quad m > m',$$
$$(m, n)\Gamma(m, n') \quad \text{if} \quad n \geqslant n'. \qquad \ldots (5)$$

However, we can prove the following

Theorem 7: *There is no ordinal measure of S based upon relation Γ.*

Let us suppose that there is an ordinal measure of S and let us write $f(x, y)$ for the measure of (x, y). Considered as a function of y, $f(x, y)$ is an increasing function; moreover, for any $x < +\infty$, we have $-\infty < f(x, y) < +\infty$. Therefore, for $x > 0$ we can put

$$h(x) = \lim_{y \to -\infty} f(x, y), \quad H(x) = \lim_{y \to +\infty} f(x, y). \qquad \ldots (6)$$

Obviously,
$$G(x) = H(x) - h(x) > 0$$

for $x > 0$. Let $I(\alpha)$ denote the set of all values of x such that (1) $0 < m < x < M$ and (2) if $x \in I(\alpha)$ then $G(x) > \alpha > 0$. There exists a δ for which $I(\delta)$ is an infinite set. Indeed, let us presuppose that $I(\alpha)$ is finite for any $\alpha > 0$, and let us consider a positive sequence $\{\alpha_i\}$ such that $\alpha_i > \alpha_{i+1}$ and $\alpha_i \to 0$ for $i \to \infty$. Every value x of the open interval (m, M) would then belong to an $I(\alpha_i)$ as well as to all $I(\alpha_j), j > i$. In this case, all values of the open interval (m, M) can be arranged in a sequence $(x_1, x_2, \ldots x_{a_1}, \ldots x_{a_2}, \ldots)$. But this is absurd, for the set represented by (m, M) is not numerable. Hence, there exists an infinite set $I(\delta)$. This set contains a strictly monotonic sequence $J = \{m_i\}$. (We need consider here only the case where $m_i < m_{i+1}$.)

Let us now choose $\varepsilon > 0$ such that $\eta = \delta - 2\varepsilon > 0$. Because of (6), for every $m_i \in J$, there is an N_i and an n_i such that
$$f(m_i, N_i) > H(m_i) - \varepsilon, \quad f(m_i, n_i) < h(m_i) + \varepsilon.$$
This yields
$$f(m_i, N_i) - f(m_i, n_i) > G(m_i) - 2\varepsilon > \eta.$$
Hence, $N_i > n_i$, and by (5) and Definition 2,
$$f(m, n) < f(m_1, n_1) < f(m_1, N_1) < \ldots < f(m_k, n_k) < f(m_k, N_k) < \ldots < f(M, N).$$
Therefore,
$$f(M, N) - f(m, n) > \sum_{1}^{k} {}_i \, [f(m_i, N_i) - f(m_i, n_i)] > k\eta,$$
for any k. Since this inequality is obviously impossible for $k \to \infty$, the theorem is proved.

This result shows that we have not yet taken into account all properties connected with cardinal measure. The missing property is connected with the name of Archimedes who first pointed out its importance for the concept of measure.

A5: The set C possesses the Archimedean property, i.e., given $a, b \in C$ and $a \neq z$, there exists a positive integer k such that $(ka) \, Gb$.[5]

The property described by A5 corresponds to the transparent fact that if the water in a reservoir, for instance, is to be measured with the aid of a pail, we must be able to empty the reservoir by removing a finite number of pails of water. It is clear that the set of Example 6 does not possess the Archimedean property: if $m > 0$, then $(m, n) \, G[k(0, 1)]$ for any k. But that set is not ordinally measurable. On the other hand, we can prove that the Archimedean property does not necessarily guarantee that an ordered set should be ordinally measurable.

Example 7: Let S be the set of all ordered pairs (m, n), where $m > 0$ and n are real numbers, and let $\perp$ be defined by (3). The set is obviously ordered by the relation R:

$$(m, n) \, R(m', n') \quad \text{if} \quad m > m',$$
$$(m, n) \, R(m, n') \quad \text{if} \quad n \geq n'.$$

---

[5] By (ka) we denote the element $a \perp a \perp \ldots \perp a$ where $\perp$ is applied $k-1$ times.

Moreover, S possesses the Archimedean property for if (m, n), $(m', n') \epsilon S$, there obviously exists a k such that (km', kn') $R(m, n)$. Yet, S is not ordinally measurable (via Theorem 7). Let us observe, however, that in this case there is no strict relationship between R and $\perp$, because S does not satisfy A4, i.e., S is not homogeneous.

We can now reach the following important result:

Theorem 8: *A set, C, possessing the properties* A1-A5, *has an ordinal measure*, meas (x), *such that if $a, b \epsilon S$ then*

$$\text{meas } (a \perp b) = \text{meas } (a) + \text{meas } (b). \qquad \ldots \text{(7)}$$

The proof of this theorem follows very familiar lines. An arbitrary $u \epsilon C$, $u \neq z$ is chosen as unit, i.e., meas $(u) = 1$. Then, for any $v \epsilon C$, $v \neq u$, a Dedekind cut, r, of the non-negative rational numbers, m/n, is obtained according to whether (mu) $G(nv)$ or $(nv)\Gamma(mu)$. It is a simple matter to show that by putting meas $(v) = r$ we satisfy the theorem. Clearly, meas $(z) = 0$.

Definition 10: A set, C, satisfying A1-A5 is cardinally measurable and any measure satisfying (7) is a cardinal measure.

Theorem 8 calls for a few remarks. First, if a cardinally measurable set, S, contains an element, u, such that uGx, $x \epsilon S$, implies $x = z$, it follows that if $a \epsilon S$, then $a = (nu)$. Indeed, if $a \epsilon S$, and $a \neq (nu)$, then by A5, we have $[(n+1)u]$ $GaG(nu)$, and by A4, $a \top (nu) = u_1 \epsilon S$. It is immediate that uGu_1 and $u_1 \neq z$, which is contrary to our premise. On the other hand, it follows that if this premise is true, S has only one cardinal measure that does not need some small print, namely, that for which meas $(nu) = n$. There are, then, structures that provide a *raison d'être* for the concept of *natural unit*.

Secondly, if meas $(b) = k$ meas (a), we may find convenient to write $b = (ka)$, even if k is not an integer. We must, however, be careful so as not to push the parallelism too far. Indeed, while it is true that if $a \epsilon S$ then $(ka) \epsilon S$ for any non-negative integer, this may no longer be true if k is not an integer; (ka) may have no meaning at all. Only in a metaphorical way can one say that $(\frac{1}{2}$ man) is an element of the set "men". This remark leads us to introduce

Definition 11: A cardinally measurable set, C, such that if $a \epsilon C$, then $(ka) \epsilon C$ for any $k \geqslant 0$ is a perfectly divisible set. (Undoubtedly, this definition corresponds to the perfect divisibility one encounters in all discussions of production factors.)

Thirdly, from the fact that a cardinally measurable set, S has no natural unit it does not follow that S is perfectly divisible.

Example 8: Let S be defined as follows: there exists a $u \epsilon S$ such that if $v \epsilon S$, then $(mu) = (3^n v)$, where m and n are non-negative integers, and conversely. The set has a cardinal measure, meas $(v) = m/3^n$, and has no natural unit; but there is no $x \epsilon S$ such that $(2x) = u$, for instance. However, any cardinally measurable set has the following interesting property:

Theorem 9: *If $a, b \epsilon C$ and $b = (pa)$, where $p = m/n$ is an irreducible fraction, then $(qa) \epsilon C$ for any irreducible fraction $q = l/n$.*

The case $m > n$ can be reduced to that where $m < n$, by observing that if $m = kn+n'$, $n' < n$, then $b = (ka) \perp b'$, aGb'. Now, if $1 < m < n$, by A4 and Theorem 8, there exists $b' \epsilon C$, such that meas $(b') = m'/n$, $m' = n-m$. If $n-m = 1$, the theorem is proved. If $n-m > 1$, let $m_1 = \min [m, n-m]$, and let b_1 be the element having m_1/n as measure. By A5, there exists an integer k such that $[(k+1)b_1] GaG(kb_1)$. Clearly, we cannot have $(kb_1) = a$, for that would mean either $km = n$ or $k(n-m) = m$, and this is impossible because m and n are mutually prime. Therefore, $a \top (kb_1) = b_2 \neq z$. And since $b_1 G b_2$, it follows that meas $(b_2) = m_2/n$, with $0 < m_2 < m_1$. If $m_2 = 1$, the theorem is proved; if $m_2 > 1$, we continue the algorithm, obtaining $m_1 > m_2 > \ldots > m_k$, with meas $(b_k) = m_k/n$. The algorithm inevitably stops when $m_k = 1$, which proves the theorem.

Fourthly, an ordinally measurable set is not, necessarily, cardinally measurable. Cardinality requires a structure with respect to which subsumption *and* subtraction should have a definite operational meaning distinct from the arithmetic manipulation of the ordinal scale. This point has an important bearing upon the prevalent view (originated by Pareto) that comparability of utility differences leads to a meaningful utility scale which leaves only the origin and the scale unit arbitrary.[6] The fact is that this latter type of measurability involves a stronger condition than comparability of differences, namely, their cardinal measurability. Undoubtedly, either this latter condition has been smuggled, as it were, in the arguments concerning the measurability of utility, or the results of such arguments have been misinterpreted. The following will make this point clear.

Definition 12: A set, S, has a weak cardinal measure if the following conditions are fulfilled:

(i) There exists a cardinally measurable set, Σ, associated with S through an external operation, $\underline{0}$, such that if $a \epsilon S$ and $\alpha \epsilon \Sigma$, then $(a \underline{0} \alpha) \epsilon S$.

(ii) For any $a \epsilon S$ and $\alpha, \beta \epsilon \Sigma$, we have $(a \underline{0} \alpha) \underline{0} \beta = a \underline{0} (\alpha \perp \beta)$.

(iii) If $a, b \epsilon S$ there exists an $\alpha \epsilon \Sigma$ such that either $a \underline{0} \alpha = b$ or $b \underline{0} \alpha = a$.

(iv) If $a \underline{0} \alpha = a$, then $\alpha = \zeta$, where ζ is the neutral element of Σ.

It can be easily proved (1) that S is ordered by the relation R defined by bRa if $b = a \underline{0} \alpha$, and (2) that S possesses an ordinal measure, meas (x), $x \epsilon S$, such that, $a \epsilon S$ being arbitrarily chosen,

$$\text{meas } (a) = 0 \quad \text{and} \quad \text{meas } (b) = \pm \text{meas } (\alpha), \qquad \ldots \text{(8)}$$

where meas (y) is a cardinal measure of Σ and the sign is set according to whether $b = a \underline{0} \alpha$ or $a = b \underline{0} \alpha$. By writing $w = x \overline{0} y$ if $x = y \underline{0} w$, we obtain

$$\text{meas } (x \overline{0} y) = \text{meas } (x) - \text{meas } (y), \qquad \ldots \text{(9)}$$

which characterizes the weak cardinal measure.[7]

---

[6]See, for instance, O. Lange, "The Determinateness of the Utility Function," *Review of Economic Studies*, i(1933/4), pp. 218-225, as well as the comments of E. H. Phelps Brown, H. Bernardelli and Lange's reply in *ibid.*, ii(1934/5), pp. 66-77.

[7]Since the proofs of the above results present no difficulty, there is no need for spelling them out here.

It is vitally important to bear in mind that $\overset{o}{\_}$ is a distinct operation from $\underline{\ \ }$: for if we fail to do so, we are apt to miss the point that S and Σ do not pertain to the same "substance" and that the operation $\overset{o}{\_}$ preceeds in *logical and operational* order the "arithmetical" difference of scale readings which appears on the right hand side of (9). No other concrete example can illustrate the points just made as well as chronological time. The latter's weak cardinal scale is not derived by adding two elements of the same substance: the elements added are neither both "chronological dates" nor both "time-intervals," but one is a "chronological date" and the other a "time-interval". The same is true of the temperature scale—another favourite example of the discussions of measurability. To a given "temperature level" we add a certain *amount* of "energy" to obtain another "temperature level." Both illustrations plainly show that the "difference" element in a weak cardinal measure must be cardinally measurable: its ordinal measurability, so much the less comparability alone, does not suffice. One may even say in general that the reason why a particular ordinal scale is singled out as having a definite operational meaning is that the scale is related in some manner to a cardinal scale. This relation is the simplest in the case of weak cardinality.

II. Divisibility, proportionality and returns

3. *Individual product and individual factor*. The very fact that we have an operational, though primary, notion of a process as a transformation of some elements into others, means that we can distinguish in a very objective way between the beginning, the *input*, and the end, the *output*. These obvious remarks justify us in laying down the following "axioms" as a first description of the productive sector:

B1: There exists a set, Π, such that if $\pi \epsilon \Pi$, then π is a concrete output and conversely.

B2: There exists a set ϕ, such that if $\varphi \epsilon \phi$, then φ is a concrete input, and conversely.

B3: To every $\pi \epsilon \Pi$ there corresponds at least one $\varphi \epsilon \phi$, and to every $\varphi \epsilon \phi$ there corresponds at least one $\pi \epsilon \Pi$.

The last proposition merely expresses the fact that a concrete output is produced and a concrete input is productive. We should explain that "concrete" is not used in the sense of "actual". "Concrete output" simply means one instance of the concept of "output".

Definition 13: A pair (π, φ) of the correspondence B3 is a concrete process.

Propositions B1–B3 in no way restrict the size or the structure of the elements π and φ: a concrete output may very well be a multi-product in the usual sense of this term. On the other hand a certain *technological horizon* must be assumed for their operationality. "Brown-and-serve rolls" became a member of Π only after we discovered how to produce them, and no φ contained a "pilot" before the invention of the airplane.

B4 : Given two concrete processes, (π', φ'), (π'', φ''), there exists a concrete process (π, φ) such that $\pi = \pi' \perp \pi''$, $\varphi = \varphi' \perp \varphi''$, where $\perp$ is a subsumptive operation.[8] This proposition states that any two concrete processes can conceivably be carried out independently of each other and at the same time. The famous objection of Pareto that "there are not two Paris cities" for Paris to be used simultaneously in two distinct processes is of no avail. Material scarcity affects only our choice, not the logical relationships of our concepts: the fact that we cannot actually draw a circle larger than the equator does not limit the properties of "circle" to that size. The following three properties hardly need any commentary:

B5 : The sets Π and ϕ have a common element, z, such that if (π, z) is a concrete process, then $\pi = z$; and if (z, φ) is a concrete process, then $\varphi = z$.

B6 : The element z is neutral for Π and for ϕ.[9]

B7 : The sets Π and ϕ satisfy A2 and A3.[10]

Obviously, it would be absurd to attribute to either Π or ϕ the homogeneity property, A4. On the other hand, we must recognize that Π (and ϕ) can be divided into cardinally measurable subsets, Π_i (and ϕ_i) having no common element other than z. The idea is that "this" particular tube of tooth paste—call it D—is a concrete instance of the cardinally measurable set $\Pi_i = \{n\ D\}$, where n is a non-negative integer. The same is true of the concrete input which includes the particular management unit of Y Inc., irrespective of whether the latter consists of a single person or a board with a sizable staff. In this way quality is clearly separated from quantity: the qualitative difference between concrete outputs or between concrete inputs is shown by the subscript of the subsets Π_i and ϕ_i to which they belong. And since qualities may be more numerous than numbers themselves, the subscripts are not necessarily numerical.

These remarks lead to

B8 : If $\pi\epsilon\Pi$ (or $\varphi\epsilon\phi$), there exists a cardinally measurable subset $\Pi' \subset \Pi$ (or $\phi' \subset \phi$) such that $\pi\epsilon\Pi'$ (or $\varphi\epsilon\phi'$).

B9 : If $\pi\epsilon\Pi$ and $\pi \neq z$, there exists a cardinally measurable subset $\Pi_k \subset \Pi$ such that if $\pi\epsilon\Pi' \subset \Pi$ and Π' is cardinally measurable, then $\Pi' \subseteq \Pi_k$ (and similarly for ϕ).

We may refer to any Π_k or ϕ_k as an *individual output* or *individual input*. Finally, we should add the transparent property

B10 : If $\pi, \pi'\epsilon\Pi$, then $(k\pi) \perp (k\ \pi') = k(\pi \perp \pi')$, and similarly for ϕ.[11]

On the basis of the above propositions we can arrive at a simple (and very general) concept of production function. Let ψ_i be the set of all concrete processes (π, φ) such that π belongs to a given individual output, Π_i. By writing $q = \text{meas}\ (\pi)$, we can describe ψ_i by a *numerical* function

$$q = q_i(\varphi). \qquad \ldots (10)$$

---

[8] Of course, this does not imply that there are no processes, (π_1, φ), (π, φ_1), where $\pi_1 \neq \pi$, $\varphi_1 \neq \varphi$.
[9] The last two propositions exclude creation and annihilation.
[10] They satisfy A1 by virtue of B4-B6.
[11] For k rational this property follows from B4, B9 and Theorem 9.

Almost needless to say, $q_i(\varphi)$ is not a single-valued function : the same φ may be used according to various "techniques" and, above all, distributed differently among the units of production.

B11 :   For a given φ, $q_i(\varphi)$ has a maximum, $Q_i(\varphi)$.

This proposition does not follow from B1-B10[12]; it is introduced only for avoiding irrelevant technicalities. A single-valued production function

$$q = Q_i(\varphi) \qquad \ldots \ (11)$$

is thus obtained. But even this form is not very helpful, either for our own problem or for most others. The individual outputs and individual inputs do not correspond to products and factors as these terms are understood both in theory and everyday business. The present Π, for instance, does not contain an individual output representing "mutton" alone. And although every concrete input includes a management unit, management by itself can produce nothing; hence, no individual input consists solely of management.

In some cases, the distinction of an individual product or factor can be based upon the fact that a *quantitative* difference exists between two *qualitatively* different outputs or inputs. If $\pi'\epsilon\Pi_i$, $\pi''\epsilon\Pi_j$, $i \neq j$, then $\pi' \perp \pi'' = \pi\epsilon\Pi$, i.e., π'' is the quantitative difference between π and π'. The relevant case, however, is that where the difference does not belong to Π : if $\pi = $ "1 lb wool and 3 lb mutton" and $\pi' = $ "1 lb wool and 2 lb mutton", the difference $\xi = $ "1 lb mutton" is not a concrete output. Unfortunately, nothing warrants that by taking all such *external* differences we shall ultimately derive every individual factor and individual product. Complementarity may prevent us from arriving at an external difference consisting of management alone. It is thus seen that it is not possible to derive the concept of individual factor (or product) from that of measurable inputs (or outputs). Qualitative distinctions cannot be derived from measure, for—as we have said in the introduction—quality is prior to quantity. Therefore, the concepts of individual factor and individual product must be introduced as primary concepts in any analysis of production. As we shall presently see, this point has not been always realized.

B12 :   There exist two classes of cardinally measurable sets, $\{P_i\}$, $i\epsilon T$, and $\{F_j\}$, $j\epsilon H$, such that if $\pi\epsilon\Pi$ or if $\varphi\epsilon\phi$, then

$$\pi = p_{i_1} \perp p_{i_2} \perp \cdots \perp p_{i_n}, \quad \varphi = f_{j_1} \perp f_{j_2} \perp \cdots \perp f_{j_m}, \qquad \ldots \ (12)$$

where $p_{i_k}\epsilon P_{i_k}$ and $f_{j_k}\epsilon F_{j_k}$. P_i and F_j are called individual product and individual factor respectively.

A concrete output (input) can then be represented as a vector in a space where each coordinate corresponds to an individual product (factor) and conversely. The

---

[12] As the instructive example in the author's paper "Leontief's System in the Light of Recent Results," *Review of Economics and Statistics*, 1950, p. 214, shows.

fact that such a space may conceivably have an infinity of coordinates raises no difficulty. In a concise form, we shall write $V(f_j)$ for the vector representing a given concrete input, where f_j, $j \epsilon H$, represent the coordinates of the individual-factor space. The production functions (10) and (11) become

$$q = q_i(V) \quad \text{and} \quad q = Q_i(V), \qquad \ldots (13)$$

respectively. They are relations between individual *factors* and a given individual *output* : therefore they are of no use in analyzing joint supply. But for our own purpose they suffice.

4. *The tautological thesis*. By this term we shall understand the thesis that "constant returns to scale is...purely a matter of definition,"[13] or in other words, that it is always possible to define the factors of production in such a way as to arrive at a production function represented by a homogeneous function of the first degree.[14] Ordinarily the argument goes no further than this assertion. Hahn, however, stands out in this respect for having tried to show how constant returns follow *tautologically* from a particular definition of factors. Consequently, his argument deserves to be analyzed in detail.

Following an old hint,[15] Hahn proposes to "classify all those units between which the marginal rate of substitution is unity into the [same] category F_i." Using our notations we can state this rule as follows: if

$$Q_i(a+V) = Q_i(b+V), \qquad \ldots (14)$$

where V represents the factors kept constant, then $a, b \epsilon F_i$. (The only way in which I can interpret Hahn's "arbitrary units 'a'" is that they are some concrete factors in our sense.) Hahn goes on to say that (14) must be satisfied "independent of the size of the input, where size is measured as a fraction or multiple of the standard unit 'a'." He fails to say anything about measure, but the last statement clearly implies that all concrete factors are cardinally measurable *before* we come to apply his equivalence rule. In conclusion, we can restate Hahn's rule as follows : if $a \epsilon F_i$, $b \epsilon F_j$, and

$$Q_i[(ka)+V] = Q_i[(kb)+V] \qquad \ldots (15)$$

for any non-negative k, then *as far as economic conduct is concerned*, $F_i = F_j$.[16] That is all, according to Hahn. However, this rule obviously fails to describe the real economic conduct adequately. To begin with, (15) may be satisfied only for a particular V. If this is the case, no producer could possibly behave as if $F_i = F_j$. We must add that (15) should be satisfied not only for any k, but also for any V; this is

---

[13] Hahn, *op. cit.*, p.137.

[14] This argument is endorsed also by McLeod, *op. cit.*, p. 130.

[15] See, for instance, N. Kaldor, "Mrs. Robinson's 'Economics of Imperfect Competition,'" *Economica*, 1934, p. 359.

[16] Clearly, this rule implies that F_i and F_j are perfectly divisible. But this restriction is not necessary. We need, however, the condition that if (ka) exists, then (kb) exists and vice versa, in other words, that F_i and F_j have the same cardinal scale.

tantamount to saying "for any size of the output." But even this amendment is insufficient: (15) may be satisfied only for some but not all products. Thus, one automobile may very well be substitutable for ten dog sleighs in Lapland: most definitely, this does not apply to Egypt. We must then add also that (15) should be satisfied for all outputs Π_i. It is clear, however, that even in this amended form the rule does not constitute a definition of individual factor. It merely imposes the condition that any distinction between the F_i's that would not matter in production should be eliminated. If, say, a certain type of worker can *always* be substituted by a certain machine without a change in output, the worker and the machine should not be listed as separate factors. But before the rule can be applied a list of F_i's, all cardinally measurable, must exist (B12).

The fact that Hahn believes that his definition tautologically leads to constant returns is due to a logical error in his proof. In passing from the factor combination (a, b) to (ka, kb) he made an incorrect use of the concept of rate of substitution by ignoring that this concept requires output to remain constant.[17] In fact, Hahn prefaces his proof with the remark that his definition does not preclude variable returns "to successive amounts ka of the category F_i." Strangely enough, it did not strike him that $(ka, 0)$ represents such successive amounts *and* proportional variations of factors as well.

Other arguments for the tautological thesis seem to run on a different line: any production function can be represented by a homogeneous function of the first degree if the ordinal scales of factors are appropriately chosen. Of course, the inverse transformation of (T), Example 1, changes the production function $q=e^{1/2}(u^2+v^2)^{-3}$ into $q = \sqrt{xy}$. However, this type of argument contains a rather unsuspected flaw, namely, the supposition that it is *always* possible to choose the scales as claimed. The following simple example exposes the fallacy.

Example 9 : Let $q = uv+u+v$ be the expression of a certain production function when the factors are measured on some arbitrarily chosen scales, u, v. The problem is to find the monotonic transformations $u = g(x)$, $v = h(y)$, such that $g(0) = 0$, $h(0) = 0$, and

$$q = \lambda(gh+g+h) = g(\lambda x)h(\lambda y)+g(\lambda x)+h(\lambda y), \quad \ldots \quad (16)$$

for all non-negative values of λ, u, v. Setting $y = 0$, we obtain the functional equation

$$\lambda g(x) = g(\lambda x), \quad \ldots \quad (17)$$

which has the well-known solution $g(x) = \alpha x$, α being an undetermined parameter. Similarly, $h(y) = \beta y$. If these solutions are introduced in (16), the latter yields $\alpha\beta = 0$. This shows that there are no ordinal scales for the factors such that the production function in question should be represented by a homogeneous function of the first degree.

In this connection we may recall Joan Robinson's tempting idea of transforming the natural units of all factors into efficiency units.[18] Obviously, this amounts to

---

[17] Hahn, *op. cit.*, p. 132. See also *infra*, fn 22.
[18] *Economics of Imperfect Competition*, p. 109 *passim*.

replacing cardinal measures by some ordinal scales chosen so as to render the production function homogeneous. The above example shows why such a proposal could but be ill-fated. This she hurried to admit at the first opportunity,[19] but it is not certain whether she realized the real drawback of her proposal: that it assumes *all* productions to be susceptible of being represented by a homogeneous function of the first degree.

To exclude the case of production functions not reducible to a homogeneous function is not a simple matter of definition, but one concerning the structure of natural laws. For after all, any production function is subject to natural laws. By a special choice of definitions and, especially, scales we may—as indeed we often are—be able to simplify the formulae by which those laws are expressed, but it is baseless to attribute to these laws a structure of our own choice without any regard for factual considerations. This, in essence, is the position from which Chamberlin's argument proceeds. His argument may be further strengthened by the following observations: The relevant shape-properties of the production function—such as constant returns, decreasing marginal rate of substitution, etc.—are characterized by the fact that they remain invariant under one single type of scale transformation and only under that type: $x = \mu u$, $y = \nu v$. This characteristic is a consequence of the fact that in the actual world products and factors are cardinally measurable (for which reason we may refer to the mentioned properties as *cardinal* properties). Now, it stands to reason that, once the factors are listed, there is no longer any room for manipulating their definitions for the purpose of arriving at a pre-selected shape of the production function.

5. *Divisibility and homogeneity.* The striking feature of all arguments advanced by the advocates of the tautological thesis is their insistence upon the perfect divisibility of factors. All sorts of fables—such as Hahn's world of ant-men and ant-machines—are designed in order to explain how from this assumption it follows tautologically that the optimum scale of production is indeterminate. But, as Chamberlin first pointed out, perfect divisibility of factors is rather a smoke screen which obscures the issue: the optimum scale may be indeterminate even though not all factors are perfectly divisible. All we need is that all scallops in Chamberlin's diagram should be tangent to the same horizontal straight line.[20] In fact, we may go even further and, for the sake of the logical probing, assume that all factors have a natural unit. In this case, the production function $q = f(x, y)$ exists only for integral values of x and y. But that does not preclude $f(\lambda x, \lambda y) = \lambda f(x, y)$ for all integral values of λ, x, y. Neither indeterminateness nor homogeneity (of a function) are concepts restricted to the case of continuous variables. Why then the insistence upon the perfect divisibility of factors?

The answer to this question is that the usual type of argument against the existence of an optimum scale confuses divisibility of *factors* with divisibility of *processes*. To wit, Hahn writes: "since perfect divisibility [of factors] is assumed, it is solely a

---

[19] Joan Robinson, "Euler's Theorem and the Problem of Distribution," *Economic Journal*, 1934, p. 402n.

[20] Chamberlin, *op. cit.*, p. 243.

matter of *subdividing any single productive process* into a large number of stages [i.e., identical processes save for size], and this by definition is possible."[21] This clearly shows that Hahn's assumption of perfect divisibility applies not only to factors but also to processes.

This tacit connection between a property of factors and one of processes implies a peculiar view of natural phenomena, to which the respective authors may or may not subscribe consciously. The view boils down to this: In nature all transformations consist of mere reshuffling of some primary substances; such reshuffling can bring about no qualities other than those inherent in the primary substances. *All processes then are indifferent to size, just as the process of measure is.* Indeed, as we have repeatedly emphasized, measuring involves some operations, i.e., it represents a process. And this process stands out as one that does not alter the quality of the element involved— its input, as it were. This is what we usually mean by "measure is indifferent to quality." The rationale of the view mentioned above seems to be that because some elements are cardinally measurable— that is, they can go through a certain process without any qualitative residual—the same must be true of all processes in which they are inputs.

Now, from the assumption that all processes can be divided or compounded without any qualitative residual it follows that all natural laws, not only economic processes, are expressed by homogeneous functions of the first degree. This *is* a tautology. It is then this assumption that is crucial for the tautological thesis. All other points upon which past controversies have spent much effort, are irrelevant. This is especially true for the assumption of "perfect divisibility of factors." Of course, the assumption is void of any meaning if factors are not cardinally measurable.

6. *Natural laws and the qualitative residual*. Examples of processes that are not indifferent to size are so abundant in natural sciences that one can only wonder how their existence may ever be ignored by other disciplines. At the microscale, organic chemistry offers innumerable examples where *new* qualities emerge after polymerization, i.e., after a certain scale is reached. For a topical illustration one may also mention the critical mass of atomic explosion. At the macroscale, in the theory of structures it is almost impossible to find a linear relation between *homogeneous and perfectly divisible* materials—iron, cement, insulation, etc—and variables expressing measures of some quality—resistance to strain, elasticity, radiation, and so on. The qualitative residual—of which we have spoken earlier —is reflected in the non-linearity of laws such as these, which relate to two distinct categories of variables: one essentially quantitative, the other pertaining to quantified qualities.

No doubt, the heat radiation from two identical spheres is twice as great as that from one such sphere. And so is the quantity of material from which they are constructed. That much everyone knows about $1+1 = 2$. But the issue of the optimum scale is only partly related with this *external* addition. The main problem concerns what happens when the *same* process is expanded or constructed. For simple but convincing illustrations we need not search the intricate world of life phenomena. To construct a sphere of radius, R, and wall thickness, d, it takes a quantity of materials

---

[21] *Op. cit.*, p. 134. Italics added.

proportional to $3R^2d - 3Rd^2 + d^3$; the heat radiated through such a sphere is proportional to R^2, while the heat that could be stored inside it is proportional to $(R-d)^3$. There is no linear relation between these expressions, and yet they all represent homogeneous variables. This example indirectly shows that the larger the furnace the less coal (proportionally) we need to keep it at a constant temperature. However, as the size of the sphere grows larger, the same wall thickness will no longer resist internal strain, and the sphere will collapse under its own weight. To prevent this we have to make the walls thicker and thicker; ultimately, any further increase in size becomes definitely uneconomical.

The illustration recalls Herbert Spencer's splendid analysis of the optimum size of a bird. As he explained, volume grows faster than the area enclosing it, and the latter, faster than its average diameter. To store energy efficiently the bird's size must be large. But a large bird is heavy and beyond a certain weight the wing bones would have to be so long that they will break under the strain of lifting the body into the air. All individual processes whether in biology or technology follow exactly the same pattern : beyond a certain scale some collapse, others explode, or melt, or freeze. In a word, they cease to work at all. Below another scale, they do not even exit.

The advocates of the indeterminateness of optimum scale, may, if they so wish, ignore all this evidence of the existence of qualitative residual in all laws. But then they must be prepared to accept another tautological conclusion, which seems to have thus far escaped notice. For if there is no qualitative residual, *there is nothing to prevent the process by which one factor is substituted for another from being indifferent to size*. Consequently, all isoquants should be linear and all production functions should be linear homogeneous expressions of factor quantities. The formula $q = \Sigma A_i f_i$ would thus represent the universal law of production.[22]

7. *Individual processes*. The fact that the authors supporting the tautological thesis mean "perfect divisibility of processes" when they say "perfect divisibility of factors", explains why they insist that every term should be defined while they fail to define "individual" process. For naturally, if processes are perfectly divisible, then *process loses all individuality* and there is nothing to define. As there is no natural unit connected with a perfectly divisible substance, there is no individual process involved in the concept of concrete process. But as we have explained in the preceding section, the issue concerning the optimum scale centers upon what happens within the *same* process. And we can speak of the same factor or the same process only in reference to an individual factor or individual process.

The difficulties connected with the concept of individual process are well-known. In spite of this, no theoretical description of reality can dispense with the primary notion of individual process : all natural laws describe in the first place an individual process. The *isolated* system of physics is nothing but such a process.

---

[22] Some points in Hahn's proof discussed earlier seem to suggest that he might have unwittingly reasoned from this position. For if the production surface is a plane the concept of rate of substitution can easily be extended to movements in any direction.

In biology and in social sciences the concept may raise some worthy issues because at times it has been applied to such troublesome categories as species, social classes or political parties. In production theory, however, we see no reason why "an ice-cream plant" should be a less legitimate notion than "tree" or "protein molecule". There is only one point that must be clearly understood. What we regard as "one" or as "many" in each situation depends upon the problem at hand. In production theory we are led to distinguish two kinds of individual processes. The first is the process for which the term plant has been imported from biology: it is an *individual physical process* which may be assimilated to a spontaneous natural process for it is governed almost exclusively by natural laws. The second kind of individual process reflects *the unit of organization* that supervises and controls production. In addition to the natural laws governing the physical processes included in it, this latter process is subject to the laws concerning the human ability to control natural phenomena for a definite purpose.

In logical order it is the plant process that comes first; it also is the more critical of the two. A most convenient and transparent way of describing it is $q = F(W)$, the relation between output and the "material" input. This relation may vary from one process to another, even if the individual output remains the same, but broadly speaking it is in each case determined by the natural laws involved.

The following propositions express some very obvious features of the situations to which we ordinarily apply the concepts just described.

C1 : There exists a class of individual factors, $\{E_i\}$, $i \epsilon M$, $M \subset H$, such that every E_i has a natural unit, m_i, and for every concrete process (π, V), the vector $V(f_j)$ has at least one coordinate $f_j \neq 0$, $j \epsilon M$.

A unit, m_i, obviously represents a management unit. Such a unit may consist of a single person or of a board together with its general staff. There is no restriction in this respect.

C2 : Given any concrete process, (π, V), there exist concrete processes $\alpha(\pi_i, V_i)$, with $\pi, \pi_i \epsilon \Pi_k$, and

$$q = q_1 + q_2 + \ldots + q_n, \quad V = V_1 + V_2 + \ldots + V_n, \qquad \ldots \text{(18)}$$

and such that for every $V_i(f_j)$ the following conditions hold

$$f_a = m_a, \; a \epsilon M \quad \text{and} \quad f_j = 0 \quad \text{for } j \neq a, \; j \epsilon M. \qquad \ldots \text{(19)}$$

It is obvious that a process $\alpha(\pi, V)$ satisfying (19) represents a process under one management. For such a process we may put $V = (W, m_a)$, where W is a vector in the space of non-management factors, and write $\alpha(\pi, W; m_a)$ instead of $\alpha(\pi, V)$.

The description of a process by $\alpha(\pi, W; m_a)$ is however incomplete, for it includes no information regarding the number of individual physical processes covered by the firm. We must add the information that

$$\pi = \pi_1 \perp \pi_2 \perp \ldots \perp \pi_u, \quad W = W_1 + W_2 + \ldots + W_u, \qquad \ldots \text{(20)}$$

where $\pi, \pi_i \epsilon \Pi_k$ and each pair (q_i, W_i) represents a concrete instance of an individual physical process. In case $u = 1$, we shall simply write $\omega(\pi, W; m_a)$ instead of $\alpha(\pi, W; m_a)$ and (20).

Definition 14 : Given Π_k, m_a, and u, the set $I_k^u(m_a)$ of all processes $\alpha(\pi, W; m_a)$ satisfying (20) is a u-plant firm. (If $u = 1$, we shall omit the upper-script.)

C3 : The set $I_k^u(m_a)$ is never empty.

C4 : If $\alpha(\pi, W; m_a) \epsilon I_k^u(m_a)$, then for the same distribution (20) and for any $b \epsilon M$, there exists a π' such that $\alpha(\pi', W; m_b) \epsilon I_k^u(m_b)$.

The meaning of these two propositions is obvious : C3 states that any product can be produced by a u-plant firm whatever u and m_a may be; C4 states that any management unit can manage any process.

C5 : If $\omega(\pi_1, W_1; m_a), \omega(\pi_2, W_2; m_a) \epsilon I_k(m_a)$ there exists a π such that $\omega(\pi, W_1 + W_2; m_a) \epsilon I_k(m_a)$.

We shall write

$$\omega(\pi_1, W_1; m_a) \oplus \omega(\pi_2, W_2; m_a) = \omega(\pi, W_1 + W_2; m_a)$$

and refer to $\oplus$ as *internal* addition.

Definition 15 : The process $\omega^+(\pi^+, W; m_a) \epsilon I_k(m_a)$ is said to be efficient if $q^+ \geqslant q$ for any $\omega(\pi, W; m_a) \epsilon I_k(m_a)$. The process $\omega^*(\pi^*, W; m_a) \epsilon I_k(m_a)$ is completely efficient if $q^* \geqslant q$ for any $\omega(\pi, W; m_b) \epsilon I_k(m_b)$, $b \epsilon M$.

From B11 it follows that every efficient process is contained in $I_k(m_a)$. The only reason for a process $\omega(\pi, W; m_a)$ not being efficient is that the respective management unit is *intentionally* either careless in choosing the best techniques according to its knowledge or negligent in running the process thus chosen. Since elimination of this whimsical aspect of the human nature does not affect the conclusions of our argument, we may assume that all processes are efficient. The difference in efficiency between two processes then is due only to the qualitative difference of their management units, which covers knowledge of techniques as well as other management abilities.

Let us observe that if all processes are efficient the set $I_k(m_a)$ can be described by a singled-value function

$$q = U(W; m_a). \qquad \ldots \text{(21)}$$

This is not identical to the usual concept of plant production function. In the first place, in $U(W; m_a)$ the unit of management is clearly set out as a qualitative unit. Secondly, in (21) the output is cardinally measurable; the formula, therefore, represents only a "slice" in the usual production function of a plant producing a multi-product. We should also add that since we shall have frequent occasions to consider only variations of W such that $W = wW_0$, W_0 being given, it is convenient to refer to w as the scale of the process.

The question of whether a completely efficient process always exists as a concrete process is more involved and will be taken up in the next section.

8. *Proportionality, returns and efficiency.* One type of argument against the existence of an optimum scale makes use of the proposition that "if every factor is homogeneous, then doubling all factors doubles the output." We shall refer to such an argument as "the proportionality thesis." It goes without saying that there can be no quarrel about the validity of the above proposition if applied to an *external* addition, as that of B4 or C2. Controversy arises only if the doubled output is obtained by an individual producer, i.e., by a firm. In examining the possible interpretations of the proposition in question it is necessary to treat the one-plant firm separately from the multi-plant firm, for the two situations do not lead to the same conclusions. But, first, we must consider the various conceivable views on the nature and the role of the management factor. These are:

(α) Management is a fictitious factor of production introduced in economic analysis by a faulty conception of the productive process. In a static process, as in any other such process, every factor has one main role, that of keeping the process going. By definition, each factor must perform its specific role if it is to be a factor at all. Consequently, a productive process does not need a special factor to control and coordinate it. If we came to talk about a management unit it is only because of the need to identify processes; a management unit then is only a name. In this case, only W is a true variable in (21), and the latter becomes

$$q^* = U(W). \qquad \ldots (22)$$

The economic process of a plant is then reduced to a purely physical process governed by technological (i.e., natural) laws.

(β) Management, even though a proper factor with a distinct role, cannot influence the relation between input and output except in a negative way by being intentionally remiss—like any other human factor. If this eventuality is excluded— as we have done— then the relation between input and output is determined exclusively by technological considerations. This leads again to (22), with the difference that m_a must now count as a cost element. We ought to add that the view just described is most unrealistic, for in a sense it attributes the quality of *perfection* to all efficient management units. We have nevertheless listed it for the sake of completeness.

(γ) Output depends upon the quality of the management unit, in the sense that each management unit can approach with various degrees of approximation the *ideal* output given by a formula such as (22). This ideal production function is determined only by technological laws and represents the individual physical process upon which the corresponding plant production rests.

In other words, q^* represents the upperbound of $U(W; m_a)$ for $a \in M$. If this upperbound is reached by $U(W; m_e)$ then m_e is completely efficient for the scale W (Definition 15). However, in order to take account, first of difference between ideal laws and actual processes, and second, of the infinite variability of human quality, it is advisable to assume that no management unit is completely efficient and that there exist infinite many management units, all different in some way from each other. In particular, given W and ε there exist infinite many management units, m_a, such that $q^* - \varepsilon < U(W; m_a) < q^*$.

We can now pass to the analysis of the two cases mentioned earlier, (1) one-plant firm and (2) multi-plant firm, for each alternative α, β, γ. (1α) In this case "doubling all factors doubles the output" can mean only that $nU(W) = U(nW)$. This leads to

$$q^* = \lambda W, \qquad \ldots (23)$$

where λ is some constant. Nothing need be added here to what has already been said in Section 6 about the non-linearity of individual physical processes in order to see that (1α) is of no avail for the proportionality thesis. (1β) This alternative, too, leads to (23) and hence is subject to the same objections as (1α).

(1γ) In this case, "doubling all factors doubles the output" may mean that

$$nU(W; m_a) = U(nW; m_b), \qquad \ldots (24)$$

where n has an appropriate value and m_b is "the n-fold of m_a."[23]

The question that arises immediately is that of the operational definition of "n-fold." There seems to be no other way to go about it than the *implicit* definition:

Definition 16: If (24) is satisfied, m_b is the n-fold of m_a and we write $m_b = n \otimes m_a$

As explained in the introductory section such a definition is useless by itself. We need an existence proposition

Z1: Given a concrete process $\omega(\pi, W; m_a)$ and any appropriate n, there exists $b \in M$ such that $\omega(n\pi, nW; m_b)$ is a concrete process.

In plain words this states that any number of one-plant firms can be replaced by a single one-plant firm without any loss of efficiency as far as the relation between the non-management input and the output is concerned. The proposition, which is

---

[23] There is absolutely no point in considering the case of a factor which would have a cardinal scale like that of Example 8: factors either are perfectly divisible or have a natural unit. The same is then true of the cardinally measurable set $\{nW\}$. Accordingly, in what follows we shall assume that $U(nW; m_b)$ exists either for all values of n, or for only such values that $nW = pW_0$, where p is an integer and W_0, the natural unit of $\{nW\}$.

far from transparent, is not of purely academic interest : it concerns the controversy over the merits of large scale production (not to be confused with large corporation.)[24]

We can easily prove that if Z1 is true, then (23) is true. Let $U(W_0)/W_0 = \lambda$ and $W = nW_0$, W being chosen arbitrarily. Let $\varepsilon > 0$ and $\varepsilon/n > \eta > 0$. Let $a\epsilon M$ be such that $U(W_0; m_a) > U(W_0) - \eta$. By Z1, there exists $b\epsilon M$ such that $nU(W_0; m_a) = U(W; m_b)$. Hence $U(W; m_b) > \lambda W - \varepsilon$. On the other hand, it is immediate that $U(W; m_b)$ is not greater than λW. Hence, (23) is proved. Therefore, (1γ) is to be rejected on the same ground as (1α) and (1β).

Before passing to the case of a multi-plant firm, let us point out a very important conclusion of the preceding argument. What refutes the proportionality thesis in the case of a one-plant firm is the fact that no individual physical process is indifferent to size. The point that something similar may hold for management organizations, if added, would make the argument all the stronger; but it is not needed. This is no longer true for a multi-plant firm. In a multi-plant firm the individual physical processes are added *externally* : if to a given W there corresponds the output q^* in a single individual process, an output of nq^* can be obtained from nW in n individual processes. The ideal output of a multi-plant firm can then be proportional to W (at least for multiples $W = nW_0$ of the optimum scale, W_0, of the individual process).

In the analysis of the multi-plant firm we shall distinguish the same alternatives (α)–(γ).

(2α) Since management is a pure fiction, a multi-plant firm, too, is just an identification label corresponding to our own subjective desire of referring in some way to n individual processes. It makes no difference whether some processes are thus grouped under one label or under several labels. The multi-plant firm becomes synonymous to "industry." Alternative (α) constitutes, therefore, a triumph for the proportionality thesis, even though in a definitely vacuous manner.

(2β) As any management unit is assumed *perfect*, any m_a can run any scale with *complete efficiency*. Therefore, if $\omega(\pi, W; m_a)$ is a concrete process, then $\omega(n\pi, nW; m_a)$, too, is a concrete process. By observing that from the last proposition, $m_a = n \otimes m_a$ (by Definition 16), we reach the conclusion that "doubling all factors doubles the output" is vindicated. However, let us point out that the optimum scale of the firm is, in this case, infinite, not indeterminate. Naturally, perfect management units—if they actually existed—should be exploited to the fullest extent.

(2γ) In this alternative, the proportionality thesis can be formulated as follows:

Z2. *If $\omega(\pi, W; m_a)$ is a concrete process, then given any $n > 1$ there exists $b\epsilon M$ such that $\alpha(n\pi, nW; m_b)$ is a concrete process of an n-plant firm. By implicit definition, $m_b = n \otimes m_a$.*

---

[24] We should point out a few further issues related to Definition 16. As it stands this definition allows for the troublesome result that $m_a = n \otimes m_b$ and $m_b = n \otimes m_a$ at the same time, even if the individual out-put and the ratios between non-management factors do not change. To avoid this, we could make the definition stronger by requiring that (24) should be fulfilled for all $W = mW_0$. But this is not needed for our immediate purpose.

Nothing we have said thus far militates against this proposition. Clearly, it allows for the existence of optimum scale in the case of one-plant firm. Nor can it be rejected on the ground that in an individual process decreasing returns must sooner of later set in if one factor is invariable and all others grow proportionately. This proposition (the so-called law of variable proportions) may be invoked to prove that a one-plant firm has necessarily an optimum scale if the management unit does not change. But in Z2, the management unit is free to change with the scale of operations. This allows for the possibility that in Chamberlin's diagram one scallop should follow another to infinity, all tangent to the same horizontal line and *each representing an individual producer operating a different number of plants*.[25]

On the other hand, there seems to be no common-sense objection against the idea that the merger of any number of firms producing the same individual output should not affect the ratio between output and non-management factors, *as long as no restriction is imposed upon the size or the quality of the new management unit*.

This does not mean that Z2 vindicates the proportionality thesis. At most it enables us to state a perfect tautology or write down an elegant formula. But it does not solve the problem of optimum scale. This is a problem of efficiency, and—as we have repeatedly emphasized —efficiency measure requires cardinal measurability. For this, $m_b = n \otimes m_a$ does not suffice: we must know how many m_a's make an m_b *in the cardinal sense*. And this is an insoluble problem in *real terms*. Indeed, not all physical elements of an m_a—presidents and commissars, consultants and advisors, generals and marshals, etc, etc—are included in every other m_b. In real terms, management units differ qualitatively, and that is the end of it. For a meaningful comparison of all management units we have no other choice than to turn to prices, provided however that in the respective economic system the latter reflect adequately the relative cost to the community. For this reason Chamberlin's idea of analyzing the problem of optimum scale with the aid of a diagram of a family of average cost curves[26] seems for more promising than using the production function and its isoquants— however more respectable the latter approach may be.

Conclusions

An analysis such as ours, using formal logic blended with elementary common sense, could not claim to settle the intricate issue of the optimum size of an economic organization, which has for long been a main target in the controversies over the merits of large corporations and of large governments as well. We can only claim to have isolated the numerous related issues.

In the first place, we have shown that the ordinary concept of efficiency (as well as other equally important concepts of production theory) has no meaning if factors and products are not cardinally measurable. The issue of cardinal measurability, therefore, is far from idle. Cardinal measurability is not merely one convenient

---

[25] This point is extremely important in view of the fallacy—heard now and then—that the existence of the optimum size for a firm follows from the law of variable proportions.

[26] Chamberlin, *op. cit.*, p. 233.

ordinal scale, but a real, specific property of numerous material variables. This property can be completely described by a series of operational propositions which are not common to all types of measurables.

Secondly, we have shown that the optimum size of an economic organization is not a technical problem like that of the optimum size of a plant. The optimum size of organization involves comparisons of quality; efficiency, therefore, belongs to the qualitative residual of the analysis of a productive process exclusively by the production function.

As a mere speculation, one may however regard efficiency as a quality of "thinking" matter. In this case, one may invoke the fact that relations between quantified qualities and true quantities are never linear and conclude that any organization has an optimum size, just as any plant has one. Clearly, the optimum size in both cases must change with the progress of our knowledge of "inert" and "thinking" matter.

CHAPTER 12

(1952)

Toward Partial Redirection of Econometrics III

The title of Mr. Orcutt's paper can hardly pass unnoticed. But whether it will arouse enthusiasm — this could be almost unlimited were it not for the qualifying term "partial" — or whether it will be met with deep skepticism is rather difficult to say. The writer's best guess is that the attentive reader, even if his anticipations were full of optimism, will ultimately wind up with a feeling of dissatisfaction. Indeed, Mr. Orcutt's Pegasus, more impatient than the mythological one, did not wait for the death of the econometric Medusa. As if realizing that the birth of his Pegasus prior to this death may be the cause of the failure to strike a new source of inspiration, Mr. Orcutt, hoping to put things in order, tries — if not to finish off the Medusa — at least to speed up the ceremony of her sacrifice.

In the introductory paragraph and first two sections of his paper, Mr. Orcutt maintains the spirit of the optimistic reader at a high pitch. Here we find formulated one by one all the dreams of an econometrician and, more especially, of a policy-maker. (Mr. Orcutt apparently thinks that the services of the economist *per se* are not worthy of sharing, even in a very modest way, the glory of collaborating to solve the problem.) While a careful listing is made of all the things which, in Mr. Orcutt's opinion, would make the life of a policy-maker very comfortable — but which also would deprive him of an inestimable glory whenever successful — the grounds for the case against econometrics are gradually built up.[1] As a start in this direction, we are told that the "econometricians have failed to attack in any force problems whose solutions not only would be useful to policy-makers but whose solutions may be more feasible," and that they have spent their time playing around with other problems whose solutions are, in Mr. Orcutt's opinion, either not feasible or less feasible. With such a horrible report-card, the econometrician is treated with the usual fatherly

[1] In his enthusiasm, Mr. Orcutt is ready to widen his target to include all physical sciences, challenging their success in the operation of control systems. This point will not be taken up in the present paper.

advice: first, that "more emphasis needs to be placed on building and testing models" and, second, that "more study of the continuity properties of economic time series is thus needed." It is not difficult to guess that the laggard schoolboy would have preferred both to have been spared the admonition and to be actually helped with some of his homework. In the latter connection, Mr. Orcutt is apparently not open to criticism since in the next two sections (III and IV) he develops a system which, in his opinion, could help the econometricians improve their grades.

One should normally go directly to the central part of Mr. Orcutt's argument and, in a written criticism, omit discussion of the earlier sections on the ground that they are introductory in character. However, some of the difficulties connected with the theme of the paper are already exhibited in the preliminary remarks. This is why one may be justified in exploring them before proceeding further.

In contrast with his strictness regarding the definitions used by other econometricians, Mr. Orcutt frequently leaves the reader in confusion as a result of the imprecision of some of his terms. More space could have been profitably diverted from obvious generalities to an explanation of what the author understands by various terms, for instance, by "more feasible solutions" and, more especially, by the "instrument by which the policy-maker may modify the course of the actual." To the very end of the paper, the reader cannot find out whether Mr. Orcutt by "instrument," (becoming successively "instrument of adjustment" and "control instrument"), means:

(a) the institutional means at the disposal of the policy-maker (e.g., the power to change the tariff, or to introduce rationing, etc.);
(b) the variables or the parameters which can be modified by the measures mentioned in (a);
(c) the theoretical (or econometric) relation existing between some variables (b) and other variables of the economic system;
(d) the actual effect of a change of the variables (b) on the other variables of the system;
(e) any other concept which Mr. Orcutt might have had in mind.

This ambiguity certainly does not help us reach a clear-cut picture of the equipment necessary to the policy-maker and the method of using it, as seen by Mr. Orcutt. And this is the origin of a haze which extends over the main part of the argument, which also considers the tools useful for the policy-maker.

The last paragraph of section I offers the first sign of Mr. Orcutt's opinion that the economic factors which "we know how to control and that we contemplate using for control purposes" are normally found among "variables" and not among "parameters." "There has been some tendency," we read, "to think of many policy actions as consisting of changes in the parameters of the econometric models," but Mr. Orcutt regards this tendency as groundless since the parameters cannot be altered without having "one or more auxiliary models [relating] to the parameter values" to the controllable variables.[2] We do not know exactly what Mr. Orcutt means by "parameters of a model," but it is clear that under the most widely accepted use of the term — that which stems originally from multiple or general equilibrium models[3] - we find some parameters among the oldest and the most preferred channels for carrying out economic policies. The outstanding examples of this are the "tariff schedule" and the "tax schedule." Furthermore, the distinction between controllable and uncontrollable factors cannot be made in the abstract, independently of the problem at hand, or by an a priori formal approach, as is the usual definition of the parameters.[4] It is not possible to know — without the help of economic analysis and its great ally, economic history — which factors are controllable and which are not. They may be parameters as well as variables, and the econometrician cannot alter their quality. He has to accept them as such and build his models accordingly.

One may heartily join Mr. Orcutt in wishing that more "models which include as exogenous variables those variables that we know how to control" be built and tested, but wishing alone will not help.

The main problem of building models is rather that of making them complete from the point of view of both economic theory and statistics.[5] If this criterion is followed, the group of exogenous variables cannot be arbitrarily set, neither can the subgroup of controllable exogenous variables be so set. In each particular case they are determined by the structure of the problem under study.[6] And this is why the same variable may be endogenous in one case, and exogenous in another. The decisions regarding the specification of exogenous variables are made, therefore, so that the problem may be handled in the best way and not, as Mr. Orcutt states, so as to "arbitrarily set[ting] the limits of the problem under consideration."[7] In order to make sense of each particular model this is the logical

---

[2] The term "control" is loosely used by Mr. Orcutt. At times, it refers to variables "we *wish to control*"—i.e., the ultimate objective of the policy-maker—while, at others, it is connected with those variables "that we contemplate *using for control purposes*"—i.e., the factors over which the policy-maker has direct control. (In both quotations, italics have been added.) It is in the latter sense that the terms "control" and "controllable" are used above by the present writer.

[3] Cf. Jacob Marschak, "Statistical Inference in Economics: An Introduction," in *Statistical Inference in Dynamic Models* (Cowles Commission Monograph, No. 10, ed. T.C. Koopmans), pp. 7-8.

[4] See, however, T. Haavelmo, "The Probability Approach in Econometrics," *Econometrica,* 12 (Supplement, July 1944), 3.

[5] See the penetrating analysis of T.C. Koopmans, "When Is a System Complete for Statistical Purposes?" (Cowles Commission Monograph, No. 10), pp. 393-409.

[6] The considerations that enter into the logical process which is involved here are basically those provided, as mentioned before, by economic analysis and economic history. They are, in Mr. Orcutt's opinion, only "some a priori knowledge of unspecified source."

[7] The present writer does not deny that a few isolated cases may be perhaps found where a certain degree of *irrationality* is present in the particular treatment of a problem, but only asserts that, if these cases exist, they do not constitute the general rule. Besides, an entire discipline—such as is the object of Mr. Orcutt's attack—cannot be made responsible for the errors committed by one of its users. One single specific example would have helped the reader to make better use of Mr. Orcutt's criticism. At the beginning of section III, he mentions *by title* some of the best known works, those of J. Tinbergen, L.R. Klein, and Colin Clark, and at no place in his paper is a connection established between his criticism and the methods used by these authors. To be more explicit, the reference by title to their works may very well be left out: the argument of Mr. Orcutt will be *in no way* affected by the omission.

procedure. Thus under certain circumstances, a Leontief model open with respect to "households" may be used, while, under others, the same model may be open to "other countries." The first would be justified under the assumption of some type of rationing, which in turn would justify the consideration of the "bill of goods" as exogenous; the second, under the assumption of foreign trade control, which may make the "exports" exogenous. Changes of this kind may prove to be at times very useful for analytical purposes.[8] They have nothing to do with models aimed at helping the policy-maker.

In section III, as the argument draws closer to Mr. Orcutt's theme — which centers upon the concept of the exogenous variable — a new accusation, far more serious than any previous one, is thrown at "economic theory (sic) and econometrics." This time it is that their "literature is far from explicit about the difference between endogenous and exogenous variables." This accusation too is not supported by any evidence.

However defective the literature may be on this point, from it Mr. Orcutt gathered a definition of exogenous variables which he seems to adopt temporarily. According to this definition, the "exogenous variables are (those) which affect the economic system but are not in turn affected by it, *or at least are only affected to a negligible degree by it.*"[9] This definition strikes a new chord, capable of deep and multiple resonances, and would have induced the writer to consider it at a great length, had this not already been done by some of the best contributors to the theory of statistical inference.[10]

Abstracting for the time being from the italicized part and taking the remainder of the above definition *ad literam*, no exogenous variables can be of an economic nature. Neither can they be "sociological, political, and psychological factors,"[11] nor factors describing the state of the arts and geographical location. If we include as endogenous all these variables, we are left with a system "open" only with respect to the initial cosmological conditions: time, the inalterable properties of matter, and its initial distribution in the space. Such a classification shades into an almost metaphysical scheme and loses all importance for any practical inquiry. It is the inclusion of the italicized phrase — a weakening condition — which makes the concept of exogenous variables useful for econometric analysis and, indeed, for all social sciences. It seems, therefore, a natural thing to formulate the definition of the exogenous variable on the basis of the weak, rather than on the strong, causal principle. For the sake of greater rigor, one should also try to make more precise the meaning of "negligible degree." Here we are confronted again with the difficulty of giving a definition to a loose concept so that it make sense as an analytical tool. Some influences work their effect quickly, others more slowly. In some cases, the effects are of a more lasting nature, almost cumulative in character, in others, the effects are short-lived, leaving no trace. Furthermore, the limit of "negligible degree" cannot be set uniformly.[12] The setting of a reasonable limit

[8] Koopmans, op. cit., p. 394.

[9] Italics added.

[10] E.g., Marschak, op. cit., p. 8; Koopmans, op. cit., pp. 393 ff. The above definition is in fact what Koopmans calls "the causal principle" (*ibid.*, p. 394).

[11] Koopmans, op. cit., p. 402.

[12] The complete failure of a similar attempt, that of the classical statisticians, to set uniformly the limit of the significance of probability at $P = .05$, is very instructive in this regard.

must be left to the model-builder or to its user. The exogenous variable is a relative and loose concept,[13] and very little indeed can be done about it.[14] Almost any model will provide a good illustration of this point. One may start with the three Marshallian models, the market, the short-run, and the long-run equilibria, characterized by three different exogenous variables, which are respectively the day supply, the size of the capital equipment, the state of the arts, and by one common to all, the tastes of a stationary population. Again, price constellations exercise their influence upon the quantities produced as well as upon the state of the arts. It is only because it takes longer for the latter than for the former to show a visible effect of this influence that treating the input-output coefficients as exogenous variables in a Leontief model is justified.[15]

The present writer is at a loss, therefore, to find a reasonable justification for Mr. Orcutt's repetitious complaints, some of which leave no room for a less strong interpretation. This is the case in the italicized statement that, "in any case, *the specification* [of which variables are exogenous] *is not subject to any test whatsoever.*" Later on, the econometricians face their n-th accusation, that they have "almost complete[ly] neglect[ed]" the "testing [of] hypotheses about which variables are wholly or partially (*sic*) exogenous to the economic system." But this time, Mr. Orcutt had consumed exactly one half of his paper in fighting econometrics, apparently for the sole purpose of preparing the ground for the "redirection." He then decides to "have another look at the definition of an exogenous variable and see what its definition means in statistical terms."

Despite the fact that the definition thus far used by Mr. Orcutt for the exogenous variable is by no means stochastic in character, no preparation is made for the turn toward statistics. The latter comes therefore as a surprise, which, however, is not to be the last.

In order to illustrate his main theme Mr. Orcutt uses a very simple, linear model (I), whose definition is given piecemeal. When the reader finally makes it out from different bits of information scattered throughout the argument, he finds, not without surprise, that Mr. Orcutt's new definition of exogenous variables is a *sui generis* interpretation of the stochastic model used by Professor Koopmans in his paper, already quoted above. Indeed, Mr. Orcutt's model consists of two sets of equations

(A)   (AI) $Y - a - bI = o$ (A2) $I = o$

which determine the solution of the entire system.[16]

---

[13] If it were otherwise, we would be in a position to answer the question that comes frequently from our sophomore classes: how long is the short run and how short is the long run?

[14] Of course, we can formalize its definition, as Professors Marschak and Koopmans did, but, while this helps tremendously to clarify our ideas and to treat some important problems of statistical inference, its use in econometrics is to help us in building the model so as not to contradict *known* facts about the nature of variables—supported by economic history, economic theory and, at times, by other tests—rather than in discovering *new* facts about these variables. *Infra,* fn. 25.

[15] For other examples, see Koopmans, op. cit., p. 394.

[16] CF. Marschak, op. cit., p. 8. The above equations correspond respectively to the systems (I.4) and (I.2). This writer feels that by confining the argument to two variables instead of two systems many fine points of the problem are obscured. It is because of this limitation that we arrive at expressions such as "*I* is exogenous to *Y*."

One point needs here a special emphasis. The definition of the exogenous variable is only interpreted by the structure of the system (A) and is not equivalent to (A). One should add to the system (A) the condition that the first equation cannot be used to explain the values of I, were the second equation to be hidden from us.[17] It is this qualitative condition that completes the definition of the exogenous variable. Whatever follows from now on is built on top of the concept. This is not, however, Mr. Orcutt's opinion, since he tries to define the exogenous variable with the help of properties other than those just mentioned.

The problem that comes naturally next is to see how we can make use of the model (A) in order to interpret a body of observed values of Y and I. For this purpose, a stochastic scheme must be introduced in (A). This can be done in various ways which, for the purpose of exposition, can be exemplified by introducing:

(a) errors in the observed values of the variables;
(b) shocks suffered by the theoretical relations (such as parallel shifts in a straight-line demand);
(c) other stochastic influences (such as changes in the slope of a straight-line demand).[18]

Mr. Orcutt chooses to consider the "simple shock model," that is, (b). This means that during the period of observation, the straight lines (A1) and (A2) suffer shocks, ϵ and η (Chart I).[19] The true position is E_0, and the observed is e. The system (A) becomes

(B)   (B1) $Y - a - bI = \epsilon$ (B2) $I = \eta$

These are particular cases of equations (3a) and (3b) used by Professor Koopmans.[20] Further, following the same author, we introduce the assumption that the random variables ϵ and η are independently distributed. Thus, if $f(\epsilon, \eta)\, d\epsilon\, d\eta$ is the distribution of ϵ and η, then[21]

(C) $f(\epsilon, \eta) = f_1(\epsilon) f_2(\eta)$.

This is introduced with the idea of keeping the two *systems* (A1) and (A2), or (B1) and (B2), free from any possible connection other than that attributed to the exogenous variables.

---

[17] E.g., if I were the rainfall and Y the crop-yield, the first equation could not offer an explanation of the rainfall in terms of the crop-yield. (It could offer, however, a method of an a posteriori estimation of the rainfall, but this is another problem.) In a note ("A Suggestion for Notation in Mathematical Economics," *Quarterly Journal of Economics*, LIV, November 1939, 165-67), Andrew W. Edson made the interesting suggestion that, in economic mathematical relations we should use an arrow, in addition to the equality sign, in order to specify the "direction" of the causal interpretation of such a relation. In other words, the arrow will show which side of the relation contains the endogenous variable.

[18] Cf. Marschak, op. cit., pp. 20-21.

[19] The relation (A2) might have been written $I - I_0 = 0$, without altering the argument.

[20] Op. cit., p. 395.

[21] *Ibid.*, p. 396, relation (4).

Because of (B2), the relation (C) becomes

(C') $$f(\epsilon, I) = f_1(\epsilon) f_2(I)$$

which brings us to Mr. Orcutt's definition that the exogenous variables are those which "are distributed independently of the excluded variables or shock terms as they are sometimes called."[22] If the term "shocks" is to be used here unambiguously, one should make clear that here only the shocks of the relations involving the endogenous variables are considered.[23]

Mr. Orcutt differs from Professor Koopmans in the fact that the former thinks that "to say that I and ϵ are independently distributed would be equivalent to saying that of the included variables I is the exogenous one." Apparently, Mr. Orcutt does not feel that this equivalence requires a proof since he does not offer one. It is not difficult, however, to see the weakness of this point which, unfortunately, serves as pivot for the positive theme of Mr. Orcutt. Indeed, it is only in the light of system (A) and of the additional specification of the

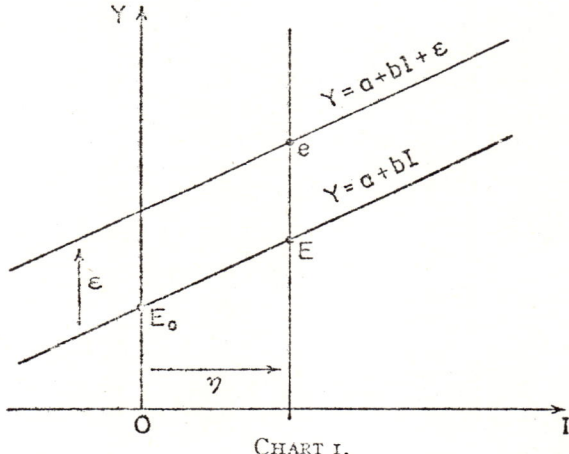

CHART I.

[22] This statement justifies completely the interpretation of Mr. Orcutt's ϵ as a *shock* term and the presentation of his model as was done above. Another alternative, available here only because of the simple structure of the model (I), would have been to interpret ϵ as determined by an *error* of Y and of I. The reader had to wait until the end of the argument to learn which interpretation was used by Mr. Orcutt.

[23] It is easy to see that from Professor Koopmans' basic assumptions regarding his shock model, i.e., from assumptions 3.1, 3.2, and 3.3, it follows immediately that the distribution of the exogenous variables $(x_{G+1}, x_{G+2}, \ldots x_n)$ is independent of that of the shocks $(u_1, u_2 \ldots u_G)$ of the relations involving the endogenous variables $(x_1, x_2 \ldots x_G)$. If $X(u_1, u_2, \ldots u_G, u_{G+1}, \ldots x_n)$ is the distribution density, then $X = f_1(u_1, u_2, \ldots u_G) \times f_2(u_{G+1}, \ldots u_n) \dfrac{\delta(u_{G+1}, \ldots u_n)}{\delta(x_{G+1}, \ldots x_n)}$

direction in which causality works that the condition (C′) acquires a meaning. The latter supplements the former, but is *not* equivalent to it.[24]

Neither can Mr. Orcutt claim that, since the condition (C′) is not a definition of exogenous variables, it constitutes at least a partial test which all such variables must fulfill. Indeed, (C′) has no meaning outside the simple shock models. Thus, the only thing that remains from the theme developed in section IV is that the distribution of the exogenous variables must — in the case of simple shock models — be stochastically independent of that of the shocks. As a consequence of this principle, we are told that "if this correlation [between I and ϵ] turned out to be significantly different from zero, then some modification of the model would seem to be called for." But Mr. Orcutt is too experienced a statistician to be unaware of the ensuing difficulty. The problem, as it stands before the econometrician, is not only to determine which variable — national income or investment — is exogenous, but to determine also a and b. And for this task, the only available information about the reality under study is a scatter formed by a number of points, such as e (Chart I). And as Mr. Orcutt rightly observes, "unfortunately, the usual fitting process ensures a selection of values for a and b such that the correlation between the variable selected to be the independent one and the obtained residuals must necessarily be exactly zero." But this is not all, however. Had the econometrician selected the national income as the independent variable, *the same scatter would have confirmed his choice of Y as the endogenous variable!*[25] While this situation may help the econometrician to maintain cordial relations with both the Keynesians and the anti-Keynesians, it reveals that, without a model accepted prior to the study of the statistical inference, the econometrician cannot solve any problem. On the other hand, the statistical inference will confirm the choice of any model provided the latter follows, be it only in general lines, the map of data. To use an analogy, an old favorite of the present writer, the situation of the statistician is such that he can prove that in every log there is a beautifully sculptured Madonna — simply because his tools are such that when he tries to get inside a piece of wood, by this very procedure, he carves the statue. Mr. Orcutt promises to put an end to this tragic fate of the modern Midas, with a subsequent paper devoted to the choice of independent variables. Until then, the econometrician is again invited, in a formal finale, to double and redouble his efforts at studying.

---

[24] Using Mr. Edson's suggestion (*supra*, fn. 17), the line connecting I and Y in Mr. Orcutt's triangular diagram should be replaced by an arrow pointing toward Y. It is the logic which justifies the direction of the arrow—the causal principle—and not the type of stochastic relation between ϵ and I that determines which variable is the exogenous one. As a matter of fact, ϵ is introduced in the diagram for stochastic, and not for fundamental, reasons.

[25] *Supra*, fn. 14.

Part IV
Pure Theory

CHAPTER 13

(1973)

VILFREDO PARETO AND HIS THEORY OF OPHELIMITY *

> Respectons l'oeuvre de nos devanciers, mais complétons–là, ne fût–
> ce qu'en donnant plus de précision aux vérités qu'ils ont découvertes.
> VILFREDO PARETO, *Cours*, § 765.

There are three reasons why I have chosen this topic for my contribution to this anniversary of the truly eminent master that Pareto was for both economics and sociology. The first is that Pareto's theory of ophelimity represents the main pillar of his momentous contribution to economic theory. The second reason is my long–standing interest in it. And the third reason is that the accidents that have punctuated the history of the theory of ophelimity illustrate better than anything else the attitude of the economic profession toward Pareto's pathbreaking ideas. In fact, they also illustrate – as we shall have occasions to see in this essay – how liable to perdurable errors even great scholars, especially economists, are.

There is no denying that Pareto's own ideas met with an incredible lack of attention from most economists during his life as well as during many years thereafter. In the last edition of Alfred Marshall's *Principles* (1920), Pareto is mentioned only once, as one of many who " have been inclined towards mathematical modes of thought ". No wonder then that for most of his life Pareto worked in isolation, a fact which earned him the famous epithet of " the hermit of Céligny ". The indifference, occasionally the opposition, toward his new ideas caused him " to withdraw into himself in a disdainful attitude, a recluse damning the damned world that never understood him " [1].

But the fate of Pareto's economic contribution is not the only famous case to show that in economics what is true, what is great, is decided by the parochial

---

* I wish to acknowledge the help I have received from Mr. Charles E. Scott during the preparation of the final version of this essay.

(1) *Lettere a Maffeo Pantaleoni*, Gabriele de Rosa ed., 3 vols., Roma: Banca Nazionale del Lavoro, 1960, II, 424. (All translations from French and Italian in this essay are mine).

temper of the day, often by the personal interests of certain powerful academic cliques. To most economists the names of Rudolf Auspitz and Rudolf Lieben are completely unknown, even though the work of these two amateur economists helped the advancement of economic theory in many important respects. And if H. H. Gossen is now known, it is rather because of the tragic history of his unique volume. If everything is kept in the proper proportions, there are some similarities between Gossen and Pareto. Like Gossen, Pareto was well ahead of his own time and also was completely aware of this fact. In a letter to Maffeo Pantaleoni, he confessed that " I am pleased with my work, which means that no one will say a good word about it. I have noticed that the things I like most are not liked by others at all " [2]. Perhaps, also like Gossen, Pareto lived in the wrong place. Who in the whole German nation in 1854 would have seen any sense in using mathematics in political economy as Gossen proposed? Who in Lausanne or in all of Switzerland during Pareto's life was interested in economic theory? And who in England could feel any incentive to propagate the ideas of an outlander who, moreover, was pushing in the opposite direction from the autochthonous school of thought? Had Pareto been associated with one of the great British universities, almost certainly economics would now be different in many respects. It would have switched to positive welfare economics much earlier and with greater elan.

There is, however, a difference between Gossen and Pareto, and a great one at that. Not only were Pareto's contemporaries already accustomed to the use of mathematics in economics but also his works were numerous, in the form of both books and articles, and written in several languages of large circulation. And although Pareto did not publish in English (apart from a few exceptions), most British economists knew of his works; a few of these works they knew even in detail. But – as Pareto lamented on repeated occasions [3] – under the spell of Alfred Marshall, to which not even F. Y. Edgeworth was immune, they chose to pay no attention to the greatest foe of the partial equilibrium analysis. Pareto had good reasons for being unhappy over this development. Walras' position, which he had endorsed and strengthened, was not receiving proper recognition from others [4]. He believed that the main cause was Marshall's inability to understand the general equilibrium theory [5].

There also were others. But the French had already proved for decades that they had no patience with any attempt to develop economics on a mathematical scaffold. As Pareto himself judged (*Lettere*, I, 419), they were interested only in

---

(2) *Lettere*, II, 376.

(3) *Lettere*, especially, I, 419, 490; II, 44; III, 63. In one letter (III, 65), Pareto even complains that Marshall " has unleashed his dogs on me ".

(4) The theme of the difference between the two schools came up repeatedly with bitter overtones in Pareto's letters to Pantaleoni (*Lettere*, II, 36; III, 60–64). In the end, Pareto decided to insert a trenchant footnote on this point in the *Manuel*, App. § 22.

(5) *Lettere*, I, 417. The accusation was certainly inept; but the truth is that Marshall was opposed even to using the idea that utilities of various commodities are interrelated (*Principles*, p. 845).

" academic doctrines ". For them, Pareto's was a lost cause *ab initio*. In the United States, the *Cours* was greeted with some rather favorable reviews, but all ended there. The revolutionary *Manuale* passed unnoticed. Only his country of birth paid attention to his ideas, at times in an utterly slavish manner, at times with strong ideological opposition. Among those who seem to have been singled out by Pareto himself, there were, first of all, Maffeo Pantaleoni, Pareto's mentor and life-long friend, as well as Benedetto Croce, Achille Loria, Enrico Barone, Luigi Amorozo, Umberto Ricci, Guido Sensini, Vittorio Rocca, Pietro Boninseign, Giuseppe Jona, Felice Vinci, and a few others of minor importance in this particular respect. A strong Pareto tradition continued in Italy for some time thereafter, its flame being kept alive by some economists of younger vintage, such as Alfonso de Pietri-Tonelli, Valentino Dominedò, and especially Eraldo Fossati and Giovanni Demaria. However, while one can discern a strong Paretoan tradition, even in Italy - Schumpeter notwithstanding - there emerged no Paretoan school " in the full sense of the word " [6]. There was no discernible " inner circle of eminent economists " not only to embrace Pareto's ideas but also to develop them in new directions [7]. Sparks of such developments are found in some works of Barone, Amoroso, and later on in those of Ricci. However, the only work to grow directly out of Pareto's work and to appear during Pareto's lifetime is the famous article by Eugen Slutsky, " Sulla teoria del bilancio del consumatore " (*Giornale degli Economisti*, 1915). Most of Pareto's contemporaries who aimed at being the standard-bearers of the new economics - Sensini, Boninsegni, as well as even the great Barone in part - only disseminated his ideas in a form that makes them guilty of glaring plagiarism [8].

The consequence of all these circumstances was that the Lausanne chair failed to attract students capable of developing Pareto's ideas - although Pareto himself blamed his enemies in Italy for keeping potential students away from him (*Lettere*, II, 448). The resident students at Lausanne singled out by Pareto himself were Léon Winiarsky, Vittorio Rocca, Pasquale Boninsegni, Giuseppe Jona, Gino Borgatta, and Giuseppe Stanislao Scalfati. Winiarsky later turned against Pareto

---

(6) JOSEPH A. SCHUMPETER, *Ten Great Economists*, New York: Oxford University Press, 1951, p. 118.

(7) Also the picture presented by JOSEPH A. SCHUMPETER (*History of Economic Analysis*, New York: Oxford University Press, 1954, pp. 858-859) is not fully accurate. He included among Paretoan economists some who shared nothing or insignificantly little of Pareto's specific thrust. He also saw in de Pietri-Tonelli, who merely published an expository manual of pure economics (*Traité d'économie rationnelle*, Paris: Giard, 1927), a hard-core Paretoan. The main explanation probably is that Schumpeter did not know of Jannaccone's paper (mentioned below), nor of *Lettere*, which appeared much later.

(8) This sad but true fact was exposed with crushing evidence in a highly interesting article by PASQUALE JANNACCONE, " Il ' Paretaio ' ", *La Riforma Sociale*, XXIII (1912), 337-368. Pareto himself noted that " no distinction can be made between Sensini's rent theory and mine: they are identical ". *Lettere*, III, 151. (Jannaccone's paper was brought to my attention by Giacomo Beccatini, who also pointed out that its title involves a pun: in Italian " paretaio " means a place where bird snares are set).

and, by some irony, in 1903 became Pareto's colleague at Lausanne [9]. Rocca and Boninsegni were at one time or other even Pareto's teaching substitutes. But Rocca was mainly a sociologist with insufficient mathematical training. And Boninsegni, who followed Pareto in the chair, proved to be no more than an uninspiring continuator. No fireworks came from the others either [10]. Among the short-time visitors was Guido Sensini, a former student of Pantaleoni, who remained thereafter in close contact with Pareto. V. Furlan, a young German economist, also spent some time in Lausanne and subsequently wrote a couple of papers related to Pareto's contribution. Luigi Amorozo, who was already a blossoming scholar, also paid a short visit to the hermit of Céligny and then remained attached both to his ideas and to him personally [11].

But the observation that outside Italy Pareto was ignored almost totally until a long time after his death is only part of the story. For about two decades, the 1930's and the 1940's, there was a fervent rediscovery of Pareto, a true Pareto vogue among Anglo-American economists to reexamine, amend, and expand his economic ideas. But since Pareto had been ignored for such a long time, to rediscover him they should have searched through the back files of an immense number of publications, not all of large circulation. On the other hand, even a middling search would have necessitated years of patient labor. Witness the fact that an important article such as Slutsky's was not discovered by the English-speaking economists before the mid-1930's. In this situation, many of Pareto's ideas still remained unknown, although many hints and associations of ideas transpired from his basic economic works, the *Manuel* and the latter-day article in the French encyclopaedia of mathematics, [12] with which almost everyone became familiar without much effort. Some of his ideas were thus rediscovered by others and recognized as Pareto's only after a long time, some only very recently. Curious though it may seem, the so-called Principle of Compensation, originally enunciated by Pareto in an 1894 paper [13], is still attributed to others, who rediscovered it some forty years later. In the same article, Pareto worked out the principle now known as "Pareto optimality", which again was used but, as John S. Chipman recently remarked [14], not credited unequivocally to Pareto until 1950 when I. M. D. Little

---

(9) *Lettere*, I, 431; II, 54, 60. Pareto's low opinion of Winiarsky's analytical abilities is vindicated by the fact that Winiarsky once devoted a long essay to the ludicruous idea that gold is "the pure personification and incarnation of the socio-biological energy". See NICHOLAS GEORGESCU-ROEGEN, *The Entropy Law and the Economic Process*, Cambridge, Mass.: Harvard University Press, 1971, p. 283.

(10) Cf. *Lettere*, II, 188, 268, 321, 353, 368, 403, 432, 445-446, 461-462; III, 163-164, 239.

(11) *Lettere*, III, 43-49, 117, 121, 151-154, 177.

(12) VILFREDO PARETO, "Economie mathématique", *Encyclopédie des Sciences Mathématiques*, Paris: Gauthier-Villars, 1911, Tome 1, vol. 4, fascicule 4, pp. 591-640 (reprinted in VILFREDO PARETO, *Oeuvres Complètes*, Geneva: Droz, 1966, Tome VIII). English translation in *International Economic Papers*, V (1955).

(13) V. PARETO, "Il massimo di utilità dato dalla libera concorrenza", *Giornale degli Economisti*, IX (July 1894), 48-66.

(14) JOHN S. CHIPMAN, "The Paretian Heritage", Paper read at the Kingston Meeting of the Canadian Economic Association, 1973 (unpublished).

did so [15]. And as pointed out by Vincent Tarascio [16], Pareto also introduced even the concept of the social welfare function as early as 1913 [17], a concept still associated mainly with his able rediscoverer, Abram Bergson. But the most piquant of all these academic piracies is that exposed recently by Jaffé [18]. The ultra-popular " Edgeworth Box " is not to be found in Edgeworth's writings; it appeared, instead, for the first time in the *Manuale* (III, § 116) [19].

Perhaps, as Sir John Hicks suggested, the modern economists were hesitant to study and criticize Pareto's works because " the sheer impressiveness of his achievements " discouraged them [20]. On the other hand, one cannot help thinking also about the peculiar code (at times explicitly defended) by which large circles of economists and perhaps of other academic servants have for some time now been guided. According to this code – which has an interesting cultural explanation if one is willing to think about it – it serves no useful purpose for one to refer to the intellectual precursors who prepared the ground for one's own inspiration. Their ideas, it is argued, are now incorporated in some better systems, more precise and better articulated. We may thus hear less and less of Pareto in the future; in fact, we already see this happening. That will be an injustice unparalleled in history for an intellectual giant of Pareto's stature. For, as John Chipman recently noted, Pareto's writings form a greater volume than the combined works of David Ricardo, John Stuart Mill, and John Maynard Keynes, each an intellectual giant in his own right. And certainly, no one can deny that, on the average, one page by Pareto contains as much intellectual substance as one by any of these other illustrious economists. When everything is taken into consideration, modern economic analysis bears Pareto's seal more distinctly than any Neoclassical economist, save perhaps Walras and Marshall, a truth that cannot be destroyed by refusing to mention his name when due.

One obstacle to the spread of Pareto's ideas should not be passed over in silence. Having begun his incredibly vast literary activity in two fields at a late age (45 years),

---

(15) See I..M. D. LITTLE, *A Critique of Welfare Economics*, 2nd ed., Oxford: Clarendon, 1957, pp. 84–87. In the 1951 *Welfare and Competition* (Chicago: Richard D. Irwin) by T. Scitovsky, Pareto is mentioned only once, by title in a bibliographical footnote! But see, for example, T. W. HUTCHINSON, *A Review of Economic Doctrines:1870–1929*, Oxford: Clarendon, 1953, pp. 224–230, 302.

(16) VINCENT J. TARASCIO, *Pareto's Methodological Approach to Economics*, Chapel Hill: The University of North Carolina Press, 1966, pp. 81–83.

(17) VILFREDO PARETO, " Il massimo di utilità per una collettività in sociologia ", *Giornale degli Economisti*, XLIV (1913), 337–341.

(18) WILLIAM JAFFÉ, " Pareto Translated: A Review Article ", *Journal of Economic Literature*, X (1972), 1190.

(19) In turn, Pareto may be suspected of having drawn inspiration from sources he failed to acknowledge. MAURICE ALLAIS (" Vilfredo Pareto ", *International Encyclopedia of the Social Sciences*, vol. 11, p. 404), for example, notes that a form of Pareto's Principle of Optimality had already appeared in F. Y. EDGEWORTH, *Mathematical Psychics*, London: Kegan Paul, 1881, pp. 23–27. Other such cases will be mentioned later on.

(20) J. R. HICKS, " Marginal Productivity and the Lausanne School: A Reply ", *Economica*, XII (1932), 300.

Pareto apparently found no time for a systematic presentation of his numerous new ideas. It is unquestionable that neither the *Cours* nor the *Manuale* is a well-composed economic treatise, as the titles announce. Both volumes appear rather as hasty arrangements of numerous notes in the form of a long sequence of short paragraphs. No wonder then that formulations are not always as precise as one would wish and that there are frequent shifts of position. Many interesting concepts and hypotheses are introduced but few are worked out fully [21]. What appear to be slips of the pen stud his main works, irritate the reader, and expose Pareto to facile attacks. A glaring example is the fact that in two places (*Manuale*, VI, § 33, App. § 45; *Manuel*, App. § 89) Pareto says that any movement from a position of " maximum " collective ophelimity necessarily benefits some members and hurts others. Yet a few short paragraphs later he admits that such a movement may also hurt everybody (*Manuale*, VI, § 37) [22].

In fairness to Pareto's great eminence, any critic of his imperfection should recognize that, had the fecund pioneer stopped to polish every one of his ideas and align them logically as he proceeded, he could not have broken as much new ground as he did. Another fact must also be considered. Being ignored by almost every competent economist, he could not benefit from any substantial cross-fertilization and constructive criticism of his ideas. As we noted earlier, he had no exceptional students, and his true followers were few and not too assiduous.

In this last respect, we should also observe that although Pareto's mathematical training was outstanding, his mathematics was not always up to the call, at times it was even mishandled. This was an important handicap in view of the extremely delicate kinds of mathematics required for the theory of consumer choice as traced out by him. Unfortunately also, the two well-qualified mathematicians who showed some interest in Pareto's work and critized it – Gaetano Scorza and Vito Volterra – were not quite able to understand the actual problems of Pareto's constructions and saddled economics with some diversionary issues [23].

II

The trend for the rediscovery of Pareto was set mainly by an Anglo-American circle of economists around 1930. At first, a few contributions pertained to pro-

---

[21] Cf. SCHUMPETER, *Ten Great Economists*, p. 123; HUTSCHISON, *A Review of Economic Doctrines*, p. 218.

[22] To be sure, the confusion had a mathematical source as well, namely, that displacements around a maximum may leave everything unchanged but ordinarily causes a decrease.

[23] One of Scorza's objections concerned the practice of simply counting the independent equations and the unknowns without inquiring whether the system of general equilibrium has several solutions or, possibly, none. As was fully recognized later, the problem thus raised is a legitimate one. However, it is still not completely solved, and its partial solutions have only little economic relevance (Cf. NICHOLAS GEORGESCU-ROEGEN, *Analytical Economics: Issues and Problems*, Cambridge, Mass.: Harvard University Press, 1966, pp. 338–340, and *Analisi economica e processo economico*, Firenze: Sansoni, 1973, pp. 185–186). Scorza's second attack involved the maximum ophelimity, on which occasion his lack of economic understanding showed up unobstructed. For

ductivity theory [24], but the real momentum came later with those focused on Pareto's theory of the consumer, viz. his ophelimity theory [25]. These were attempts at smoothing some rough corners and at settling some issues which the writings of Pareto left open.

One contention of this writer is that some of these issues were in essence, but not in form, spurious from Pareto's basic standpoint. The confusion was due in part to Pareto's over-reaction to Volterra's censure about the integrability of a total differential equation, and in part to an unsympathetic interpretation of Pareto's presentation of his position – a fact justified by the already mentioned shortcomings of his economic writings. A second contention is that some confusion still prevails and, unfortunately, it concerns perhaps the most vital issue of consumer theory, namely, the consistency of the various analytical frameworks of choice. A final contention is that Pareto, at one stage in his conceptual development, reached the only correct position on ophelimity as a coordinate of choice, but soon thereafter departed from it under the influence of various old residuals and new suggestions.

The position to which I have just referred is best expressed in a long, enlightening letter of December 28, 1899, to Pantaleoni (*Lettere*, II, 287-293):

> " Edgeworth and others start from the concept of the final degree of utility and *arrive* at the determination of the curves of indifference (as in fact I have myself done in the articles of the *Giornale*) [26]. I now leave completely aside the final degree of utility and *start* from the curves of indifference. In this lies the whole novelty. It is curious that this step has not been made before. The reasons, I believe, are two: 1) the mania of always going *beyond* experience; 2) the science [of economics] began by considering the final degree of utility: all have continued on the same track. I do not believe that the first reason had any influence on myself when I wrote the articles of *Giornale*, where the curves of indifference are precisely discussed. The second cause is probably the one which did operate.
>
> Finally, be it as it may, up to now the principles of pure economics have been founded

---

the controversy Scorza-Pareto, see CHIPMAN, " The Paretian Heritage ". Volterra's case will be amply discussed below.

(24) E. g., HENRY SCHULTZ, " Marginal Productivity and the General Pricing Process ", *Journal of Political Economy*, XXXVII (1929), 505-555, and " Marginal Productivity and the Lausanne School ", *Economica*, XII (1932), 285-296; J. R. HICKS, " A Reply, " *ibid.*, 297-300.

(25) E. g., RAGNAR FRISCH, " Sur un problème d'économie pure ", *Norsk matematisk forenings skrifter*, Series I, No. 16, 1926, pp. 1-40 (English translation in JOHN S. CHIPMAN, et al., *Preferences, Utility, and Demand*, New York: Harcourt Brace Jovanovich, 1971, pp. 386-423); R. G. D. ALLEN, " The Foudations of a Mathematical Theory of Exchange ", *Economica*, XII (1932), 197-226; J. R. HICKS and R. G. D. ALLEN, " A Reconsideration of the Theory of Value ', *Economica*, New Series, I (1934), 52-76, 196-219; OSKAR LANGE, " The Determinateness of the Utility Function ", *Review of Economic Studies*, I (1934), 218-225; E. H. PHELPS BROWN, " Notes on the Determinateness of the Utility Function ", Part I, *Review of Economic Studies*, II (1934), 66-69; NICHOLAS GEORGESCU-ROEGEN, " Note on a Proposition of Pareto ", *Quarterly Journal of Economics*, XLIX (1935), 706-714, and " The Pure Theory of Consumer's Behavior ", *Quarterly Journal of Economics*, L (1936), 545-593. (The article by Sir John Hicks and Sir Roy Allen was in fact a rediscovery of Slutsky rather than of Pareto).

(26) Reference is made to Pareto's series of articles " Considerazioni sui principii fondamentali dell'economia politica pura " published in five installments in the *Giornale degli Economisti* during 1892-1893. These articles later constituted the basis for Pareto's *Cours*.

on the final degree of utility, the *rareté*, the ophelimity, etc. Well! This is unnecessary. One can start from the curves of indifference, *which are a direct result of experience* ".

The same thought is summarized in the " Sunto " of Pareto's new views published a few months thereafter [27], and later repeated in several notes (*Manuale*, ch. III, § 54; Appendix, § 2; " Economie mathématique ", note 25). The last notes have " utility " instead of " the final degree of utility ", a change which suggests that Pareto did not see clearly that the crux of the matter turns on the latter, not on the former notion [28].

Naturally, Pareto had to denounce as completely senseless the interpersonal comparability of utility (*Manuale*, IV, § 32), a possibility to which others, such as Bentham and Edgeworth, subscribed explicitly [29]. The position of Bentham that without the addibility of individual utilities " all political reasoning is at a stand " (*Works*, I, 304) constitutes the tacit rationale of all those economists who use community indifference curves or community consumer's surplus for any economic argument – and they are legion. While Pareto thus refused to follow this road, he did not want to do away with the individual preference map as well. He seemed to have intuited that, as demonstrated by Gustav Cassel and Enrico Barone [30], demand and supply schedules are all we need (*Manuel*, App. § 43) for the analysis of the economic process. Yet, Pareto, the sociologist, was loath to remove all considerations of welfare from pure economics.

III

To mark the novelty introduced by Pareto, let us recall that Edgeworth assumed that the utility of any combination of goods $x = (x_1, x_2 \ldots, x_n)$ can be measured with the aid of some " hedonimeter ", in whose possible construction he strongly believed [31]. In this case, once the unit of utility is chosen, the amount of utility is a perfectly determined function $u(x) = u(x_1, x_2, \ldots, x_n)$. With some additional assumptions (about differentiability and shape), Edgeworth arrived at the now

---

(27) " Sunto di alcuni capitoli di un nuovo trattato di economia pura ", *Giornale degli economisti*, Part I, XX (March 1900), 216–217.

(28) See note 42, *infra*.

(29) *The Works of Jeremy Bentham*, 11 vols., John Browning ed., Edinburgh: William Tait, 1838–1843, I, 16; EDGEWORTH, *Mathematical Psychics*, pp. 7, 59–60.

(30) GUSTAV CASSEL, " Grundriss einer elementaren Preislehre ", *Zeitschrift für die gesamte Staatswissenschaft*, LV (1899), 395–458; ENRICO BARONE, " Il ministro della produzione nello stato colletivista ", *Giornale degli Economisti*, XXXVII (1908), 267–293, 391–414. English translation in *Collectivist Economic Planning*, F. A. Hayek ed., London: Routledge & Kegan Paul, 1935, pp. 245–290.

(31) EDGEWORTH, *Mathematical Psychics*, p. 101. Bentham (*Works*, I, 304) also believed in the possibility of a " moral thermometer ". And curious though it may seem, as late as 1926 we find F. P. RAMSEY (*The Foundations of Mathematics and Other Logical Essays*, New York: Humanity Press, 1950, p. 161) still visualizing the possibility of measuring utility by some kind of " psychogalvanometer ".

ultrafamiliar map of indifference curves (varieties) (32). But to pinpoint the difference between Edgeworth's and Pareto's positions, we should call these varieties *iso–utility* (33) rather than *indifference* varieties, for to represent equal *amounts of utility* is their fundamental property. The fact that the individual would be indifferent between any two combinations of the same iso–utility variety is a corollary of the definition of such varieties. However, for this framework to be useful in analyzing consumer behavior one must add the hedonistic axiom:

H_1. *The individual always chooses the combination of highest utility among those available to him whenever such a combination exists* (34).

Without this axiom, which is often ignored (35), there is no possible connection between utility theory and consumer behavior.

If we now turn to examine carefully the rest of Pareto's letter to Pantaleoni, quoted above, as well as the other passages of his writings in which he discussed the concept of ophelimity and the indifference curves, we can arrive at only one possible interpretation of his position concerning consumer theory.

In saying that he leaves measurable utility aside and starts from the indifference curves – conceived as the loci of all commodity combinations between which the individual is indifferent – *Pareto could have meant only that the existence of these curves constitutes his fundamental assumption*. And as evidenced by the graph which he immediately used (*Lettere*, II, 291; *Manuale*, III, § 53), Pareto implicitly assumed that the indifference varieties have the same shape properties as those implicitly attributed by Edgeworth to his iso–utility curves. That is, indifference varieties are nonintersecting, continuous and convex toward the origin; moreover, each variety separates the commodity space into two disjoint domains, one consisting of the combinations preferred, the other of the combinations nonpreferred relative to those on that variety (36). Moreover, Pareto also made good his claim that the indifference varieties are given by experience directly by indicating how they could possibly be constructed from the individual's answers to binary choices (*Lettere*, II, 289-290; *Manuale*, III, § 52).

---

(32) In fact, the current map was introduced in the unjustly ignored volume of R. AUSPITZ and R. LIEBEN, *Untersuchungen über die Theorie des Preises*, Leipzig: Dunker and Humblot, 1889, and became popular through its adoption by Irving Fisher (1892), *Mathematical Investigations in the Theory of Value and Prices*, New Haven: Yale University Pres, 1925. In *Mathematical Psychics*, Edgeworth has the coordinate axes differently oriented.

(33) W. E. JOHNSON ("The Pure Theory of Utility Curves", *Economic Journal*, XXIII (1913), 481) first referred to them as "constant utility curves", instead of "indifference curves".

(34) A combination of maximum utility may not necessarily exist even if $u(x)$ is a continuous function. A simple clarifying example is that in which the set of available combinations is represented by $x_1 + x_2 < 1$ and the iso-utility curves have the usual shape.

(35) A notable exception is Gossen, who began his 1854 *Entwickelung der Gesetze des menschlichen Verkehrs und der daraus fliessenden Regeln für menschliches Handeln*, Berlin: Prager, 1889, by proclaiming that man's aim in life is "to obtain the highest posible life pleausre", which term meant "utility" for that author. EDGEWORTH (*Mathematical Psychics*, p. 6) also mentions this assumption. See note 39, *infra*.

(36) See also "Sunto", Part II, pp. 536–537; *Manuale*, App. § 10; *Manuel*, App. § 2, 44; "Economie mathématique", § 16.

By present standards, Pareto should have cast his argument into an axiomatic form. But given the way even the best mathematical works were written at that time, one can hardly impugn him for not having done so. It was thus left to later writers to complete this task on the basis of what Pareto said in a few brief paragraphs about how measurable utility may be replaced by an ordinal series of indices (*Manuale*, III 52-57; App. § 2) [37].

Yet Pareto can rightly be faulted on two other counts. First, he did not see that in order to connect his indifference map with the behavior of the consumer, one needs some axiom equivalent to H_1. The point was perceived by V. Furlan, who proposed to replace " utility " by " ophelimity " in that axiom [38]. But an axiom more in the spirit of pure theory of choice is:

H_2. *Given a set of alternatives S, the choice of the individual is an alternative $a \in S$ such that no $b \in S$ is preferred to a (provided that such a exists).*

Perhaps Pareto thought that such an axiom is unnecessary. For he seems to have adopted a position similar to that of Irving Fisher, who instead of any hedonistic axiom chose as his basic point of departure the notion that " *Each individual acts as he desires* " [39]. But this is a bare tautology which tells us nothing about what the individual desires.

The second reason, already hinted above, is that Pareto did not remain continuously faithful to the idea that consumer theory needs only the map of the indifference varieties, not measurable utility [40]. Even before he came to explain his theory of choice, Pareto unequivocally stated that since it is not possible to prove that utility is measurable, " from now on, when we shall speak of ophelimity, one should always understand that we simply wish to denote one of the systems of indices of ophelimity " (*Manuale*, III, § 36) [41].

---

[37] As a matter of fact, the first attempt (GEORGESCU-ROEGEN, " The Pure Theory of Consumer's Behavior ") at providing Pareto's position with an axiomatic basis came only in 1936 and, moreover, was not completely adequate. See, however, the subsequent improvement in HERMAN WOLD, " A Synthesis of Pure Demand Analysis ", *Skandinavisk Aktuarietidskrift*, XXVI (1943), Part II, 221-223.

[38] V. FURLAN, " Cenni su una generalizzazione del concetto d'ofelimità ", *Giornale degli Economisti*, XXI (1908), 259-260. See also Axiom II in WOLD, " A Synthesis ", Part II, p. 221.

[39] FISHER, *Mathematical Investigations*, p. 11. Cf. *Manuale*, II, § 29.

[40] We should not fail to observe that this conclusion, in a slightly different form, appeared for the first time in IRVING FISHER, *Mathematical Investigations*, pp. 98-89: " Thus we may dispense with the total utility density and conceive the ' economic world ' to be filled merely with the lines of force or ' maximum directions ' ". Pareto read Fisher's contribution soon after it appeared (*Lettere*, I, 291, 301), that is, seven years before he communicated his own idea to Pantaleoni. One cannot help suspecting a connection of which, curiously, Pareto makes no clear mention. Among the several references to Fisher in the *Manuale*, only in one is Fisher connected with the issue of measurable utility, and then in a highly vague way (*Manuale*, III, § 35). See note 59, *infra*.

[41] It is for this reason that in the special literature (as in this essay) " ophelimity " came to mean " ordinal utility ". As TARASCIO (*Pareto's Methodological Approach*, pp. 78-80) observed, Pareto retained in his sociological writings the term " utility " to denote the level of social welfare derived from all causes not only from those of an economic nature.

But probably even Pareto could not completely ban the idea of measurable utility from his mind. That idea had been around for such a long time that, as has happened with many other notions, it was hard to get rid of all its deep-seated roots. Thus, Pareto immediately after making the statement just quoted turned to identifying the indifference curves with the contour lines of " the hill of the indices of pleasure "; in the end, he came to speak only of " the hill of pleasure ", the height of which represents the total ophelimity (*Manuale*, III, § 58; IV, § 32). Explicitly, he proposed then to treat ophelimity as " a quantity ", asserting that " it will be easy to modify the argument so as to have recourse only to the concept of ophelimity indices " (*Manuale*, IV, § 33). With the last statement Pareto embraced an error of which he never became aware thereafter. It did not occur to him, as even nowadays it does not occur to the overwhelming majority of economists, that a sound concept of " marginal utility " requires that both utility *and* the corresponding commodity be cardinally measurable entities [42].

Since from any ophelimity index, φ, we can derive another by the operation

(1) $$\varnothing = F(\varphi),$$

where F is an arbitrary but increasing function, then (as in *Manuale*, App. §§ 6-7; *Manuel*, App. §§ 3-4)

(2) $$\varnothing_x = F'\varphi_x,$$

where the usual subscript notation for partial derivatives is used [43]. Pareto therefore produced the simple evidence that $\varnothing_x$ depends on an arbitrary factor F' and hence " marginal utility " is not a determined quantity [44], but failed to derive the elementary conclusion from it.

The persistence to the present day of the error of speaking of marginal utility without specifying any restrictions may be due to several factors. First, the fact that the sign of $\varnothing_x$ is always positive may prompt the idea that some quantitative " more " applies to the margin of ophelimity. Second, there is the notion that the individual can compare not only the pleasures derived from different combinations, but also the differences between these pleasures. But if this is the case, then – Pareto and a long list of authors after him maintained – one can establish a grid of equal differences and hence a cardinal scale for pleasures [45]. Such authors

---

(42) NICHOLAS GEORGESCU-ROEGEN, " Utility ", *International Encyclopedia of the Social Sciences*, New York: Macmillan and The Free Press, vol. 16, p. 239.

(43) In passing, we may note the circular reasoning (one of several examples which mar Pareto's economic exposition) by which relation (1) is justified in the *Manuale*.

(44) A similar argument applies to the case in which the commodity is only ordinally measurable and hence x may be replaced by an arbitrary increasing function $v(x)$.

(45) *Manuale*, IV, § 32. See also A. OSORIO, *Théorie mathématique de l'échange* (with an introduction by V. Pareto), Paris: Giard, 1913, p. 312; A. L. BOWLEY, *The Mathematical Groundwork of Economics*, Oxford: Clarendon, 1924, p. 2; RAGNAR FRISCH, " On a problem ", pp. 387–395; OSKAR LANGE, " The Determinateness of the Utility Function ", pp. 222–223; W. E. ARMSTRONG, " The Determinateness of the Utility Function, " *Economic Journal*, XLIX (1939), 453–467; to list only a representative sample.

have not realized that cardinality requires more than that. To put it in the most direct way, cardinality requires, above all, that if the individual finds that the utility differences between four alternatives A_1, A_2, A_3, A_4, are such that

(3) $$\Delta(A_1, A_2) = \Delta(A_3, A_4) < \Delta(A_1, A_3),$$

then he should also find that

(4) $$\Delta(A_1, A_3) = \Delta(A_2, A_4).$$

One must therefore introduce this or an equivalent condition as an additional assumption [46].

But Pareto wanted to continue to speak of decreasing marginal ophelimity (i.e., of the third glass of wine producing less pleasure than the second) without returning to Edgeworth's measurable utility. To avoid this, Pareto accepted the power of the individual to know whether two differences are different but not whether they are equal. Obviously, this is a specious solution for it assumes that the individual is a perfect scale when he has to compare combinations but no longer so when differences are involved. The inconsistency is reminiscent of what I have termed " the ordinalist's fallacy ", which consists precisely of assuming that there is indifference between " sure " alternatives but not between " sure " and " risk " alternatives [47].

If, like Pareto, we believe that the individual cannot detect equal differences, to be consistent we should also deny him the faculty of finding equal (indifferent) combinations. And there is no danger at all in doing away with the case of indifference altogether, as Little suggested [48]. Only, we need in this case to justify to the traditional school the absence of indifference. This is achieved if we return to the older idea that choice is based on the hierarchy of wants. On this basis, all combinations can be ordered in a chain. Yet indifferent combinations may not necessarily exist; two combinations may always be discriminated by some particular want or another. This scheme is not only entirely sufficient for the behavior of the individual, it also reveals that the meaning of the principle of decreasing marginal utility is nothing other than that of the hierarchy of wants, which goes back to Plato and which was so wittily described in Carl Menger's famous fable about the way an isolated farmer uses successive bags of corn. Last, but not least, the same scheme provides an *objective* basis for many welfare issues, which otherwise would require either the assumption of interpersonal comparison of utility or a *subjective* welfare function [49].

[46] By a formal and more involved argument, the necessity of an additional assumption was pointed out by PAUL A. SAMUELSON, " The Numerical Representation of Ordered Classifications and the Concept of Utility ", *Review of Economic Studies*, VI (1938), 65–70.

[47] NICHOLAS GEORGESCU-ROEGEN, *Analytical Economics*, pp. 190, 206, and *Analisi economica*, pp. 144–145.

[48] LITTLE, *A Critique*, 1st ed., p. 22. Also Axiom G, *infra*.

[49] GEORGESCU-ROEGEN, *Analitical Economics*, pp. 192–103 (where references to Carl Menger and Plato also are to be found), and " Utility ", pp. 262–265.

IV

Pareto apparently believed that his hybrid position on the comparability of differences entitled him to attribute a definite significance also to the second partial derivatives of any index–function. I say "apparently" because he offered no justification for his using these derivatives (and even those of higher order), nor did he mention the point at all. Everything seems to indicate that Pareto was not aware of the fact that even the feeble invariance (as to sign) enjoyed by the first partial derivatives disappears for the second order partial derivatives. It is hard to conceive why he did not go one step further beyond relation (2) to see that

(5) $$\varnothing_{xx} = F''\varphi_x^2 + F'\varphi_{xx}, \qquad \varnothing_{xy} = F''\varphi_x\varphi_y + F'\varphi_{xy},$$

and, hence, that $\varnothing_{xx}$ and φ_{xx} (or $\varnothing_{xy}$ and φ_{xy}) do not necessarily have the same sign. Sure of himself, Pareto proceeded to establish a series of inequalities intended to express various properties of the indifference varieties and of the hill of pleasure in terms of any ophelimity index–function φ, inequalities which all involved the second and higher order partial derivatives of φ (*Manuale*, App. § 11–14, 49; *Manuel*, App. § 46–49, 124; "Economie mathématique", § 17). Pareto showed no sign of doubting that most of the formulae he established have no operational value unless ophelimity is cardinally measurable. Yet, even subject to this restriction, Pareto's formulae broke entirely new ground.

In the related argument, Pareto made use of a *factual* assumption which he introduced without marking the move sufficiently – a characteristic blemish of his exposition. The assumption, namely, that the principle of decreasing marginal utility applies to baskets of commodities as well (*Manuale*, IV, § 42; "Economie mathématique", § 17) is completely reasonable, but it is an *assumption* nevertheless. In a mathematical form, it states that if

(6) $$dx_i dx_k \geq 0 \text{ for any } i, k = 1, 2, \ldots, n,$$

then

(7) $$d^2\varphi < 0.$$

We may refer to it as the *weak* principle of decreasing marginal utility and also note that it allows saddle points in the hill of pleasure [50]. For two commodities, it is equivalent to

(8) $$\varphi_{xx} < 0, \qquad \varphi_{yy} < 0, \qquad \varphi_{xy} < \sqrt{\varphi_{xx}\varphi_{yy}}.$$

As these relations show and as Pareto carefully proved for the general case (*Manuale*, App. § 49; *Manuel*, App. § 47), for competitive and independent com-

---

[50] NICHOLAS GEORGESCU-ROEGEN, "A Diagrammatic Analysis of Complementarity", *Southern Economic Journal*, XIX (1952), p. 3.

modities the weak principle results from purely mathematical considerations. In case $\varphi_{xy} \geq 0$, relations (8) are equivalent to

(9) $$\varphi_{xx} < 0, \qquad \varphi_{xx}\varphi_{yy} - \varphi_{xy}^2 > 0,$$

as Pareto has (*Manuale*, App. § 12; *Manuel*, App. § 48; " Economie mathématique ", § 17). Interpreted differently, this means that for complementary or independent commodities the weak principle of decreasing marginal utility entails [51]

(10) $$d^2\varphi \leq 0 \quad \text{for any} \quad dx_i,$$

i.e., it entails the *strong* principle of decreasing marginal utility [52]. However, as Pareto also proved, the same proposition is not true if all commodities are competitive. To be sure, his mathematical proofs are exasperatingly muddled in places – as was not uncommon with Pareto – but in exchange we are offered a real mathematical treat by the proof of the above propositions for the general case (*Manuale*, App. § 49; *Manuel*, § 124) [53]. Perhaps Pareto thought that he did not need to be as careful in the lateral expositions of the same conclusions.

For many, the preceding considerations may present only a marginal theoretical importance. Be this as it may, the undeniably important consequence of Pareto's failure to perceive the fundamental difference between cardinal and ordinal utility concerns his classification of a pair of commodities into complementary, independent, and competitive according to whether [54]

(11) $$\varphi_{xy} >, =, < 0.$$

In this he followed Auspitz and Lieben (*Untersuchungen*, pp. 154, 482) [55]. However, Pareto's analysis is of a still unsurpassed insight (*Manuale*, IV, § 9–15, 34–42).

---

[51] This restriction perhaps misled Pareto into believing that also the weak principle is valid only for complementary or independent commodities. Otherwise, the thought seems without foundation.

[52] GEORGESCU–ROEGEN, " A Diagrammatic Analysis ", p. 5.

[53] The accusation of Allais (" Vilfredo Pareto ", p. 402) that Pareto's proof is " rather obscure and unconvincing ", is therefore a good example of the unsympathetic attitude of even some of the selfstyled heads of a Paretoan school (see *ibid.*, p. 407).

[54] *Manuale*, App. § 12; *Manuel*, App. § 47 (where obvious typographical errors reversed some signs).

[55] The same work had inspired Fisher (*Mathematical Investigations*, pp. 65–71). Fisher's work was already known to Pareto (note 40, *supra*), but Fisher did not use definition (11). The *Untersuchungen* also were known to Pareto, but to what extent is a moot question. Pareto lamented that he did not know " one single word of German ", but he obtained a translation of that volume (or of parts of it) from Pantaleoni (*Lettere*, I, 163, 196). In any case, there is no basis for crediting Edgeworth with the innovation, as is done by many, including the present writer (GEORGESCU–ROEGEN, " Utility ", p. 252). In the *Mathematical Psychics*, pp. 34, 108, Edgeworth asserts only that $\varphi_{xy} < 0$ because of a principle akin to that which requires $\varphi_{xx} < 0$. Something similar to (11) appears only in an 1897 paper of his, by which time he was quite familiar with the *Untersuchungen* (F. Y. EDGEWORTH, *Paper Relating to Political Economy*, 3 vols., London: Macmillan, 1925, I, 117; II, 295).

What is more, he offered a practically equivalent but more transparent definition, which can rightly be called Pareto's own. Within a certain domain, two commodities are complementary, independent, or competitive according to whether increments of them yield together a greater, equal, or smaller utility than when used *separately* [56]. This definition lends itself to a very simple and useful grahical representation. Take a grid of equally distant iso–utility curves, $u, u + \Delta u, u + 2\Delta u$, andco nstruct the rectangle ABCD as shown in Fig. 1. The situation (a), (b), (c) represent complementarity, independence, and competitiveness respectively [57].

In fairness to Pareto's position on complementarity one should ponder over the fact that some thirty years had to go by before Sir John Hicks and Sir Roy Allen discovered that, because of (5), one cannot classify commodities according to (11) if utility is only ordinally measurable [58]. They thought up a new definition which is extremely complicated and which, later, was shown to reduce to the general formula established by Slutsky in 1915. Yet the conception is *formally* identical to that of Pareto's. If instead of movements on the utility surface, we now consider movements on an indifference surface for the two goods and money, then on a grid of equally distant contour lines for money, $m, m - \Delta u, m - 2\Delta u$, the classification is made according to the same rule of Fig. 1. The difference is that

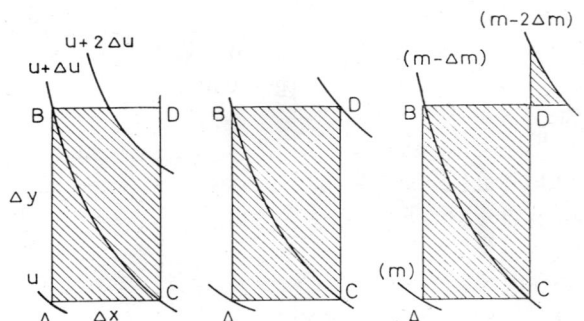

Fig. 1

in Pareto's definition valuation is made in terms of amounts of utility, in that of Allen–Hicks in terms of amounts of money. One could hear a Marshall redivivus pressing the point that, after all, money is the measure of all motives and pleasures. And if the marginal utility of money were constant in some absolute sense (which is impossible), then the two definitions would be completely, not only formally, identical [59].

(56) Curiously, Pareto did not realize the blow inflicted on this definition by his denial of the individual's faculty to ascertain whether two increments of ophelimity are equal and of the cardinality of ophelimity.
(57) GEORGESCU–ROEGEN, " A Diagrammatic Analysis ", pp. 6–7.
(58) HICKS and ALLEN, " A Reconsideration ", pp. 60, 197.
(59) GEORGESCU–ROEGEN, " A Diagrammatic Analysis ", pp. 11–12. The above graphical device has the additional merit of providing an extremely simple topological criterion for recognizing whether an indifference map can be represented by an ophelimity index function of the form $u = f(x_1, x_3, \ldots, x_n) + g(x_2, x_3, \ldots, x_n)$, in which variables x_1 and x_2 are separated and $n \geq 2$.

V

There are mathematical models of choice galore, yet few have paid attention to the issue of the connection between the paper–and–pencil construction and the actual situations in which the individual must make a choice. There is ample evidence to think that the model Pareto had in mind *ab initio* was complete in this particular sense, but he failed to arrive at a clear, explicit picture of it. Only a most general model of choice will help settle the issue.

Any *analytical* model of choice must begin by affirming the fact of choice. To this effect, we must consider the *universe* U of all possible alternatives, which need not consist only of baskets of commodities. For a reason to become clear presently, U must be allowed to contain " identical " alternatives in any number whatever. The fact of choice in the most general form may then be stated as follows:

A. THE GENERAL AXIOM OF CHOICE: *To any element I of a set P and any set* $X \subseteq U$, *there corresponds a nonnull subset* $c_I(X) \subseteq X$, *called the choice set of* X.

In relation to actuality, I represents some particular individual of the population P, and $c_I(X)$ represents the set of all alternatives that might be chosen from X by I. The fact that we have allowed U to contain identical alternatives eliminates the impossibility of choice portrayed by Buridan's ass, which starved to death between identical piles of hay. Let us also observe that we can simplify notations by doing away with the subscript I; all arguments in this domain implicitly assume that I remains the same throughout.

There are good reasons, however, to think that A is much too strong. Whenever we see an individual choosing, he chooses from some particular type of set. We may therefore prefer the following axiom:

A_1. THE RESTRICTED AXIOM OF CHOICE: *There exists a family* $\mathscr{F}$ *of subsets of* U *such that to each* $X \in \mathscr{F}$, *there corresponds a nonnull choice set* $c(X)$.

On the vasis of this postulate we may introduce the following relations for $x, y \in X$, $X \in \mathscr{F}$ (60):

The vertex D of the rectangle ABCD constructed as in Fig. 1 on any two indifference curves in any plane (x_1, x_2) must describe another indifference curve. The criterion introduced later by G. DEBREU (" Topological Methods in Cardinal Utility Theory ", in *Mathematical Methods in the Social Science*, K. J. Arrow et al. eds., Stanford: Stanford University Press, 1960, pp. 20–25) applies only if $n \geq 3$.

Incidentally, the problem of separability is related to Pareto's argument that, if commodities are independent, ophelimity is cardinally measurable (*Manuale*, III, § 35; App. § 8; *Manuel*, App. § 10) – an argument hich proves without any doubt that Pareto was not aware of the fact that the notion of independence, $\varphi_{xy} = 0$, requires cardinality. CHIPMAN (*Preferences*, pp. 325–326) errs in saying both that the argument goes back to Irving Fisher *and* that Pareto acknowledged this fact in the *Manuale*, III, § 35.

(60) The logical symbols, $\wedge$, $\vee$, $\Rightarrow$, $\Leftrightarrow$, and $\sim$ are used with their usual meanings – " and ", " or/and ", " implies ", " is equivalent to ", and " it is false that ".

(1): $x S_X y \Leftrightarrow [x \in c(X)] \wedge [y \in X]$;
$x P_X y \Leftrightarrow [x \in c(X)] \wedge [y \notin c(X)]$;
$x E_X y \Leftrightarrow [x \in c(X)] \wedge [y \in c(X)]$;
$x N_X y \Leftrightarrow [x \notin c(X)] \wedge [y \notin c(X)]$.

The point that these relations are only notations cannot be emphasized sufficiently [61]. Of course, one could label them selection, preference, equivalance, and noncomparability, respectively. But that would not be wise. The connotations that these terms carry with them in the ordinary language may lead us astray at this juncture.

Obviously, A_1 does not preclude the relation between x and y from being different in two different sets. With an eye on the living individual, we should naturally like to eliminate this "inconsistency". To this effect, we may adopt a new axiom, which is in fact a generalization of Samuelson's Weak Axiom of Revealed Preference [62].

B_1. THE AXIOM OF CONSISTENCY OF CHOICE [63]: $\sim [(x S_X y) \wedge (y P_Y x)]$. This axiom does not exclude

(12) $\qquad (x E_X y) \wedge (y E_Y x)$,

i.e., $c(X)$ may have more than one element. In fact, B_1 yields

(13) $\qquad (x E_X y) \Rightarrow (y E_Y x) \vee (y N_Y x)$.

Instead of B_1, we may adopt:

B_2. THE STRICT AXIOM OF CONSISTENCY OF CHOICE: *If $x \neq y$,* $\sim [(x S_X y)| \wedge (y S_Y x)]$.

This entails that, for $x \neq y$, $\sim (x E_X y)$. Therefore, B_2 may be decomposed into B_1 and

C. THE SINGLETON CHOICE AXIOM: *The choice set $c(X)$ is a singleton.*
From B_1, it easily follows that if X denotes any set of $\mathscr{F}$ to which a given pair x, y belongs, there are only four situations:

(a) $(x P_X y) \vee (x N_X y)$;
(b) $(y P_X x) \vee (y N_X x)$;
(c) $(x E_X y) \vee (x N_X y)$;
(d) $(x N_X y)$.

---

[61] It is useful to note that S_X is reflexive, P_X is asymmetrical (hence irreflexive), and E_X, N_X symmetrical and reflexive.
[62] PAUL A. SAMUELSON, "A Note on the Pure Theory of Consumer's Behavior", *Economica*, N. S., V (1938), 65; HIROFUMI UZAWA, "Note on Preference and Axioms of Choice", *Annals of the Institute of Statistical Mathematics*, VIII (1956), 35–40; KENNETH J. ARROW, "Rational Choice Function and Orderings", *Economica*, N. S. XXVI (1959), 121–127.
[63] Here and herafter, X, Y, Z, . . denote sets belonging to the *same* $\mathscr{F}$.

Hence, we can now introduce some relations analogous to (I), but *independent of* X (not of $\mathscr{F}$!):

(II): $\quad xPy \Leftrightarrow xP_Xy$ for some $X \in \mathscr{F}$;

$\quad\quad\quad xEy \Leftrightarrow xE_Xy$ for some $X \in \mathscr{F}$;

$\quad\quad\quad xSy \Leftrightarrow (xPy) \lor (xEy)$;

$\quad\quad\quad xNy \Leftrightarrow \sim [(xSy) \lor (ySx)]$ [64].

It is only at this juncture that we are entitled to relate these abstract relations to the actual acts of choice. For it is only now that we have some basis for saying that by his acts of choice the individual reveals to us some uniformity of preference and indifference. For example, if in one situation he chooses x and rejects y, he will always either do so or reject both of them. Hence, we may refer to P, for example, *as revealed preference*, but without forgetting to add " on the basis of $\mathscr{F}$ ".

A new set of relations may now be *defined* as follows:

(III): $\quad xPt_1Pt_2 \ldots t_nPy \Leftrightarrow xP_ty$;

$\quad\quad\quad xSt_1St_2 \ldots t_nSy \Leftrightarrow xS_ty$;

$\quad\quad\quad xEt_1Et_2 \ldots t_nEy \Leftrightarrow xE_ty$;

where t stands for the chain $t_1, t_2, \ldots, t_n$. A special form of transitivity applies here *by definition*:

$$xP_tyP_uz \Rightarrow xP_{t,u}z,$$

(14) $\quad\quad xS_tyS_uz \Rightarrow xS_{t,u}z,$

$$xE_tyE_uz \Rightarrow xE_{t,u}z,$$

as well as the mixed relations of the form

$$xP_tyE_uz \Rightarrow xS_{t,u}z,$$

(15) $\quad\quad xEy_tS_uz \Rightarrow xS_{t,u}z,$

$$xP_tyS_uz \Rightarrow xS_{t,u}z.$$

This tautological form of transitivity should not be confused with the true form. For, obviously, xP_ty does not warrant xPy; actually, between x and y there may exist any one of the relations (II). Therefore, at this juncture, we cannot directly connect the definitions (III) with any act of choice defined by A_1.

We are now in a similar situation to that created by (I). The new relations (III) also are tied to some set, t instead of X. Hence, we may raise again the question of consistency in the same form, namely, is it possible that xP_ty and yP_ux?

---

[64] It stands to reason that xNy includes the case in which no X contains both x and y.

The axiom that eliminates this possibility generalizes the one advanced by Jean Ville, rediscovered independently by H. S. Houthakker [65], but at times called Samuelson's Strong Axiom of Revealed Preference.

D_1. THE AXIOM OF REGULARITY OF CHOICE: $\sim [(xS_t y)| \wedge (yP_u x)]$.

Obviously, this axiom generalizes B_1, which is obtained if t and u are null chains. Like B_1, D_1 allows $xE_t y$ for t nonnull and $x \neq y$. B_2 is generalized by

D_2. THE STRICT AXIOM OF REGULARITY OF CHOICE: For t nonnull and $x \neq y$, $\sim [xS_t y) \wedge (yS_u x)]$.

Like B_2, D_2 eliminates equivalence, except the trivial $xE_X x$. For this reason, it is well to operate with D_1 and introduce C only when required by the structure in point.

If we write T_t for any $S_t \neq E_t$, D_1 may be replaced by an equivalent but more telling axiom.

D. THE CYCLE AXIOM: $\sim (xT_t x)$.

According to this axiom, if $xP_t y$, for example, for any other u either $xT_u y$ or $xN_u y$, the last relation meaning that $\sim [(xT_u y) \vee (yT_{u'} x)]$, where u' is u in reserve order [66]. On the basis of this regularity, we may, just as in the earlier case, classify the possible relations between any x and y into four classes and define some new relations [67]:

(IV): $\quad xP^* y \Leftrightarrow xP_t y$ for some t;

$\quad\quad\quad\quad xE^* y \Leftrightarrow xE_t y$ for some t;

$\quad\quad\quad\quad xS^* y \Leftrightarrow xS_t y$ for some t;

$\quad\quad\quad\quad xN^* y \Leftrightarrow \sim [(xS^* y) \vee (yS^* x)]$.

Recalling that P_t, E_t are defined by paper and pencil operations, we may refer to these relations as PAP (paper-and-pencil) preference, equivalence, selection, and noncomparability. This terminology is justified by the fact that PAP preference does not always correspond to an act of choice. That is, although $xP^* y$, the individual may never be confronted with some $X \in \mathscr{F}$ such that $xP_X y$. If he is not, we have no way of knowing how he would treat x in comparison to y. Briefly, D establishes a particular structure of choice, which helps us in making some predictions about the individual's choices. For example, if $xP^* y$, we know that $\sim ySx$.

---

[65] JEAN VILLE, " Sur les conditions d'existence d'une ophélimité totale et d'un indice du niveau des prix ", *Annales de l'Université de Lyon*, IX (1946), Sec. A(3), 32–39; English translation in *Review of Economic Studies*, XIX (1951–52). H. S. HOUTHAKKER, " Revealed Preference and the Utility Function ", *Economica*, N. S., XVII (1950), 159–174.

[66] All this is reminiscent of Pareto's open and closed cycles. But as we shall see in detail in Section VIII below, Pareto's approach differs in several important aspects.

[67] P^* is asymmetric and transitive; E^* is reflexive, symmetric and transitive; S^* is transitive; N^* is irreflexive and symmetric.

But D does not affect the lack of information that may be inherent in a consistent choice over $\mathscr{F}$.

A highly important, yet apparently neglected issue that presents itself in connection with the above model is the consistency or the rationality of the individual in an entirely different sense than that considered by axioms B_1 and D. Let us assume that according to the choice over $\mathscr{F}$, we learn, say, that $x\mathrm{P}y$. If the same I chooses over a different family of sets, $\mathscr{F}^1$, would he still reveal to us that he prefers x to y, i.e., that $x\mathrm{P}^1 y$? Nothing in pure logic warrants that $x\mathrm{P}y \Leftrightarrow x\mathrm{P}^1 y$. On the other hand, if for the same individual both $x\mathrm{P}y$ and $y\mathrm{P}^1 x$ were true, we would be entitled to think that the individual is *irrational* [68]. An axiom formally similar to B_1 and D_1 will establish rationality in this particular sense.

E. THE RATIONALITY AXIOM: *If $\mathscr{F}$, $\mathscr{F}^1$ are any two families of subsets of U, then* $\sim [(x\mathrm{P}y) \wedge (y\mathrm{S}^1 x)]$.

In this case, it is still possible to have $x\mathrm{P}y$ and $y\mathrm{N}^1 x$. The general case, therefore, is one in which the relations N and N^1 overlap, i.e., neither $\mathrm{N} \subset \mathrm{N}^1$ nor $\mathrm{N}^1 \subset \mathrm{N}$. However, if $\mathrm{N} \subset \mathrm{N}^1$, we shall say that the choice on $\mathscr{F}$ covers the choice on $\mathscr{F}^1$, for then $x\mathrm{P}y \Rightarrow x\mathrm{P}^1 y$, and $x\mathrm{E}y \Rightarrow x\mathrm{E}^1 y$. Obviously, if $\mathscr{F}^\circ$ is the family of *all* binary subsets of U, then $\mathscr{F}^\circ$ covers any other $\mathscr{F}$. This is the main reason for the unique position the binary choice occupies in the theory of the consumer.

This theory now appears as a particular case of the foregoing axioms A_1 and D, provided that we also adopt the following axiom:

F. THE COMPLETENESS AXIOM: *Axiom A_1 applies to the family $\mathscr{F}^\circ$ of all binary subsets of U.*

To be sure, one may adopt a weaker axiom, which asserts the same thing only for a subset of $\mathscr{F}^\circ$, in which case some pairs of alternatives may be intrinsically incomparable. Reasons may be offered for doing so. One such reason is that demand theory does not require complete comparability. However, this argument alone is not too weighty. Ockham's razor does not prevent walking creatures from possessing a greater than necessary number of legs. But in view of the fact that we wish to follow Pareto's line of thought, the adoption of F is here in order. Indeed, as evidenced by the presentation in the *Manuale* (III, § 52) and especially by the detailed explanation in the already mentioned letter to Pantaleoni (*Lettere*, II, 287–293), complete binary comparability is the first pillar of Pareto's revolutionary theory of the consumer.

Pareto used several other assumptions, which, again, he failed to state explicitly but which can be detected without much difficulty. Only one of these assumptions, the assumption of transitivity of choice, seems to have been taken completely

---

[68] If threshold is introduced in choices, it is possible that the individual may be irrational in this sense. For an example in which $\{x\} = c\{x, y\}, \{x\} = c\{x, z\}$, but $\{y\} = c\{x, y, z\}$, see GEORGESCU-ROEGEN, *Analytical Economics*, p. 239.

for granted by Pareto. From all we can judge, Pareto, like a host of others after him, did not realize that transitivity is a factual property that may be falsified. Within the general framework developed in this section, the transitivity assumption is a direct corollary of D (or of D_1) and F, a connection worth retaining for further use. With self-explanatory notations, from D we have

$$x P° y S° z \Rightarrow \sim z S° x \Leftrightarrow x P° z;$$

(16) $$x S° y P° z \Rightarrow \sim z S° x \Leftrightarrow x P° z;$$

$$x E° y E° z \Rightarrow \sim [(z P° x) \lor (x P° z)] \Leftrightarrow x E° z.$$

As is well-known by now, D and F suffice to order all alternatives in a chain, but do not warrant the existence of an ophelimity index [69]. In case all alternatives are combinations of cardinally measurable commodities – hence representable by the points of the nonnegative ortant Ω_n of an n-dimensional Euclidean space – a nicely-behaving ophelimity function (specifically, a continuous function over Ω_n) can be constructed with the aid of the following simple axiom.

G. THE INDIFFERENCE AXIOM: *If $yP°x$ and $xP°z$, on the segment uniting z to y there exists a point $w = \alpha y + (1 - \alpha)z$, $0 < \alpha < 1$, such that $wE°x$* [70].

The axiom is tantamount to what is ordinarily known as the Continuity Axiom. There can be no doubt that continuity was assumed by Pareto in his demonstration of how indifference curves may be constructed from the individual's answers to binary choices. To set our ideas in good order, however, we should note that an ophelimity index may be constructed even if G is not true [71].

All we need to add now are the two axioms of shape.

J. THE CONVEXITY AXIOM: *If $yS°x$, $zS°x$, and $w = \alpha y + (1 - \alpha)z$, $0 < \alpha < 1$, then $wS°x$.*

This is one way of saying that the indifference varieties are convex. It is the only axiom to which Pareto paid some attention in that he justified it by invoking what later became known as the Principle of Decreasing Marginal Rate of Substitution (*Manuale*, IV, § 45; App. § 10).

K. THE AXIOM OF THE DOMINANCE OF SATURATION. *In the commodity space, at a finite or at an infinite distance, there exists at least one saturation combination,*

---

[69] The factually relevant illustration of the last point is the scheme of the hierarchy of wants. GEORGESCU-ROEGEN, *Analytical Economics*, pp. 198–201, and "Utility", pp. 262–264.

[70] Cf. Postulate A in GEORGESCU-ROEGEN, "The Pure Theory", p. 136; Axiom V in WOLD, "A Synthesis", Part II, p. 223.

[71] Take the case of $\varphi(x, y) = xy + 1$, if $xy \neq 0$, and $\varphi(x, y) = x/(x + 1) + y/(y + 1)$, if $xy = 0$.

s, such that $\sim xP°s$, for any x. If $w = \alpha s + (1-\alpha)x$, $0 \leq \alpha \leq 1$, is a saturation point only for $\alpha = 1$, then $wP°x$ for $0 \leq \alpha < 1$ [72].

In the standard case, considered by Edgeworth and Pareto, all points at infinity are saturation points. K then simply says that the individual always prefers to have more of any commodity.

VI

We have thus come to a clear axiomatic interpretation of the way in which Pareto thought about the connection between binary choice and indifference varieties. In addition, we have seen that two fundamental axioms of this theory – the axioms of completeness and transitivity – fit into a far more general theory of choice. But one additional point deserves unparsimonious emphasis. If we do not look at Pareto's economic writings in a narrow manner, taking each line by itself, but instead consider these writings as a presentation of a unified conception, we have to come to the conclusion that Pareto viewed the consumer not only as a consistent agent (in the sense of Axiom B_1 and D_1), but also as a rational one (in the sense of E). In other words, although he did not say so explicitly, Pareto's main line of thought rested on the belief in an alternative form of H_2 (of Section III, above).

H_3. THE AXIOM OF BINARY SUPREMACY: *If $\mathscr{F}$ is a family for which A_1 is true and $X \in \mathscr{F}$, then relations P_X and E_X are equivalent to $P°$ and $E°$ respectively.*

This means that regardless of the set from which the individual *can* make a choice, that choice cannot differ from his binary choices. In other words, for any individual there exists a regular indifference map. On the basis of this map and on it alone the individual chooses whenever he is in a situation in which he can make a choice. This position is visible only in very few places in Pareto's writings – for example, where he argues that, for an individual who has to choose the optimal combination of a budget with fixed prices, it is always better to move to a position with " a greater ophelimity " (*Manuale*, III, § 97). This argument has ever since been one of the few hard rocks not only of textbook economic theory but of the ordinary literature as well. Actually, the writers who, like Pareto, used that argument without being aware of the logical necessity to adopt H_3 explicitly are legion.

It was only because Pareto did not cling to his initial position continuously, that he began encountering difficulties. And although the reason why he shifted his position appears in retrospect extremely simple, there still are many untold yet important details of the intricate story.

The merit of the theory of binary choice as a simple and transparent theory is undeniable. However, in order to use that theory for discovering the preference

---

[72] From J, it follows that the set of saturation points is convex. The points at infinity may be handled by considering the projective equivalent of the commodity orthant, which is a simplex. GEORGESCU–ROEGEN, " The Pure Theory ", pp. 142-143.

field of an individual, we must submit the individual to a "laboratory" experiment, as that described by Pareto in his earliest discussion of the subject. We must ask the individual "What do you prefer? 10 cherries and 10 dates or 9 dates and 11 cherries?" (*Lettere*, II, 289) [73]. But such an experiment, though conceivably feasible, is time-consuming beyond all reason. On the other hand, we can observe any individual directly as he chooses the optimal distribution of his budget. Instead of submitting the individual to the lengthy interrogation required by Pareto's approach, we may simply observe his behavior in the market and base our theory on these observations. This simple and at the same time highly appealing idea goes back to G. B. Antonelli (1866) – who, like Pareto, was trained as an engineer [74]. It reappeared in Pareto's writings first as a hint in a 1902 article for the German Encyklopädie [75], and later in full force in the *Manuel* [76]. More recently, Samuelson revived it by offering an entirely new axiomatic basis intended to do away with "the last vestiges of utility analysis" ("A Note", p. 62). It is this last claim that brings to the surface some vital issues about this particular branch of the economic science.

The new approach, now known as Revealed Preference, boils down to constructing a consumer theory only on the choices pertaining to the family $\mathscr{F}^B$ of all possible budgets B $(p; m)$, a budget being a subset of the commodity space satisfying the condition

$$(17) \qquad b(x, p) - m = \sum_{l}^{n} p_i x_i - m \leq 0,$$

where $p = (p_1, p_2, \ldots, p_n)$ represents uniform prices, and $m \geq 0$ is the individual's income. In relation to our theoretical framework (Section V), this means that A_1 is true for $\mathscr{F}^B$. And if the last vestiges of utility analysis are to be dropped out, we cannot assume that A_1 is true for binary choices as well, since such an assumption may lead us back to ophelimity.

But when one theory competes with an old one for supremacy, one important issue is whether the truths accepted by one remain true for the other. Now, if

[73] Simple experiments of this type have been attempted by many psychologists. They are of little, if any, practical value. For a relevant answer, the individual must first experiment with each alternative before being sure that his answer represents what he will choose in the end. On this point, see GEORGESCU-ROEGEN, "The Pure Theory", p. 163, and "The Theory of Choice and the Constancy of Economic Laws" (1950), reprinted in *Analytical Economics*, o, pp. 171–183.

[74] G. B. ANTONELLI, *Sulla teoria matematica della economia politica*, Pisa: Tipografia del Folchetto, 1886. English translation in CHIPMAN *et al.*, *Preferences*, pp. 333–360.

[75] V. PARETO, "Anwendungen der Mathematik auf Nationalökonomie", *Encyklopädie der mathematischen Wissenschaften mit Einschluss ihrer Anwendungen*, 1902, Band I, pp. 1094–1120. Reprinted in *Oeuvres Complètes*, vol. VIII, 126–152. Italian translation in *Giornale degli Economisti*, XXXIII (1906), 424–453.

[76] Whether Pareto drew inspiration from Antonelli's work is a thorny issue. Pareto not only knew of that work but he must have already read it with sufficient care in order to judge in a letter of December 14, 1891, to PANTALEONI (*Lettere*, I, 121) that "Antonelli certainly went up into the clouds". Antonelli's volume is also listed in the bibliography of the *Encyklopädie* article.

one accepts Axiom H_2, which is a particular case of the Rationality Axiom E, then Binary choice also explains how the individual distributes his budget. That is, if A_1 is true for $\mathscr{F}°$, it is true for $\mathscr{F}^B$ as well. We can then observe the individual on the market and still reason according to the theory of binary choice. All we need is that the validity of H_2 and of the other axioms on which that theory rests be verified for a sufficient number of individuals. Indeed, in no science whatsoever is every theoretical principle verified for every instance.

Accepting the validity of the theory of binary choice *does not preclude us* from taking into consideration the fact that to determine the indifference map of any given individual by binary choices is a highly tedious task. Absolutely nothing stops us from using a simpler method, the simplest one for that matter, if such a method exists. To be sure, to determine the magnitude of the angle under which the earth is seen from the moon, one does not have to travel to the moon and measure that angle *directly*. The simplest way to reach the same result is to calculate that angle *indirectly*, by trigonometry, from observations made from the earth.

It is in this manner, I contend, that Pareto viewed the problem at first. In other words, he assumed the existence of a family of indifference varieties represented by the equation

(18)
$$\varphi(x_1, x_2, \ldots, x_n) = \text{const.},$$

which, some conditions being granted, yields the total differential equation

(19)
$$\varphi_1\, dx_1 + \varphi_2\, dx_2 + \ldots + \varphi_n dx_n = 0.$$

On the other hand, from the market equilibrium conditions (*Manuale*, App. § 30),

(20)
$$\frac{\varphi_1}{p_1} = \frac{\varphi_2}{p_2} = \ldots = \frac{\varphi_n}{p_n}, \quad \Sigma p_i x_i = m,$$

one may derive the demand functions

(21)
$$x_i = D_i(p; m),$$

and straightforwardly the inverse demand functions

(22)
$$p_i = m\varphi_i / (\Sigma x_k \varphi_k).$$

In the *Manuale*, however, Pareto was content to use a diagram, an ultra familiar one by now, to show how from *a given* indifference map, one can derive graphically the geometrical representation of the demand functions when only p varies. He referred to these lines variously as " exchange lines ", " supply lines ", or " demand lines ", (*Manuale*, III, § 97, § 180). It is very important to note, in addition, that he explained further that " One may cover the [commodity] plane with a great number of exchange lines; one may thus obtain a representation of the hill of ophelimity indices, *which would be analogous to that obtained by covering the plane with indifference lines*" and which could serve the same purpose as the other lines [77]. All this is sufficient proof that, at that stage in the evolution of his ideas, Pareto considered that all demand functions are based on some indifference map. The statement in the *Manuale* (App. § 6) that by integrating (19) we can obtain the ophelimity–index function, foreshadowed his subsequent position, namely, that from market data we can arrive at a total differential equation

$$(23) \qquad q^1(x)\,dx_1 + q^2(x)\,dx_2 + \ldots + q^n(x)\,dx_n = 0$$

equivalent to (19). Hence, this equation must have an integral which represents the family of all possible ophelimity functions (*Manuel*, App. § 13, 42) [78]. As long as one abides by Pareto's theory of binary choice, this position is faultless. And in truth, Pareto never abandoned it completely. As late as 1911, we see him insisting that "From the mathematical viewpoint, it is immaterial for the determination of equilibrium to know the actions of the individual through the supply and demand functions or through the index–functions. The choice depends on experimental opportunities" ("Economie mathématique", § 3).

What is the situation if one adopts the revealed preference approach? Beforehand, there are three possible alternatives. First, it may happen that the PAP relations of such a theory completely order all commodity combinations and that this ordering coincides with that established by the actual experiments with binary choices. In this case, the old vestiges are only hidden behind the curtains of some new assumptions. Second, the order of the PAP relations is not complete, but it still agrees with actual binary choices. In this case should the new approach be considered superior although it could not account for *all* the facts? Perhaps one may submit that not all pairs are actually comparable. Indeed, it is by no means unreasonable to believe that we can compare only alternatives which do not differ too much. In this case, we may arrive at the most preferred alternative not like a bird diving directly on its chosen prey, but rather like a worm which at each moment decides where to move by comparing only infinitesimally close situations [79]. But this rejoinder would not do. Revealed Preference begins by admitting that even far apart combinations are comparable by S_B. Besides, the point simply reminds us that the old vestiges are the essence of the manner by which the individual arrives at $c(B)$. Before any individual is ready for the check–out line of a self–service store, he has made one binary choice after another, each time pondering whether it would be better to have one dollar less and five apples or one dollar less and a dozen eggs. Finally – and this is the greatest stumbling block of Revealed Preference – its PAP relations, to some extent even S_B, may be completely void of any operational sense. This surprising point will form the object of the concluding section of this essay. Meanwhile, there are some other observations

---

[77] *Manuale*, III, § 97; App. § 19. My italics.
[78] See Section VIII, *infra*.
[79] NICHOLAS GEORGESCU–ROEGEN, "Choice and Revealed Preference", *Southern Economic Journal*, XXI (1954), 119–130, reprinted in *Analytical Economics*, pp. 216–227.

worth making about the problem of Binary Choice versus Revealed Preference.

VII

The brute fact that the individual is able to spend a given budget in some way (invariable in the long run) rather than starve is the cornerstone of Revealed Preference. But one must be careful about how this idea is formulated. We may adapt the axiom

P_1. *The demand functions are singled–valued, nonnegative functions $x_i = D^i (p; m)$, defined for all $p > 0$ and $m \geq 0$,*

without specifying anything else about this form [80]. Even the familiar axiom

P_2. THE EXHAUSTIBILITY AXIOM: *The demand functions satisfy the identity*

(24) $$\Sigma p_i b_i (D, p) = m.$$

is not innocuous. (It denies the existence of a saturation point at a finite distance). And if all vestiges of Paretoan theory are to be discarded, one may like to introduce also

P_3. *The demand functions are homogeneous of the first degree in prices and income,*

which again is not innocuous, not even always factually true. This axiom eliminates money illusion as well as the complex of conspicuous consumption. In the Paretoan framework, it is hidden under the assumption that the ophelimity function does not depend on m or p, or that choice is not influenced by these economic coordinates. If φ depends only on x, a simple look at (20) establishes P_3 as a theorem.

P_1 and P_2, together, constitute Samuelson's Postulate I (" A Note "), and P_3 is his Postulate II. His Postulate III states that

(25) $$[b(D^2, p^1) \leq b(D^1, p^1)] \Rightarrow [b(D^1, p^2) > b(D^2, p^2)],$$

where the meaning of the superscripts is self–explanatory. Translated into the terminology of Section V, this is tantamount to

(26) $$(xS_X y) \Rightarrow [y \in c(Y) \Rightarrow x \notin Y],$$

which is an equivalent form of B_2. Hence, III implies that the demand functions are single–valued, but Samuelson's proof (" A Note ", pp. 353–354) that it also

---

[80] It is conceivable (and in some cases perhaps true) that demand functions may be many–valued. The restriction simplifies the argument without affecting its object–lessons.

implies the other part of I is incorrect. If D is not defined, (25) makes no sense. And if (24) is not assumed, one cannot write $b(D^1, p^1) = b(D^2, p^1)$, where D^2 corresponds to $p^2 = \lambda p^1$.

For an alternative basis we may begin by adopting A_1 for $\mathscr{F}^B$. A_1 excludes from the outset money illusion and conspicuous consumption. It also implies that demand functions are homogeneous of the first degree since $B(p; m) = B(\lambda p; \lambda m)$. We may then add axioms P_2 and B_2 to reach a basis equivalent to Samuelson's.

The question now is to see what we can build on this basis in addition to the properties of demand already mentioned. For this we must turn to B_2. This axiom establishes a convex structure, which may be shown in several ways. But for the present circumstances let us first introduce the assumption that the demand functions (21) can be inverted to yield the positive, singled-valued functions defined over the entire commodity space [81]

$$(27) \qquad p_i = mL_i(x),$$

which, by P_3, satisfy

$$\Sigma x_i L_i(x) = 1.$$

This means that any x is chosen in some unique budget expressed in real terms by

$$(29) \qquad \Sigma x_i L_i - 1 \leq 0.$$

In this case, the directions Δx, with the constant provision that $\Delta x \neq 0$, may be divided into two categories, *nonpreference* and *preference*, according to whether

$$(30) \qquad \Sigma L_i(x) \Delta x_i \leq \text{ or } > 0.$$

It is convenient to divide the nonpreference directions into *limiting* and *antipreference* directions according to whether

$$(30a) \qquad \Sigma L_i(x) \Delta x_i = \text{ or } < 0.$$

An alternative but equivalent form of Axiom B_2, alias Samuelson's Weak Axiom, is the following:

B_4. THE PRINCIPLE OF THE PERSISTENCE OF NONPREFERENCE: *If Δx is a nonpreference direction at x, it is an antipreference direction for any $y = x + \rho \Delta x$, $\rho \geq 0$, such that $y' = x + \rho' \Delta x$ belongs to Ω_n for some $\rho' > \rho$* [82].

---

[81] Since (22) is no longer available, the inversion must now be postulated. The existence of these functions over the entire commodity space also can be falsified. In relation to an indifference map, the assumptions mean that indifference varieties are strictly convex everywhere, save on the boundary of Ω_n.

[82] GEORGESCU–ROEGEN, "The Pure Theory", pp. 138, 141, and "Choice and Revealed Preference", pp. 219–220.

With this change, we pass from Samuelson's Revealed Preference to the theory of Directional Choice, which preceded Samuelson's and, in addition, presents many analytical advantages. However, this fact does not diminish in the least Samuelson's great merit of having stated the same truth as B_4 in a new form which is not only plainly transparent but also cast in finite terms [83].

As can be seen without much ado, if we start with the Paretoan framework, B_4 is equivalent to the (strict) Principle of Decreasing Marginal Rate of Substitution, which establishes the convexity of indifference varieties. As happened time and again, Pareto did not fail to perceive the connection and noted that, if along a straight line ophelimity decreases at one point, it will keep decreasing constantly thereafter (*Manuale*, IV, § 70). For B_4 says that, if $(x + \Delta x) \in \Omega_n$, then

(31) $$\Sigma L_i(x)\Delta x_i \leq 0 \Rightarrow \Sigma L_i(x + \Delta x)\Delta x_i < 0.$$

This yields [84]

(32) $$\Sigma L_i \Delta x_i = 0 \Rightarrow \Sigma \Delta L_i \Delta x_i < 0.$$

Obviously, (32) may also be written as

(32a) $$\Sigma p_i \Delta x_i = 0 \Rightarrow \Sigma \Delta p_i \Delta x_i < 0.$$

a form which expresses the restrictions imposed upon the demand functions by B_4. In case ophelimity exists and φ is differentiable, (32a) is equivalent to

(32b) $$\Sigma p_i dx_i = 0 \Rightarrow \Sigma dp_i dx_i < 0,$$

which expresses the condition for the *strict* convexity of the indifference varieties [85]. Sir John Hicks rightly argued that (32b) constitutes the most general law of demand [86].

For an even more conspicuous association of B_4 with convexity, let us examine first the legitimacy of the terminology introduced by (30) and (30a). The use of "nonpreference" is immediately understood: instead of saying that x is the only choice from the budget (29), it is natural to say that the direction Δx from x to $y = x + \rho \Delta x$ is a nonpreference direction if $y \neq x$ belongs to that budget.

We should note that B_4 represents the strict formulation, which is in accord with (25), Samuelson's formulation. In a nonstrict formulation, B_3, "antipreference" should be replaced with "nonpreference".

(83) Cf. GEORGESCU–ROEGEN, "Utility", p. 257.

(84) GEORGESCU–ROEGEN, "The Pure Theory", pp. 139–141; SAMUELSON, "A Note", p. 67.

(85) Paul A. SAMUELSON, "The Empirical Implications of Utility Analysis", *Econometrica*, VI (1938), pp. 345–347; J. R. HICKS, *Value and Capital*, 2nd ed., Oxford: Clarendon, 1946, pp. 311–329.

(86) Of course, (32) represents this law regardless whether or not ophelimity exists. For some interesting analytical differences between the two cases, see GEORGESCU–ROEGEN, "The Pure Theory", pp. 139–141, 165–170.

In fact, we can refer to the entire oriented segment xy (not necessarily including y) as a nonpreference direction and write xy^-. In this case, the opposite direction $(-\Delta x)$ at any point of the segment yx (*except perhaps at* x) is a preference direction. If the individual is offered the budget distribution y, he will reject it and choose x. We may therefore call the oriented segment yx a preference direction and write yx^+. With this understanding,

$$(33) \qquad xy^- \Rightarrow yx^+.$$

At x however, Δx may be a limiting direction, in which case $(-\Delta x)$ is a nonpreference (limiting) direction at x.

The preference direction – as Pareto noted for an indifference map – does not necessarily persist; it may turn into a nonpreference one. But this is not all in our general case. The axioms adopted so far do not exclude the possibility that Δx may be a preference direction at x and a nonpreference direction for every $y = x + \rho \Delta x$, $\rho > 0$, $y \in \Omega_n$. Curiously, *this discontinuity does not exist for two commodities.*

Indeed, if $\Delta x_1 \geq 0$, $\Delta x_2 \geq 0$, it is immediate that $\Sigma L(y) \Delta x > 0$ for any ρ [87]. And if $\Delta x_1 < 0$, $\Delta x_2 > 0$, the choice z of the budget determined by the line (x, y) must be such that zx^-, hence xz^+. For two commodities, therefore, the preference direction also persists but only over a possibly limited segment open at one end. However, the following proposition is true for any number of commodities.

THEOREM I. *Let x, $y \in \Omega_n$ and let $w = \alpha x + (1 - \alpha)y$, $0 < \alpha < 1$. Then either xw^+ or yw^+.*

For this we need only observe that either $\Sigma x L(w) \leq 1$ or $\Sigma y L(w) \leq 1$ [88].

The following corollary, which is straightforward, is extremely important for the framework under discussion.

COROLLARY 1: *If relations (III) refer to the choice on $\mathscr{F}^B$, then the set*

$$(34) \qquad P(x) = \{y \mid y P_t x \text{ for some } t\}$$

is convex [89].

The next theorem brings to the surface the factors that clouded for a long time, and often still cloud, our understanding of the power of Samuelson's Weak Axiom of Revealed Preference and of the integrability issue.

---

[87] This result, that if $y \geq x$, $y \neq x$, then xy^+, is true in general and, in substance, is tantamount to Axiom K.

[88] Of course, we may have both $\Sigma x L(w) = 1$ and $\Sigma y L(w) = 1$, but the existence of such a w is not warranted except in the case of two commodities, when we may have at the same time $w = x$ or $w = y$. We may make sure that such a w exists always by postulating that $L_i(x)$ is a continuous function (Cf. GEORGESCU–ROEGEN, "Choice and Revealed Preference", p. 220). What is highly desirable is an example to prove that the discontinuity of the preference direction cannot be eliminated without an additional postulate.

[89] This generalizes proposition (v) in GEORGESCU–ROEGEN, "Choice and Revealed Preference", p. 224.

THEOREM II: *For two commodities, Axiom B_2 – the Weak Axiom of Revealed Preference – implies Axiom D_2 – the Strong Axiom of Revealed Preference in its strict form.*

Indeed, in *two dimensions*, for any nonnull closed chain $(x, t_1, t_2, \ldots, t_n, x)$, we must have $t_j \geq t_{j+1}$, for some j (with $t_0 = x$ and $t_{n+1} = x$). But then $t_j \overline{t_{j+1}}$, which is incompatible with $xP_t x$.

Since, as we shall prove in the next section through a simple, direct example, this theorem is not true for more than two commodities, at this juncture we may pass to the general case by *explicitly introducing* the Strong Axiom in its strict form, i.e., as D_2. On this basis, we can easily obtain a series of interesting results. To this purpose, let $K(x)$ stand for the closure of $P(x)$ and $k(x)$ for the boundary of $K(x)$.

THEOREM III: $x \notin P(x); x \in k(x)$.

THEOREM IV: *The hyperplane $\Sigma_i x_i L_i(x) = 1$ is the unique supporting plane of $K(x)$ at x.*

COROLLARY IV: *The map of the $k(x)$'s coincides with that of the integral varieties of the equation*

$$\Sigma L_i(x) dx_i = 0. \tag{35}$$

THEOREM V: *If $yP_t x$, then $P(y) \subset P(x)$.*

An extremely important conclusion is that two combinations x and y may not be comparable by PAP relations, i.e., neither xP^*y nor yP^*x is true. In this case $K(x)$ and $K(y)$ overlap, and since at any of their common points there is one

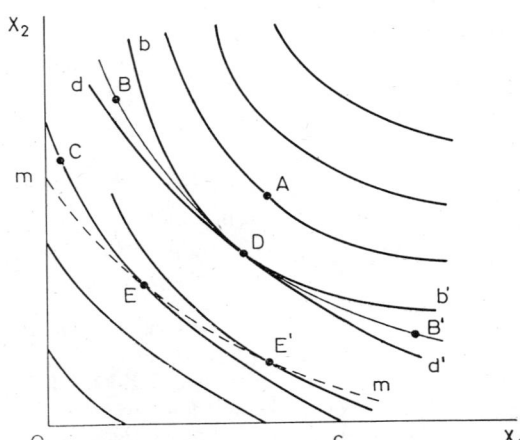

Fig. 2

and only one supporting plane of both, it follows that $k(x)$ and $k(y)$ are tangent at any of their points of intersection. The qualitative analysis of the integral lines leads to the illustration of Fig. 2. There are, first, points such as A, through which passes only one integral line that, in addition, does not intersect any other integral line. In this case, the integral line coincides with $k(A)$, and A is PAP comparable with any $x \in \Omega_n$, except if $x \in k(A)$. In other words, either xP^*A or AP^*x. Second, there may be points such as B or C, through which passes only one integral line, but this integral line touches one or several other integral lines. These cases are related to what in the theory of differential equations are called singularities. D is a singular point through which pass many integral lines, all having the same tangent. It is known at times as a *node*, at times as a *pole*. E also is a singular point – through it pass two integral lines; we may refer to such a point as a *simple pole*. The integral line mm' is another type of singularity; all its points are simple poles. The result is that the integral lines have no "individuality", as it were. We have no ground to say whether DB and DB' belong to one and the same integral line. However, let bDb' be the highest integral line passing through D and dDd', the lowest. Then $k(B)$ is BDb' and $k(B')$ is $B'Db$, which is the same as saying that $K(B)$ and $K(B')$ overlap. The points within the curved triangle bDd are PAP noncomparable not only with those on B'D but also with any point within the curved triangle $b'Dd'$. Similarly, C is noncomparable with any point within mEc.

VIII

The foregoing theorems and the qualitative analysis of the integral lines expose in a simple, palpable manner the roots of several errors that have been besetting the theory of Revealed Preference and the issue of integrability [90]. Perhaps the most important point is that the integral varieties of (35), whether for two or more commodities, do not necessarily represent indifference varieties, unless the existence of a regular indifference map is already assumed by the approach [91] – as, according to my contention, Pareto did. The confusion, very common among economists, began with Vito Volterra's famous criticism of Pareto for having not taken into account in the Appendix of the *Manuale* the fact familiar to all students of calculus that a total differential equation in two variables, $Xdx + Ydy = 0$, is as a rule integrable, whereas one in more than two variables need not be so [92]. With this remark, Volterra shifted the attention of theoretical economists from more

---

[90] Except Theorem II, which very likely is new, all other results were presented for the the first time in GEORGESCU–ROEGEN, "Choice and Revealed Preference". The assumption of the continuity of $L_i(x)$, made at that time, is now seen to be unnecessary. The case of the pole and its bearings on the PAP pseudo measure of utility were rediscovered later by L. HURWICZ and M. RICHTER ("Revealed Preference Without Demand Continuity Assumptions", in CHIPMAN et al., *Preferences*, p. 62). The highly interesting case of the singular integral line was introduced into the picture by W. M. GORMAN ("Preference, Revealed Preference, and Indifference", *ibid.*, p. 85).

[91] GEORGESCU–ROEGEN, "The Pure Theory", pp. 148–151, 164, and "Choice and Revealed Preference", pp. 219, 225.

[92] VITO VOLTERRA, "L'economia matematica ed il nuovo manuale del prof. Pareto", *Giornale degli Economisti*, XXXII (1906), 296–301. English translation in CHIPMAN et al., *Preferences*, pp. 365–369.

basic issues to that of the simple existence of an integral of a total differential equation. Integrability thus became the target of a long series of endeavors.

Samuelson was the first to argue in his masterly *Foundations* that his Weak Axiom alone suffices to derive a map of indifference varieties from a complete set of market data [93]. Later, Little revisited the problem and claimed to arrive at the same result by a more economical procedure [94]. Then, Samuelson followed with a more technical essay pertaining to two commodities and assuming, in addition, the classical Cauchy–Lipschitz condition for regular integrability [95]. But this last essay of Samuelson's is of little value, for it merely repeats the mathematical textbook argument about this condition. It does reveal, however, that at the time Samuelson still believed that the Weak Axiom is not only a convexity condition, but it also has some important bearings on integrability. In retrospect it is highly curious that both authors ignored a fact proved long before, namely, that convexity alone is not sufficient for integrability.

That proof pertained to a framework of Directional Choice of the kind outlined in the preceding section. The idea of Directional Choice, which goes back to Pareto (*Manuel*, App. § 13), is to determine by experience the limiting directions at every x and thus arrive at a total differential equation

$$(36) \qquad dx_1 + M_2(x)dx_2 + \ldots + M_n(x)dx_n = 0,$$

equivalent to (35) [96]. If one begins by assuming – as Pareto appears to have done – that an ophelimity function necessarily exists, then (36) certainly has a regular integral which represents the family of ophelimity indices itself. But if we adopt only B_4 or only Samuelson's Weak Axiom, all we know is that the M's satisfy the " convexity " condition

$$(37) \qquad dx_1 + \sum_2 M_i(x)dx_i \leq 0 \Rightarrow \sum_2 \Delta M_i \Delta x_i < 0.$$

And this condition alone does not warrant the integrability of (36) for $n < 2$. The reason is revealed by the proof of Theorem II, which establishes that, for $n = 2$, B_4 implies D_2. It can hardly be overstressed that even for $n = 2$, the proof of Theorem II collapses if a saturation point exists at a finite distance; for in that case, $x \leq y$ no longer implies xy^+. The same theorem collapses for $n > 2$, because in that case there are closed chains $(x = t_0, t_1, \ldots, t_n, t_{n+1} = x)$ such that

---

[93] PAUL A. SAMUELSON, *Foundations of Economic Analysis*, Cambridge, Mass.: Harvard University Press, 1947, pp. 150–154.

[94] I. M. D. LITTLE, " A Reformulation of the Theory of Consumers' Behavior ", *Oxford Economic Papers*, N. S., I (1949), 90–99, and his *A Critique*, Appendix, II, pp. 283–289.

[95] PAUL A. SAMUELSON, " Consumption Theory in Terms of Revealed Preference ", *Economica*, N. S., XV (1948), 243–253.

[96] Pareto, just like this writer in his proof, erred by believing that if x and $x + \Delta x$ are two indifferent combinations, and Δx is very small, Δx is a limiting direction. Yet the fact that the limiting directions can be determined experimentally is obviously true.

$t_j \geq t_{j+1}$ for no j. These points are easily established by considering the simplest, yet relevant, case of $n = 3$.

From A_1, P_2 and B_4, we know that the choice $x°$ of the budget $p/m = (1, M_2°, M_3°)$ is unique and satisfies the equation

(38) $$x_1 + M_2° x_2 + M_3° x_3 = x_1° + M_2° x_2° + M_3° x_3°$$

where $M_i(x°) = M_i°$. The corresponding "budget simplex", $B(x°)$, is the triangle obtained by the intersection of the budget plane (38) with Ω_3. Let now $y \in B(x°)$, $y \neq x°$. The intersection of the budget plane

(39) $$x_1 + M_2(y) x_2 + M_3(y) x_3 = y_1 + M_2(y) y_2 + M_3(y) y_3$$

with (38) represents a limiting direction at y. Any such intersection may be regarded as a "budget simplex" – in this case, a budget segment – within $B(x°)$. In order to facilitate the analysis of this structure, we may project it on the plane $x_2 O x_3$ (Fig. 3). In the projection simplex, $a_2 O a_3$, the family of the budget segments is represented by the differential equation

(40) $$(\bar{M}_2 - M_2°) dx_2 + (\bar{M}_3 - M_3°) dx_3 = 0,$$

where

(41) $$\bar{M}_i = M_i[x_1° + M_2° (x_2° - x_2) + M_3° (x_3° - x_3), x_2, x_3].$$

Simple algebra shows that, if the budget equations (39) satisfy the Principle of the Persistence of Nonpreference, so does the family (40). It is also seen that

(42) $$(\bar{M}_2 - M_2°)(x_2° - x_2) + (\bar{M}_3 - M_3°)(x_3° - x_3) \geq 0,$$

the equality prevailing if and only if $x = x°$. This means that $xx°$ is always a preference direction and $x°x$ a nonpreference direction. Within $B(x°)$, $x°$ is a point of saturation. One cannot, therefore, maintain that saturation points and their properties are irrelevant issues for Revealed Preference.

Now, equation (41) – as Volterra said in his review – is integrable, but only if "integrable" is interpreted in a broad sense. It is in this broad sense that we may say that $ydx - xdy = 0$ is integrable, the integral lines being represented by $y - kx = 0$. However, $x = 0$, $y = 0$ constitutes a singular point (usually called a *vertex*), because through it passes an infinity of integral lines that have no common tangent. In a strict, very strict, sense of the term, the same equation is *not* integrable at $x = 0$, $y = 0$. The crucial point which Volterra failed to see in relation to the problem considered by Pareto is that the integrability (in the broad sense) of (42) is not sufficient to establish the existence of a PAP ophelimity index over $B(x°)$. And if such an index does not exist, certainly no PAP ophelimity index exists over Ω_n. The point overlooked by Volterra – and, as a result, ignored by mathematical economists for a long time thereafter – is that if the integral solution of (42) has some particular kind of singularities, no PAP ophelimity index may be constructed over $B(x^\circ)$. And in fact, (x_2°, x_3°) – hence, x° itself – is a singular point; for $x_2 = x_2^\circ$, $x_3 = x_3^\circ$, the equation (42) no longer determines dx_2, dx_3.

To determine the pattern of the integral lines around a singularity in the general case is one of the most intricate problems of qualitative analysis. Among all possible cases, there is that in which (x_2°, x_3°) is a *focus*. This means that the integral lines are asymptotic spirals around the focus. Of course, the same pattern, shown

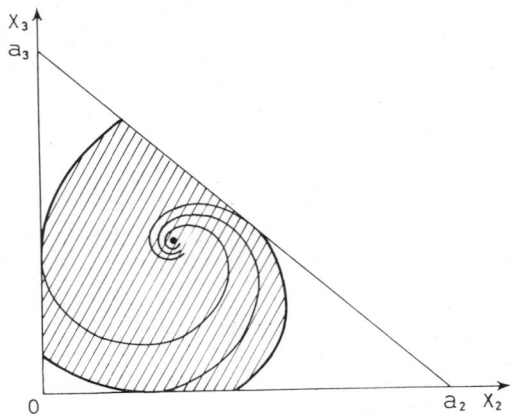

Fig. 3a

in Fig. 3a, portrays the situation in $B(x^\circ)$. The Principle of the Persistence of Nonpreference obviously applies, and, as is easily seen, Corollary I also is true. For any point z of the shaded area (the curved boundaries *excluded*), the corresponding $P(z)$ is that area itself, which *is* convex. This proves that $z \in P(z)$, which means that Theorem III, based on the Strong Axiom of Revealed Preference, is not valid in this case. And if z' is another point of the shaded area, then both zP_tz' and $z'P_uz$ are true for some chains. Therefore, no PAP order, implicitly, no PAP ophelimity index can be constructed.

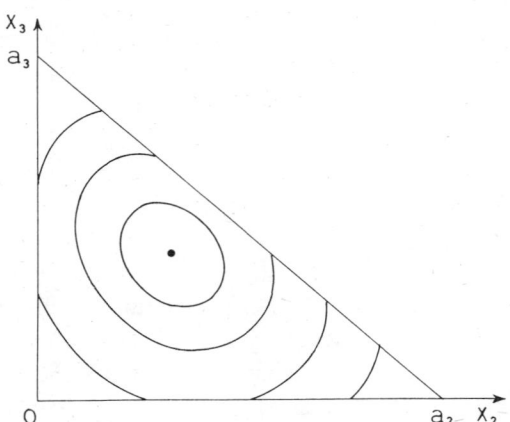

Fig. 3b

However, if the singular point (x_2°, x_3°) is a *center*, instead of a focus, and *all integral lines have no point in common* (as shown in Fig. 3b), then a PAP ophelimity index can be constructed immediately over $B(x°)$. And if this pattern prevails for every $x°$, there exists a PAP ophelimity index over the entire commodity space [97].

The foregoing results were revived when Ville and Houthakker came forth with the so–called Strong Axiom of Revealed Preference, which states that D_2 is true for $\mathscr{F}^B$. For the effect of that axiom is, in the first place, to eliminate the cases in which the singular point represented by the combination chosen from a budget is either a focus or a vertex([98]). As we have seen in Section VII, the ultimate effect is rather impressive: The Strong Axiom guarantees the existence of integral lines for (35), alternatively, for (36). However, Samuelson again set his hopes too high. For he acclaimed the Strong Axiom as placing us " in a position to complete the programme begun a dozen years ago [in his 1938 " A Note "] of arriving *at the full empirical implications for demand behavior of the most ordinal utility analysis* " [99]. Apparently, he believed that that axiom finally removed all obstacles, with the result that on the basis of it we can always arrive at a map of the same pattern as that of Pareto, except for the fact that its varieties could be called " indifference varieties " only by courtesy. Some function ψ could then be constructed over Ω_n, such that for any $x, y \in \Omega_n$,

(43) $$[\psi(x) < \psi(y)] \Leftrightarrow y P^* x,$$

$$[\psi(x) = \psi(y)] \Leftrightarrow x N^* y.$$

Such a function may be referred to as a *quasi ophelimity index* [100]. But, as subsequent results showed (Section VII), the Strong Axiom does not exclude a map like that of Fig. 2. In this case, all we can do is to construct a *pseudo ophelimity index* $\chi(x)$, for which only

(44) $$y P^* x \Rightarrow [\chi(x) < \chi(y)]$$

is true. This means that if $x N^* y$, all cases, $\chi(x) \lesseqgtr \chi(y)$, are possible.

At present, the difficulty of constructing a true ophelimity index by paper–and–pencil operations stems from the existence of poles. Of course, one could eliminate all singularities by openly assuming from the outset the Cauchy–Lipschitz condition. The drawback of this plan is that there is no practical way of testing the Cauchy–Lipschitz condition. The problem before us, therefore, is to find an additional axiom formulated (as the Strong Axiom is) in finite terms and capable of eliminat-

---

[97] For further details, see GEORGESCU–ROEGEN, " The Pure Theory ". The case of $x P_t x$ may be encountered in still another pattern, in which $x°$ is a *vertex* through which passes a pencil of convex half–curves, as shown in GEORGESCU–ROEGEN, " Utility ", Figure 10.

[98] The only remaining pattern, other than one with a center, is that in which $x°$ is a saddle-point. But this pattern is eliminated by the convexity condition, the Weak Axiom.

[99] PAUL A. SAMUELSON, " The Problem of Integrability in Utility Theory ", *Economica*, N. S., XVII (1950), 369.

[100] We should be reminded that we have been assuming all along that the choice set is a singleton; hence, no nontrivial equivalence exists.

ing the poles. But given that, as we shall argue in the following sections, the PAP ophelimity has no economic relevance unless the demand functions or the limiting directions are derived from an existent, albeit unknown, indifference field, such an axiom may present a greater interest in mathematics than in economics.

IX

There are indications that Pareto was disconcerted by Volterra's objection. The footnote in which Pareto refers to Volterra's objection (*Manuel*, App. § 12), is quite telling. On the one hand, he dogmatically, even arrogantly, asserts that " The criticisms of the literary economists have no value "; on the other hand, in an obviously deferent tone, quite unusual for him, Pareto goes on to say that " the observations and the criticisms coming from a savant such as Mr. Volterra have a great value and are precious for the progress of science ". Another indication is the fact that immediately after the publication of Volterra's review, Pareto produced a related article which appeared in the very next issue of the same journal [101] and was reproduced in part in the *Manuel* (App. § 14–21).

The article was an obviously hurried attempt at face-saving and did not come to grips with the precise issue raised by Volterra. Instead, it shifted the problem to a psychological ground – an idea which represented a reversal of an earlier position of Pareto's. In the *Manuale* (IV, § 7), Pareto had mentioned that even though the pleasure derived from consumption varies with the order in which the goods – dishes at a meal, for example – are consumed, economics avoids this thorny and irrelevant problem because it simply considers only the quantities at the disposal of the individual, who is supposed to choose the order that suits him best. This time, however, Pareto turned to the order of consumption in the hope of explaining the difference between integrability and nonintegrability. But he also had to turn to another idea even more damaging to his novel theory of ordinal utility: he assumed that ophelimity is cardinally measurable, thus reverting to Edgeworth's position.

The basic step in this new endeavor is the assumption that, if the actual consumption of one of the goods increases by dx_i, the ophelimity – Pareto actually meant " the amount of pleasure " – increases by $Q_i dx_i$.

Pareto (" L'ofelimità nei cicli chiusi ", *passim*; *Manuel*, App. § 14–15) also considered only the case in which the order of consumption is set for one entire quantity after another. The formula for the amount of pleasure for the order $x_1, x_2, \ldots, x_n$ is

(45) $$u(x) = \int_0^{x_1} Q_1(x, 0, 0, \ldots, 0)\, dx + \int_0^{x_2} Q_2(x_1, x, 0, \ldots, 0)\, dx +$$
$$\ldots + \int_0^{x_n} Q(x_1, x_2, \ldots, x_{n-1}, x)\, dx.$$

---

[101] VILFREDO PARETO, " L'ofelimità nei cicli non chiusi ", *Giornale degli Economisti*, XXXIII (1906), 15–30. English translation in CHIPMAN *et al.*, *Preferences*, pp. 370–385.

Even in this form, it is obvious that $u(x)$ changes with the order, unless the " marginal utilities ", $Q_i(x)$, satisfy certain conditions. But here, as in many other situations, the general case leads us more directly to the solution of the problem at hand. Instead of Pareto's (45), we may let the order of consumption follow any nonbackward-bending line L, uniting the origin of the corrdinates to any given x. In this case, the amount of pleasure is given by the line integral

$$(46) \qquad u(x; L) = \int_L \Sigma Q_i dx_i.$$

Let us now consider an infinitesimal extension of L along the sides of the rectangle determined by x and $x + dx$, where $dx = (dx_1, dx_2, 0, \ldots, 0)$, and assume that the Q_i's have continuous derivatives over Ω_n. For $u(x + dx; L + dL)$ not to depend on the order in which the sides are followed, we must have

$$(47) \qquad \frac{\partial Q_2}{\partial x_1} dx_1 dx_2 = \frac{\partial Q_1}{\partial x_2} dx_2 dx_1,$$

and hence, in general,

$$(48) \qquad \frac{\partial Q_i}{\partial x_k} = \frac{\partial Q_k}{\partial x_i}, \quad i, k = 1, 2, \ldots, n.$$

This means that the expression under the integral (46) is an exact differential; that is, there is a function $Q(x)$ such that

$$(49) \qquad dQ = \Sigma Q_i dx_i.$$

It is curious that Pareto did not see that at least for $n = 2$ and $n = 3$, conditions (48), hence (49), follow from the Green–Stokes formulae, with which an engineer of his training must have been quite familiar. To recall, if the derivatives involved are continuous, Green's classical formula is

$$(50) \qquad \int_\Gamma (Q_1 dx_1 + Q_2 dx_2) = \iint_A \left(\frac{\partial Q_2}{\partial x_1} - \frac{\partial Q_1}{\partial x_2} \right) dx_1 dx_2,$$

where A is a finite domain and Γ is its boundary. In Pareto's terminology, a cycle Γ is closed, if after following it, the amount of pleasure returns to its initial value (*Manuel*, App. § 20). By this definition, the line integral in (50) must be zero for any Γ, if all cycles are to be closed. The double integral then yields (48).

The reason why Pareto did not come to grips with Volterra's objection is that he focused his attention on the situation in which (48), hence (49), is satisfied. (" L'ofelimità ", § 14; *Manuel*, App. § 17). This problemconc erns both the case of $n = 2$ and of $n = 3$. But Volterra had referred to an entirely different problem, that of whether the total differential equation (23) of Section VI,

(51) $$\Sigma\, q^i\, dx_i = 0$$

has a solution R, which must be such that

(52) $$\frac{\partial R}{\partial x_i} \Big/ \frac{\partial R}{\partial x^n} = \frac{q^i}{q^n}.$$

In fact, Pareto never came to mention the classical condition(s) for the integrability of (51).

However, Pareto's effort still bore witness to his analytical intuition. For, as we have seen in the preceding Section, the issues of the existence of a PAP ophelimity index depends on what happens along a cycle. Only, it does not concern the amount of the ophelimity (pleasure). Instead, it concerns the kind of the direction in which the cycle proceeds (preference or nonpreference). If this direction is of the same kind all along the cycle, there is no PAP ophelimity index, not even a pseudo index.

X

Even after his excursion into the problem of closed and open cycles, Pareto never abandoned his position of the *Manuale*. We see this best in the " Economie mathématique ", where he did not separate the picture of the exchange lines from that of the indifference lines. It seems in order, therefore, to believe that Pareto never felt that a complete theory of the consumer capable of subserving positive welfare economics could be constructed from market data alone. Such a theory must be based on the fundamental assumption that any individual can always choose between two alternatives in the manner implied by his theory of Binary Choice, i.e., in a manner that leads to the construction of a typical indifference map. Market data can only provide us with an indirect means for determining the indifference map, simply because they fit that map *ex hypothesi* (Section VI) [102].

This view of Pareto's probable standpoint leads us to a question not yet raised by those who, like the present writer, have been too busy trying to improve the PAP arguments of Revealed Preference theory along the tracks initiated by Antonelli and adopted, although somewhat on the margin, by Pareto himself. The question is this. Does this theory prove Pareto wrong? In more precise terms, do the PAP relations P* derived by this theory in its strongest form – i.e., in the form in which all singularities are excluded – have any operational binary meaning? Surprising as this may seem, the answer to this question is a definite " No ". The PAP relations P*, to the extent to which P* is not reduced to a simple P, have no meaning outside the paper-and-pencil operations. They can have an operational

---

[102] Probably because of this belief, Pareto took many things for granted – some even not true – in his mathematical derivation of some relations concerning the demand functions or the differential equation of the indifference map. Cf. GEORGESCU-ROEGEN, " Note on a Proposition of Pareto ".

meaning only if the Binary Choice theory is already assumed, in which case xP^*y must coincide with $xP°y$.

As an analytical illustration of this point, let us consider the case in which the demand of a community satisfies the usual axioms of Revealed Preference. As is well-known, the demand of a community does not, in general, satisfy even Axiom A_1. The reason is that the demand functions of the community depend on the income *distribution* as well. There is, however, a situation, in which there exists a PAP quasi measure of ophelimity, $\Psi(x)$, and hence, a PAP " indifference " map, $\Psi(x) =$ constant. Although realistically absurd, this situation will serve our analytical purposes to the point.

Let us begin by noting that, conceivably, the entire national income, M, may be allocated to any one individual. All individuals must therefore have the same demand, for all prices and incomes. And, the first commodity being taken as *numéraire*, we have

(53) $$\Delta(p_2, p_3, \ldots, p_n; M) = D(p_2, p_3, \ldots, p_n; M),$$

where Δ is the community demand and D the typical demand of every individual. The individual demands being identical, we must also have

(54) $$D(p_2, p_3, \ldots, p_n; M) = \Sigma_k D(p_2, p_3, \ldots, p_n; m_k)$$

for any distribution $M = \Sigma m_k$. And if D is assumed continuous in M, (54) yields

(55) $$D(p_2, p_3, \ldots, p_n; m) = md(p_2, p_3, \ldots, p_n).$$

This means that all Engel curves are straight lines through the origin [103]. If we assume now that each x corresponds to only one budget $B(1, p_2, p_3, \ldots, p_n; m)$, the functions

(56) $$x = md(p_2, p_3, \ldots, p_n),$$

can be inversed, and since

(57) $$d_i/d_1 = x_i/x_1, \quad i = 2, 3, \ldots, n,$$

the inverse functions are homogeneous of degree zero:

(58) $$p_i = M_i \left(\frac{x_2}{x_1}, \frac{x_3}{x_1}, \ldots, \frac{x_n}{x_1} \right), \quad i = 2, 3, \ldots, n.$$

---

[103] The problem of the indifference varieties for a community was treated, almost at the same time, by W. M. GORMAN, " Community Preference Fields ", *Econometrica*, XXI (1953), 63–80, and A. NATAF, " Possibilité d'aggrégation dans le cadre de la théorie du choix ", *Metroeconomica*, V (1953), 22–30, who arrived at a different result than the one reached here and by a more involved procedure.

But by Corollary IV, equation (36) is integrable. If we put $x_i = w_i\, x_1$, (36) becomes

$$(59) \qquad (dx_1/x_1) + (\Sigma M_i dw_i) / (1 + \Sigma M_i w_i) = 0.$$

This proves that the integral of (36) is reducible to the form

$$(60) \qquad \Psi(x) = x_1\, G\left(\frac{x_2}{x_1}, \frac{x_3}{x_1}, \ldots, \frac{x_n}{x_1}\right),$$

and the integral varieties are homothetic with respect to the origin of the coordinates.

The map defined by $\Psi(x) =$ constant is the indifference map of every individual and also the map of behavior varieties of the entire community. Yet the meanings of these two maps are totally different. For the individual we may very well assume that the theory of Binary Choice is true and hence, even though (60) has been derived by paper-and-pencil operations from his market behavior, $\Psi(x) \geq \Psi(y)$ means that $x S° y$ *factually*. Obviously, $\Psi(x) > \Psi(y)$ does not necessarily mean that the community prefers x over y. The community will disagree even on how to use x. The only case in which $\Psi(x) > \Psi(y)$ has *some* operational meaning is when x is chosen from a budget that contains y. But since this situation is part of the informational basis, it constitutes no novelty.

The upshot is clear: along with the paper-and-pencil operations, the builders of the Revealed Preference theory have tacitly attached to these operations a gratuitous operational significance. This is a logical error which cannot be imputed to Pareto's handling of the same problem, however faulty that handling may be from other viewpoints.

Our journey has been rather long, yet it has proved definitely worthwhile. For it is important, I think, to discover that Pareto's theory of ordinal ophelimity based on Binary Choice is the only right way to theorize fruitfully about the problems of the consumer. Pareto's visible reluctance to depart from this idea, even after Volterra's review seemed to have somewhat intimidated him, may bear the best witness to his genial intuition as a theoretical economist.

REFERENCES

1. ALLAIS, MAURICE, "Vilfredo Pareto", *International Encyclopedia of the Social Sciences*, New York: Macmillan and The Free Press, Vol. 11, pp. 399–411.
2. ALLEN, R. G. D., "The Foundations of a Mathematical Theory of Exchange", *Economica*, XII (1932), 197–226.
3. ANTONELLI, G. B., *Sulla teoria matematica dell'economia politica*, Pisa: Tipografia del Folchetto, 1866. English translation in [12], pp. 333–360.
4. ARMSTRONG, W. E., "The Determinateness of the Utility Function", *Economic Journal*, XLIX (1939), 453–467.
5. ARROW, KENNETH J., "Rational Choice Functions and Orderings", *Economica*, N. S., XXVI (1959), 121–127.

6. AUSPITZ, R. and LIEBEN, R., *Untersuchungen über die Theorie des Preises*, Leipzig: Dunker and Humblot, 1889.
7. BARONE, ENRICO, " Il ministro della produzione nello stato collettivista ", *Giornale degli Economisti*, XXXVII (1908), 267–293, 391–414. English translation in *Collectivist Economic Planning*, F. A. Hayek ed., London: Routledge and Kegan Paul, 1935, pp. 245–290.
8. BENTHAM, JEREMY, *The Works of Jeremy Bentham*, 11 vols., John Browning ed., Edinburgh: William Tait, 1838–1843.
9. BOWLEY, A. L., *The Mathematical Groundwork of Economics*, Oxford: Clarendon 1924.
10. BROWN, E. H. PHELPS, " Notes on the Determinateness of the Utility Function ", Part 1, *Review of Economic Studies*, II (1934), 66–69.
11. CASSEL, GUSTAV, " Grundriss einer elementaren Preislehre ", *Zeitschrift für die gesamte Staatswissenschaft*, LV (1899), 395–458.
12. CHIPMAN, JOHN S.; HURWICZ, LEONID; RICHTER, MARCEL K.; and SONNENSCHEIN, HUGO F., *Preferences, Utility, and Demand*, New York: Harcourt Brace Jovanovich, 1971.
13. CHIPMAN, JOHN S., " The Paretian Heritage ", Paper read at the Kingston meeting of the Canadian Economic Association, 1973 (unpublished).
14. DEBREU, G., " Topological Methods in Cardinal Utility Theory ", in *Mathematical Methods in the Social Science*, K. J. Arrow et al. eds., Stanford: Stanford University Press, 1960, pp. 16–26.
15. EDGEWORTH, F. Y., *Mathematical Psychics*, London: Kegan Paul, 1881 (Reprinted, New York: August M. Kelley, 1953).
16. — —, *Papers Relating to Political Economy*, 3 vols., London: Macmillan, 1925.
17. FISHER, IRVING, *Mathematical Investigations in the Theory of Value and Prices*, New Haven: Yale University Press, 1925 (First published in 1892).
18. FRISH, RAGNAR, " Sur un problème d'économie pure ", *Norsk matematisk forenings skrifter*, Series I, No. 16, 1926, pp. 1–40. (English translation in [12], pp. 386–423).
19. FURLAN, V., " Cenni su una generalizzazione del concetto d'ofelimità ", *Giornale degli Economisti*, XXI (1908), 259–265.
20. GEORGESCU-ROEGEN, NICHOLAS, " Note on a Proposition of Pareto ", *Quarterly Journal of Economics*, XLIX (1935), 706–714.
21. — —, " The Pure Theory of Consumer's Behavior ", *Quarterly Journal of Economics*, L (1936), 545–593. Reprinted in [25], pp. 133–170.
22. — —, " The Theory of Choice and the Constancy of Economic Laws ", *Quarterly Journal of Economics*, LXIV (1950), 125–138. Reprinted in [25], pp. 171–184.
23. — —, " A Diagrammatic Analysis of Complementarity ", *Southern Economic Journal*, XIX (1952), 1–20.
24. — —, " Choice and Revealed Preference ", *Southern Economic Journal*, XXI (1954), 119–130. Reprinted in [25], pp. 216–227.
25. — —, *Analytical Economics: Issues and Problems*, Cambridge, Mass.: Harvard University Press, 1966.
26. — —, " Utility ", *International Encyclopedia of the Social Sciences*, New York: Macmillan and The Free Press, Vol. 16, pp. 236–267.
27. — —, *The Entropy Law and the Economic Process*, Cambridge, Mass.: Harvard University Press, 1971.
28. — —, *Analisi economica e processo economico* (with an Introduction by Giacomo Becattini), Firenze: Sansoni, 1973.
29. GORMAN, W. M., " Community Preference Fields ", *Econometrica*, XXI (1953), 63–80..
30. — —, " Preference, Revealed Preference, and Indifference ", in [12], pp. 81–113.
31. GOSSEN, H. H., *Entwickelung der Gesetze des menschlichen Verkehrs und der daraus fliessenden Regeln für menschliches Handeln*, Berlin: Prager, 1889. (First edition, 1854).
32. HICKS, J. R., " Marginal Productivity and the Lausanne School: A Reply ", *Economica*, XII (1932), 297–300.
33. HICKS, J. R. and ALLEN, R. G. D., " A Reconsideration of the Thoery of Value ", *Economica*, N. S., I (1934), 52–76, 196–219.

34. HICKS, JOHN R., *Value and Capital*, 2nd ed., Oxford: Clarendon, 1946. (First edition, 1939).
35. HOUTHAKKER, H. S., "Revealed Preference and the Utility Function ", *Economica*, N. S., XVII (1950), 159–174.
36. HURWICZ, L. and RICHTER, M. K., "Revealed Preference Without Demand Continuity Assumptions ", in [12], pp. 59–76.
37. HUTCHISON, T. W., *A Review of Economic Doctrines: 1870–1929*, Oxford: Clarendon, 1953.
38. JAFFÉ, WILLIAM, " Pareto Translated: A Review Article ", *Journal of Economic Literature*, X (1972), 1190–1201.
39. JANNACCONE, PASQUALE, " Il ' Paretaio ' ", *La Riforma Sociale*, XXIII (1912), 337–368.
40. JOHNSON, W. E., "The Pure Theory of Utility Curves ", *Economic Journal*, XXIII (1913), 483–513.
41. LANGE, OSCAR, " The Determinateness of the Utility Function ", *Review of Economic Studies*, I (1934), 218–225.
42. LITTLE, I. M. D., " A Reformulation of the Theory of Consumers' Behavior ", *Oxford Economic Papers*, N. S., I (1949), 90–99.
43. — —, *A Critique of Welfare Economics*, 2nd ed., Oxford: Clarendon, 1957. (First edition, 1950).
44. MARSHALL, ALFRED, *Principles of Economics*, 8th ed., London: Macmillan, 1920. (First edition, 1890).
45. NATAF, A, " Possibilité d'aggrégation dans le cadre de la théorie du choix ", *Metroeconomica*, V (1953), 22–30.
46. OSORIO, A., *Théorie mathématique de l'échange* (with an introduction by V. Pareto), Paris: Giard, 1913.
47. PARETO, VILFREDO, " Considerazioni sui principii fondamentali dell'economia politica pura ", *Giornale degli Economisti*, IV (1892), 389–420, 486–512; V (1892), 119–157; VI (1893), 1–37, VII (1893), 279–321.
48. — —, " Il Massimo di utilità dato dalla libera concorrenza ", *Giornale degli Economisti*, IX (July 1894), 48–66.
49. — —, *Cours d'économie politique*, 2 vols., Lausanne: F. Rouge, 1896–1897. Also in *Oeuvres complètes*, Vol. I, Geneva: Droz, 1964.
50. — —, " Sunto di alcuni capitoli di un nuovo trattato di economia pura ", *Giornale degli Economisti*, XX (March, June 1900), 216–235, 511–549.
51. — —, " Anwendungen der Mathematik auf Nationalökonomie ", *Encyclopädie der mathematischen Wissenschaften mit Einschluss ihrer Anwendungen*, 1902, Band I, pp. 1094–1120. Reprinted in *Oeuvres complètes*, Vol. VIII, 126–152. Italien translation in *Giornale degli Economisti*, XXXIII (1906), 424–453.
52. — —, *Manuale di economia politica pura, con una introduzione alla scienza sociale*, Milan: Società Editrice Libraria, 1906.
53. — —, " L'ofelimità nei cicli non chiusi " *Giornale degli Economisti*, XXXIII (1906), 15–30. English translation in [12], pp. 370–385.
54. — —, *Manuel d'économie politique*, Paris: Giard & Brière, 1909.
55. — —, " Economie mathématique ", *Encyclopédie des Sciences Mathématiques*, Paris: Gauthier–Villars, 1911, Tome 1, Vol. 4, fascicule 4, pp. 591–640. (Reprinted in Vilfredo Pareto, *Oeuvres complètes*, Geneva: Droz, 1966, Tome VIII. English translation in *International Economic Papers*, V (1955).
56. — —, " Il massimo di utilità per una collettività in sociologia ", *Giornale degli Economisti*, XLIV (1913), 337–341.
57. PIETRI–TONELLI, ALFONSO de, *Traité d'économie rationnelle*, Paris: Giard, 1927.
58. RAMSEY, F. P., *The Foundations of Mathematics and Other Logical Essay*, New York: Humanity Press, 1950.
59. ROSA, GABRIELE de, ed., *Lettere a Maffeo Pantaleoni*, 3 vols., Roma: Banca Nazionale del Lavoro, 1960.
60. SAMUELSON, PAUL A., " The Empirical Implications of Utility Analysis ", *Econometrica*, VI (1938), 344–356.

61. — —, " A Note on the Pure Theory of Consumer's Behavior ", *Economica*, N. S., V (1938), 61–71, 353–354.
62. — —, " The Numetrical Representation of Ordered Classifications and the Concept of Utility ", *Review of Economic Studies*, VI (1938), 65–70,
63. — —, *Foundations of Economic Analysis*, Cambridge, Mass.: Harvard University Press, 1947.
64. — —, " Consumption Theory in Terms of Revealed Preference ", *Economica*, N. S., XV (1948), 243–353.
65. — —, " The Problem of Integrability in Utility Theory ", *Economica*, N. S., XVII (1950), 355–385.
66. SCHULTZ, HENRY, " Marginal Productivity and the General Pricing Process ", *Journal of Political Economy*, XXXVII (1929), 505–551.
67. — —, " Marginal Productivity and the Lausanne School ", *Economica*, XII (1932), 285–296.
68. SCHUMPETER, JOSEPH A., *Ten Great Economists*, New York: Oxford University Press, 1951.
69. — —, *History of Economic Analysis*, New York: Oxford University Press, 1954.
70. SCITOVSKY, T., *Welfare and Competition*, Chicago: Richard D. Irwin, 1951.
71. SLUTSKY, EUGEN, " Sulla teoria del bilancio del consumatore ", *Giornale degli Economisti*, LI (1915), 1–26. English translation in *Readings in Price Theory*, G. J. Stigler and K. E. Boulding eds., Homewood, Ill: Richard D. Irwin, 1952, pp. 27–56.
72. TARASCIO, VINCENT J., *Pareto's Methodological Approach to Economics*, Chapel Hill: The University of North Carolina Press, 1966.
73. UZAWA, HIROFUMI, " Note on Preference and Axioms of Choice ", *Annals of the Institute of Statistical Mathematics*, VIII (1956), 35–40.
74. VILLE, JEAN, " Sur les conditions d'existence d'une ophelimité totale et d'un indice du niveau des prix ", *Annales de l'Université de Lyon*, IX (1946), Sec. A (3), 32–39; English translation in *Review of Economic Studies*, XIX (1951–52), 123–128.
75. VOLTERRA, VITO, " L'economia matematica ed il nuovo manuale del prof. Pareto ", *Giornale degli Economisti*, XXXII (1906), 296–301. English translation in [12], pp. 365–369.
76. WOLD, HERMAN, " A Synthesis of Pure Demand Analysis ", *Skandinavisk aktuarietidskrift*, XXVI (1943), Part I, 86–118, Part II, 220–263, XXVII (1944), Part III, 70–120.

CHAPTER 14

(1952)

A DIAGRAMMATIC ANALYSIS OF COMPLEMENTARITY*

Introduction.—I. Interpretations of the principle of decreasing marginal utility.—II. First graphic interpretation of Edgeworth-Pareto complementarity.—III. Second graphic interpretation of Edgeworth-Pareto complementarity.—IV. Translation of the preceding interpretation into Allen-Hicks schemata.—V. Derivation of Edgeworth-Hicks diagrams.—VI. Limiting cases.—VII. A new interpretation of the basic postulates used in consumer's theory.

On the theoretical and, even more, on the pragmatic usefulness of the concept of complementarity, opinions seem to be thus far strongly divided. Even the latest milestones in the literature of economic theory take opposing views on the matter. On the one hand, Professor Samuelson, in his *Foundations* (p. 183), expresses the view that "the problem of complementarity has received more attention than is merited by its intrinsic importance." On the other hand, in the second edition of his *Value and Capital,* Professor Hicks persists in using the concept of complementarity as one of the main theoretical backbones of his economic analysis.

Not only the world's best economists, but thousands of other serious students of economics have spent, and are still spending, "considerable time and energy" in the study of Professor Hicks' book.[1] A great deal of this time and energy is occasioned, I believe, by the absence of a link between the verbal argument of the main text and the mathematical proofs presented independently in the Appendix; for, unlike Marshall or Pareto, Hicks does not offer an adequate diagrammatic analysis to bridge the gap between the two types of argument. The primary purpose of the present paper is to provide such a bridge.[2] The non-mathematical reader will find here a new diagrammatic analysis which, it is hoped, will help him to handle almost any problem involving complementarity.[3] This diagram-

---

* The presentation of this paper benefited from the criticism made by Professors Jesse W. Markham, William H. Nicholls and Paul A. Samuelson.

[1] Samuelson, [23], p. 184. Also Machlup, [17], p. 297. (The list of references appears at the end of the paper.)

[2] The basic results presented in this paper, especially those of Sections I–VI, have been used by the author since 1949 in his graduate classes at Vanderbilt University in place of verbal or algebraic proofs (Cf. Hicks, [11], pp. 1 and 45). The paper was read, for the first time, before the Economics Seminar of Vanderbilt University on April 19, 1951.

[3] The concept of complementarity considered in this paper is of a purely static nature, i.e., it presupposes constant tastes. But complementarity has other aspects, even more

matic interpretation[4] applies equally well to the analysis of complementarity in either the Edgeworth-Pareto[5] sense or the Allen-Hicks[6] sense. This shows that the two definitions are not *basically* different. The Edgeworth-Pareto definition will be considered first. For despite the fact that many no longer regard it as a part of modern orthodox theory—for the reason that it requires a measure of utility[7]—this definition provides a more intuitive background for many aspects of complementarity. It will also be shown that the point of view adopted by Johnson,[8] in classifying the possible relationship between two commodities, complements, rather than conflicts with, the Edgeworth-Pareto definition.[9] As a natural consequence, it will be seen that Johnson's point of view, if embodied in the Allen-Hicks approach, helps to clarify many an obscure point which one finds in the applications of complementarity. In fact, without this addition, the Allen-Hicks theoretical rationale could not possibly fully serve its own purpose.

The problem of the nature of the relationship between commodities is intimately connected with that of the shape of the indifference varieties. This connection—which certainly would not surprise the reader but which would be difficult to find explicitly expressed in the economic literature—explains the necessity of dealing first with some shape properties of the indifference curves (Section I), even at the risk of appearing to duplicate the extensive contributions already available on the subject. A natural continuation of the analysis of complementarity in the manner adopted in this paper led to a new interpretation of the basic postulates of consumer's behavior. These results are presented in Section VII.

---

important for economic theory. They have been pointed out by Machlup [17], p. 286, Lange [15], pp. 59–60, Li [16]. The dynamic aspects of complementarity have been considered by Lange [15] and Ichimura [12]. Since there are yet fundamental objections to be raised in connection with determining the static field of preference (Georgescu-Roegen, [9]), it is natural that the parametric representation of the change in tastes, although throwing interesting light upon the subject, constitutes, at this stage of our knowledge, a daring procedure.

[4] An attempt (perhaps the first) at a diagrammatic analysis of the concept of complementarity was made by Professor Hayek [10]. Professor Hayek's approach, which is completely different from that of the present paper, uses the properties of indifference curves of X and Y for $Z = const$ in order to describe the relationship between X and Z. No attempt was made to use this description as an analytic tool for further theoretical results.

[5] Edgeworth, [5], Vol. I, p. 117 ff. Pareto, [20], §94 and 974; [21], Ch. IV, §8–19, 39–41, Appendix, § 47; [22], pp. 611–12.

[6] Allen and Hicks, [3], pp. 69ff. and pp. 211–217. Hicks, [11], Ch. III. (See, however, *infra*, fn 38).

[7] However, this standpoint appears shaken since v. Neumann and Morgenstern [19], pp. 15–29, offered a procedure for measuring utility. Their method is a reversal of that used by Daniel Bernoulli to answer a paradox of gambling (Todhunter, [29], pp. 213ff., especially nos. 384–386). Evidently, the point of view that measurability of utility is not necessary anyhow for a theory of consumer's equilibrium still largely prevails. It is, however, very weak scientifically. Could we refuse to take account of animals with more than two feet, on the ground that only two feet are needed for walking?

[8] W. E. Johnson, [14], p. 496.

[9] *Cf.* The controversial argument of Edgeworth, [5], Vol. II, pp. 464–466. Also Allen [2]; Schultz [26], Ch. XIX.

I

As is known, the principles of decreasing marginal utility and decreasing marginal rate of substitution are two different things.[10] But, to the author's knowledge, the exact difference between the two has never been fully investigated.

In the usual interpretation, the principle of decreasing marginal utility means that successive *additional* equal amounts of one commodity will add successive decreasing amounts of utility[11] or that in order to add successive equal amounts of utility U, successively increasing amounts of the commodity X are needed.[12] This will be referred to as the *crude* principle of decreasing marginal utility. Its validity, with a few and irrelevant exceptions, is hardly subject to question.[13]

Since, from the consumer's point of view, there is no objection to considering baskets of identical commodity composition as identical units of a new commodity, it is only natural to extend the above principle to successive additions of such composite commodities.[14] Therefore, if, as in Fig. 1, we draw a series of indifference curves *equally distant in utility*, they will intercept on any line of positive slope (i.e., along which no commodity would decrease in amount) a series of increasing segments $A_{-2} A_{-1}$, $A_{-1} A$, $A A_1$, $A_1 A_2$, ...[15]

The formulation just described will be referred to as the *weak* principle of decreasing marginal utility.[16] It does not assure the convexity of the indifference curves, i.e., it does not imply the principle of decreasing marginal rate of substitution. Indeed, the weak principle guarantees the upward convexity of the utility surface along the directions $\overrightarrow{Aa}$ but not necessarily along $\overrightarrow{Aa'}$. The utility surface may therefore be like a *saddle*, concave in some directions and convex in others.[17]

---

[10] Allen and Hicks, [3], p. 57. Hicks, [11], p. 16.

[11] Jevons, [13], pp. 54ff. and 62. Edgeworth, [4], pp. 34, 61, 108. Pareto, [20], p. 19; [21], IV, §32–33, 69, Appendix, §46. Marshall, [18], pp. 93–94 and Note I, p. 838. Johnson, [14], p. 492.

[12] Johnson [14], p. 485.

[13] In terms of theory of production, this corresponds to the law of diminishing returns of a single factor (*cf*. Marshall, [18], IV, iii) or to that of variable proportions (*cf*. Stigler, [27], p. 116).

[14] Pareto, [21] IV, §42, Appendix, §47–48. Pareto was apparently the first to see the possibility of making this extension, but, as in many other instances, he surrounded this new way of looking at things by some theoretical inconsistencies and mathematical slips. Thus, Pareto seems to restrict the extension of the crude principle to baskets that contain complementary commodities only. Also, the passing from conditions (66) to (67) of the Appendix is not completely justified until a later stage (§124).

Independent of all this, on p. 576 there are evidently some manuscript errors, namely the signs of the inequalities (63) and (64) which must be reversed.

[15] The double notation on the graphs used here will become clear subsequently; the notation in terms of money (m) should be ignored while using the graphs for discussing the Edgeworth-Pareto scheme, while that in terms of utility (u) should be disregarded when analysing the Allen-Hicks approach.

[16] On the production side, the principle of decreasing returns to scale (*cf*. Stigler, [27], pp. 128ff.) is a particular case of the weak principle. This remark helps us to understand why we cannot derive *analytically* the weak principle from the crude one.

[17] The possibility of the indifference curve's being concave towards the origin, if only

We know by now that all we need for a coherent theory of exchange is the principle of decreasing marginal rate of substitution.[18] Thus not only in early writings but even in later ones, we find this principle assumed jointly with the weak principle of decreasing marginal utility, at times without any support, at others, painfully and only partially justified.[19] The explanation of this hybrid

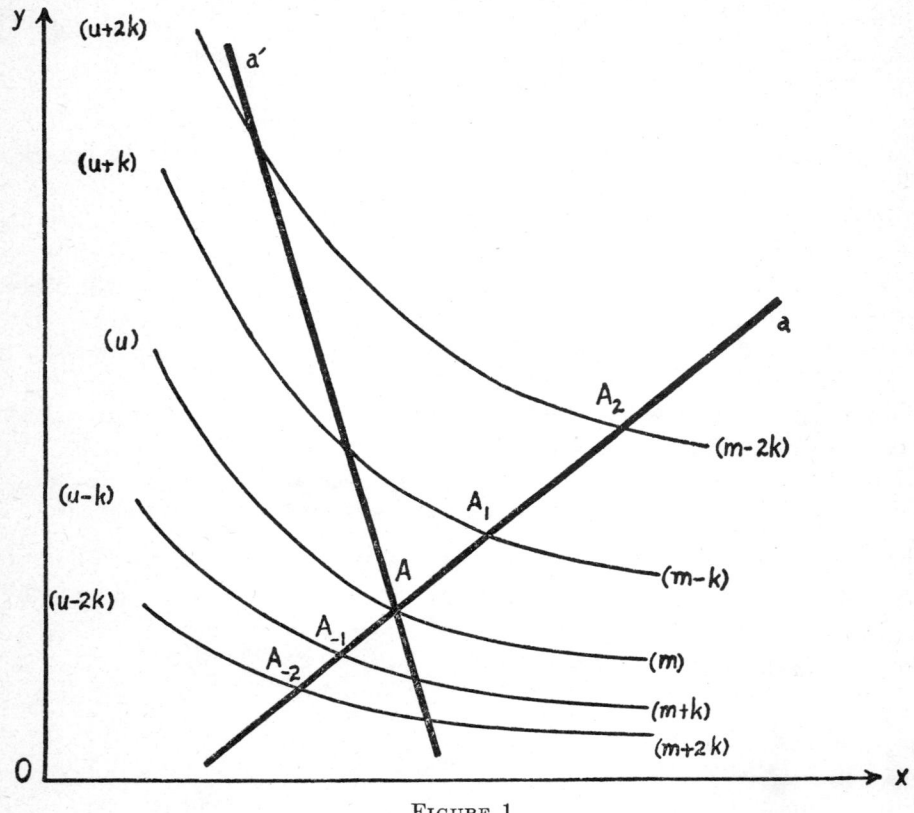

FIGURE 1.

axiomatic approach is, perhaps, the fact that satisfactory evidence in support of the convexity of the indifference varieties is not yet available.[20] By keeping the

---

some form of a weak principle is assumed, is mentioned in Pareto, [21], p. 654. A numerical example illustrating this *analytical* possibility is offered by the utility function, $U = 54(x+y) - (2x^2 + 5xy + 2y^2)$, where $x, y \leq 6$.

[18] Professor Allen was the first to present clearly this point intuited and hinted at by both Fisher and Pareto. Fisher [6], pp. 67, 69, 88 and, later, Pareto [21] Appendix, §6, proved that what is essential for economic theory is the existence of indifference directions. Professor Allen's first formulation [1], pp. 203, 218–219, differs in form from that reached later in collaboration with Professor Hicks [3].

[19] Pareto, [21], Appendix, p. 572, (55), p. 575, (59). Johnson, [14], p. 485, (a), p. 486 (c). Edgeworth, [4], pp. 36 and 108.

[20] Hicks, [11], p. 25. The replacement of the geometric condition of convexity by the principle of decreasing marginal rate of substitution is a mere change of vocabulary and

two postulates together, it was hoped that one day the principle of decreasing marginal rate of substitution could eventually be related to that of decreasing marginal utility.

Now, if a start is made with some sort of decreasing utility, additional assumptions, other than those contained in the weak principle, must be made in order to arrive at the convexity of indifference curves. This can be achieved, for instance, by introducing the strong principle:

The marginal utility decreases along any preference direction.

In other words, the grid of equi-distant indifference curves of Fig. 1 will intercept segments of increasing size not only on $\overrightarrow{Aa}$, but also on $\overrightarrow{Aa'}$ so long as the latter remains a preference direction.[21] The difference between the weak and the strong principle is therefore clear.[22] The strong principle assures the over-all convexity of the utility surface and, implicitly, that of the indifference curves.

Conversely, however, the principle of decreasing marginal rate of substitution does not necessarily imply any form of the principle of decreasing marginal utility.[23] Intuitively, one can see that the utility surface can be contracted or expanded like an accordion without changing the shape of the contour lines (i.e., the indifference curves). This will, however, alter the concavity of the utility surface in different directions, or, in other words, will affect the variation of marginal utility.[24]

The strong principle of decreasing marginal utility and that of decreasing marginal rate of substitution are much more intimately connected than thus far appears. Their common basis is evident if we compare their respective meanings,

---

carries with it no explanatory element. This is another example of implicit theorizing to be added to the list of Professor Leontief ("Implicit Theorizing: A Methodological Criticism of the Neo-Cambridge School," *Quarterly Journal of Economics*, February, 1937, pp. 337–351). The preceding remark is not intended to impair in any way the importance of Professor Allen's contribution regarding the other side of the problem, mentioned in note 18, above.

[21] At a given point, a preference direction is any one which leads to a higher utility variety (Georgescu-Roegen, [8], p. 551).

[22] The strong principle is hard to find in the literature except in its mathematical form as a condition for the maximization of utility (*Cf.* Pareto, [21], Appendix, §122–123; Edgeworth, [5], I, p. 117.) or for the minimization of long-run average cost (*Cf.* Hicks, [11], p. 87; Stigler, [27], p. 130n).

[23] For the case of decreasing returns of a single factor of production, see Samuelson, [23], p. 62.

[24] In mathematical terms, if $u(x)$ is the utility function in vector notation, we have

(1) The crude principle means $\dfrac{\partial^2 u}{\partial x_i^2} \leq 0$

(2) The weak principle means $d^2u \leq 0$ whenever $(dx) \geq 0$ or $(dx) \leq 0$.

(3) The strong principle means $d^2u \leq 0$ for $du > 0$ or $du < 0$.

(4) The decreasing marginal rate of substitution is expressed as $d^2u \leq 0$ for $du = 0$.

Thus (4) follows from (3), but not conversely. In fact, even (2) and (4) do not suffice to ensure (3). Many points of theory of production are intimately connected with this analytical hierarchy.

for the former, in the case of a utility surface involving two commodities, and for the latter, in the case of an indifference surface involving three commodities. Both principles express the same property: the convexity of each of the two surfaces.

If the curves in Fig. 1 are assumed to be the intersections of *one* and the *same* indifference surface of X, Y, M with equi-distant levels of M, we can use a phrasing similar to that of the strong principle of decreasing marginal utility in order to make more precise the verbal definition of the decreasing marginal rate of substitution for the case of more than two commodities.[25] Under this new interpretation, the segments intercepted on the lines $\overrightarrow{Aa}$, or $\overrightarrow{Aa'}$, will increase in the direction of decreasing M. This last remark underlies, as will be shown later, the complete *formal* identity of the Allen-Hicks definition of complementarity and that of Edgeworth and Pareto under the assumption of the strong principle of decreasing marginal utility. For this reason, despite the fact that the Edge-

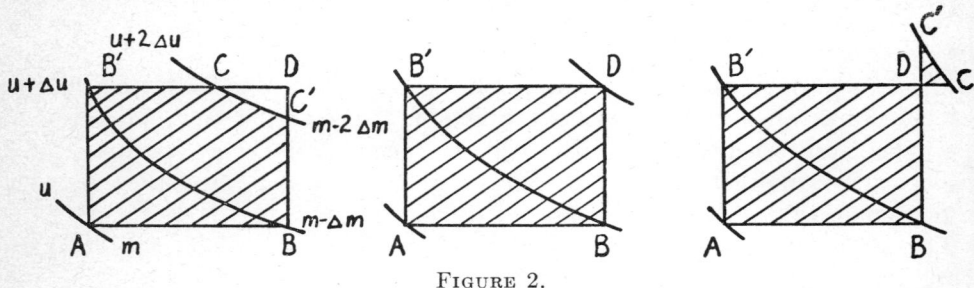

FIGURE 2.

worth-Pareto definition makes sense even without it, the latter principle will be postulated in the greater part of the subsequent analysis.

II

Let us now consider two indifference curves corresponding to the utility levels u and $u + \Delta u$. Let A be a point on the first indifference curve (Fig. 2). By increasing separately X and Y, we obtain on the second indifference curve the points B and B'. If the increment of utility Δu is very small, so will be the increments $\Delta x = AB$, $\Delta y = AB'$. As the marginal utility of X is the ratio between the increase in utility Δu caused by Δx and the selfsame Δu, we have

(1) $$\Delta u = (muX)_A \, \Delta x = (muY)_A \, \Delta y$$

where $(muX)_A$ represents the marginal utility of X at A. This relation shows that

(II.1) *The increments Δx and Δy between pairs of indifference curves equidistant in utility vary inversely with the corresponding marginal utilities.*

This remark is fundamental to the diagrammatic analysis of complementarity. Let us complete the rectangle $ABB'D$ and call it the *complementarity rectangle*. If we draw now the indifference curve corresponding to the utility level $u + 2\Delta u$, we are confronted with three alternatives, each corresponding to one of

---

[25] Hicks, [11], p. 25. However, on p. 87, fn. 2, we find an interpretation which is equivalent to the *weak* form of this principle.

the three possible relationships between X and Y as defined by Edgeworth and Pareto. It is clear that

(a) *If the indifference curve* $(u + 2\Delta u)$ *intersects the complementarity rectangle,* X *and* Y *are complementary (Fig. 2a);*

(b) *If the indifference curve* $(u + 2\Delta u)$ *only touches the complementarity rectangle,* X *and* Y *are independent (Fig. 2b);*

(c) *If the indifference curve* $(u + 2\Delta u)$ *has no common point with the complementarity rectangle,* X *and* Y *are competitive. (Fig. 2c).*

This is the graphic interpretation of complementarity as defined by Edgeworth and Pareto. A second interpretation of (a)–(c) is the well known fact that the utility of two complementary (competitive) commodities is greater (smaller) than the sum of the utilities represented by each commodity separately.

As an immediate application, the graphic interpretation just described provides a self-evident proof of the reversibility of the relationships between X and Y. However, a closer examination of the competitive case reveals that it presents an asymmetry which becomes evident as soon as other relevant aspects of the problem are taken into consideration. With the help of (II.1) and the crude principle of decreasing marginal utility we arrive at the following symmetrical property of complementary or independent commodities:

$$(2) \qquad (muX)_B < (muX)_{B'}, \qquad (muY)_{B'} < (muY)_B$$

This may no longer be valid if X and Y are competitive. Indeed, in the latter case, we may also have

$$(3) \qquad (muY)_{B'} > (muY)_B$$

This means that the increase of X (along an indifference curve) *decreases* the marginal utility of Y, or, in other words, that Y must be *highly competitive* with X. But if this is so, X cannot be *highly competitive* with Y, i.e., this particular relationship is *not* reversible. This can be shown with the help of Fig. 3, where, according to (3), we must have $BC' > B'C'''$. The slope of $C'C'''$ is therefore smaller than that of BB'.[26] But as the slope of $C''C$ must be smaller still then that of $C'C'''$, it follows that $B'C < BC''$. Therefore, we must by necessity have $(muX)_B < (muX)_{B'}$.

In fact, since the slope of CC'' is smaller than that of BB', it follows that the point on the curve $(u + 2\Delta u)$ where the marginal rate of substitution is the same as at B lies to the left of C'. This means that X is *inferior* with respect to Y. Therefore, Y being highly competitive with X and X being inferior to Y are equivalent concepts. The basis of irreversibility appears now clearly.

Summing up, we have the important result:

(II.2) *If one commodity is substituted for another along the same indifference curve, its marginal utility decreases, except when the first commodity is highly competitive with the latter.*[27]

[26] For sufficiently small intervals, the indifference curves can be assimilated to segments of straight lines.

[27] It is worth emphasizing that the results of this section require only the crude principle of decreasing marginal utility and that of decreasing marginal rate of substitution.

III

Complementarity can be analyzed with greater efficiency with the help of tools other than those introduced thus far. Let us consider the *loci* of all combinations of commodities for which the marginal utility of one commodity, say X, remains constant. These *loci* are in fact new iso-curves; along them the *subjective evaluation* of X, measured by the marginal utility of X, is the constant. Because the evaluation of X is in this case based upon the measure of utility, we shall refer to these loci as the *utility isotimetics* of X.[28]

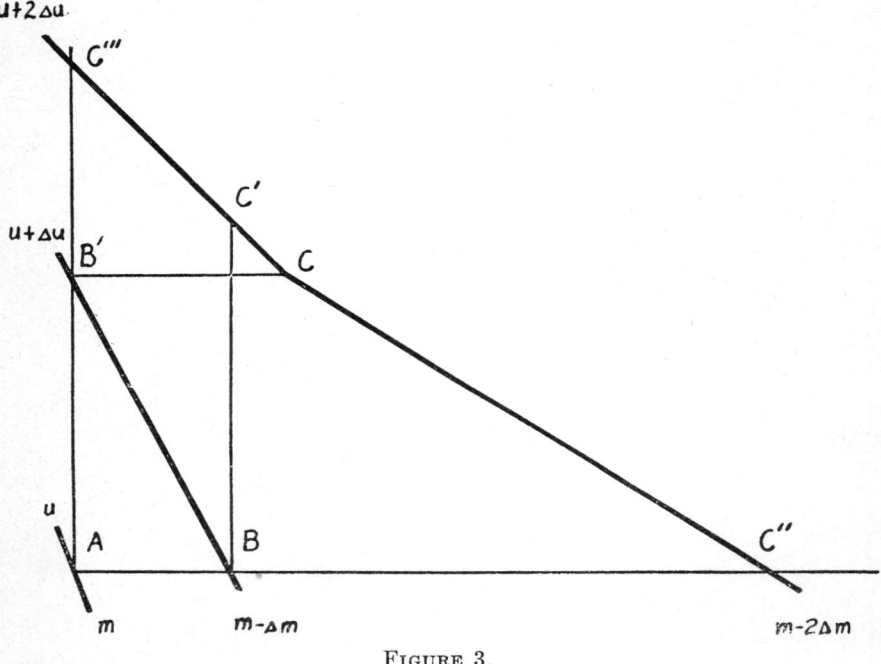

FIGURE 3.

In order to avoid formal repetition, we shall confine our detailed analysis for each proposition to a single illustration, although for a rigorous argument each type of relationship between X and Y should be considered individually. Following the pattern of the proof of the illustration, the argument can be readily reconstructed for all the omitted cases.

In line with this rule, let us consider first the case of X and Y mildly competitive. From the definition of competitive commodities, it follows that $(muY)_B < (muY)_A$. Furthermore, according to (II.2), the point B_1 of $(u + \Delta u)$, where (muY) is the same as in A (i.e., $AB' = B_1C_3$) must lie to the right of B (Fig. 4). Consequently, the utility isotimetic of Y is *negatively* inclined. Further, AC_2 being a preference direction, *according to the strong principle of decreasing marginal utility*, we have $AB_1 < B_1C_2$. Hence, by comparing the triangles

[28] From τιμητικοσ = evaluating, estimating. The assistance of Prof. F. W. Mitchel in coining this word is acknowledged.

AB_1B', $B_1C_3C_2$, we conclude that $B_1C_1 > AB$, i.e., $(muX)_A > (muX)_B$. (For the purpose of facilitating the comparison, the triangle $B_1B_1'C_3$, equal to AB_1B' has been introduced in Fig. 4.)

Results of the same nature can be obtained for all other types of relationships between X and Y and can be interpreted graphically as shown in Fig. 5 (in the same order of letters as below). Denoting by $[X]$ the utility isotimetic of X, these results can be presented systematically as follows:

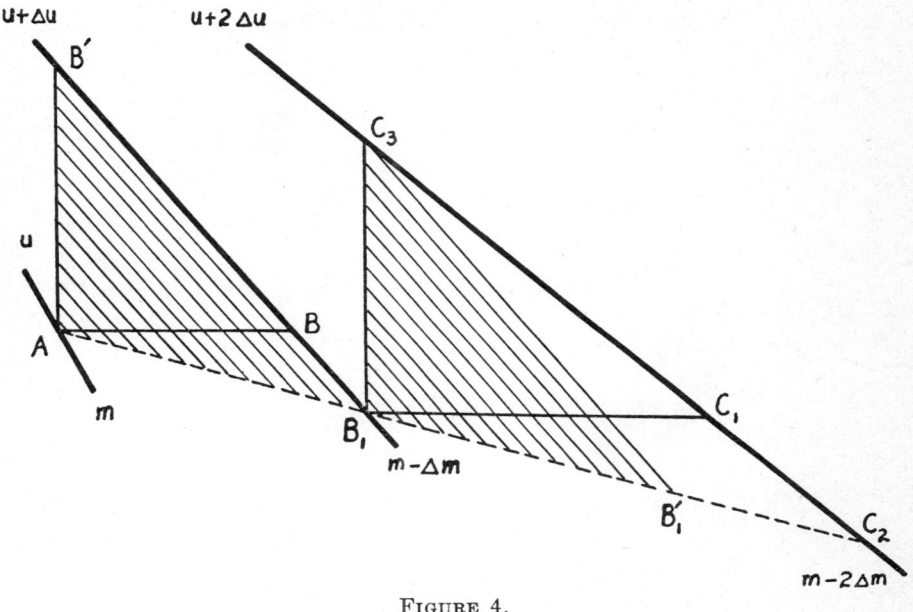

FIGURE 4.

(III a). *If X and Y are complementary, we have at A*

$$0 < \text{slope of } [Y] < \text{slope of } [X]$$

(III b). *If X and Y are independent, $[X]$ and $[Y]$ are straight lines parallel respectively to Oy and Ox.*

(III c). *If X and Y are mildly competitive, at A we have*[29]

$$0 > \text{slope of } [Y] > -(mrsX) > \text{slope of } [X]$$

(III d). *If Y is highly competitive with X, at A we have*

$$-(mrsX) > \text{slope of } [Y] > \text{slope of } [X]$$

Fig. 5 represents a new diagrammatic interpretation of complementarity and constitutes an excellent analytical tool for treating further problems in the field.

As a first application, let us assume that the marginal utility of money is

[29] Here $mrsX$ stands for the marginal rate of substitution of X for Y and, therefore, $-(mrsX)$ is the slope of the indifference curve.

constant and examine the effect a fall in the price of X has upon the demand for Y, while the price of Y remains constant. It is well known that the conditions of this problem lead to the following results: (a) X will increase, (b) muY will remain constant.

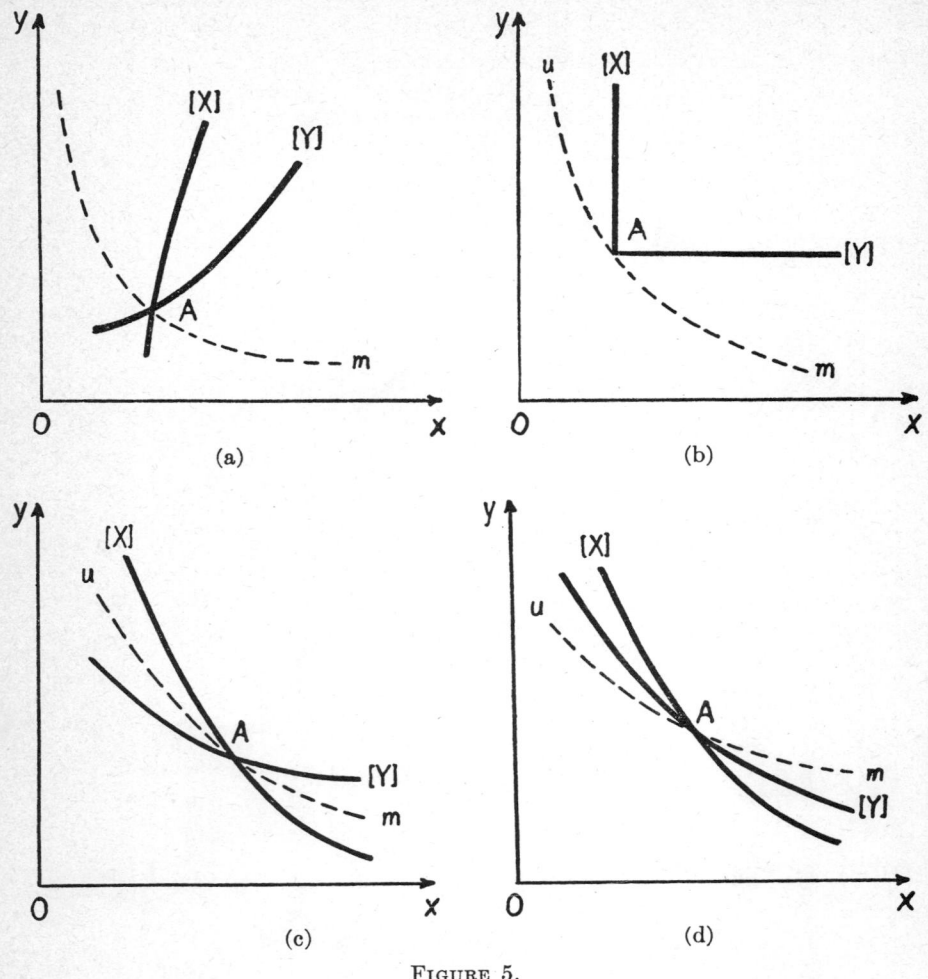

FIGURE 5.

A brief inspection of Fig. 5 shows that Y will increase if X and Y are complementary and will decrease if they are competitive.[30]

Bearing in mind that a decrease in the marginal utility of a commodity means

[30] *Cf*. Hicks, [11], p. 42. On this point the argument of Professor Hicks seems defective. Indeed, the increase of the marginal utility of Y following the increase of the demand for X cannot be an immediate consequence of the definition of complementarity (the constancy of Y with a fall in the price of X is incompatible with complementarity and constant marginal utility of money). Since a fall in the price of X will necessarily bring about a change in Y, the argument as a verbal proof is clearly elliptic.

a shift of the respective utility isotimetics in the direction away from the origin, an important consequence can be derived directly from Fig. 5. This is an alternative definition of complementarity:

(III.1) Let muY be constant and muX decrease, then
(a) *If Y increases, X and Y are complementary;*
(b) *If Y remains constant, X and Y are independent;*
(c) *If Y decreases while U increases, X and Y are mildly competitive;*
(d) *If Y decreases together with U, Y is highly competitive with X.*

IV

The background of all diagrams used so far consisted of a grid of indifference curves equi-distant in utility. The concept of complementarity was defined and analyzed in terms of this grid. The grid, derived from the utility surface, is no longer available if we discard the assumption of measurability of utility. Some other surface has to be used as a base for a new grid. For their definition of complementarity, Professors Allen and Hicks decided to resort to the indifference surface of X, Y, and M, where M stands for all other goods lumped together, i.e., for Marshall-Hicks money.[31] In short, the consumer of Edgeworth-Pareto walks on a utility surface, that of Allen-Hicks, on an indifference surface. Otherwise, the rationale is identical in both approaches.

With the help of an indifference surface between X, Y and M we can obtain in the plane $x0y$ a family of iso-curves by intersecting this surface at constant levels of M. By considering equally distant levels of M, we obtain a grid similar to that used in Sections II and III, with the only difference that as X (or Y) increases the level of M decreases. Moreover, the principle of decreasing marginal rate of substitution imposes upon the indifference surface the same shape properties that the strong principle of decreasing marginal utility imposes upon the utility surface.[32] The only difference between the two cases is a question of the direction of the convexity of each surface.

Therefore any of the Figs. 1–4 can be relettered so as to represent the new situation. The only changes are that: (1) u, $u + \Delta u$ and $u + 2\Delta u$ must be replaced respectively by m, $m - \Delta m$ and $m - 2\Delta m$ and (2) the marginal utility of X (or Y) must be replaced by the marginal rate of substitution of X (or Y) for M, when Y (or X) is kept constant. With these changes Figs. 2a, 2b, 2c can be used also for a diagrammatic interpretation of complementarity, independence and competitiveness in the Allen-Hicks sense.[33]

It goes without saying that, this time, $[X]$ represents the *price isotimetic* of X, i.e., the loci of the combination for which the marginal rate of substitution of X for money, p_x, (the marginal price) is constant.

As a first application, Fig. 2 offers an immediate non-mathematical proof of

---

[31] Allen and Hicks, [3], pp. 69–71. Hicks, [11], pp. 44ff.
[32] *Supra*, Section I.
[33] Allen and Hicks, [3], p. 71. Hicks, [11], p. 44. Professor Hicks attaches little importance to the case of independence.

the reversibility of the relationships there described.[34] It shows also that, if X and Y are complementary (competitive), the individual would pay a greater (smaller) price for Δx and Δy taken together than the sum of the prices which he would be willing to pay separately for Δx or Δy.

Further, if we assume both the measurability of utility and the constancy of marginal utility of money, it follows that utility can be measured in money units. The isoquants in X and Y for which $m =$ constant and u varies can be considered as isoquants in X and Y for which $u =$ constant and m varies. The grids used for the Edgeworth-Pareto or for the Allen-Hicks definitions are therefore identical. In this particular setup, both criteria give the same answer.[35]

At this point of the argument it is useful to point out that price isotimetics are intimately connected with important economic concepts found in the current literature. If the third dimension is money, M, the price isotimetic $[Y]$ is nothing else than one way of representing the demand law as interpreted either by Professor Friedman—assuming a constant real income—or by Professor Hicks—abstracting for the income effects.[36] This demand, regarded as a relationship between X and p_x can be derived from $[Y]$ exactly as the Walrasian demand law can be derived from the price-consumption curve. If, on the other hand, the third dimension is the product produced with the help of X and Y, $[Y]$ is the demand of the entrepreneur under the assumption of constant prices of Y and of the product.

Some familiar properties of these demand laws are obtained immediately by a simple inspection of Fig. 5:

(IV.1) *Under the assumption of constant real income, if p_x falls while p_y is kept constant then*:

(a) *Y increases, if X and Y are complementary,*

(b) *Y remains constant, if X and Y are independent,*

(c–d) *Y decreases, if X and Y are competitive.*

This is, in fact, a second definition of complementarity.[37] The preceding argument shows that, within the framework of Hicksian analysis, this definition is equivalent to the first (Fig. 2).[38]

In *Value and Capital* (p. 46), Professor Hicks uses a third definition which is

---

[34] This proof is not given in Hicks [11]. The proposition establishing reversibility of the relationship between X and Y should not be considered as trivial. One may be convinced of the necessity of the proof by referring to the case when Y is highly competitive with X.

[35] Hicks, [11], p. 44.

[36] [7], p. 465, [11], p. 32.

[37] Allen and Hicks, [3], p. 70; Hicks, [11], Appendix, p. 311.

[38] This equivalence had been challenged by Professor Samuelson, [23], p. 184–186. He has since retracted his criticism (Samuelson, [25], p. 379n), but not without simultaneously emphasizing that the Allen-Hicks definition differs fundamentally from that of *Value and Capital*. (The same point appears in Ichimura, [12], p. 179.) From the short footnote containing his retraction one may gather that Professor Samuelson had realized that money in Professor Hicks' sense was not a Walrasian *numéraire*. But by this very argument the difference between the two definitions mentioned above should disappear also. For even in Allen and Hicks, [3], we find that the background of comparison is not a *third* commodity, one of the *others*, but *all* the other commodities lumped together. "The second (component) will depend on how far the substitution in favor of X takes place at the expense of Y rather

based not only upon the changes in the quantities of X and Y, but also upon the change in the amount of money. This third definition is not equivalent to the other two. By proving this, we shall implicitly explain why, though he considers only two relationships—complementarity and competitiveness—Professor Hicks is forced to speak of *"three* possible cases."[39]

The argument of Section 3, on p. 46, of *Value and Capital* is at times loosely worded and, because of this, we are never sure of what elements are kept constant. Speaking of complementary commodities, Professor Hicks certainly cannot mean by "a substitution of X for money," *any* such substitution, for if this were the case, Fig. 5a would show immediately that we can substitute X for money without "a parallel substitution of Y for money." If, on the other hand, the substitution of X for money follows a fall in the price of X only, then the conclusions regarding the competitive case are no longer always true. Indeed, with the exception of the case described by Fig. 5d (Y highly competitive with X), a fall in the price of X (only) will always cause a substitution against money. The case of Y being highly competitive with X constitutes therefore a special case of particular importance in that it leads to an increase of money with a fall of p_x. It is the only case when we can speak of "a substitution in favor of money and against Y" and—since it is not a dominating feature of competitiveness—we cannot, even by way of an attempt at a realistically simplified description, attribute the latter property to all cases of competitive pairs of commodities.

We see, therefore, why the case of Y highly competitive with X has to be considered as an important separate case. The reasons are similar to those which in the case of the Edgeworth-Pareto definition led to the Giffen paradox. Thus, (IV.1) has to be supplemented by

(IV.2) *A fall in the price of X, with that of Y being constant, will cause a decrease of M, if X and Y are mildly competitive, and an increase of M, if Y is highly competitive with X.*

The relationship between Y and X in the latter case of (IV.2) is no longer reversible. Thus, a fall in p_y, with p_x constant, will cause a substitution against money.

At this point, the question of whether the difference between high and mild competitiveness leads to distinctions other than (IV.2) may be raised. Along this line, it is easy to see, on Fig. 5d, that if M remains constant and X is increased, the marginal rate of substitution of M for Y (the reciprocal of p_y) increases. According to (III.1), this means that X and M are complementary. Now, leaving out the simple case of independence, we may represent the triangular relationship between X, Y and M as in Fig. 6. This shows in a schematic way the role of the three relationships and the deeper reason for the irreversibility of high competitiveness.

So long as the third dimension is not assigned a special role, the last case of

---

than of the other goods (Z)" (p. 69). The only reason for a different impression regarding this point may be the fact that unfortunately the mathematical part of the paper presents the formulae in terms of only three commodities.

[39] Allen and Hicks, [3], p. 70 (italics added). Also Hicks, [11], pp. 68 and 93.

Fig. 6 does not raise any particular problem. This is no longer true if the third dimension does not occupy an inter-changeable position with X and Y. Then the symmetry of the mutual relationship breaks down. This is especially conspicuous in the case where the third dimension is the output in a production function using X and Y as factors.

We may find in this an explanation of why, in *Value and Capital*, the high competitiveness appears only as a peculiarity specific to the equilibrium of the firm. On this occasion, Professor Hicks even introduces a new term, that of a *regressive factor*,[40] proving that he was not aware of the fact that the essence of this latter concept is identical with that of a highly competitive commodity or with that of an inferior good.

With the help of the preceding diagrammatic analysis, this point of view is grasped in its entire generality. Professor Hicks' cases (1), (2), (3)—distinguished

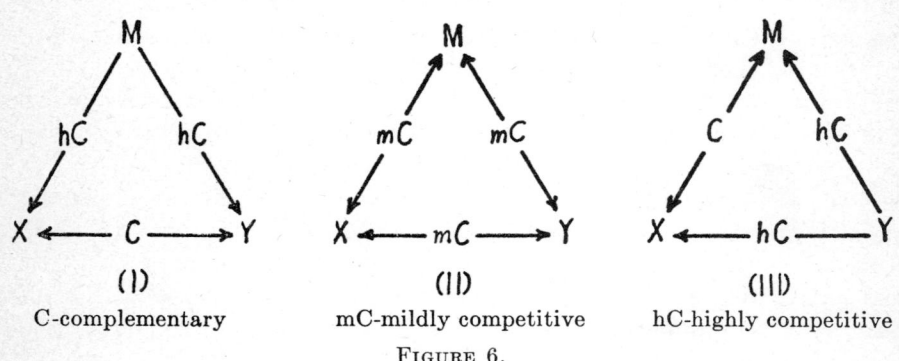

FIGURE 6.

on p. 93 of *Value and Capital*—correspond respectively to (c), (a), and (d) of our basic Fig. 5 or to (II), (I), (III) of Fig. 6. It is therefore seen that there is nothing mysterious about a regressive factor, which would make "the reader.... rub his eyes." A regressive factor is nothing but an "inferior" factor of production, i.e., one whose input will decrease with an increase of output (the conditions of equilibrium implying the constancy of the prices p_x, p_y, ... exactly as in the case of an income-consumption curve).[41] Professor Hicks needs a whole section (5, pp. 94–96) of the most difficult-to-follow verbal arguments in order to reach the conclusion that X (in his notations, A) is a factor "suited for small scale production." He does not identify it positively either with the inferior consumer's good, or with the only form of relationship which would make his quasi-definitions of p. 46 consistent with the other ones.[42]

[40] Hicks [11], p. 93.
[41] Examples of inferior factors of production could be easily procured (hand-operated machinery, small power generators, even labor, or at least some types of it, etc.). Therefore, inferior factors do not seem to be at all "grossly improbable." (Hicks, [11], p. 94.)
[42] Professor Hicks is forced, nevertheless, to single out the highly competitive case in Fig. 16, [11], p. 68, without pursuing its significance further.

V

One of the other high-lights of Hicksian analysis, namely the stability of multi-exchange equilibrium,[43] can be more easily described in terms of our basic diagrams, Fig. 5. As in the preceding sections, we shall confine the detailed analysis to one type of relationship. For this illustration, let us assume, this time, that Y is highly competitive with X and let us find the relationships between p_x and p_y expressing the conditions of partial equilibrium for the exchange market of each commodity.[44]

A graph derived from Fig. 5d will be used in parallel with an Edgeworth-Hicks diagram.[45] Let us assume that equilibrium on both exchange markets is reached at A and let us represent the corresponding pair of equilibrium prices

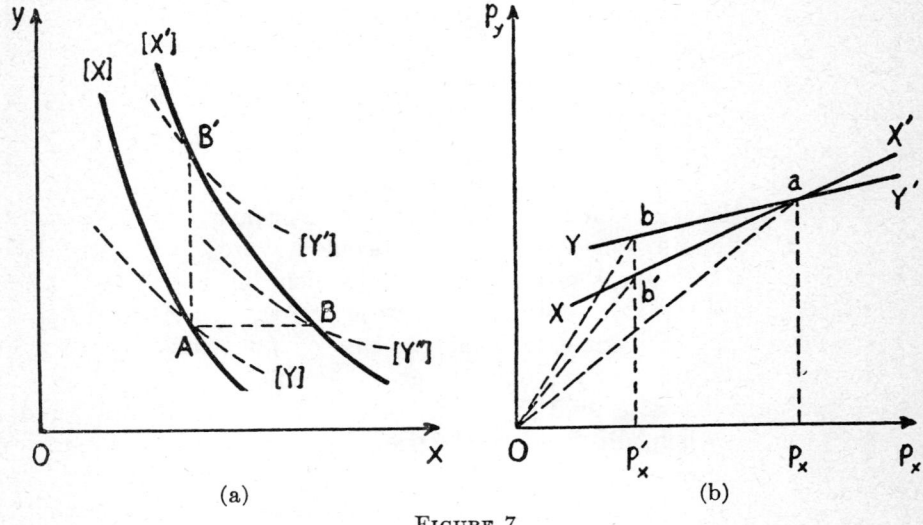

FIGURE 7.

at a (Fig. 7). A fall in the price of X only, from p_x to p'_x is represented by a shift of $[X]$ to $[X]'$, in the direction away from 0. Let B and B' respectively be the new separate equilibrium positions of the exchange market for Y and X (with $y = const$ and $x = const$, respectively). Since Y is assumed competitive with X, the equilibrium price of Y will be smaller at B and B' than at A, and smaller at B' than at B. This follows directly from Fig. 5d by drawing the price isotimetics $[Y]'$, $[Y]''$ of a negative slope smaller in absolute value than that of $[X]'$ at B and B'. We have further to remark that the slope of $0a$ is the ratio between the equilibrium prices and it is, therefore, equal to the marginal rate of substitution of Y for X at A. If, at this point of the argument, we take into consideration that Y is highly competitive with X and that this implies that the

[43] Hicks, [11], Ch. V.
[44] With Professor Hicks, we assume that market supplies x_0, y_0 are fixed and that income effects are ignored.
[45] Edgeworth, [5], Vol. I, pp. 129–130; Hicks, [11], p. 68.

marginal rate of substitution of Y for X is smaller at A than at B' and smaller at B' than at B, we obtain the new pair of equilibrium prices at b and b', as shown in Fig. 7b. From this, we determine the relative slopes of XX', YY' representing the relationship between p_x and p_y when the X and Y exchange markets are *separately* in equilibrium. Fig. 7b is identical with the last graph of Professor Hicks' Fig. 16, [11], p. 68. The necessity for his using a third commodity, T, now appears clear. This can be dispensed with if the title of his last graph is changed to: Y highly competitive with X. To be consistent throughout, the title of his first graph should be changed to: X and Y mildly competitive.

VI

Provided we leave out of the picture the limiting cases of full complementarity and of full competitiveness, we can be sure that Fig. 5 contains all possible cases which the relationship between X and Y can reveal. Indeed, it will suffice to go over the argument relating to Fig. 4 in order to see that, if the principle of decreasing marginal rate of substitution or the strong principle of decreasing marginal utility is assumed, under no circumstance can the absolute slope of the isotimetic $[Y]$ be steeper than that of $[X]$ at the same point. In other words, the blades of the scissors formed by $[X]$ and $[Y]$ could not cross past each other.

Furthermore, to each graph of Fig. 5 there corresponds a unique Edgeworth-Hicks diagram which can be obtained following the same steps used in the preceding section. Consequently, ignoring the case of independence, as well as other limiting cases, the analysis of multi-exchange equilibrium can lead only to one of the three diagrams used by Professor Hicks in his Fig. 16, [11], and to no other. With their help, Professor Hicks proved in a simple and elegant way the extremely important result regarding the stability of multi-exchange equilibrium.

At this point, it becomes almost a necessity to examine what would happen if the limiting cases are considered also. This leads us to examine all limiting and border-line cases.

The extreme limit of complementarity—full complementarity—and that of competitiveness—full competitiveness—correspond to the case in which the isotimetics $[X]$ and $[Y]$ coincide, the only difference between the two extremes being in the signs of the slopes of the isotimetics. The other two special cases, mentioned here for the sake of unity, are those of independence and of quasi-constant marginal utility (of quasi-constant marginal rate of substitution). The former is the border-line case between complementarity and competitiveness, while the latter lies between mild and high competitiveness. The interpretation of each special case presents no difficulty.

Let us refer first to full complementarity in the Edgeworth-Pareto sense. Obviously, this includes the text-book example of the key and lock—the rigid complementarity—but the latter is only a particular case. As $[X]$ and $[Y]$ coincide, each $[X]$, or $[Y]$, is also an income-consumption curve for the appropriate set of prices. In this case, along an income-consumption curve the marginal utility of money is constant. Conversely, if the marginal utility of money is constant along an income-consumption curve—i.e., the marginal utility depends only upon prices—this curve must coincide with [X] and [Y].[46] It is easy to see that the income comsumption curve will be a straight line, not necessarily passing through the origin. Indeed, if an increase in income $\Delta\mu$ brings about an increase of Δu in utility and the increases Δx, Δy of X and Y respectively, the next additional increases equal to Δx, Δy will necessitate a new increase in income equal to $\Delta\mu$ and, in the present scheme, will cause the utility to increase exactly by Δu. Full complementarity means therefore that any additional income (not necessarily beginning at the level zero) will be spent in the same proportions on X and Y, provided prices remain constant. Examples approximating this extreme case may be found in that "conspicuous" consumption, when any additional income above a certain level is divided in a constant ratio—varying with the price ratio—between, say, servants and decorative dwellings.

The case of full competitiveness presents the same properties as that of full complementarity except that the income-consumption curve is negatively inclined. One may think of X being a nutritious but not tasty food, and of Y being less nourishing but more palatable than X. Any additional increase in income will very probably bring about an increase Δy in the consumption of Y and a proportionate decrease Δx, in that of X, the ratio between Δy and Δx varying with the price ratio of the two commodities.[47] Full competitiveness includes the case of *perfect substitutability* when [X] and [Y] coincide also with the indifference curves, which, in this case, must be straight lines. Finally, the border-line case between mild and high competitiveness (Y highly competitive with X) is that in which [Y] coincides with the indifference curve (u). This is the case we referred to above as the quasi-constant marginal utility (of Y). The term seems justified by the parallelism (in the direction of Y) of all indifference curves.

As an illustration of the results which can be obtained along the same line, if the third dimension in Fig. 5 is Marshall-Hicks money, let us refer to the extreme case of full complementarity. In this case, it is seen that money is a perfect substitute for combinations of constant proportions of X and Y which is another transparent property of extreme complementarity. But the relevant conclusion is that the knowledge of p_x, p_y will fail to determine the equilibrium of the consumer. Since [X] and [Y] coincide, it means that if the marginal price of X is given, that of Y is implicitly determined. The curves XX', YY' in Professor Hicks' notations, [11], p. 68, must therefore coincide if X and Y are fully complementary (as well as if X and Y are fully competitive). Equilibrium of the multi-exchange market becomes *indifferent* and, therefore, *indifferently stable*. What we usually represent in theory by curves is very often a simplified representation of a more complicated and less sharply defined phenomenon—

[46] The constancy of marginal utility of money in this sense as well as in the sense of Marshall-Hicks, should not be confused with either of the two definitions given to this concept by Samuelson [24], despite his statement to the contrary (*ibid.*, p. 80).

[47] This is the extreme case of Y highly competitive with X.

368   Energy and Economic Myths

which would be better pictured as a penumbra.[48] Consequently, one may ask whether the initial intuition of Professor Hicks regarding "extreme complementarity" [11], p. 77, should not be reconsidered, at least in relation to the degree of stability.

It should be observed, however, that the preceding analysis shows clearly that in Professor Hicks' theoretical setup there is no place for a degree of instability greater than that of the indifferent equilibrium. Indeed, in order to have a clear-cut case of instability, the relationship between X and Y should be described by graphs other than those of Fig. 5; but this, as already mentioned, is prevented by the principle of decreasing marginal rate of substitution.

VII

It has become clear by now that the impossibility of the isotimetics $[X]$ and $[Y]$ crossing past each other is only another form of the strong principle of decreasing marginal utility or of that of decreasing marginal rate of substitution. But this impossibility is equivalent to a more transparent property:

(VII.1) *If the marginal rate of substitution for Y for money (or the marginal utility of Y) remains constant while X increases, the marginal rate of substitution of X for money (or the marginal utility of X) decreases.*

We reached by a different way a well known theorem,[49] which is, however, a particular case of a more general one.[50] The approach followed in this paper enables us to offer for the latter the following economic interpretation:

(VII.2) *If X increases, while either the amount or the evaluation of every other commodity is kept constant, then the evaluation of X decreases.*

The word *evaluation* stands here either for "marginal utility" or for "marginal rate of substitution for money." This formulation covers both types of evaluation—subjective and objective—revealing thus their intrinsic similarity. Under the first interpretation, (VII.2) is equivalent to the strong principle of decreasing marginal utility; under the second, to that of decreasing marginal rate of substitution. Either of these principles finds in (VII.2) an intuitive economic interpretation.[51]

Looked at from another angle, (VII.2) constitutes a straightforward generalization of the crude principle of decreasing marginal utility. In order to grasp the

---

[48] Taussig, [28], p. 407. Also Georgescu-Roegen, [8], pp. 583, 587.
[49] Hicks, [11], p. 309; Samuelson, [23], p. 109, Relation (72).
[50] Hicks, [11], pp. 52 and 311; Samuelson, [23], p. 109, Relation (71).
[51] The proposition (VII.2) is nothing else than the economic interpretation of the maximization conditions found in the early writings only as purely mathematical inequalities:

$$\varphi_{xx} < 0, \quad \begin{vmatrix} \varphi_{xx} & \varphi_{xy} \\ \varphi_{yx} & \varphi_{yy} \end{vmatrix} > 0, \quad \begin{vmatrix} \varphi_{xx} & \varphi_{xy} & \varphi_{xz} \\ \varphi_{yx} & \varphi_{yy} & \varphi_{yz} \\ \varphi_{zx} & \varphi_{zy} & \varphi_{zz} \end{vmatrix} < 0, \cdots$$

Cf., Pareto, [20], p. 577. For Pareto's scheme, φ is the utility function. If the interpretation of φ is changed so that $m + \varphi = 0$ is the indifference variety, the above conditions express the general principle of the decreasing marginal rate of substitution.

essence of this generalization, one should recall that, in the crude principle, the constancy of the amounts of the commodities other than X is intended to eliminate the influence of the other commodities upon the change in utility due to a change in X. However, when the crude principle was first introduced into economics, its originators were not aware of the inter-dependences of utilities. Under the assumption of independent commodities, keeping the amount of a commodity constant and keeping its marginal utility constant are one and the same thing. These two conditions necessarily acquire a distinct individuality if the "classical" assumption of independent commodities is dropped. A generalization of the crude principle applied to X must naturally consider the amounts as well as the marginal utilities of all commodities other than X. This is what (VII.2) does.

Now we are in a position to perceive the economic meaning of the difference between the strong principle of decreasing marginal utility and the classical hybrid set consisting of both the crude principle of decreasing marginal utility and the principle of decreasing marginal rate of substitution. The difference must crystalize itself into some exception to (VII.2). This is the possibility of the isotimetics crossing past each other, i.e., going beyond the limiting cases of full complementarity and full competitiveness. In other words, the hybrid set allows for the existence of excessive complementarity or excessive competitiveness. These excessive relationships require that

The marginal utility of X should increase with X if the marginal utility of Y is kept constant.[52]

The possibility of such excessive relationships becomes therefore the test case between the strong (or the generalized) principle of decreasing marginal utility and the classical hybrid set of postulates. The decision as to which one of these two rationales describes better the consumer's behavior depends largely upon whether some evidence can be offered for the possible existence of excessive complementarity or excessive competitiveness.

Frequently, in both economic and econometric models, we either postulate explicitly that utility is measurable or we reason as if this were so.[53] Therefore, the question of what type of postulate is adopted for describing the decreasing marginal utility is not of negligible importance since the admissibility of excessive complementarity and excessive competitiveness depends upon this choice.

REFERENCES

1. ALLEN, R. G. D. "The Foundation of a Mathematical Theory of Exchange," *Economica*, May 1932, pp. 197–226.
2. ——— "A Comparison between Different Definitions of Complementary and Competitive Goods," *Econometrica*, 1934, pp. 168–175.
3. ALLEN, R. G. D., AND HICKS, J. R. "A Reconsideration of the Theory of Value," *Economica*, February 1934, pp. 52–76 and May 1934, pp. 196–219.
4. EDGEWORTH, F. Y. *Mathematical Psychics*.

---

[52] This is easily seen on the graphs obtained from Fig. 5a and 5d, by crossing [X] and [Y] past each other.

[53] *Cf. Supra*, fn 7.

5. —— *Papers Relating to Political Economy*, Vols. I & II.
6. FISHER, I. *Mathematical Investigations in the Theory of Value and Prices*, (1925 Reprint).
7. FRIEDMAN, M. "The Marshallian Demand Curve," *The Journal of Political Economy*, December 1949, pp. 463–495.
8. GEORGESCU-ROEGEN, N. "The Pure Theory of Consumer's Behavior," *Quarterly Journal of Economics*, August 1936, pp. 545–593.
9. —— "The Theory of Choice and the Constancy of Economic Laws," *Quarterly Journal of Economics*, February 1950, pp. 125–138.
10. HAYEK, F. A. "The Geometric Representation of Complementarity," *The Review of Economic Studies*, 1943, Vol. X, No. 2, pp. 122–125.
11. HICKS, J. R. *Value and Capital*, (2nd ed.).
12. ICHIMURA, S. "A Critical Note on the Definition of Related Goods." *The Review of Economic Studies*, 1950–51, No. 47, pp. 179–183.
13. JEVONS, W. S. *The Theory of Political Economy*, (1871).
14. JOHNSON, W. E. "The Pure Theory of Utility Curves," *Economic Journal*, December 1913, pp. 483–513.
15. LANGE, O. "Complementarity and Interrelations of Shifts in Demand," *The Review of Economic Studies*, October 1940, pp. 58–63.
16. LI, CHOH-MING "A Note on Professor Hicks' Value and Capital," *The Review of Economic Studies*, November 1941, pp. 74–76.
17. MACHLUP, F. "Professor Hicks' Statics," *The Quarterly Journal of Economics*, February 1940, pp. 277–297.
18. MARSHALL, A. *Principles of Economics*, (8th ed.).
19. NEUMANN, J. v., AND MORGENSTERN, O. *Theory of Games and Economic Behavior* (2nd ed.).
20. PARETO, V. *Corso di Economia Politica*, Vols. I and II, Torino, 1948.
21. ——, —— *Manuel d' Economie Politique*.
22. ——, —— "Economie Mathématique," *Encyclopédie des Sciences Mathématiques*, Tome 1, Vol. IV, Fasc. 4, pp. 591–640.
23. SAMUELSON, P. A. *Foundations of Economic Analysis*.
24. ——, —— "Constancy of the Marginal Utility of Income," *Studies in Mathematical Economics and Econometrics: in Memory of Henry Schultz*, pp. 75–91.
25. ——, —— "The Problem of Integrability in Utility Theory," *Economica*, November 1950, pp. 355–385.
26. SCHULTZ, H. *The Theory and Measurement of Demand*.
27. STIGLER, G. *The Theory of Price*.
28. TAUSSIG, F. W. "Is Market Price Determinate?," *The Quarterly Journal of Economics*, May 1921, pp. 394–411.
29. TODHUNTER, I. *A History of the Mathematical Theory of Probability*.

Index

Abel, Niels Henrik, xxvii
Abelson, Philip H., 35
Abstinence vs. waiting, 251
Accumulation, 184, 249. *See also* Decumulation
Adams, John Couch, 4
Agents of production, 64
    idleness of, 66, 68, 90, 225
    idleness and internal arrangements, 90-91. *See also* Funds
Agrarian communism, 215
Agrarian Question, the, 109-110
Agrarian reform, 137
    and north-east Brazil, 145
Agrarian structure
    in USSR, 142-143
    in Poland, 143-144
Agrarianism, 103, 135
    vs. Marxism, 137
Agrarians, xii-xiii, 112-113, 136, 220, 223
    and optimum farm size, 138. *See also* Narodniki
Agriculture
    and economic development, 135
    and Entropy Law, 99
    and idleness of agents, 68-69
    and Industrial Revolution, 28
    vs. industry, 99-100, 107-108
    mechanized and economy of resources, 28, 58, 100
    organic vs. mechanized, xv, xix, 34, 229
    and population pressure, 28-29
Allais, Maurice, 311, 320, 346
Allee, W.C., 207, 229
Allen, R.G.D., xxvi, 76, 313, 321, 346-347, 352-354, 361, 363, 369
Amorozo, Luigi, 309-310
Analysis, 238
    and Change, 40-41. *See also* Dialectics, Science
Anaxagoras, 62
Anderson, James, 228
Antonelli, G.B., 329, 344, 346
Archimedean Axiom and ordinal measure, xxiv, 279
Arensberg, C.M., 221, 224-225, 229
Aristotle, 244
Arithmomorphic models, 61. *See also* Dynamic models, Models
Armaments, production prohibited, 33
Armstrong, W.E., 317, 346
Arrow, Kenneth J., xii, xxvii, 19, 125, 323, 346
Artin, Tom, 35
Aumann, R.J., 235, 253

Auspitz, Rudolf, xxvi, 308, 315, 320, 347
Axiom
    Binary Supremacy, 328
    Completeness, 326
    Convexity, 327
    Cycle, 325
    Demand Homogeneity, 332-333
    Dominance of Saturation, 327-328
    Exhaustibility, 332
    Indifference, 327
    Rationality, 326
Axiom of Choice
    consistency of, 323
    general, 322
    hedonistic, 315-316
    restricted, 322
    singleton, 323
Axiom of Revealed Preference, xxv, 323-324
    and indifference varieties, 341
    and poles, 341-342
    strong and focus, 341
    strong, and Regularity of Choice, 325
    strong, for two commodities, 336
    and vertex, 341
    weak, 336
    weak, and integrability, 335-338
    weak, and Principle of Non-Preference Persistence, 333-334
Ayres, Eugene, 60

Bach, G.L., 4, 60, 236, 253
Baden-Powell, B.H., 202, 204, 206, 210, 213, 215, 218-221, 223, 229
Baer, Werner, 164, 171, 173-174, 178, 186
Bancroft, A.L., 222-223, 229
Barnett, Harold J. 18, 35
Barone, Enrico, 131-132, 309, 314, 347; on Socialist state, 136
Baumol, W.J., 237, 242, 253
Beccatini, Giacomo, 309
Beckerman, Wilfred, 3, 14, 21, 35
Behavior directions, 333-335. *See also* Directional choice
Behavior, hedonistic, 128-129, 259
Behavior varieties vs. preference map, 346
Bentham, Jeremy, 157-158, 172-173, 186, 314, 347
Bergson, Abram, 311
Bernardelli, H., 281
Bernoulli, Daniel, 352
Bernstein, E.M., 186
Billings, Josh, xxiv
Binary choice, 326, 328-330
    and demand integrability, 338
    vs. market data, 344. *See also* Opheli-

372 Index

mity, Revealed Preference
Bioeconomic program, 33-35
    its obstacles, 59
Bioeconomics, 25-35
    and demand vs. supply, 33
    and durable goods, 34
    and fashion, 34
    and gadgetry, 34
    and high yield varieties, 39
    integenerational economics, 30
    and leisure, 34
    and mechanized agriculture, 29
Biological evolution, 25, 203, 227
Biology and physics, 30
Black, John D., 110, 202, 229
Blin-Stoyle, R.J., 35, 60
Bloch, Marc, 204, 208, 212, 214-216, 220-221, 229
"Blueprint for Survival," 22, 30, 35
Blueprint — models vs. simile — schemas, 118, 257
Bo-Aires, 220
Bohr, Niels, 62, 105
Boninsegni, Pietro, 309-310
Boon, Gerald K., 100
Borgatta, Gino, 309
Borgstrom, Georg, xv
Borlaug, Norman E., 29
Bormann, F.H., 14, 24, 35
Boulding, Kenneth E., xv, 9, 12, 23-24, 32, 35, 61, 73-75, 80-81, 95
Bowley, A.L. 61, 73, 317, 347
Bray, Jeremy, 17, 35
Breeder reactor, 16, 26
    its risks, 27. See also Entropy Law
Bresciani-Turroni, C., 160, 162, 164, 168, 172, 185-186
Breshkovskaia, Katerina, 112
Bridgman, Percy W., ix, 3, 15, 35, 60-61, 200, 229
Brown, D.J., 235, 253
Brown, E.H. Phelps, 281, 313, 347
Brown, Harrison, xv, 35
Brown, Lester R., 35
Bruton, Henry J., 173, 180, 186
Bucharest Conference, 32
Burnet, J., 62
Burnet, Macfarlane, 19
Bye, R.T., 60

Campos, Roberto de Oliveria, 165, 173, 180, 186
Cannon, James, 35
Cantillon, Richard, 200
Capital intensiveness, 249
Capital maintained constant, 64, 83, 87
Capital-labor ratio and statistical data, 46, 69
Capitalism
    calvary of, 134
    theory of, 106. See also Feudalism
Cardinal measure, 274-282, 295
    and efficiency, 295
    and ophelimity index, 327
    weak, 281-282. See also Measure

Carlson, Sune, 61, 74, 81
Carnot, Nicolas Sadi, 7-8, 11, 54
Carson, Rachel, 20
Cassel, Gustav, 314, 347
Chamberlin, E.H., 51-52, 271, 287, 295
Chaos, 8
Chenery, Hollis B., 75, 77-78, 141
Chicherin, B.N., 204, 210, 215, 229
Childe, V. Gordon, 135
Chipman, John S., 310-311, 313, 322, 337, 342, 347
Choice, see Binary choice, Directional choice, Ophelimity
Choice and cultural matrix, 128-129, 260
Civil society, 260
Clark, Colin, 29, 58, 60, 299
Clark, John Maurice, 130
Clausius, Rudolf, 8
Cloud, Preston, 19, 35
Club of Rome, 20
Coaration, 221
Commodity fetishism, 40-41, 64
Commodities
    highly competitive, 357, 359, 361, 364
    inferior, 357, 364
    mildly competitive, 359, 361
    vs. processes, 251. See also Competitiveness, Complementarity
Commoner, Barry, 30, 35
Communism, primitive, 210
Communist Manifesto, 107, 220
Comparability of differences
    and cardinal scale, 317-318
    and measurability, 281
Competitiveness
    excessive, 369
    full, 364, 367. See also Commodities, Complementarity
Complementarity, xxvi-xxvii, 319-321, 351-370
    Allen-Hicks, 321, 356, 361-362
    excessive, 369
    Edgeworth-Pareto on, 352, 356, 362
    extreme, 368
    full, 367
    and ordinal utility, 320-321
    of Pareto and of Allen-Hicks, 321
    topological rectangle of, 356
    and utility measurability, 361
Conspicuous consumption, 332-333
Cost
    fixed, 46
    and flows and funds, 47
    total, 46-51
    variable, 46. See also Working day
Cost curves, 50
    vs. isoquants, 295
Cost of environmental factors, xx, 13, 30-31
Coulanges, N.D. Fustel de, 205, 230
Cramer, D.L., 60
Croce, Benedetto, 309
Culbertson, John M., 35
Curran, Peter F., 36
Curve fitting, 266-267

Dagum, Estela M. Bee de, 153
Daly, Herman E., 23-24, 35
Daniels, Farrington, 26, 35
Debreau, Gerard, xii, xvii, 19, 125, 322, 347
Declining state, 25
Decumulation, 184. 249. *See also* Accumulation
Delewski, J., 112
Demand
   and economic development, 60
   of future generations, 30-32, 59
   and idleness of agents, 90
Demand schedule, 332
   and binary complementarity, 326
   of community, and Axiom of Choice, 345
   Marshallian, 362
   and PAP "indifference," 345-346
   Walrasian, 362. *See also* Isotimetics
Demaria, Giovanni, 309
Denman, D.R., 208, 210, 221, 230
Deuterium, 11, 27
Development, *see* Growth, Innovations
Diagram of reproduction, 4, 53
   flows vs. funds in 85. *See also* Transformation problem
Dialectics, 40, 244
   vs. analysis, 40
Directional choice, 333-335, 338
Distribution and institutional patterns, 129
Distribution *per stirpes*, 215
Divisibility of processes, 288
Divisibility of commodities
   and homogeneity, 287-289
   and scale, 282
Dodd, J.H., 60
Domar, E.D., 236
Dominedo, Valentino, 309
Dorfman, Robert, 39, 52, 85
Draft animals, a burden for farmer, 100
Dual economy, 140
Duration *See* Process
Dynamic equilibrium and waiting, 252
Dynamic models, 21
   and lags, 251-252
   and mere growth, 244-250
   producer vs. consumer goods in, 248-250
   reversibility of, 237. *See also* Models

Eccarius, J.G., 111
Ecological regulations, quantitative, 33
Econometric models, xxiii, 298-300
   and the Madonna effect, 267, 304. *See also* Models
Economic conflict over income flows, 216
Economic development
   and luxury goods, 182-183
   and wage goods, 185-186. *See also* Working day
Economic dynamics and differential equations, 250-251
Economic process
   changing, ix-x
   circular, 42, 53, 56, 226, 236
   and enjoyment of life, 9, 56, 227
   an entropic transformation of natural resources into waste, xv, 8-10, 97
   and the environment, 4, 6
   irreversible, 9, 237
   as a mechanical analogue, 4
   and nature, 241-242
   and waste, 9. *See also* Thermodynamics, Entropy, Entropy Law
Economics
   analytico-physiological, xxiv, 236-237
   and commodity, 80
   and ecology, 30
   mathematico-imaginative, xxiv, 235
   mechanico-descriptive, xxiv, 236, 243
   truth in, 307. *See also* Mechanistic epistemology
Economists, opportunism of, ix, 3, 200
Edgeworth, F.Y., xxvi, 12, 128, 308, 311, 314-315, 320, 328, 347, 352-353, 355, 357, 365, 369-370
Edgeworth-Hicks diagram, 365-366
Edson, Andrew W., 302, 304
Effective demand and employment, 220
Efficiency, its theoretical limits, 11
Efficient processes, 291-292
Ehrlich, Paul, 20
Einstein, Albert, 35
Einstein equivalence, 11, xvii
Elementary processes
   arrangements in series, in parallel, in line, 239
Ellis, H.S., 170
Energy
   accessible, xvii, 10-12
   available, 7-8
   conversion into matter, 11, 25
   geothermal, 27
   not net, xvii
   of thunderbolts, 11
   tidal, 27
   unavailable, 7-8, 54-55. *See also* Solar energy
Engel curves, 345
Engels, F., 87, 96-97, 108-110, 112, 120, 137, 204-206, 210, 220, 228, 230
Enjoyment of life, 9, 97
Entropic deficit, 10, 55
Entropic indeterminateness, 9
Entropic problem, 15-19
Entropy, 7, 8, 54-55
   and disorder, 54
   and the economic process, 54
   and economic value, 60
   and information, 60
Entropy Law, 7-9, 55-56, 98, 102, 227
   and breeder reactor, 15
   and economic process, 8-10, 55, 97-98
   and economics, 8
   and life phenomena, 8
   and living organisms, 55
   and matter, 8, 12
   and natural laws, 9
   and recycling, 12
   and statistical interpretation, 15

and statistical thermodynamics, 8
the taproot of economic scarcity, 9
vegetable growth, 228
Ercelawn, Aly Alp, 149
Eris, Ibrahim, 149
Eucken, Walter, 201, 230
Event at an instant of time, 63
Evolution, irreversible, 59
Evolutionary factors and arithmomorphic models, 267
Exogenous variables, 299-304
Exosomatic addiction, 28
Exosomatic evolution, xix-xx, 25
and value of labor, 211. *See also* Social conflict

Factors
controllable vs. uncontrollable, 299
limitational, 49, 119
limitative, 119
Factory, 89-90, 239
and agents' idleness, 227
and chicken farms, 227
an economic innovation, 68, 240
and fund coordinates, 92
its process-fund, 45
and processes in line, 92
and waiting, 94, 252
and working time, 94-95. *See also* Farming, Industrial Revolution
Factory process, analytical representation of, 89-96
Farm economics, 202
Farming vs. factory system, ix-xiv, 42-43, 227, 240
Fertility in peasant communities, 224
Feudalism vs. capitalism
and distribution, 126
and Marx's surplus value, 128
Finsterbusch, Gail, 35
Firm vs. plant, 293. *See also* Management unit
Firth, Raymond, 201
Fisher, Irving, 86, 97, 128, 259-260, 262, 315-316, 320, 347, 354, 370
Fisher, R.A., 224, 230
Fisher, W.D., 240
Flows, 41, 84-86, 239
and the Conservation Law, 67
dimension of, 94
vs. flow rates, 38, 61-62, 86
vs. services, 42
vs. stocks, 24-25, 86. *See also* Funds
Flow complex, 86
Flurzwang, 215, 221
Foner, Philip S., 134
Food, from crude oil, xvi, 59
Forest, in early communities, 212
Fossati, Eraldo, 309
Friedman, Milton, 151, 153-154, 183, 186, 362, 370
Frisch, Ragnar, 21, 37-38, 52, 61-62, 313, 317, 347
Functional, 43, 238
degenerate, 44

vs. Dirichlet function, 88
Functional space, 40, 65, 238
Funds, 41, 84-86, 239
dimension of, 94
vs. flows, 64-65
as inventories, 92
in productive processes, 87-88
service of, 66. *See also* Agents, Process, Process fund, Production function
Furlan, V., 310, 316, 347

Gabor, Denis, 22
Galton, Francis, 264
Gamow, George, 60
Gatti, G., 109, 231
Gee, W., 202, 230
Giffen paradox, 363
Gillette, Robert, 35
Gini, Corrado, 255-257, 268-269
Glaser, Peter E., 35
Glasson, E.D., 205, 230
Gleaning, 127, 214
Goeller, H.E., 35
Gofman, John W., 35
Gold, as socio-biological energy, 310
Goldschmidt, R., 245, 253
Gomme, G.L., 217, 222, 230
Goods, producer vs. consumer, 248-250
Gorman, W.M., 337, 345, 347
Gossen, H.H., 71, 308, 315, 347
Green plants and Entropy Law, 9
Growth
vs. development, 19, 242-245
limits to, 20-22
multiple senses of, 19
and underdeveloped economies, 32. *See also* Innovations, Labor, Working day
Gunton, G., 134

Haar, D. ter, 36, 60
Haavelmo, T., 299
Hague, Douglas C., 37, 52, 73
Hahn, F.H., 19, 39, 52, 271, 285-288
Hailstones, T.J., 60
Haldane, J.B.S., 5, 207, 230
Hammond, Allen L., 36
Hammond, R. Philip, 14, 26, 36
Hansen, Alvin, 151, 183, 187
Hardin, Garrett, xvi, 36
Harrod, R.F., 236, 243-244, 252-253
Hasek, C.W., 60
Havens, R.M., 60
Haxthausen, August F.L.M. von, 202, 215, 230
Hayek, F.A., 132, 314, 352, 370
Heat, 7
Heat Death, 8
Hedonimeter, 314
Hegel, G.W.F., 201, 272
Hegelianism, 109
Heilbroner, Robert L., 4, 236, 253
Henderson, J.S., 60
Herakleitos, 79
Herera, Philip, 36
Herzen, Alexander, 112, 202

Hibbard, Walter R., 36
Hicks, J.R., xxvi, 61, 74, 311, 313, 321, 334, 347-348, 351-356, 360-370
Hierarchy of wants and Principle of Decreasing Marginal Utility, 318
Hoffer, C.R., 202, 230
Holdren, John, 8, 36
Homo capitalisticus vs. *Homo agricola,* 130
Homo oeconomicus, strictly hedonistic, 130
Homogeneity, fallacy of general, 39
Hoover, Calvin B., 138
Hotelling, Harold, 31, 36
Houthakker, H.S., 325, 341, 348
Hubbard, Leonard E., 115
Hubbert, M. King, 26, 36
Hume, D., 172-173, 187
Hurwicz, Leonid, 337, 347-348
Hutschison, T.W., 312, 348

Ichimura, S., 352, 362, 370
Income, contractual and noncontractual, 161
Indifference varieties
 and binary choice, 315
 and isoutility varieties, 315. *See also* Integral varieties
Industrial capacity
 needed for agriculture, 29, 100
 redistribution of, xix, 32
Industrial Revolution and factory, 43, 68, 240
Industrialization
 in agricultural countries, 136
 excessive, 145
 the myth of, 101
Infeld, Leopold, 35
Inflation, xx, 149-197
 and balanced growth, 181-183
 and economic development, 172-181
 and extra profit, 171, 174
 and forced saving, 173
 and income shift, 176-177
 and investment, 174
 and labor's share, 178-179
 losers and winners, 166-172
 and monetary policy, 183-186
 and structural lock, 183-186
 and taxation, 163-164
Innovations
 and development, 20, 243-245, 251-252
 types of, 18
Inorganic domain, 203-204
Input, 37, 40, 63, 238, 282-283
Integrability
 in broad sense, 339
 and order of consumption, 342-343
Integral varieties, 336-337
 vs. indifference varieties, xxv, 337
 and PAP ophelimity index, 339
 for two commodities, 337
Interplanetary travel, 11
Investing vs. disinvesting, 237
Irrational numbers and empirical observations, 257
Isotimetics, xxvi-xxvii, 358, 366, 368-369
 and demand law, 362
 and indifference map, 358, 361
Istock, Conrad A., 36

Jaffé, William, 311, 348
Jannaccone, Pasquale, 309, 348
Jevons, W. Stanley, x, 3, 36, 53, 60, 71, 226, 236, 253, 353, 370
Jona, Giuseppe, 309
Johnson, Harry G., 8, 13-15, 21, 36
Johnson, W.E., 315, 348, 352-353, 370
Johnston, Bruce F., 135
Junker, Louis J., 252

Kafka, Alexandre, 172, 177-178, 187
Kahn, A.E., 141
Kaldor, N., 242, 253, 285
Katchalsky, A., 36
Kautsky, Karl, 105, 107-110, 119, 131, 135, 219, 222, 230
Kaysen, Carl, 17, 21, 32, 36
Keynes, John Maynard, xxvi, 151-152, 157, 159-160, 163-164, 187, 200, 229, 257, 311
Kimball, S.T., 221, 224-225, 229
Klein, L.R., xxi, 299
Kneese, Allen, 36
Knight, Frank H., 220, 230, 256
Koentjaraningrat, R.M., 210
Koopmans, T.C., 62, 78, 80-81, 87, 95, 299-303
Kovalevskii, M., 210-211, 213-215, 218, 221, 230
Kroeber, A.L., 203, 230
Kuenne, R.E., 259
Kuznets, Simon, xxiii, 132

Labor
 and growth, 248
 share in feudal system, 127
 supply of, 120
Labor
 and efficiency, 139
 marginal productivity of, 117-118, 121-123, 126-128, 134
 and overpopulation, 124-125
Labor time and measure of value, 68, 96
Lafargue, A.L., 204, 211, 230
Land
 as agent, 82
 individual property of, 204-205
 utilization of, 100
Lange, Oskar, 281, 313, 317, 348, 352, 370
Laplace, Pierre Simon de, 3, 36
Laveleye, Emile L.V. de, 210-211, 213, 215, 224, 230
Law of concentration, 108
Leftwich, Richard H., 37, 52, 73
Leibenstein, Harvey, 115
Leisure
 and development policies, 69, 101-102
 and national income, 131-133
 in peasant communities, 222
 in strictly overpopulated economies, 131, 133
 unwanted, xii, 133-134
Lenin, V., 111-112
Leontief, W.W., 4, 19, 21, 36, 80, 85, 87, 101, 236, 246-248, 250-251, 253, 355

Leverrier, Urbain, 4
Levi, Carlo, 226
Lewis, W. Arthur, 117, 160, 163-164, 173, 177, 187
Lexicographic order and ordinal measure, 278
Li, Choh-Ming, 352, 370
Lieben, Rudolf, xxvi, 308, 315, 320, 347
Liebig, Justus von, 33
Life
    amount of, and terrestrial resources, xv, 23, 58
    cannot violate natural laws, 9, 55
Limitationality, 67
Linear thinking, 14
Lithium-6, 27
Little, I.M.D., 310, 318, 338, 348
Loria, Achille, 309
Lotka, Alfred, xv, 25
Lovering, Thomas S., 36
Luxemburg, Rosa, 112

McCarthy, Terence, 105
McCulluch, J.R., 222, 230
McDonald, Gordon J.F., 36
McLeod, A.N., 39, 52, 271, 285
McManus, M., 250, 253
Mach, Ernst, 272
Machlup, F., 351-352, 370
Maddox, John, 17, 19, 21-22, 36
Madgearu, V., 115
Mahalanobis, P.C., xxiv
Maine, H.J. Sumner, 202, 206-207, 209-210, 214-215, 217-218, 220, 226, 230
Maitland, F.W., 203, 206, 211, 218, 230
Malthus, Thomas, 4, 12, 22, 127
Malthus's error, 242
Man, the limiting biological nature of, 57
Management unit, 76, 283-284, 290-291
    and quality, 295
Mandeville, Bernard, 182
Manu Law, 213
Marginal pricing
    and employment, 219-220
    in feudalism, 126-127
    and optimum welfare, 136
Marginal productivity in Communist regime, 137
Marginal rate of substitution and price ratio, 48
Marginal utility and cardinality, 317
Market mechanism and ecological crises, 32
Markham, Jesse W., 351
Marschak, Jacob, 299-302
Marshall, Alfred, xxiv, 6, 36, 41, 60, 159, 187, 201, 222, 226, 230, 236-237, 245, 253, 307-308, 311, 321, 348, 351, 353, 370
Marx, Karl, 36, 40-42, 60, 68, 82-83, 85, 87-88, 95, 101-102, 105, 108-109, 111-112, 114-115, 120, 127-128, 131, 134, 137, 157, 180, 201, 205, 219-220, 228, 230, 237-238, 246, 248
    a marginalist, 211
    on natural resources, 5
    and the peasant economy, 107
    on services, 65
    on the working day, 96. *See also* Diagram of reproduction
Mathematical formalism, 256-259
Mathematics and science, 61
Matter
    accessible, xvii, 10-12
    a crucial item, xviii
    not net, xvii
    subject to entropic degradation, xvii-xxviii, 12, 59. *See also* Energy, Entropy Law
Meade, J.E., 242, 245, 253
Meadows, Donella H., 23, 30, 36, 237, 253
Measure, 271-296. *See also* Cardinal measure, Ordinal measure
Mechanical processes, reversible, 7
Mechanics
    and Conservation Principle, 6
    and locomotion, 6
Mechanistic epistemology, x, 3-4, 236, 243
Menger, Carl, 220, 231, 318
Methodenstreit, 200
Metz, William D., 36
Meyers, A.L., 171
Micromutations and evolution, 244
Migration, freedom of, 34
Mill, John Stuart, 23-24, 36, 154, 164, 187, 311
Mir, 204
Mirrlees, J.A., 242, 253
Mishan, E.J., xv, 36
Mitrany, David, 109-113, 130, 201, 220, 231
Models, Leontief, 101, 245-251
    open, 246
    and reversibility, 250. *See also* Dynamic models, Working day
Money, 150
    constant marginal utility of, 359-360
    Marshall-Hicks type, xxvi, 361, 367
    quasi-constant marginal utility of, 367
Money fetishism, 158-162
Money illusion, 158-159, 332-333
Moore, W.E., 114-115, 138
Morgan, L.H., 97, 201
Morgenstern, O., 352, 370
Morse, Chandler, 18, 35
Multi-exchange equilibrium, 365, 367
Multi-plant firm, 291
Multi-process, analytical representation of, 240-241
Murdoch, William W., 36
Murdock, G.P., 204, 208, 221, 231
Myths
    of exponential technology, 17
    of immortality of human species, 5
    of industrialization as cure-all, 101, 228
    and man, 5

Nardi, G. di, 141
Narodniki, 202, 220
    and Agrarians, 112
    and Marxism, 112
Nasse, E., 213-214, 217, 231
Nataf, A., 345, 348

National income in overpopulated economies, 132
Natural laws expressed by linear functions, 288-289
Natural resources
    and amount of life, 23, 58
    fossil fuels, 26
    and Great Migration, 56
    and history, 6
    and interest rate, 31-32
    limit, 16
    low entropy, 97
    terrestrial are stocks, 98
    and time horizon, 31-32
    vs. waste, 56. *See also* Cost of environmental factors, Standard economics
Neoclassical theory of production, 37-51
Neumann, John von, 15, 81, 240, 352, 370
Newcomb, S., 86
Nicholls, William H., 96, 351
Nicol, Hugh, xv
Notestein, Frank, 11
Nothing, 275-276
    vs. zero, 275
Novelty vs. arithmomorphic structures, 237
Novick, Sheldon, 36
Nuclear fuels, 11, 26
Nuclear garbage, 13
Numéraire, 149-150
Nunn, Charles, 144
Nurkse, Ragnar, 228, 231

O'Curry, E., 213, 215, 220-221, 231
Oncken, Auguste, 127
Onsager, L., 23
Onsager's conditions, 23-24
Open fields, 213-214
Ophelimity, 307-349
    and PAP index, 339
    and Pareto's cycles, 325, 342-344
    and prices and income, 332
    psuedo index of, 341
    quasi index of, 341. *See also* Paretoan ophelimity
Optimal budget and relative saturation, 339
Optimum size, 39, 75, 271-296
    in agriculture, 107, 225
    and proportionality, 292-295
    of village community, 209
Orcutt, G.H., 298-304
Order and comparability, 277
Ordinal measure, 39
    vs. cardinal measure, 273, 281
    and economic principles, 274. *See also* Lexicographic order, Measure
Ordinalist's fallacy, 318
Organic domain, 203-204
Organs
    endosomatic, 25
    exosomatic, 25
Orwin, C.S., 204, 208, 212, 216, 228, 231
Osborn, Fairfield, 20
Osorio, A., 317, 348
Output, 37, 40, 63, 238, 282-283
Overcapitalization in farming, 68
Overpopulation, 96, 113-114, 126
    in agriculature, 123

    in Brazil, 145
    and bureaucracy, 133
    in Communist countries, 145
    and investment criteria, 141
    and marginal pricing, 140
    and occupation ratio, 133
    and profit maximization, 135
    strict, 123, 127
    and underdevelopment, 116, 133. *See also* Labor, Village community

Pantaleoni, Maffeo, 307-309, 313
PAP operations, xi, 325
Pareto, Vilfredo, xxiv-xxv, 72, 128, 236, 253, 260, 307-346, 348, 351-355, 357, 368, 370
    and Edgeworth box, 311
    on final degree of utility vs. indifference curves, 313
    on the hill of pleasure, 317
    and Principle of Compensation, 310
    and social welfare function, 311
Papi, Ugo, 144
Paretoan ophelimity
    and commodities, 260
    and rational choice, 223
Paretoan optimality, 310
Patel, I.G., 186
Patinkin, Don, 38, 42, 240
Peasant and the town, 225-226
Peasant communities, 199-231
    institutions in, 129. *See also* Village community
Perpetual motion
    of the first kind, 5
    of the second kind, 5
    and themodynamics, 7
Perroux, Francois, 152
Petty, Sir William, 6, 53, 60, 228
Pietri-Tonelli, Alfonso de, 309, 348
Pigou, A.C., 4, 36, 61, 69, 74, 76, 86, 95
Planck, Max, 61
Plato, 63, 318
Plutonium-239, 13
Poincare', Henri, 159
Pollution, 13, 33, 57
    a surface phenomenon, 33. *See also* Cost of environmental factors
Population
    explosion, xiii, 99
    and land, 242
    optimum size, xv, 29-30, 58, 114
    superfluous, 115. *See also* Agriculture
Prebisch, Raul, 184, 187
Prices and ecology, 10
Prigogine Principle, 24
Principle of Complementarity, 105
Principle of Conservation of Matter-Energy, 53
Principle of Decreasing Marginal Rate of Substitution, 327, 334, 353-355, 366, 368-369
    and cardinal measurability, 38
Principle of Decreasing Marginal Utility, 366
    crude, 353, 355, 369
    and hierarchy of wants, 318

vs. Principle of Decreasing Marginal Rate
    of Substitution, 353-355
  strong, 320, 355, 368-369
  strong, and Principle of Decreasing Marginal
    Rate of Substitution, 361
  weak, 319-320, 353, 355
Principle of Nonpreference Persistence, 340
Process
  in agriculture, 37
  and analysis, 63
  analytical, 40
  an analytical tangle, 39-40
  and analytical boundary, 40, 62-63, 79-80,
    82
  its analytical difficulty, 41
  its analytical representation, 6, 78-89
  arranged in line and agriculture, 68
  arrangement in line, 42, 66
  arrangement in parallel, 42-43, 66
  arrangement in series, 43, 66
  and Change, 39
  and dialectics, 62
  and duration, 40, 51, 63, 65
  elementary, 41-42, 65-66
  and flow rates, 80
  and gross output, 85
  in manufacturing, 37
  reproducible, 64-65
  and stocks, 81
  its structure and economic development, 43
  subsumptive, 283
  unit of 289-290
  and vulgar philosophy, 62
Process analysis, 37-51
Process fund and qualitative change, 67, 93.
    See also Factory
Production capacity of plant, 45
  utilization (at a moment vs. overtime), 46,
    67
Production function, 38-46, 61-68, 72-78, 238,
    283
  and duration, 44, 65
  engineering, 77
  and flows, 44, 47, 67
  and flow coordinates, 45
  and flow rates, 44
  flow rates vs. quantities in, 38, 62,
    74-75, 258
  as functional, 65
  and funds, 47, 67
  general form, 43
  homogeneity of, 38, 46, 62, 67,
    285-286
  limitational, 44
  linearity of, 38
  management in, 292-293
  and natural resources, 65
  and process-fund, 45
  and quality, 75-76
  and services, 95
  vs. utility function, 71-72, 273
  and waste, 48, 65
  See also Agents of production, Factory
    production function
Profit vs. tithe, 125
Property, 210-211

Quesnay, F., 127, 135, 200, 237
Qualitative residual
  and efficiency, 296
  and natural laws, 288
  and quality quantification, 271-272
Quality, 271-296
  lack order, 245
  vs. quantity, 271, 283

Randomness vs. irregularity, 261
Ramsey, F.P., 314, 348
Recycling
  not complete, xviii, 12
  not free, 57
  not perpetual, 12
Regression line vs. true relations,
    xxiii-xxiv, 262-266
Regressive factor, 364
Revealed preference
  vs. binary choice, xxv, 331-332
  and operational binary choice, 344-345
  See also Axiom of Revealed Preference
Revelle, Roger, 16, 19, 29, 36
Ricardian land, 6, 238
  a fund, 64
  vs. natural resources, 56
Ricardo, David, 157, 163, 172-173, 176,
    187, 201, 311
Ricci, Umberto, 309
Richardson, George, xii
Richter, Marcel K., 337, 347-348
Ridker, Ronald, 36
Roberts, L.H., 112
Robertson, D.H., 157, 187
Robinson, A., 235, 253
Robinson, G.T., 204, 208, 231
Robinson, Joan, 39, 52, 286-287
Rocca, Vittorio, 309-310
Roosevelt, Theodore, 201
Rosa, Gabriele de, 307, 348
Rostow, W.W., 242, 253
Rousseau, Jean Jacques, 211
Run-rig, 215
Ruopp, Phillips, 206, 231
Rutherford, Ernest, 11

Sameness, 40, 64
Samuelson, Paul A., xxv, xxvi-xxvii,
    4, 37, 39, 49, 52, 60-61, 72, 74,
    76, 78, 85, 236, 238, 253, 318,
    323, 329, 334, 338, 341, 348-349,
    351, 362, 367-368, 370
Sanderson, D., 202, 231
Saving and development, xxiv, 242-250
Sawhill, John, xvi
Say, J.B., 86
  on economics, 80
Scalfati, Giuseppe Stanislao, 309
Scatter and its kernel, 262-263
Schmelev, G., 142-144
Schneider, Erich, 61, 74
Schrödinger, Erwin, 9, 36
Schultz, Henry, 313, 349, 352, 370
Schultz, Theodore, 116
Schumpeter, Joseph, xi, xxiv, 19, 170,

Index   379

185, 200, 220, 222, 231, 237,
241, 243-244, 251, 253, 309,
312, 349
   on economic development, 245
Science
   and mathematics, 61
   oversold, 57
   theoretical, and analysis, 79
Scitovsky, T., 349
Scorza, Gaetano, 312
Seebohm, F., 202, 208, 210, 214-215, 221, 231
Senior, Nassau, 95
Sensini, Guido, 309-310
Separability, topological criterion of, xxvii, 321-322
Services, 88
   See also Funds
Set
   homogeneous, 277
   ordered, 273
   perfectly divisible, 280
   subsumptive, 275-277
Silk, Leonard, 32, 36
Simonsen, Mario Henrique, 171, 174, 176-177, 187
Singularities of integral varieties
   center, 340
   focus, 340
   pole, 337
   vertex, 339
Slavophiles, 202
Slutsky, Eugen, 309-310, 321, 349
Smith, Adam, 37, 201, 212, 237-238
Social conflict and exomatic evolution, 25
Social evolution, 203
Solar energy, 25-28, 57, 60, 98-99
   and dictatorship of the present, 33
   a flow, 25, 98
   its intensity, 26-27
   and pollution, 28
   and Ricardian land, 25, 42
   vs. terrestrial energy, 27, 29, 57, 245
Solo, Robert A., 8, 14, 16
Solow, Robert M., 8, 12, 17, 20-21, 32, 36, 85, 236
Sonnenschein, Hugo, 347
Sorel, G., 220, 226, 231
Spencer, Herbert, 289
Spengler, Joseph J., xv, 22, 36
Spengler, Oswald, 107, 208, 218, 231
Sprout, Harold, 27, 36
Sprout, Margaret, 27, 36
Stahl, H.H., 203-205, 208, 212, 214-215, 218, 226, 231
Stalin, J., 110
Standard economics
   and feedbacks, 4
   and natural resources, 4-5
   and production coordinates, 46
   See also Mechanistic epistemology
Starcs, P., 222, 226, 231
Static process, 62, 64
   and analytical difficulties, 83
   analytical representation of, 84
   and fund coordinates, 87

   See also Stationary state
Stationary state, 4-5, 69, 86, 237, 239, 242
   production of commodities in, 251
Statistical tests
   and normality, 262
   and randomness, 260-261
Steady state, 22-25
   and cosmological hypotheses, 8
   and pollution, 24
   and technology, 24
   See also Stationary state
Stigler, George, 62, 74-75, 80, 95, 353, 355, 370
Stockholm Conference, 32
Stocks, *See* Flows
Stonier, Alfred W., 37, 52, 73
Substitutability, perfect, 367
Substitution of materials, 17-19
Subsumption
   and measurability, 274-275
   vs. subtraction, 278
Summers, Claude M., 36
Superorganic domain, 203-204
Sweezy, Paul M., 85, 112
Swift, Jonathan, 228

Tang, Anthony M., xxvii
Tarascio, Vincent J., 311, 316, 349
Taussig, F.W., 368, 370
Technical idleness of agents, 42
   See also Agents
Technical inventories, 45
Technological horizon, 136, 282
Technology, *See* Myths
Technology, the power of, xv, 16
Teller, Edward, xix
Tepicht, J., 143-144
Terrestrial resources
   a finite stock, 25
   See also Natural resources
Theorizing in vacuum, 200
Theory
   a logical file, 104-105, 272
   and reality, xi, 106-111
Thermodynamics, 7
   and the economic process, 53, 97-98
   and irreversibility, 24
   and life, 227
   physics of economic value, 54
   and statistical interpretation, 15, 60
   See also Entropy Law
Thermonuclear energy, xix, 11, 27
Thurnwald, Richard, 139, 222, 231
Tinbergen, J., 299
Tithe, maximization of, 126-127
Todhunter, I., 352, 370
Torr, Dona, 109
Town vs. countryside, 102, 110-111
Toynbee, Arnold, 210, 231
Transformation problem, 85
Transitivity
   falsifiable, 327
   tautological form of, 324
Tritium, 27

Tschajanow, A.V. (Chayanov), 113, 129, 202, 209, 229
Tschuprow, A.A., (Chuprov), 202, 206, 229

Universe, an hourglass, xvi
Uranium-235, resources, 26
Used equipment and accounting, 83
Utility
    and community welfare, 128
    and interpersonal comparability, 314
    measurability of, 272-273, 316-317
    measured in money units, 362
    See also Ophelimity
Uzawa, Hirofumi, 323, 349

Variability, biological and social, 203
Veblen, Thorstein, 108, 119, 207
Vieira, Dorival Teixeira, 170
Villaça, Maria J., 168
Village community, xxii
    control of income flows, 209-210
    distribution in, 216
    equality vs. opportunity, 211, 219-221
    and gregarious instinct, 207
    and kinship, 206
    land in, 209
    and Marxism, 220
    oral tradition, 209, 216-217
    and overpopulation, 226
    and peasant sociology, 205
    physiology of, 211
    See also Peasant communities
Village, 203-205
    optimum size, 207, 209
    an organic territory, 208-209
Ville, Jean, 325, 341, 349
Vinci, Felice, 309
Viner, Jacob, 51
Vinogradoft, Paul, 204, 206, 209, 212, 231
Volterra, Vito, xxv, 312-313, 337, 339, 342-343, 346, 349

Waiting, 240, 251-252
Wald, Abraham, 125
Wallich, Henry C., 36

Walras, Léon, 72, 83, 86, 131, 154-156, 187, 226, 308, 311
Walrasian system, critique of, xii, 125
Warriner, Doreen, 115-116, 138, 226, 231
Waste, 12, 53, 88, 93, 238-239
    its flow rate, 45
    high entropy, 98
    irrevocability of, 57, 98
    material output of economic process, 97
Weber, Max, 151
Weinberg, Alvin M., 14, 26-27, 30
Welfare state, 130
Westfield, F., xxvii
White, L.D., 116
Whitehead, Alfred North, 63, 210, 223, 231
Wicksell, Knut, 95
Wicksteed, Philip, 37-38, 61, 65, 71-73, 238, 253
    on production function, 37-38
Wilson, M.L., 202, 226, 231
Wilson, Woodrow, 202
Winiarsky, Léon, 309
Wittgenstein, Ludwig, 235
Wold, Herman, 316, 327, 349
Wolstenholme, G., 60
Working day, 44-45, 51, 259
    and daily production, 45
    and economic development, 44, 101-102, 229, 246
    and natural resources, 246
    and Neoclassical economics, 96
    and the production function, 45, 68, 94-95
    and total cost, 46-48, 94
    in Leontief model, 246
Worley, J.S., xxvii
Woytinsky, E.S., 103
Woytinsky, W.S. 103, 134

Xenophon, 222

Zadruga, 226
Zasulich, Vera, 109
Zeman, Jiří, 25